SUSTAINABLE GOVERNANCE OF WILDLIFE AND COMMUNITY-BASED NATURAL RESOURCE MANAGEMENT

This book develops the Sustainable Governance Approach and the principles of Community-Based Natural Resource Management (CBNRM). It provides practical examples of successes and failures in implementation, and lessons about the economics and governance of wild resources with global application.

CBNRM emerged in the 1980s, encouraging greater local participation to conserve and manage natural and wild resources in the face of increasing encroachment by agricultural and other forms of land use development. This book describes the institutional history of wildlife and the empirical transformation of the wildlife sector on private and communal land, particularly in southern Africa, to develop an alternative paradigm for governing wild resources. With the twin goals of addressing poverty and resource degradation in the world's extensive agriculturally marginal areas, the author conceptualises this paradigm as the Sustainable Governance Approach, which integrates theories of proprietorship and rights, prices and economics, governance and scale, and adaptive learning. The author then discusses and defines CBNRM, a major subset of this approach. Interweaving theory and practice, he shows that the primary challenges facing CBNRM are the devolution of rights from the centre to marginal communities and the governance of these rights by communities, a challenge which is seldom recognised or addressed. He focuses on this shortcoming, extending and operationalising institutional theory, including Ostrom's principles of collective action, within the context of cross-scale governance.

Based on the author's extensive experience this book will be key reading for students of natural resource management, sustainable land use, community forestry, conservation, and development. Providing practical but theoretically robust tools for implementing CBNRM it will also appeal to professionals and practitioners working in communities and in conservation and development.

Brian Child is an associate professor in the Department of Geography and Center for Africa Studies at the University of Florida, Gainesville, USA and Professor Extraordinary in the School of Public Leadership at the University of Stellenbosch, South Africa. As a wildlife practitioner from southern Africa, he previously managed the CAMPFIRE programme in Zimbabwe and acquired experience in private conservation and park management.

Earthscan Studies in Natural Resource Management

Forest Management Auditing
Certification of Forest Products and Services
Edited by Lucio Brotto and Davide Pettenella

Agricultural Land Use and Natural Gas Extraction Conflicts
A Global Socio-Legal Perspective
Madeline Taylor and Tina Hunter

Tropical Bioproductivity
Origins and Distribution in a Globalized World
David Hammond

The Commons in a Glocal World
Global Connections and Local Responses
Edited by Tobias Haller, Thomas Breu, Tine De Moor, Christian Rohr, Heinzpeter Zonj

Natural Resource Conflicts and Sustainable Development
Edited by E. Gunilla Almered Olsson and Pernille Gooch

Sustainable Governance of Wildlife and Community Based Natural Resource Management
From Economic Principles to Practical Governance
Brian Child

Sustainability Certification Schemes in the Agricultural and Natural Resource Sectors
Outcomes for Society and the Environment
Edited by Melissa Vogt

For more information on books in the Earthscan Studies in Natural Resource Management series, please visit the series page on the Routledge website: www.routledge.com/books/series/ECNRM/

SUSTAINABLE GOVERNANCE OF WILDLIFE AND COMMUNITY-BASED NATURAL RESOURCE MANAGEMENT

From Economic Principles to Practical Governance

Brian Child

from Routledge

First published 2019
by Routledge
2 Park Square, Milton Park, Abingdon, Oxon OX14 4RN

and by Routledge
52 Vanderbilt Avenue, New York, NY 10017

Routledge is an imprint of the Taylor & Francis Group, an informa business

British Library Cataloguing-in-Publication Data
A catalogue record for this book is available from the British Library

Library of Congress Cataloging-in-Publication Data
Names: Child, B. (Brian), author.Title: Sustainable governance of wildlife and community-based natural resource management : from economic principles to practical governance / Brian Child.
Description: Abingdon, Oxon ; New York, NY : Routledge, 2019. |
Series: Earthscan studies in natural resource management | Includes bibliographical references and index.Identifiers: LCCN 2018061154 (print) | LCCN 2019009021 (ebook) | ISBN 9781315211152 (eBook) | ISBN 9780415793278 (hbk) | ISBN 9780415793292 (pbk) | ISBN 9781315211152 (ebk)
Subjects: LCSH: Community-based conservation. | Conservation of natural resources–Citizen participation. | Natural resources–Management.
Classification: LCC S944.5.C57 (ebook) | LCC S944.5.C57 C45 2019 (print) | DDC 333.72–dc23
LC record available at https://lccn.loc.gov/2018061154

ISBN: 978-0-415-79327-8 (hbk)
ISBN: 978-0-415-79329-2 (pbk)
ISBN: 978-1-315-21115-2 (ebk)

Typeset in Bembo
by Swales & Willis, Exeter, Devon, UK
Printed and bound by CPI Group (UK) Ltd, Croydon CR0 4YY

This book is dedicated to my father, Graham Child, and his life's work as a field ecologist, wildlife administrator, and pioneer of the new paradigms of wildlife conservation discussed in this book. This book is also dedicated to my mother, Diana Child, who was an avid and active influence on both my father's and my own philosophy and work, and who provided unending support to us both.

CONTENTS

List of figures   ix
List of tables   xii
List of boxes   xiv
Acknowledgements   xvi
Preface: some personal reflections   xix

1 Introduction: poverty, conservation, and governance   1

2 The emergence of man, governance, and rules   11

3 A brief history of man's impact on the planet   32

4 Institutions and ungoverned spaces   51

5 Proprietorship   70

6 Economic principles for wildlife governance   97

7 The institutional history of wildlife and its governance   116

8 Changing the game   134

9 Assessing the economics of wildlife: tools and lessons   158

10 The sustainable governance approach   180

11 Kindling CBNRM: history and lessons from CAMPFIRE in Zimbabwe   200

12 Does it take a village? Is there a difference between participatory and representational governance? 226

13 The global emergence of CBNRM practice and theory 250

14 The application of theories of micro-governance to CBNRM 268

15 Implementing CBNRM 287

16 Participatory governance and revenue distribution in practice 315

17 Conclusions 341

References *357*

Index *378*

FIGURES

2.1   A model for prosperity and sustainability   28
3.1   Drivers of the environmental Kuznets curve   34
3.2   The 'closed' economy and ungoverned spaces   35
3.3   Exceeding the safe operating space for humanity in ungoverned spaces   35
3.4   Comparison of returns from land with weak and strong institutions   42
3.5   Expanding out of the agricultural sweet spot into drylands and forests   43
3.6   The value of ecosystem services relative to agriculture across a rainfall gradient   47
3.7   The financial and economic prices of wild and domestic production systems   49
5.1   Why private goods have non-problematic characteristics for economic
      allocation   72
5.2   Internalising costs and benefits by matching resource fugitiveness and
      jurisdictional scale   81
5.3   Definition matrix for private, common pool, club, and public goods   83
5.4   Wild resources are private or common pool goods and becoming more so   84
5.5   The interaction between resource fugitiveness, rights, owners, and exclusion   88
5.6   The different purposes of collective action in natural resource management   90
5.7   CBNRM and decentralisation   94
5.8   Rights and the governance of resources in political or economic market places   95
6.1   Biodiversity loss in high, middle, and low-income countries over time   99
6.2   Economic model of the wildlife economy and the sustainable use approach   102
8.1   The process of change in southern Africa   148
8.2   Trends in wildlife populations in Africa (1970–2005)   151
8.3   Trends in wildlife numbers in Kenya   151
8.4   Comparing wildlife and livestock populations in southern Africa with the
      Kenyan counter-factual   152
8.5   Changing wildlife policies and threats   155
9.1   Fence line photo comparing game and cattle ranches (1984)   161
9.2   A financial comparison of wildlife and livestock in Zimbabwe in 1984   164
9.3   Map of Buffalo Range cattle and wildlife sections   167

9.4 Comparing cattle productivity (calving rates) and range condition on Buffalo Range 168

9.5 Commercial and communal cattle populations in Zimbabwe (1910–1988) 170

9.6 Using beef parity pricing to assess if commercial beef production was taxed or subsidised 171

9.7 Total economic value and vulnerability pyramid for South Luangwa National Park 177

9.8 Dematerialising production: replacing the agro-extractive economy with the bio-experience economy 179

10.1 Private values determine land use outcomes 181

10.2 Goal of sustainable governance approach is to lift financial curve as far as possible to reflect total economic value 182

10.3 The interactions between price and proprietorship, and different wildlife economies 186

10.4 Distribution of the costs and benefits of wildlife and domestic species across scale 190

10.5 Comparing control hierarchies for wildlife and farming 190

10.6 Polyvalent governance and the allocation of roles across scale 194

10.7 Collaborative adaptive management 196

10.8 Developing learning communities: using action to catalyse cognitive change and agency 197

11.1 Using simple databases to optimise the marketing of hunting in CAMPFIRE 210

11.2 Participatory negotiation of hunting concessions 211

11.3 Participatory activity-based budgeting 215

11.4 Each person gets their whole share in cash 216

11.5 Money is paid back into buckets for projects (tax) 216

11.6 Allocation of CAMPFIRE revenues over time 221

12.1 Comparison of old and new policies used in community training in Luangwa 232

12.2 Allocation of expenditure in 43 Village Action Groups in Lupande GMA (1996–2001) 236

12.3 Comparing performance of top-down (1989–1995) and bottom-up (1996–2001) phases in CBNRM in Luangwa 237

12.4 Comparison of trust in small and large communities 240

12.5 Relationship between size of community and allocation of wildlife income to community benefits 242

12.6 Comparison of financial allocations and satisfaction with participatory and representative budgeting processes 243

12.7 Participatory and representational governance, scale, and public goods 247

13.1 The four steps in CBNRM: proprietorship, price, micro-governance, and resource management 264

13.2 An expanded model of CBNRM and the sustainable governance approach 265

13.3 The relationship between governance and management in CBNRM 266

14.1 An expanded explanation of Ostrom's principles 273

14.2 The process of building micro-governance and common property management through CBNRM 283

15.1 Four stages for implementing CBNRM at a new site 288

15.2 Examples of data from livelihood surveys: food vulnerability and sources of
      household production   303
15.3 Examples of data from livelihood surveys: associational and social capital   304
15.4 Example of dashboard data from Mangalane community in Mozambique   305
16.1 The revenue distribution ceremony   329
16.2 Community financial summary   331

TABLES

2.1	Chronology of the evolution of man	14
5.1	The four categories of right owners	75
5.2	Defining rights as a bundle of entitlements and obligations	78
5.3	Defining proprietorship in terms of use rights	78
5.4	Challenges to using simple cadastral property regimes for wild resources	80
5.5	The unhealthy political economy of wildlife	87
5.6	The confusing use of the terms 'public' and 'private' in natural resource governance	88
5.7	Checklist to illustrate a strategic approach for regulating the governance of wild species and spaces	91
7.1	The changing political ecology of wildlife	117
7.2	Tropical forest tenure	130
8.1	The sustainable governance approach as a policy experiment	150
8.2	Trends in numbers of rare wildlife species in South Africa	153
9.1	Changes in wildlife ranges in the Midlands (1975–1984)	163
9.2	Simple financial calculations for comparing the profitability of wildlife and livestock	163
9.3	Meat production from cattle and wildlife reflects range condition on Buffalo Range	167
9.4	Return on investment from cattle and wildlife enterprises in Zimbabwe 1989/1990	169
9.5	Policy analysis matrix	172
9.6	Summary of PAM results for eight wildlife properties in South Africa in 2010	173
9.7	Trends in wildlife and livestock populations on private land in southern Africa	176
10.1	The alignment of goods, rights, and owners in public and sustainable governance approaches	184
10.2	A comparison of the sustainable governance approach with old-style public conservation	184
10.3	The structure of cross-scale governance regimes in Zimbabwe	193
11.1	Historical events leading to CAMPFIRE	201

11.2   The rate of growth of the early CAMPFIRE programme   218
11.3   Phases in the devolution of proprietorship   219
11.4   Measuring progress towards fiscal devolution by 'following the money'   220
12.1   Policy guidelines for community-based wildlife management agreed at
       LIRDP's Inter-Ministerial Review and Policy Committee in 1996   231
12.2   A comparison of attitudes and understanding of community finances in
       different phases of the community programme in Luangwa   239
12.3   Performance of 20 CBNRM communities in southern Africa   241
15.1   Example of checklist approach for a financial and economic assessment of
       CBNRM resources   290
15.2   Checklist to assess the forces for and against participatory governance   296
15.3   Structure of quantitative livelihood survey   299
15.4   Results chain and indicators for a community wildlife programme   307
16.1   Inclusive and extractive institutions   317
17.1   Practical guidance for allocation functions across hierarchical scale   351
17.2   Conservation addictions and their antidotes   355

BOXES

2.1	Institutions, organisations, and governance	14
2.2	Language, human reciprocity, and exchange	16
2.3	Human group size and Dunbar's number	18
2.4	The origins of land tenure	21
2.5	Early democracy in Athens and Rome	22
2.6	Extractive and inclusive forms of governance	25
2.7	The newness of nation states and democracy	27
3.1	Farming models	39
3.2	African savannas: ecological complexity and institutional challenges	44
3.3	The extent and losses of tropical forests and drylands	45
4.1	A brief introduction to 'budget-funded institutions'	53
4.2	The role of state agencies in conservation	55
4.3	The perpetuation of communal lands in South Africa	57
4.4	17th-century English rhyme that stands the test of time: 'The Goose and the Commons'	58
4.5	Four levels of social and economic analysis	65
5.1	A southern Africa perspective on rhino regulation and policy	92
5.2	Defining terms used with decentralisation	93
6.1	The concept of comparative advantage	103
6.2	The false economy of subsistence use and alternative livelihoods	109
6.3	Economic versus financial analysis	111
7.1	Thinking beyond moral certainties and simplistic solutions	129
8.1	Zimbabwe's Intensive Conservation Areas (ICAs) and Natural Resources Board (NRB)	140
8.2	Graham Child: game changer	142
8.3	Wildlife economics and the incoherence of 'demand reduction'	153
9.1	A brief history and geography of Zimbabwe	159
9.2	How does safari trophy hunting work?	165
10.1	Parks in transition	185
10.2	Price, proprietorship, and regime shifting	188

10.3   A quick guide to the oral and grey literature informing the sustainable
           governance approach   198
11.1   Useful documentaries for training   206
11.2   The CAMPFIRE principles (1991)   208
11.3   Reducing elephant offtakes through high-value sustainable use   212
11.4   Role play illustrating the differential ownership of cattle and wildlife and its
           financial consequences   213
12.1   Certification of VAG compliance and approval of release of funds   235
13.1   Property rights and social movements in community forest management in
           the Americas   253
13.2   Participatory forest management in Asia   254
13.3   Community fisheries management in South East Asia   255
13.4   CBNRM and REDD in Tanzania   256
13.5   CBNRM and wildlife in Namibia   257
14.1   Ostrom's design principles for long-enduring common property resource
           institutions   270
14.2   Murphree's principles for viable communal property regimes   271
14.3   Murphree's laws for scaling   272
14.4   Dahl's five criteria for democracy   276
14.5   Institutional conditions for effective collective action and CBNRM
           governance   284
16.1   Tools for inclusive CBNRM governance   318
16.2   Standard agenda for general community meetings   320
16.3   CBNRM principles for beginner communities   322
16.4   The 'rules of the money'   323
16.5   Description of roles and responsibilities in CBNRM communities and
           cross-scale agencies   324
16.6   Explaining the source of the money   326
16.7   Calculating individual dividends and shares for each village   326
16.8   Participatory activity-based budgeting   327
16.9   Format for a monthly/quarterly financial variance analysis   331
16.10 Recommended financial cycle for managing community finances   332
16.11 Community cash as tax   334
16.12 Monitoring form for CBNRM general meetings   335
16.13 Checklist of governance compliance criteria for inclusive CBNRM   336
16.14 A simple client orientated monitoring system   338

ACKNOWLEDGEMENTS

It is rare to be given an opportunity to thank the many people that have contributed to your life's mission and that have shaped your experiences and the way you think, who have assisted in your career, with whom you have worked with and for, and who have inspired, directed, and taught you lessons, both personal and professional. This book is a step in a journey that has taken many years, and to which many people have contributed, none more so than my late father, Graham Child. He absorbed me in the wonder of ecology, wildlife, and its observation; the global importance of parks and a deep understanding of the history and practice of wildlife governance and administration; the value of listening to and empowering all people, rich and poor alike, to live to their potential; the conceptualisation of problems by attempting to solve them; and the value and enjoyment of working with rural people. Growing up as the son of a field man, ecologist, and administrator, my father is the architect of the many ideas around which this book is built. I recount some sage advice he gave me as we investigated university opportunities together. Despite our shared passion for wildlife and ecology, his advice was that if I really wanted to make a difference, I needed to become a wildlife economist. This led to a BSc Honours in Agricultural Economics at the University of Zimbabwe, one of the best decisions I ever made, where passionate teachers like Kay Muir-Leresche and others showed me how academics could and should engage in real world challenges.

My earliest dream was to join the Department of National Parks and Wildlife Management in Zimbabwe. This book is deeply informed by the culture and philosophy of this remarkable agency. My colleagues and friends in the Department framed my thinking. Despite challenging times, including civil war and political transitions, they showed me that there is always a way to get things done, better, cheaper, differently, and that if norms were not delivering they needed to be challenged, thoughtfully and boldly. There was also no shortcut to 'doing things properly', and wildlife conservation required that clear policies and quality administration and systems were in place; intentions and opinions would never be enough. In this highly individualist department of rugged men, and a few thoughtful ones too, we were given clear authority and were expected to deliver the goods somehow, by making a plan, which is why I have so much confidence in the power of devolution, done properly. To this day being part of this brotherhood leaves me with a sense of pride and honour. Coloured by hilarious tales of hairy moments in the bush, these underpaid, hardworking, dedicated rangers and ecologists managed parks and wildlife to an exceptional standard, and pioneered wildlife conservation on private land and in communities. As conservationists, they played off the front foot, very much in contrast to conservation more generally, perhaps the true dismal science. Through bold policy and competence they expanded wildlife's footprint and an attitude of caring about the environment across the landscape and into the minds of many Zimbabweans.

My early career and my DPhil on the economics of wildlife and livestock in drylands owes so much to the many farmers and ranchers who were pioneering a new economy based on intact habitats and the utilisation of wildlife, carefully, profitably, and in preference to the range degradation and overstocking that was almost unavoidable with cattle. Men like Clive Style, Bob Swift, Clive Stockil, and many others hosted me, entertained me, and shared deep wisdom about the realities of managing wildlife and livestock, hunting and tourism. Even today, I use a film of Clive Stockil to demonstrate to my students that practical men think as deeply and clearly, and often more realistically, than we academics.

Perhaps the greatest privilege and fastest learning curve of my life was to be entrusted with coordinating the CAMPFIRE programme by my boss, Rowan Martin. Few people have been as lucky as I have to be able to participate in such an all-inspiring initiative and its courage for self-criticism and innovation. Here I thank an incredible community of practice for many ideas and hours of debate, enjoyment, and inspiration that are only partly reflected in this book, and not least for their dedication in recognising and empowering the rural people of Africa to manage and benefit from their own wildlife: Rowan Martin (National Parks) never one to be scared of breaking intellectual chains; the guruship, wisdom, and delight of Marshall Murphree (University of Zimbabwe); the clarity, grace, and professionalism of David Cumming and Russell Taylor (WWF); the energy of Simon Metcalfe, Rob Monro, Liz Rihoy, and others in the Zimbabwe Trust; the excitement of Richard Diggle as he took on life and the philosophical challenges of rural CAMPFIRE, my energetic introduction to actually working with and for black Africans by the swashbuckling politicians, Tap Mavheneke and Ephraim Chafesuka (CAMPFIRE Association); and dedicated community leaders and administrators like Bongani Sibanda, Jerry Gotora, Amos Mulaudzi, and many others, and the wonderful international volunteers like Cherry Bird and Bjarne Kaulberg that believed in this experiment called CAMPFIRE in the remote communities they lived in. I must extend similar thanks to the incredible team that was doing parallel work in Namibia – Chris Weaver, Chris Brown, Brian Jones, Richard Diggle, Greg Stuart-Hill, Patricia Skyer, Maxi Louis, Garth Owen-Smith, and Margie Jacobson to name but a few.

Much of this learning was brought together in the marvellous company and debates we shared in 'Sausage' – the IUCN Southern African Sustainable Use Specialist Group (SASUSG) – promoting the economy and democratisation of wildlife, often in direct opposition to global norms and pressures.

I cannot possibly name all these friends and colleagues, but foremost among these was my late friend, Ivan Bond, who shared so many campsites and community workshops with me.

In Zambia, I thank Rodgers Lubilo and his team for six years of inspiration and hard work as we learned together how to manage villagers to a high standard, and to the Norwegian officials who supported us so strongly, especially Ambassador Jon Lomoy and Gundbrand Stuve who was so tragically killed in the Yemen.

I thank my bosses Leonardo Villalon and Pete Waylen at the University of Florida for making such an unusual hire, and their adamancy that my role was to bridge scholarship and practice, incredible support that continues through Jane Southworth and Brenda Chalfin who give me the flexibility and encouragement to work so much in Africa. As a practitioner at heart, my time away from the coalface of conservation is made bearable and fulfilling mainly through the wonderful students who pass through my classes from all corners of the world, and the conversations and feedback over many semesters as I have struggled to shape and communicate the materials in this book. Not least I acknowledge the friendship and support of the CBNRM 'marines' – Drs Patricia Mupeta, Shylock Muyengwa, and Rodgers Lubilo, I am proud to say – who have worked so tirelessly in the field, who have helped introduce so many students to rural Africa, and who share my passion about community governance.

This book was made possible through a sabbatical at the Stellenbosch Institute for Advanced Studies (STIAS), through further experience gained working for The Nature Conservancy, and, more recently, the Peace Parks Foundation, and through the time, resources, and material preparation made possible through a partnership between the University of Stellenbosch, Copperbelt University, the Norwegian University of Life Sciences (NMBU), and the Southern African Wildlife College, financed through a NORHED project.

Natasha Bolognesi was a wonderful editor, and Sarah de Beer patiently adjusted and readjusted the figures for me. At Earthscan/Routledge I especially thank Tim Harwick who helped us steer five books into publication, as well as Hannah Ferguson and Amy Johnson, who have so patiently and professionally managed the process of completing this book, and Nicholas James Fox for copyediting the manuscript.

Finally, to the many colleagues and friends that have shared and shaped this journey in idealism, practicality, fieldwork, and academia over so many years – please excuse me if I have not listed you personally. I hope I have done some justice to the many ideas and experiences that I have shared with you all over more than thirty years and have pushed the boundary just that little bit further. I take accountability for any errors of thought or recollection in this book.

It is to my family, Adrienne and Thomas, that I offer my final thanks, for the many hours I have taken from them to complete this marathon.

PREFACE: SOME PERSONAL REFLECTIONS

For reasons of history, politics, norms, and ignorance, most drylands and forests in the tropics lack appropriate rules of governance and economy, resulting in extreme inefficiency and neglect, and a reinforcing cycle of poverty and environmental degradation. One would not enjoy playing cricket or rugby or football without sound, clear, or fair rules, because these games would quickly disintegrate into rancour, influence peddling and dysfunction. Why would forests and drylands and the wild species in them be any different? The central hypothesis of this book is that forests and drylands represent de-institutionalised (or ungoverned) spaces and species. Rural poverty and the loss of biodiversity are rooted in rules that are failing, and institutions that disempower local people and disadvantage the economy of wild species. We cannot make progress until we address these deep and underlying causes of poverty and biodiversity conservation.

Every day, article after article brings home the enormous damage that we are imposing on our planet: the depletion of soils and species by industrial agriculture, the loss of insects and birds, hurricanes and heat waves, and new data that shows the enormity of the scale at which we are squandering precious, life-giving biodiversity. On my own continent, Africa, populations of elephants and rhinos have plummeted in my lifetime, and the ordinary species that are not in the limelight have usually declined by even more. I would be surprised if we have as much as 10%, or even 5%, of our original wildlife, and I fear that the situation is actually worse than the media portrays. While the press gives so much space to the illegal wildlife trade, my personal horror is the loss of wild spaces to bad land use, the cut and burned trees and unproductive fields where wildlife used to roam, and the abuse of ecosystems and their functionality. Wildlife has declined and their habitats have been taken over by people and their domestic plants and animals, who have also harvested this wildlife without thinking. Yet there is no winner here, as these people remain desperate, often unable to feed their children every day, and at the mercy of the whims of climate and politics.

I have written this book because I believe that there are ways to avoid this catastrophe if we are bold and sensible enough. When I started my career in wildlife three decades ago, cattle fences stretched from one horizon to the other across southern Africa's savannas. Over-grazed savannas left me anxious and distressed about how much we are damaging fragile ecosystems and our disrespect for natural systems. Yet, following the insightful actions of a few astute men, I have witnessed and participated in a major rewilding of the sub-continent, and the restocking of vast landscapes with its original suite of incredible wildlife. This has not been perfect, but in the big picture it is a vast improvement on the dust and degradation associated with man's misplaced confidence in his ability to control nature, and the destructive influences of production systems based on exotic monocultures of domestic plants and animals in ecosystems to which they are not suited. Of course, we need to feed the world, but just as industrial agriculture and factory farming raise environmental and ethical questions in developed countries, in Africa the assumptions that we will solve rural poverty by promoting farming in agriculturally marginal environments needs to be deeply questioned.

I have also witnessed in very personal ways the transformative power of community-based natural resource management (CBNRM). The vital engagement of rural people, the hope in their eyes, and the recovery of wildlife leaves me in no doubt that CBNRM, properly done, is an antidote to the hopelessness of poverty and environmental destruction.

My doubts are not about whether CBNRM works or not, but why we do it so poorly. I fully acknowledge that fighting for the rights of rural people, and insisting on good governance, is often exceedingly difficult to achieve, especially where states are weak and unable to provide public leadership. I also accept that the CBNRM programmes that I draw on in the text are associated with rare combinations of people and circumstances. Nevertheless, these examples provide what businesses call 'proof of concept', easing the way for replication in less ideal circumstances.

These positive experiences gave me the confidence to write this book, and evidence for how to do so. Our goal is not only to stop the gradual erosion of the wildlife, but to proactively make policies and take actions that achieve the vision of a vastly expanded natural world built on the economic contributions that it can make to us all. I am the son of a man who dedicated his life to parks, and built park systems in three countries, and I have lived and worked in and around parks all my life. But parks are not enough, even if they one day reach the global goals of protecting 17% of the earth's land and sea. I am completely empathetic with E. O. Wilson's call to protect half of the earth. However, I don't believe that this will be achieved through public conservation, because these standard approaches to conservation are reaching the limits of what they can accomplish. It is only when we develop new and complementary paradigms for conservation, including the sustainable governance approach and CBNRM, that we will have the tools to make this happen. The fact is that public financing and management can only do so much; over much of the planet, wild species and habitats will have to pay for themselves, and to be managed by local communities and landholders. This is quite possible, and even a better approach, if we can get the underlying institutions of property, markets, and governance right.

This will not be easy. We are currently dealing with a crisis in political leadership. In a capitalist world, we are forgetting that the essence of free markets is transparency and competition. We are also forgetting the importance of public service and of public servants with the professionalism to shape the markets and institutions that guide unfettered competition towards the public good. In an increasingly centralised and transactional world, we are losing the attributes of community, while allowing special interests and celebrity politics with an illusion of understanding to hold sway over the management of complex systems (Sloman & Fernbach, 2019). Worryingly, there is evidence that ignorance is the very fuel that enables the strongly held and extreme views that capture the media limelight (Sample, 2019). Getting conservation right in a crowded and complex world requires systems thinking and institutional design similar to an engineering project. We would never build a rocket on the basis of public opinion or a celebrity tweet – why would we expect wildlife conservation to work this way? Perhaps the greatest challenge is the impending demographic tidal wave, which is difficult to contemplate, especially in Africa, where it is combined with weak governance, environmental degradation, and climate change. Indeed, IPBES (2018) warns that environmental challenges may result in the migration of as many as 700 million people. Given the destabilising effects of a far lower number of migrants on political systems globally, each and every one of us has a vested interest in resolving these challenges, even before we consider the contribution of biodiversity to the health and well-being of the planet. The consequences of not taking on the challenges represented by CBNRM are dire and global. Without effective CBNRM, where will the tragedy of ungoverned spaces end? Can the world cope with the accelerating degradation of forests and drylands on a vast scale, and what will happen to the hundreds of millions of people living in these places? As citizens of the world, can we afford to leave these problems in the hands of political leaders who appear not to have the best interests of the planet and their people foremost on their minds? Or do we need to insist that rural people in tropical countries have the same rights that have given us two centuries of escalating prosperity and freedom. If we don't, their pathway out of poverty and environmental collapse lies in vast urban slums or in migrating to the places where people do have freedoms and rights.

Having seen with my own eyes, more than once, the power of CBNRM to address the many angles of multi-dimensional poverty, I am puzzled by how rarely it is done properly. I have just re-read the introduction to Elinor Ostrom's *Governing the Commons* and Marshall Murphree's paper about scale. Both papers leave me with a sense of awe at the elegance, intellect, and sheer beauty of their arguments. Together, they provide a cogent theory for common property management, and the challenge of managing scale, that is directly applicable to the forests and drylands of the planet. Yet, on the ground, problems for which we have powerful theoretical justification and several practical examples of success are not being solved. Why?

Part of the problem is that many policy-makers and practitioners have never heard of Ostrom or Murphree, and most conservation and development projects act in ignorance of these theories. The way that we have structured the political economy of these systems is also a direct impediment to transforming them, creating a problem of reinforcing circularity, a vicious cycle if you will. The nationalisation and globalisation of wild resources means that local people are disempowered and local solutions are not emerging. Ironically, these failures lead to a further centralisation of power, and the further alienation of local peoples, although we know in our hearts that prosperity comes from empowering, not disabling, all people to live to their full potential. The symptoms of this are clear to see. Unlike my father's and my grandfather's generations, the people with the will and passion to address these problems are no longer based in rural areas working with rural people. Their skills sets have also changed. They have lost a deep understanding of rural systems and operational practicality. They live in air-conditioned offices, where they are rewarded for honing their skills to operate in a political market-place, skilfully spinning stories and writing slick proposals to capture ever more centralised dollars, with remarkable little accountability to the realities and resolution of problems on the ground. Thirdly, addressing rural poverty and environmental decline requires sustainable financing, well used. These challenges will not be solved through short termism and quick fixes.

The term CBNRM appears to have originated from a concern about natural resource management, tacking on the term 'community-based' in the 1980s as the centrality of communities in the sustainable management of wild systems was gradually recognised. In 1990, Elinor Ostrom highlighted the importance of governance. Strictly speaking, this book is about community governance, and should be titled CBNRG, but the acronym is even uglier and harder to pronounce than CBNRM. Ostrom's writings and those of her colleagues have led to a growing normative acceptance of the importance of rights-based approaches to conservation and development, supported by an expanding body of theory. The language of decentralisation and devolution, community empowerment, proprietorship, participation, and governance is gradually seeping into the language of development agencies, NGOs, and conservation and development projects, though it is difficult to argue that this narrative has been translated into practice.

While the intensions behind CBNRM are gradually being normed, the underlying theory remains incomplete and needs to be extended and tested in the rigours of real life implementation. Moreover, we will not succeed in operationalising community conservation and governance through good intentions; as in many complex systems, the design of CBNRM institutions and systems really matters, often in subtle but critical ways.

To add value to this field, consequently, I took two approaches. First, I wove the development of conceptual ideas together with practical experiences, hopefully strengthening both. Second, I recognised that CBNRM is a sub-set of what I have come to call the 'sustainable governance approach'. It takes time for many conservationists to grasp this, because it runs contrary to many long held beliefs or dogmas. The sustainable governance approach emerged from common property theory and the recognition that most wild resources are partly excludable and partly rival. They are best categorised as private goods with some common pool properties, not as public goods. This miscategorisation of wild resources means that there is a mismatch between the public management and ownership of wild resources, and their economic characteristics. The implication is that public ownership, except in specific circumstances such as parks and protected areas, is unfit for purpose more often than not. The hegemony of public conservation, it seems, may be an outcome of the success of the somewhat idiosyncratic and economically peculiar North American Model and public trust doctrine, rather than the conceptual integrity of this approach. If we were to design the institutions for wild resources theoretically in a laboratory, unencumbered by deep-seated norms and ideologies, we would be much more likely to come up with the sustainable governance approach than with current practice. I have grown up watching the positive transformation on wildlife and people in southern Africa of the devolution of rights and the development of wildlife markets and products. I have called the institutional reform that underlies this transformation the 'sustainable use approach' or the proprietorship-price hypothesis. My contribution in this book is to frame these ideas in economic and institutional history and theory.

Having been deeply involved in community wildlife management in southern Africa for three decades, and having read about parallel experiences in community forests and fisheries on other continents, perhaps my biggest surprise in writing this book is the lack of a cogent, transdisciplinary theory that provides a comprehensive framework for operationalising CBNRM. There are major gaps in understanding the economics of wild resources upon which communities depend. I was doubly surprised to see how little the internal governance of communities has been addressed. Even where there is excellent theory about key components of CBNRM, such as common property theory, it tends to be ignored in implementation. It is surprisingly rare, for instance, to find a community programme being evaluated against Ostrom's principles. My contribution, again compiled by weaving theories and practice into a single narrative, is to provide a comprehensive hypothesis about what CBNRM is, together with suggestions for how to implement it.

As a practitioner who has grown up working with rural people and wildlife in the drylands of southern Africa, I have learned how much talent and capability there is in many of these villages and, but for a quirk of birth, it could easily have been me struggling to eke out an existence in such disadvantaged circumstances. These people may benefit a little from projects, and even less if we do things for them in ways that are paternalistic and disempowering. What they really need is redesigned rules that encourage and enable them to unlock their own significant potential, together with some recapitalisation of social and natural systems in ways that reinforce dignity and self-responsibility. This book will, I hope, contribute at least a little to new processes, rules, capacities, and ways of thinking that enable rural people to uplift themselves, bringing back healthy forests and grasslands full of wildlife and biodiversity, because they depend on them, legally, economically, and spiritually.

Currently, we are doing a lot of things which are destroying wild environments and their enormous potential to finance themselves. As an agriculturist by training, I recognise the importance of food and farming, but we are unthinkingly promoting farming practices that are environmentally destructive. As a wildlife biologist at heart, my greatest pleasures are sharing wild places with Africa's incredible wildlife. However, I am deeply worried about the direction in which the field of wildlife conservation is moving. From where I sit, the narrative is being captured by armchair conservationists who may be increasingly connected, but are also increasingly disconnected from a complex world of people, livelihoods, wildlife, and politics, over-simplified by the narrow glimpse provided through televisions and twitter feeds. The focus on single species and even single animals underestimates the complexity of natural systems and the politics and economies that they are embedded in. We need to think more deeply and pragmatically, including about democratic choice and financing, if we are to succeed in the massive challenge of maintaining Pleistocene wildlife amongst Holocene people and their agricultural land.

I have tried to lay out the reality that the central challenge is to ensure that wild species can outcompete domestic species in the intense competition for land. Although it is controversial, I have often discussed the importance of hunting for wildlife conservation because, currently, hunting is a critical tool for wildlife conservation in many ecosystems because, ultimately, hunters pay the most. I am not a hunter, but the most intense wildlife experiences can come from hunting, as do some of the most passionate conservationists. In a field as impassioned as wildlife conservation, there is an unfortunately tendency to fixate on the means rather than the ends, and to think fast (and emotionally) rather than slow and systematically. Thus, many people who care deeply about wildlife are consumed by the need to ban wildlife trade and trophy hunting, and in taking their eyes off the ultimate goal and in sidelining the people who live with wildlife, may pose a far greater threat to wildlife than the illegal wildlife trade. If we are distracted from the reality that the root cause of the loss of wildlife on nearly 90% of the terrestrial planet outside parks is incomplete rights for wildlife, and the undervaluing of wildlife resources, we will fail in our mission.

Philip Lymbery's (2017) revealing book lays out the ethical challenges and environmental consequences of factory farming and industrial agriculture in shocking detail. For the health of ourselves and for our planet, we should be switching to wild foods and using wild products wherever we can. We cannot continue to cut down our forests, poison our habitats with chemicals and hormones, subject our health to food-like substances, or lock sentient animals in tiny cages. At the risk of committing conservation heresy, let me suggest that Chinese consumers who pay so much for rhino horn, and American executives who pay so much for wildlife trophies, are the pathway to the future. They give tremendous value to wild products in ways that markets for biodiversity, carbon, or water have so far failed to do, and it is our job as conservationists to shape the institutional playing field so that these values do good and not harm.

Forests and drylands have been the realm of hunter-gatherers for millennia. We face a stark choice between replacing them with farmers and farmland, or adapting these age-old practices of utilising wildlife to the modern and global world so that they are economically competitive and sustainable. The deluge of environmental horror stories is proof enough that business as usual won't do. In a crowded world, I will suggest that there is a window of opportunity to reverse the wholescale destruction of wildlife through disruptive innovation. This will require radical new institutions and consumer norms that encourage us to use wild products and services and to pay for them in full, provided these payments are reinvested directly back into wild-based production and the communities that own it. Indeed, I have suggested that these guiding principles would enable the Convention for the International Trade in Endangered Species (CITES) to regulate wildlife trade proactively, and with far greater positive impact – by encouraging trade wherever 100% of the proceeds are reinvested in the producer communities and landholders, and by questioning trade where these principles are not in place. I have written this book to bear witness to this alternative approach, and to explain why and how I think it works. I believe it has widespread relevance because wild species and spaces have enormous value, which we have not yet learned to translate into land use outcomes despite the urgency of these lessons.

In writing this book my goal is to describe and provide technical credibility to a new approach that seeks more wildlife in healthier habitats, where rural people are safe and food secure with the political and economic space to be resourceful, and where the use of wildlife is as natural, humane, respectful, and as efficient and non-wasteful as possible. I owe so much to the legacy of my late father who pioneered these ideas and implemented them. In many ways, I have simply added experiential depth and theory to the courageous policy experiment led by these pragmatic field men as they set out a simple but profound goal: to maximise the value of wildlife to the people who live with it. To quote my father, 'If wildlife is permitted to contribute meaningfully to their welfare, people will not be able to afford to lose it in their battle for survival' (Child, 1995).

1

INTRODUCTION

Poverty, conservation, and governance

LEARNING OBJECTIVES

This chapter introduces a broad overview of why current policies for wild resources may be delivering the wrong outcomes. A general hypothesis for a different approach is offered. You will learn:

1. That poverty and biodiversity are intertwined.
2. That humans are the only species that trades and makes rules.
3. The definition of wild resources or wild life and how this differs from domestic plants and animals.
4. That weak institutions and market failure affect wild resources.
5. About the management of wild resources as public goods, and why this might not be the best system.
6. That no one owns wild resources, and that they are being replaced globally by domestic plants and animals because people do own these.
7. That economic progress depends on security of person and property, and on markets and democracy. Do we need to apply these same ideas to wild resources?
8. That institutions have never been specifically designed to govern wild resources, and therefore that we may need to design new economic rules for them.
9. That one new strategy is the sustainable governance approach which depends on the price-proprietorship hypothesis.
10. A quick definition of sustainability.
11. A brief outline of CBNRM.

The challenge

We face the pivotal conundrum that the world's most spectacular animals live with the world's poorest people. This means that we cannot conserve global biodiversity without also taking on the challenge of enabling rural people and communities to take up their rights as citizens of humanity and as custodians of this biodiversity. The facts that bombard us daily, including recent dire reports that we have wiped out 60% of mammals, birds, fish, and reptiles since 1970 (WWF, 2018), suggest that the way we govern the interface between man and nature is not working well. These are the facts in front of our nose, and we should not delude ourselves by doing more of the same.

I have been fortunate to work with some exceptional initiatives and people who have addressed these complex challenges in public protected areas, on private land, and in rural communities in southern Africa. In working diligently to discern the facts through actionable science and evidence-based decision-making, these initiatives have, over the past seven decades, developed and tested a new paradigm of wildlife management. These new paradigms – the sustainable governance approach and community-based natural resource management – are based on the democratisation and promotion of the wildlife economy on private and community land respectively. The use of wildlife, and its democratisation, are both controversial. Nonetheless, they have rolled back the inevitability of wildlife decline and have led me into the fields of wildlife economy and governance, an arena often clouded in confusion and disagreement. This book is an attempt to analyse and share these lessons. While the specifics are local, the emerging principles are global and can be adapted to different wild resources and communities elsewhere in the world. My central argument is that wild life is no longer a public resource best managed through public agencies. Tackling the dual challenge of conservation and the upliftment of rural society will require an institutional and economic approach. We need new rules, or institutions, to govern wild life on a crowded planet if we intend to reverse the bleeding.

The success of *Homo sapiens* lies ultimately in the ability to cooperate in bands, clans, tribes, nations, and as a global civilisation. We are bound together in common enterprise through uniquely human myths, norms, and rules (Harari, 2014; North, 1990). As the only species with the ability to trade, we have created enormous wealth through innovation, specialisation, and trade (Smith, 1776). We are, indeed, economic man. But our success lies in the making, following, and enforcing of the rules that shape and enable this collective power. This book is about these formal and informal rules that structure human behaviour – known as institutions (North, 1990). We will not solve the intertwined problem of poverty and environmental loss by denying the power of the global economic system. The question is, rather, can we reshape the rules that govern our interactions with nature to get better outcomes for us all?

The biological rules of natural selection and evolution, unfortunately, do not apply to institutions. Institutions often become stuck in unhealthy configurations, held in a state of non-evolution by the people who are benefiting from the status quo, for decades, centuries, and even millennia (North, 1990), sometimes to the point of collapse (Diamond, 2005). As a species, we do not always get these rules right, as evidenced by the violence and inequity in the world, but nonetheless we seem, slowly, to be moving in the right direction (Fukuyama, 1992; Pinker, 2012). However, at the interface of environment and poverty in forests and drylands, we are patently getting the rules wrong: these systems and the plants, wildlife, and fish they support are depleted, while the people living in them are disproportionately vulnerable to violence, hunger, and deprivation. The United Nations reports that in drylands alone the forces driving environmental degradation create social instability and even violence, and may force as many as 700 million people to migrate to new homes (IPBES, 2018). This will, inevitably, disrupt the stability of the planet as we know it, and is the direction in which we will continue to travel, until we change the current rules and norms.

In this book I hypothesise that wild resources, and the people who depend upon them, are locked in institutional configurations that evolved on the frontier of the industrial revolution, were designed for a different age, and are no longer fit for purpose. Only by governing wild life and wild spaces differently, by aligning economic rules and forces in new ways, can we address the wicked problem of poverty and environmental decline. The hope that public ownership and management could control how wild species are used on the private and community lands that make up 85% of the land area outside protected areas is widely misplaced. The long hand of the law cannot possibly reach out into all the places where it needs to protect wildlife, and is ineffective where it is not welcomed (Wade, 1987). Neither will it be welcomed until the people who live with wildlife, and their age-old livelihood practices of hunting and gathering, are legitimised (and modernised) and they have a genuine seat at the rule-making table from which they have been so long excluded. Without this, we cannot overcome the deep resentment and resistance (Scott, 1985) that outsiders have caused when they impose foreign environmental norms on millions of rural people. The imperiousness of conservation policy continues to this day; somehow species like elephants and rhinos are so important that, in deciding their future, we do not need to listen to the opinions and wisdom of the local people and farmers that live with them.

Daniel Kahneman (2011) suggests that we solve problems in two different ways. Usually by thinking fast, emotionally, and repeating the patterns that we know. And occasionally, by thinking slow, carefully analysing systems, and inventing truly innovative solutions. My plea, in this book, is to think slow; to have an intelligent (slow) conversation, not an ideological one, and to open-mindedly consider that the solution to the problems that we share may be very different from our current norms and practices, and require very different models, principles, skills, and approaches.

We have all, at some stage in our lives, played or witnessed a game where unclear and poor rules have led to conflicts and arguments, turning something that should be fun into a mess. When rules are fair, and clear, by contrast, there is a lot of fun to be had, and you can focus on the game and not the rules or the politics and conflicts surrounding the rules. This book, essentially, is about rules and how they shape the relationship between man, nature, and the economy.

The study of rules, or institutions, is called new institutionalism, or institutional economics. It is gaining currency as an explanation for the difference between affluent and vulnerable societies (Acemoglu & Robinson, 2012), and some of the best known economic historians use the theory of institutions to explain why some societies prosper and others do not (North & Thomas, 1973). Historically, prosperous societies are rare. Living in an age of unprecedented freedom and prosperity, it is easy for us to forget that, until very recently, the conditions of life for most humans were bleak. The European Dark Ages, for example, lasted nearly a thousand years after the collapse of the Roman Empire, a period in which human populations, ravaged by war, famine, and disease, hardly grew. Then, at a critical juncture around the time of the Glorious Revolution in England, a series of transformations occurred. These included the Enlightenment, the Scientific Revolution, and the Industrial

Revolution. The governance of society was radically altered when the rights of man replaced the divine right of kings, and when the purpose of property was transformed to protect individuals rather than as a means of extraction by the ruling elite. 'Warre of every man upon every man' (Hobbes, 1651) was superseded by security of person and property, and the way political systems and the economy functioned was revolutionised. This was the time of Adam Smith. Great wealth was created as value was added to raw materials through the economic processes of specialisation, diversification, and exchange.

The Industrial Revolution led to the colonisation of much of the planet by Western society and ignited the flames of environmental destruction, which is where our story starts. Paradoxically, the response to environmental destruction was, in institutional terms, almost the exact opposite of the conditions that led to prosperity. The future of wild resources were placed in the hands of the state, not the people. Wildlife, writ large, became a public good, not the property of the local people who had lived with it since time immemorial. Indeed, wildlife governance failed to adopt the principles of ownership, self-governance, specialisation, diversification, and exchange upon which prosperity was being built. Rather, leaders like Roosevelt suggested that wildlife was destroyed by human 'greed'. As powerful bureaucrats, they took the future of wildlife into their own hand and nationalised it, banning commercial uses and taking wildlife out of the marketplace. This laid the foundation for conservation ideology for more than a hundred years. My hypothesis is that the underlying problem was incorrectly analysed. It was not human greed and markets. Rather, it was the lack of enlightened rules and local controls for channelling and controlling this 'greed'.

This era also saw the birth of national parks, an idea that has stood the test of time. Parks are an example of effective public management that is well aligned with the needs of society. Parks work – provided of course that the considerable benefits they provide to society are recognised in sustainable funding and quality management, and where public management is accountable to society.

However, governing wildlife as a public good on private or community land outside parks is an idea that has failed, and has failed catastrophically. The pressures of demography and a global economy shine an ever harsher light on the underlying deficiencies of the misfit between public goods on non-public land. Wild life and wild spaces are being replaced at an unrelenting scale by domestic species and farming, even when this farming is hardly viable, does little to reduce poverty, and is environmentally hazardous (Chapter 3). Wild species are also threatened by over-exploitation in the absence of institutions to protect them. Looking back, has taking wildlife out of the economy, and placing it under public management, been a mistake? Do we need to take the hard-headed approach of leveraging the high value of wildlife to pay for and protect itself? And if we do, can we achieve two things at once – reduce poverty by conserving biodiversity and using it wisely?

The interconnected problems of poverty, demography, and environmental loss

Demographic growth is fastest in rural areas where people are poor and women are disempowered. Urban slums in Africa, Latin America, and Asia have absorbed some of this demographic growth. However, this pressure is beginning to overwhelm agriculturally marginal forests and drylands, where livelihood options fade as desperate people cut down and burn habitats and scrabble to harvest wild resources like forests, forest products, fisheries, and wildlife. The poorest people on the planet are living with, or expanding into, our remaining intact habitats, creating unprecedented threats to global biodiversity and ecosystem services. Yet, despite high environmental opportunity costs, these people and communities are in no way getting rich. They are hamstrung from using wild resources legally or efficiently, so are forced to depend ever more on low-value agriculture and low-value uses of wild resources. They have become locked in a reinforcing cycle of poverty, demographic growth, and environmental loss.

Wild resources and wildlife: a brief analysis

Throughout this book I use the terms 'wild resources', 'wild life' (two words), 'wild spaces', 'forests', and 'drylands', interchangeably, to represent indigenous forests, wildlife, fisheries, rangelands, non-timber forest products and other non-domestic species that occur in largely intact habitats in forests, drylands, mountainous areas, and even the high altitude tundra. The important point is that wild resources and wild life have similar economic characteristics, but differ fundamentally from domestic plants and animals, and from habitats that have been converted into simple agricultural systems or plantations (Chapter 3).

Wild resources are not public goods, although they have been managed as such since the progressive era began in the early 1900s. Being largely excludable and subtractable – terms elucidated in Chapter 5 – they are usually private goods with some common-pool attributes that also provide global values. This introduces the root of the problem. There is a serious mismatch between the private and common-pool nature of wild resources and the public systems governing them.

It also turns out (Chapter 6) that wild resources and intact habitats may be economically more valuable than many of the domestic species that are replacing them, especially outside the sweet spot for farming. Losing species that have so much value does not make economic sense. Are these losses a direct consequence of governance failures and therefore avoidable? Wild resources are certainly more difficult to govern because they are complex and fugitive. However, lumping most forests, wildlife, and fisheries as public goods, rather than more correctly as private or collective goods, has ultimately under-valued them and made governance even more difficult. Moreover, they are part of complex systems – and complex systems are not suited to centralised systems of management or, for that matter, to agricultural simplification.

Centralisation, the mistrust of the abilities and aspirations of local people, and the emasculation of long-acting systems for governing local commons, was exacerbated by the authoritarian colonial state and the post-colonial one-party socialist state. These inclinations continue to be reinforced by the acceptance that the public management of local wild resources is normal. Moreover, people have expanded into marginal areas in large numbers only in very recent times, overwhelming the capacity of resources, but also the capacity of society to evolve new rules for governing these resources. Central systems are slow to evolve, but the local systems that might have done so have been rendered impotent. Many forests and dryland people, especially in Africa and America, have been traumatised by slavery, colonial subjugation, socialism, state capture, and conflict. Their social capital has been greatly diminished, they have lost their rights to land and resources (Chapter 4), and urbanisation and the modern economy has further disrupted community and family structures. At a time of great need these social ecological systems have been deprived of their adaptive capacity, and the ability and resourcefulness of people and communities to solve their own problems (Wall, 2017).

I am personally sympathetic to E. O. Wilson's (2016) call that half the earth should be set aside for wildlife and wild spaces. Where I differ is the means of getting there, and who we intend as the beneficiaries of this half-earth. I refer to this half as 'ungoverned spaces', because the rules in use are so bad that the problems of demography, poverty, and environmental decline feed voraciously on each other. As a conservationist, I am appalled by the loss of wild species and even more so by the damage to ecosystem processes. As an economist my horror is with the wastefulness and inefficiencies of the system. As a human, I am empathetic to the hardship and unfulfilled dreams and potentials of the people who live there.

It is neither possible, nor moral, to simply remove poor people from these environments in search of the 'greater good'. The case for 'alternative livelihoods', as currently conceived, is even worse. Tweaking domestic production systems is extremely difficult, and doing so will only further displace wild species. Surely, the alternative to the unsustainable use of wildlife is the sustainable use of wildlife, combined with the educational opportunities to enable people to seek lives away from the drudgery of marginal agriculture. Thus the pathway to half-earth may not lie in stopping the use of wild species and spaces, or in declaring them public protected areas. It may lie, alternatively, in deliberately governing and using them in better and more sustainable ways, while empowering local people as the custodians and primary beneficiaries of ecosystems that are rightfully theirs.

Sustainability: more, from less, for more, forever

There may be much overlap between the quest for sustainable development and the dreams of half-earth. As a proponent of the concept of sustainability I define it with the succinct ditty, 'more, from less, for more, forever'. This includes three fundamental ideas. First is the notion of economic efficiency and dematerialisation; the quest to generate more output from less input, or better livelihoods from a smaller ecological footprint. Second is the pursuit of economic justice, ensuring that more and poorer people benefit as much as the rich do. Finally, we need to maintain and expand the stock of environmental capital, living only off the flows that can be harvested sustainably.

We misidentify the problem when we blame it on the use of wild resources. After all, we evolved as hunter–gatherers, using, and deeply in touch with, our environment. Indeed it is the urban critics, supposedly so connected, that are disconnected from the realities of rural lives and environments, or their own environmental footprints. More correctly, the problem is with use that is inefficient, inhumane, distant, and unaccountable, and that harms the capital stock. Unfortunately the rules that we currently apply guide us towards a wild life economy characterised as 'less from more'. Not only are wild resources over-utilised, with a huge environmental footprint, but this use is economically inefficient, favouring low-value, wasteful uses, and leaving people hardly better off. The nub of the problem is that priceless wild resources (Costanza et al., 1997) are made worthless to the people who live with them, or are extracted to fuel the modern economy without reinvesting in the resource base or benefiting local people.

This situation is avoidable. Both wild resources and rural communities are the victims of weak institutions and the process of 'de-institutionalisation' (Chapter 4). Weak institutions allow colonial states, predatory elites, and the modern economy to profit by extracting resources from rural people and environments without paying the full cost of them, thereby de-capitalising wild spaces. 'Rent seeking' and 'free-riding' disproportionately enrich those in power, providing the incentive for perpetuating failing configurations of rules in the global modern economy.

The institutions for governing wild resources are so egregious that we waste resources while over-utilising them. We are far away from the boundary at which conservation and development are optimised. Consequently, conservation and development are only a zero-sum game when we ignore the role of institutions. With institutional reform, it is quite possible to have a lot more of both. My argument is that carefully crafted 're-institutionalisation' can conserve wild resources by making their use more efficient and profitable. By placing the control of wild resources democratically in the hands of landholders and communities, wild resources can provide a foundation for local livelihoods. If wildlife contributes meaningfully to the livelihoods of local people, they will not be able to afford to replace it with domestic species, or over-utilise it (Child, 1995).

Two hypotheses: price-proprietorship and face-to-face governance

This leads to the central hypotheses of this book. The first half develops what I have called the 'proprietorship-price hypothesis'. This suggests that better conservation-livelihood outcomes arise where people fully own wild life as individuals or groups (proprietorship) and can utilise it to best advantage including through global markets (price). The goal is to maximise the value of wildlife to the people who live with it. If wildlife contributes meaningfully to livelihoods, people will not be able to afford to lose it in their battle for survival (Child, 1995). This bold hypothesis lies at the core of the 'sustainable governance approach' (Chapter 10) which is controversial, at least to people who have normalised the public nature of wildlife (especially if they live in urban rather than rural areas).

The second half is about community-based natural resource management, or CBNRM. The core principle is that we need to entrust the people who live with wildlife to make the rules to govern it, by empowering the whole community – not just the elite – through deep, informed democracy. This is the same message that runs through the work of Elinor Ostrom (Wall, 2017). Finally, in complex systems we need constantly to adapt, by monitoring, learning, and problem-solving our way to a better future. Thinking slow, with judgements based on evidence rather than ideology, will bring us to our goals more surely than thinking fast.

Before we specifically address CBNRM we need to confront some of the assumptions that hamstring conservation, and to present and clarify some theory. Thus, Chapter 2 traces the institutional history of humankind. I identify the Glorious Revolution of England as the tipping point in the emergence of a society that is prosperous and fair, and wonder if this has important lessons for our cause. Chapter 3 reviews the impact of humans on the planet and highlights the institutional differences in the way we govern wild and domestic species. I suggest that the loss of intact forests and drylands is caused by low-value agriculture and exploitation, because the institutions for governing wild species are missing. Wild economies may well be more suited to complex ecosystems that simplified agricultural ones, and current outcomes may be neither economically wise nor ecologically sustainable. Our challenge is to develop new economic institutions for owning, using, and trading wild species, so that the high economic value of wild species is translated into land use outcomes – a sustainable economy built on high-value, wild, and intact ecosystems, not ecosystems converted to low-value domestic species. Transformation will require new economic institutions for owning, using, and trading wild species.

Chapter 4 suggests that we should not take the governance systems for wild species as given. The governance of wild resources has already progressed rapidly through several stages and is changing, but it may need to change a lot more. The underlying challenge is that the forests and drylands of the tropics have become 'ungoverned spaces' with severe weaknesses in their governing institutions, for reasons described in this chapter.

The next two chapters tackle the classification of wild resources in terms of their governance and economic characteristics. Being excludable and rival to a significant extent, most wild resources are, if we are to be accurate, private or common-pool goods. The result is an institutional misfit between the characteristics of wild resources and the historical norm of public management (Chapter 5). This institutional misfit may well be causing conservation to malfunction at the very time we need it most. Public ownership has undermined local resourcefulness, responsibility, and rule-making. Additionally, wild resources have been excluded from competing in the market on an even playing field (Chapter 6). Consequently, they have become exceedingly vulnerable to replacement by domestic species and to over-exploitation under open access conditions, especially where the public capacity to finance and protect wild species declines.

To understand the misalignment at the heart of the governance of wild resources, we illustrate how and when, historically, wild resources came to be publically governed (Chapter 7). Public governance of wild resources arose in the much emptier world of the late 1800s and early 1900s. It was a response to the threat of over-exploitation in the absence of rule at the frontier of the Industrial Revolution. Wild species were effectively nationalised by a few powerful conservationists in the era of progressive governance and colonial expansion. The world is now much fuller, and the threats are different and more intense. These threats usually take two forms – over-exploitation on the one hand, and neglect, under-investment, and replacement on the other. Over-exploitation was the catalyst for contemporary public management, and it still occurs where rights to resources are ill-defined and unprotected, including the rising scourge of the illegal wildlife trade. Nonetheless, the greater threat is habitat replacement, mainly by poor people desperately seeking a living in environments that can hardly support them (IPBES, 2018; Ripple et al., 2015). Our responses are informed almost entirely by the old challenge of over-exploitation, but the world is different now, and these responses accelerate the threat of habitat loss. We have failed to develop governance regimes that can manage both threats simultaneously.

The public management of wild resources on non-public land greatly under-prices wild species and, predictably, is not succeeding. Yet, such is the hegemony of the Rooseveltian and colonial public conservation governance model, that examples of economic approaches to wildlife conservation are rare. An important outlier is the wildlife economy in southern Africa, where the recovery of wildlife lends considerable credibility to an alternative approach based on reintegrating wildlife into the global economy under private and community ownership.

Therefore in Chapter 8 we begin to assemble the tools to simultaneously address the challenges of habitat replacement and over-exploitation. This learns from the practical policy reforms that led to the substantial recovery of wildlife over large parts of southern Africa. These policy reforms sought to get the economics of wild resources right, by reframing the rules for the wildlife economy. They maximised the value of wildlife and ensured that these values were internalised by landholders in the marketplace, so that they make the right choices about land use and switched to wildlife. Economics was perceived, broadly, as living wisely on a planet with limits and scarcities, where values are not only financial, but include altruism, a sense of ownership and responsibility, and cooperation and concern for the planet and the fellow species that occupy it.

Chapter 8 provides an example of how thinking differently about the governance of wildlife led to massive rewilding of land and economic growth. This is supported by Chapter 9, which describes some of the economic tools used to understand and promote the wildlife economy, and Chapter 10, which conceptualises the institutional lessons for governing wild resources and scale as the 'sustainable governance approach'. Given the high value of many wild resources and ecosystem services, this is highly likely to advantage natural systems compared to systems modified by farming and over-use to a much larger extent than is the case today.

The second half of the book discusses the theory and practice of transferring the sustainable governance approach to the rural communities that occupy most drylands and forests through CBNRM. Private conservation justifies the valorisation and ownership of wildlife by landholders. CBNRM is a (large) sub-set of the sustainable governance approach that addresses the governance of wild resources by rural communities. With communities (rather than individual landholders) acting as production units, it takes on the additional challenge of governance inside these communities.

CBNRM emerged spontaneously for forests, fisheries, and wildlife, in Africa, Asia, and Latin America, under various labels after the 1970s (Chapter 13). By and large, CBNRM has not lived up to its promise. There is a compelling and convincing theoretical case for CBNRM (Kellert et al., 2000), but seldom has it been 'done properly' in practice. Like Ostrom, I believe that there are enough examples to show, with confidence, that CBNRM works. Like Ostrom, I believe that the essence of CBNRM is deep democracy and entrusting wild resources to local people provided they adhere to democratic structures for self-governance (Wall, 2017). However, on the back of such excellent scholarship, my goal extends beyond studying systems that are already in place, towards proposing how to build and manage new ones.

There is, as yet, no clear understanding of what CBNRM is. Scholarship has teased out many of the aspects of it in great detail, but it has never developed a comprehensive model to link these together, with the largest gaps being economics and the valorisation of wild resources, the importance of scale, and the significance of inclusive governance. I have used my own experiences managing CBNRM programmes in CAMPFIRE, Zambia, and southern Africa in general to inductively address these gaps. CAMPFIRE (Chapter 11) illustrates how an understanding of CBNRM emerged through bold policy experiments and adaptive management over several decades. Chapter 12 digs down into the challenge of micro-governance (or deep democracy; Wall, 2017). I use the experience of CBNRM in the Luangwa Valley in Zambia to quantify the contrast between participatory and representational governance. I then broaden this analysis to southern Africa to test if these differences are as profound as they seem to be.

This adds a second hypothesis to the proprietorship-price hypothesis of the sustainable governance approach: that face-to-face inclusive governance is profoundly different to representational forms of governance, and lies at the very core of CBNRM. Wall (2017) suggests that this was the essence of Ostrom's thinking, which was radical, and that Ostrom believed 'that the more power people had to make decisions that affected their lives the better such decisions were likely to be'. CBNRM that ignores these insights pays large opportunity costs in terms of informed participation, equitable benefit sharing, elite capture, and the development of the trust and cooperation necessary for successful human endeavour. With this empirical background, in Chapter 13 I review the literature and history of CBNRM in broad terms and propose an operational model based on proprietorship, price, and inclusive micro-governance. The theoretical case for micro-governance is explained in Chapter 14, both in terms of what happens within the community and the external framing of these conditions.

'Doing CBNRM properly' requires the proper application of both the sustainable governance approach and of participatory governance within the community. Thus, CBNRM has four foundational conditions: the devolution of rights for wild resources to communities; maximising prices for wild resource; inclusive, face-to-face governance; and attention to the need to build local systems and capacities through adaptive learning. Chapter 15 provides practical advice for implementing CBNRM, while Chapter 16 elaborates on practical tools for establishing inclusive governance. Both these chapters emphasise the co-construction of knowledge and systems through the active participation of communities and the polyvalent layers of governance that service or constrain them. CBNRM is not about doing things for communities; it is about unlocking the capacities and capabilities of communities to do things for themselves.

A description of the institutional history of the current age of prosperity (Chapter 2) may seem out of place in a book about the contemporary challenges of wild life and communities. However, I have found that understanding where we have come from as a species, and the deep characteristics of our minds and the rules we depend upon for cooperation, to be immeasurably useful for interpreting some of the problems we face today. The emphasis on the Glorious Revolution that enabled humankind to step into the light of prosperity and self-fulfilment (Pinker, 2012; Sen, 1999) hints at the enormous power of the new principles it introduced – the primacy of individual rights over the divine right of kings, the wisdom of individual and collective self-determination, security of person and property, and the protection of these rights through inclusive governance. It is not coincidental that these very same principles are embedded in the sustainable governance approach and CBNRM: the right to self-governance, the application of the principles of proprietorship and price to wild resources, and the importance of inclusive governance, informed participation, and equitable benefit sharing at the local level.

CBNRM is unashamedly an economic approach to conservation and development. However, before we denigrate it with ideological terms like neoliberalism or commodification, we need to be clear about what kind of economics we are talking about. Laissez-faire capitalism certainly deserves little credibility, as the powerful make rules to benefit themselves, causing inequity and environmental harm. By contrast, allowing local people and wild resources to compete fairly and justly in the global economy is a very different matter. It is preventing them from doing so that is unjust, inequitable, and counter-productive.

2

THE EMERGENCE OF MAN, GOVERNANCE, AND RULES

LEARNING OBJECTIVES

1. The way we think determines the rules we make, and rules determine conservation outcomes much more than many of the issues we currently focus on.
2. Man has his origins in Africa, and we have cognitive capacities that enable us to work collectively. We are the only species that creates wealth through exchange.
3. Our ability to work in self-accountable small groups reflects our hunter-gather past and is defined by Dunbar's number.
4. The Agricultural Revolution led to large, sedentary societies that were usually stratified and 'extractive'. Ordinary farmers were usually worse off than hunter-gatherers.
5. Widespread prosperity only followed the Glorious Revolution, when the divine right of kings and feudal-type systems of land tenure were replaced by a belief in the rights of ordinary people to security and property.
6. We are concerned about the loss of 'ungoverned' spaces and species, where institutions reflect pre-Glorious Revolution conditions.
7. Solving the interlinked problems of biodiversity decline, land degradation, and poverty will depend on transforming these institutions – wild life's own 'Glorious Revolution'.
8. This will require major changes in the way we structure conservation, especially devolved approaches that utilise markets.
9. Whether we succeed depends, in large part, on the way we think.

Norms and institutions

The way we think

The way we think as a species determines the rules that govern us, and these rules and norms (or institutions) are reflected in the success or otherwise of human societies (North, 1990; Williamson, 2000). Around 70 000 years ago, Harari (2014) suggests that sapiens underwent a cognitive revolution that developed our unique ability to cooperate in large groups and even societies through governing myths and rules. Together with man's already formidable abilities to imagine and build alternative futures and things, and as the only species able to create wealth through trade, cooperation allowed man to dominate the planet. However, the way we cooperate, economically and institutionally, has enormous implications for the sustainability of the planet.

For most of the 200–300 000 years that *Homo sapiens* has roamed the planet, the basic building block of human society has been the band – groups of some 150 individuals with a considerable predisposition for accountability, cooperation, and reciprocity (Dunbar, 1993). We will return often to Dunbar's number – about 150 to 220 individuals – because of its importance for designing community institutions to govern local commons.

With the Agricultural Revolution, however, small egalitarian bands were replaced by hierarchical, authoritarian, and extractive societies in which the majority toiled in sickness and subservience for ten thousand years. These societies were held together by two ideas. First, the divine rights of kings – with all-powerful monarchs subject to no earthly authority – in which property served largely as a means of controlling people and extracting wealth from them. It was only when this was superseded by the idea that all men are created equal that we began the journey to today's unprecedented prosperity and freedom (and the enormous impacts on the planet; see Chapter 3) in a period of massive societal change, including the Reformation, the Scientific Revolution, and the Industrial Revolution. Such is the power of ideas and ideologies, and the institutional changes that emerge from them.

In this chapter I argue that human prosperity stems from an Institutional Revolution that unlocked the potential of ordinary people and their abilities to cooperate at an unprecedented scale. I set up the hypothesis that these changes have bypassed tropical drylands and forests, with the environmental problems that we face today being a hang-over of quasi-feudal systems where wild species and spaces are plundered and wasted in the absence of rules (institutions) that protect the persons, property, and wild species that live in these environments.

Our concern is with the world's local commons and marginal lands – the forests, drylands, mountains, and wetlands inhabited by people who are poor and marginalised. These are defined as 'ungoverned' spaces for two reasons. Economically, resources are unowned and unpriced (or mispriced) and therefore misallocated, while rural people have few political rights despite a history of trauma and abuse. This leads to the suggestion that empowering rural people, reflecting the experience of the Glorious Revolution, may be the only route to poverty reduction and environmental sustainability, through a combination of personal and economic rights maintained and shaped through inclusive political systems.

This explains why a book about communities and the sustainable use of wildlife begins with a foray into human evolution and 17th-century European history. Indeed, England's

Glorious Revolution may provide the institutional template for rebuilding local commons in ungoverned drylands and forests if combined with the attributes of reciprocity and cooperation and collective action that evolved in bands of hunter-gatherers.

In much of the rest of this book, I argue that the rules and norms governing wild resources, established in the early 20th century, are no longer fit for purpose; and I make the case that new ways of thinking and new rules are necessary to govern wild life on a crowded planet. Rules reflect the way we think. Daniel Kahneman (2011) suggests that we think in two ways. System 1 thinking is fast, unconscious, automatic – and error prone. System 2 thinking is slow, conscious, effortful – but necessary for complex decisions; it is also reliable. In the 1950s, the way we think about wildlife reached a major fork in the road. Enraptured by the story of Elsa the Lion, the environmentalists Joy and George Adamson began to anthropomorphise wildlife conservation into dramatic, single species, and even into single individual (Cecil the Lion) responses (Martin, 2012). Policymaking was gradually captured by an energetic, passionate, well-meaning, urban elite, ever more connected but perhaps also increasingly disconnected from rural realities, thinking fast, emotionally, and stereotypically, with quick, automatic (but error prone) responses to the threats to wildlife. Directing much of the money flowing into conservation, this side-lined old-fashioned, dependable, dispassionate, scientific management, and slow systems thinking, especially by field managers, game rangers, and rural people who live in the field. The end result is unhealthy bipartisan conflict between those who love animals and those who seek to save them, which has not been good for conservation (Martin, 2012).

In this book, wildlife is a proxy for the problems facing wild life – all life that is wild, including the wild spaces and species in forests and drylands. The starting paradox is that wildlife is priceless, but worthless. Wildlife is important, loved, and valuable to rich and poor alike. It is deeply ingrained in our psyche. It feeds our physical and spiritual needs. As a species, we cannot survive on a dead planet, or even on one that is highly domesticated. So why then is wild life, which is valuable in every way we think about it, disappearing so fast? This does not happen with iPhones or crops, cattle or cars, goats or golf courses, indeed with almost anything else we value highly. Wildlife is different. Is the way we think about wild life wrong, and is this why we have got the institutions for governing wild life so terribly wrong, especially in the tropics?

What are institutions?

Institutions are the 'rules of the game', analogous to the rules in sport (North, 1990). They consist of written and formal rules, and unwritten and informal norms. Institutions don't arise on their own, but are shaped by the cultures, histories, and norms of societies (Williamson, 2000). Indeed, they are embedded in the very way we think. They are strongly path dependent, and change only slowly and incrementally, except in rare circumstances (one of which is the subject of this book). Institutions differ from organisations, which are the groups of players engaged in purposive activity. Both reflect the ideologies and mental models of the societies concerned (North, 1992). Governance is the process by which the rules (institutions) are formed and stewarded (Hyden and Court, 2002) (Box 2.1). Outcomes, including environmental outcomes, reflect institutions, so if we want that better outcomes for wild life we will require better institutions for governing it.

BOX 2.1 INSTITUTIONS, ORGANISATIONS, AND GOVERNANCE

- Institutions: the formal and informal constraints for governing society; the rules of the game.
- Organisations: the players.
- Governance: the formation and stewardship of the rules.

It is well known that the differential performance of economies and societies is strongly related to the quality of its institutions (North, 1990). Rules may be invisible, but like Voldemort in Harry Potter, they are immensely powerful and ever present, and bad rules and bad outcomes have a remarkable power of persistence. Unlike species, institutions do not evolve inexorably towards providing the greatest good for the greatest number. Poorly performing institutions persist, because rulers devise and maintain them in their own interests. The governance of wildlife is no different. We cannot rely on natural selection to remove unhealthy institutions from the population. The opposite is more the case. Unhealthy institutions survive, for hundreds and even thousands of years – think of the misery of the feudal Dark Ages. The terrible news we hear daily about the demise of life that is wild is a reflection of our failure to adapt institutions to our rapidly changing and ever more crowded world. However, before we analyse the institutional governance of wild life in the past century, we need to understand why man is so peculiar in his reliance on institutions, and how these have evolved (Table 2.1).

TABLE 2.1 Chronology of the evolution of man

4 m years ago	Bipedal gracile hominids (Australopithecines) emerge as nomadic scavengers in East Africa.
2.8 m–1.5 m years ago	Hominid brain size begins to increase, and 'Handy Man' *Homo habilis*, the first of the genus *Homo*, begins to use and make tools.
1.9 m–70 000 years ago	'Upright man' *Homo erectus* invents more sophisticated tools, spreads out over much of Asia, lives in groups or bands that characterise hunter-gatherers today, uses language and fire, and shows evidence of imagination and the ability to trade. *Erectus* is probably driven to extinction by a catastrophic volcanic event, or perhaps *Homo sapiens*.
200–300 000 years ago	'Wise man' *Homo sapiens* evolves in Africa.
70 000 years ago	The Cognitive Revolution. Man develops an unprecedented capacity to work cooperatively, and rapidly colonises the planet.
12 000 years ago	The dawn of agriculture and of stratified, hierarchical societies.
1542–1687	The Scientific Revolution (Copernicus to Newton).
1640–1705	The Institutional Revolution. Birth of liberalism: John Locke argues that each man has a natural right to life, liberty, and property, protected by the state. Inclusive institutions emerge in England following the English Civil War 1642–1651, the Glorious Revolution (1688), and the Bill of Rights (1689).
1760–1840	The Industrial Revolution.
50 years ago	The Great Acceleration threatens planetary boundaries.

Sources: Bethell (1998), Harari (2014), Murrell (2017) and Reader (1999).

The evolution of man and his institutions

Nomadic scavenger and hunter-gatherer

Africa's mammalian fauna transformed from the ancient forms romanticised in the film *Ice Age* into the familiar Pleistocene array between 4 and 11 million years ago. During this period proto-humans evolved from baboon-sized quadrupeds – which lived in trees and ate fruit near Lake Victoria some 18 million years ago – into the primates and bipeds that we recognise today as apes or humans (Reader, 1999). The search for the missing link reflects the intriguing 10-million-year blank in the fossil record between species like Proconsul and the appearance of hominids in Tanzania about 4 million years ago.

Earth entered a climatically cooler and drier phase around 10 million years ago. As the African forests retreated, proto-humans emerged onto the savannas, where walking on hind legs (bipedalism) provided a distinct evolutionary advantage. The transition from quadruped to nomadic biped marks the evolutionary step to true hominid and may have been quite rapid, with apes already carrying 60% of their weight on their hind legs. Reader (1999) suggests that the association between numerous fossils in East Africa and the great migrations of the East African plains is more than coincidental. Did hominids leave the forest to exploit a niche provided by the great East African migrations? Predators and scavengers, such as lions and hyaenas, were territorial and tied to their offspring in dens. However, 'gracile' Australopithecines (similar to 'Lucy', *Australopithicus afarens*, found in Ethiopia) could carry their young and follow the migration, supplementing their omnivorous diet with the unexploited fat in the marrowbones and braincases of the carcasses left behind. For nearly 3 million years (150 000 generations) Australopithecines were the only hominids in existence. They were probably widely dispersed nomads with a generalist lifestyle able to adapt to long-term changes in climate, leaving behind footprints like those of three small-brained gracile hominids who walked across the east African plains nearly 4 million years ago through the volcanic ash of Laetoli in the East African Rift Valley (Reader, 1999).

The first human species, 'handy man' (*Homo habilis*) appeared in East Africa and began to master Oldowan stone flake tools some 2.8 million years ago. He was followed by 'upright man' (*Homo erectus*, 1.9 million to 70 000 years ago), who used much more sophisticated Acheulean stone tools, and tools like needles for sewing skin clothes, bark rope for tying blades to shafts, and so on. This bigger-brained hominid obtained specific types of stone for specific uses from distant locations, perhaps through trade (Reader, 1999). For the first time, a species on earth could imagine something beyond what they see in front of them, set out to make it, or even to trade it.

Homo erectus used fire and cared for each other – features that require communication and language – and was probably the first hominin to live in bands like the hunter-gatherers with which we are familiar today. *Homo erectus* was highly adaptive and successful, occurring widely in Africa and migrating as far as China and Indonesia in Asia. After surviving for more than a million years, they died out suddenly 70 000 years ago. Their extinction coincides with global cooling caused by the giant explosion of the Toba volcano, but also with the expansion of *Homo sapiens* beyond Africa.

Out of Africa

Homo sapiens emerged some 200 000–300 000 years ago in Africa. Humans are so genetically homogeneous that they probably declined to very low numbers, with our species surviving by a thread, perhaps living off shellfish in caves on the southern coasts of Africa during periods of extreme climatic cold that transformed much of Africa into harsh drylands about 70 000 years ago. The modern humans first dispersed out of Africa some 130 000–115 000 years ago via modern-day Israel, but were unsuccessful and never established themselves. However, a small group of humans crossed the Red Sea about 70 000 years ago, reaching South Asia by 50 000 years ago and Australia 46 000 years ago, quite possibly filling a void left by the extinction of *Homo erectus* at about this time. Humans migrated northwards into Europe about 43 000 years ago and interbred with but replaced the Neanderthals which disappeared some 40 000 years ago. About 15 000 years ago, humans crossed the land bridge between present day Russia and Alaska and within a thousand years had reached the southern tip of South America. Humans may be fragile, but they are intelligent and highly adaptable, while attributes of language and cooperation gave them considerable evolutionary competitiveness. This competitiveness enabled humans to utilise, displace, control, and dominate other species. The dark side to this success is the extinction of all our hominin cousins, and most species of large animals. Modern man is now alone, but exceeding the ecological bounds of the small planet on which we live.

The rapid domination of the planet by *Homo sapiens* may be a consequence of a new set of characteristics that Harari (2014) calls the cognitive revolution. Most apes and monkeys live in groups, cooperate, and strengthen group bonds through mutual grooming and alliances. Humans have taken this to a much higher level, with language and myth becoming a powerful tool to support reciprocity, social sanctions, and mutual altruism in much larger groups and, later, societies (Box 2.2). Richard Wrangham, a British anthropologist, watching both humans and chimpanzees around an elephant carcass, suggests that human language and cooperation bring order to the chaos and excitement around the carcass far beyond what chimpanzees exhibit. Humans are uniquely able to work collectively in an economy, through reciprocity, sanctions, and exchange. Indeed, humans are the only species able to make the exchanges through which so much wealth is created. We are, indeed, economic man. Ostrom's refrain 'beyond markets and states' recognises the importance of market exchange, the role of the state in writing the rules governing these exchanges, but also the considerable capacity of humans to make subtle but important non-monetary trade-offs, especially when they live in small groups. These are the capabilities that we depend upon when we bring humans together to reduce transaction costs in firms (Coase, 1937), and it seems logical to use these same capabilities to manage complex environmental trade-offs and externalities by empowering local collective action (Chapter 8).

BOX 2.2 LANGUAGE, HUMAN RECIPROCITY, AND EXCHANGE

Primates are highly social animals, which strengthen social bonds through physical grooming. However, grooming is time consuming, and human language and gossip may have evolved as a much more efficient means of maintaining social cohesion, as well as for coordinating group activities. These advantages are illustrated by a comparison of chimpanzees and humans tackling the carcass of a large animal

(Reader, 1999). Both humans and chimps are excited by large quantities of meat, with a great deal of noise and chaos. However, the strongest chimpanzees get the most meat in a selfish free-for-all. Beneath the pandemonium, however, humans operate very differently, using language to divide tasks, settle debts, make deals, and maintain a semblance of fairness. Feelings of guilt, reciprocity, obligation, and altruism also play their part. Reader (1999, pp 104–105) illustrates this by quoting Wrangham:

> The significance of language struck Richard Wrangham [a renowned primate biologist] most forcefully on an occasion when a group of [Mbuti] hunters had killed an elephant … Excitement was intense, and appeared dangerously volatile as the animal was skinned and dismembered. In terms of activity and noise, the scene matched anything of a comparable nature that Wrangham had observed among chimpanzees … The noise was cacophonous, but amid the din patterns of negotiation became discernible. The hunters and those with immediate rights to a share of the carcass were told to honour the obligations of kinship and give meat to their relatives. Old debts and favours were settled in exchange for meat; new pledges were contracted. The talking went on for hours, doubtless reinforcing a long-standing web of reciprocal obligations that was fundamental to the social order of the region.
>
> Chimpanzees in a comparable situation would have gone berserk. They would have screamed and squabbled and physical strength ultimately would have determined the distribution of the meat, and there probably would have been some violence between competing individuals. There may have been some bad feeling among the Mbuti too, but aggressive tendencies were constrained by the intervention of other individuals. They could talk about their differences, and bring in the issues of what happened in the past and what might happen in the future. In short – they could negotiate. Talking reduced the fighting.

This quote illustrates the emergence of three things that are essentially human – reciprocity, social sanctions, and trade. Social sanctions, shame, guilt, and even altruism provide the mechanism for human cooperation and reciprocity. Moreover, humans are the only species able to create wealth through the process of economic exchange (Ridley, 2010); a human with two apples will trade off with a human that has two bananas, but a chimpanzee will not.

For 95% of our existence, humans lived in bands of about 150 individuals. Built into our genes are the abilities to work cooperatively and hold ourselves to account in groups of 150–220. This is called Dunbar's number (Box 2.3), defined as the cognitive limit to the number of people with whom one can maintain stable social relationships (Dunbar, 1993; Mccarty et al., 2001). Small groups are the basic building blocks of society, and must not be ignored when designing local collective action where scale is a critical factor (Olson, 1965). The principles of informed participatory democracy that I promote later in this book match closely with observations about the scale and function of early human society. When we get these wrong, and human groups are larger, they become vulnerable to extractive governance where the few benefit at the expense of the many.

BOX 2.3 HUMAN GROUP SIZE AND DUNBAR'S NUMBER

In the 1990s, the British anthropologist, Robin Dunbar, noticed that the size of primate groups was correlated with the size of their brain, and quite possibly their ability to communicate. Using regression data for 38 primate families (or genera), he predicted that the mean size of human groups able to maintain stable relationships without too many restrictive rules and enforcement was 148. This was also the approximate size of New Stone Age (or Neolithic) farming villages; conservative religious communities in America (Hutterites), which 'split' into two groups once they exceed 150 members; the basic unit of the Roman army; and so on. Russell Bernard and Peter Killworth used network analysis to provide a slightly larger number of approximately 290 (McCarty et al., 2001). It is generally accepted that cohesion in social groups occurs at a size somewhere between 100 and 250 actors.

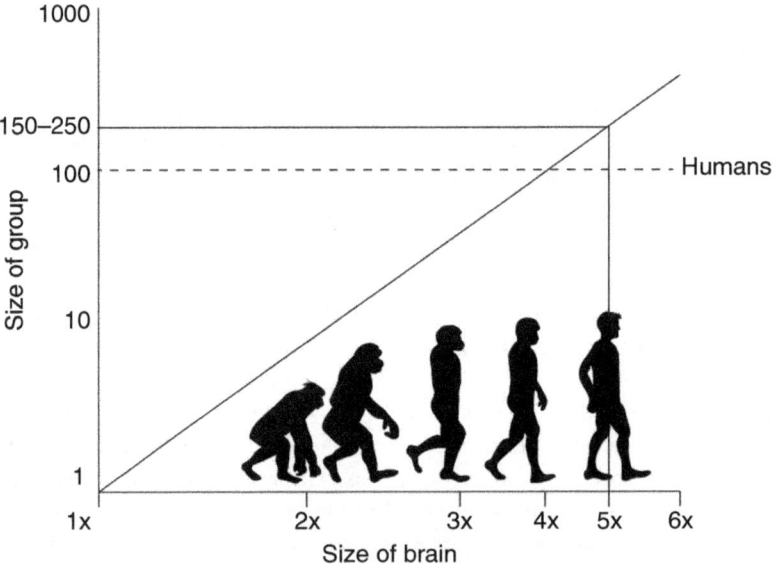

The relationship between brain size and group size in primates
Source: adapted from (Dunbar, 1998)

The fundamental unit of human society has, for thousands of years, been the band or group, but this contrasts significantly with modern lifestyles, economic theory, and social practice, which places much more emphasis on the individual. Beginning with the Agricultural Revolution, about 10 000 years ago, humans began to live in much larger groups as societies. Harari (2014) suggests that humans' ability to tell stories and make myths took cooperation to a new level, claiming that the ability to create unifying myths is our most powerful, distinguishing characteristic as a species (Gerson, 2015). Unifying myths, and especially religion, has been an essential force in binding together huge swathes of humanity in common norms – such as Christians and Muslims. Quite recently, humanity was bitterly separated into two blocks by deep beliefs, one built on the norms of communism and central planning, and the other around norms of the free market and democracy. However, myths alone cannot keep large groups of people together. For this we need norms and rules, or the concept of institutions articulated by Douglass North and others. Thus, as we began to settle in groups, institutions gained in importance.

Settling down: the dawn of agriculture, cadastres, and hierarchical society

As the climate warmed and stabilised 10 000 years ago ('the long summer'), nomadic humans began to live in more permanent settlements, often in large, fertile river valleys. They gradually domesticated the plants that they gathered, as well as some animals. The famous Russian botanist, Nikolai Vavilov, identified eight centres of origin for domestic plants (including Egypt and the Middle East, the Indus Valley, China, the Andes, and Meso-America), which coincide with the origins of sedentary human settlement, agriculture, and civilisation.

The Fertile Crescent and 'cradle of civilisation' encompasses the alluvial soils of the Nile River and delta, the Tigris and Euphrates rivers in Mesopotamia, and part of the Levant along the eastern Mediterranean where, around 12 000 years ago, nomadic people began to established permanent villages. They gradually domesticated wheat and barley and at least six other species of plant over the next 2 000 years, along with four of the five most important domestic animals – cattle, goats, sheep, and pigs. Agricultural and animal husbandry produced surpluses, and a significant expansion of population, including the first cities of tens of thousands of people about 5 000 years ago – namely Memphis (now Cairo) and Uruk (later Babylon) on the lower Euphrates in current-day Iraq. This transformed the way people are organised. As production increased through diversified roles and specialisation (Fukuyama, 2011), surpluses fed political centralisation and require-ments for administration, bureaucracy, and record keeping. Social stratification replaced egalitarian bands of hunter-gatherers with Big Men like kings, pharaohs, emperors, and chiefs living lives of immense affluence. Urbanisation encouraged the process of specialisation, diversification, and exchange; technological innovation including tools, manufacturing, cera-mics, copper implements, and weapons; as well as organised warfare, literature, and art. Complex societies were associated with extractive institutions that allowed political elites to prosper and extract a surplus from the majority of people ('subjects') (Acemoglu and Robinson, 2012) – perhaps the normal human social condition for much of our existence. Egyptian societies learned to survey flooded lands, marking them off as cadastres, a mechanism of tenure that allowed the elite to control farmers and extract a surplus from them. Cities were often held together through religion and monarchy, and perhaps this is the origin of the concept of the divine right of kings. Writing eased the recording of ever more complex transactions.

Ironically, the Agricultural Revolution may have left the majority of people worse off, locking farmers in a system of drudgery and narrow diets based around cereals and simple carbohydrates. Coupled with a higher disease burden, farmers were shorter in stature than hunter-gatherers, giving up the comparative leisure, excitement, and freedom of this lifestyle (Harari, 2014). As Thomas Malthus said in *An Essay on the Principle of Population*, humankind has 'a tendency … to increase, if unchecked, beyond the possibility of an adequate supply of food' so that farming communities often increased to the point of food insecurity. This lock-in continues to today, with a trade-off between food production and wild environments. Modern society has satiated itself with technical agriculture and factory farming. Declining productivity on 20% of the world's croplands and a 76% decline in insects in Germany (Hallmann et al., 2017) are warnings that technical agriculture may be a more serious threat to the planet than climate change (Monbiot, 2017), while, in the global south, and especially in Africa, we cannot assume that we can continue to hide poor people away in rural environments without causing extreme hardship and environmental devastation.

The need for rules and the emergence of formal institutions

In large societies, the face-to-face interactions of hunter-gatherer societies no longer ensure the fairness of interactions. More formal rules, laws, administrative systems, and written records are necessary. However, for most of the 10 000 years of 'civilisation', these rules have been hierarchical, personalised, and extractive (Acemoglu and Robinson, 2012; Fukuyama, 2011; North et al., 2009), written by the elites, to serve the interest of the elites. Olson (2000) uses the metaphor of the 'stationary bandit' to explain why rulers have an interest in stabilising and increasing the production of farmers for the self-interested purpose of monopolising the extraction of surplus from these farmers. Property boundaries and standardised measures initially emerged for very different reasons to their purpose today – to enable elites to govern and tax (Scott, 1999). Early Egyptian civilisations developed surveying, cadastres, and even writing to enable taxation (Box 2.4), while European feudal systems required peasants to pay taxes and provide (military) labour to their lords and peasants, with these ideas being perpetuated, for example, as the hut taxes and indirect rule of the British colonialists in Africa.

Centralised administrative systems are necessary to govern city-states, and warfare. Chinese civilisations emerged in the fertile Yellow River plain roughly 4 000 years ago. These dynasties, with a 'mandate from heaven', developed highly meritocratic bureaucratic systems in support of a supreme state, including professional civil service exams (Fukuyama, 2011). In India, the Indus Valley Civilisations (from 5 300 years ago) also developed writing, centralised authority, urban sewer systems, trade networks, and domesticated millets and legumes. Europeans were relative latecomers to this game.

Land tenure

One of the most important institutions is property rights. The purpose of early systems of land tenure was to provide a stable base for agriculture and other forms of production, so that the elite could extract a surplus (Box 2.4). Not only is this the antithesis of the modern interpretation of property, at least in liberal democracies, but the importance of property rights as the foundation for economic exchange and growth is so well known that it is often not even mentioned in economic text books (Chapter 6). However, property rights are invariably central to the problems of human destitution and environmental degradation. As human constructed systems, property rights often entrench inequities. Less recognised is the centrality of property rights to inclusive governance, including democracy. Indeed, Hayek (1944) argued that private property was essential to guarantee individuals freedom against a totalitarian state and suggests that the protection of property rights is a critical ingredient of liberal democracy:

It is only because the control of the means of production is divided among many people acting independently that nobody has complete power over us, that we as individuals can decide what to do with ourselves. If all the means of production were vested in a single hand, whether it be nominally that of 'society' as a whole or that of a dictator, whoever exercises this control has complete power over us.

(1944, pp. 103–104)

It is only when the rights of individuals and their property is protected that they can engage in the political action necessary to protect and deepen the democratic state. Thus in African communal lands and the European feudal system, individuals do not own land. They are obligated to the lord of the manor or the chief on whose land they live, and they cannot challenge authority without risking their access to land and hence their livelihoods.

Thus the institution of property, or its absence, plays an important role in modern society. Enlightenment philosophers, especially Thomas Hobbes and John Locke, intensively debated the purpose of land and government, not least during the turbulent English Enclosure movement when common land began to be titled to individuals before and during the Tudor period (1548–1603). Thus, Thomas Hobbes's (1588–1679) famous treatise, *Leviathan* (1651), written during the English Civil War, argues that strong central government (in his case, the monarchy) is necessary to protect life and property, and to prevent 'warre of all against all'. Without central authority, people would simply take each other's property by force in an endless cycle, making the life of man 'solitary, poor, nasty, brutish and short'. John Locke (1632–1704), by contrast, discusses the transition from the 'natural state' in which every person had an equal right to use natural resources, to his 'labour theory of property' which suggests that people acquire property by applying their labour to an unowned natural resource. Both philosophers defined the purpose of the state as protecting person and property. Property then changes hands through voluntary exchange, and voluntary exchange allows for peaceful coexistence by avoiding the acquisition of property by brute force. (As an aside, even today, land acquisition often requires that land is physically 'developed' (e.g. by clearing natural vegetation) to establish ownership. The wrinkle in this logic is that uncleared land and intact ecosystems may now be more valuable than land which is modified by people's labour.)

BOX 2.4 THE ORIGINS OF LAND TENURE

A cadastre is a register of real estate boundaries that describes the ownership, tenure, and precise location of individual parcels of land. The etymology (origin) of 'cadastre' has roots in the idea both of a tax and of a register. The early cadastre or title deed was a measure imposed on the land by rulers to enable them to extract a surplus from peasant farmers in the form of a tithe. Indeed, Scott (1999), in *Seeing Like a State*, argues that maps and measures are a mechanism that make local situations eligible to outsiders, including for the collection of data, control of people, and extraction of taxes. This is very different from the modern interpretation of property rights, which empowers landholders to record and protect their land rights.

Systems of land ownership have deep roots. Ancient Egypt's system of land administration dates as far back as 3 000 BC, including land surveys, titles, and taxes. The pharaoh owned the land, but delegated its use to deserving officials, temples, and individuals who were required to pay tax. Land boundaries were surveyed with primitive instruments and boundaries were marked with inscribed stones ('stelae'). Information similar to titles of today was recorded on tomb walls and papyri using hieroglyphics.

Roman Emperors used cadastral maps to appropriate land to the state from individuals. Similarly, the English Domesday Book records the Great Survey of much of England and Wales ordered by William the Conqueror in 1086. It established reliable boundaries for land, all owned by the king, and established knight's fees (or fiefs) for valuation and taxation. In the feudal systems that dominated much of Europe during the Dark Ages, land was held by lords and nobles. In exchange for permission to use land, and for protection, vassals paid the lord through tithes on output. However, services, including military services, largely collapsed after the Black Death when people became scarce relative to land, and the lords lost their hold over the peasants.

The forerunner of modern land administration is the comprehensive cadastral system developed for France. Napoleon replaced complex, contradictory, and illiberal royal and feudal laws with a revolutionary and codified civil legal system (the Napoleonic Code) that, amongst many other things, over threw feudal land ownership and liberalised property laws.

Source: Larsson (1991).

Democracy and the classical period

The emergence of humane and productive societies are often linked to the direct democracy of ancient Athens (c.508–146 BC), the representational democracy of the Roman Republic (c.509–27 BC), and even the benign dictatorships of the Pax Romana. Athens provided the 'classical literature' of Socrates, Plato, and Aristotle, and laid the foundation for Western philosophy and science, while Rome provided the model for modern representative democracy adopted by the United States of America (Box 2.5). The Roman Empire was successful for several hundred years under some exceptional emperors, but eventually it succumbed to a succession of weak dictators and leadership struggles. The decline of the Roman Empire laid the foundation for feudalism, and with its fall Europe entered the Dark Ages (from the 6th to the 14th century), a time of warfare and starvation. For a thousand years, emperors, kings, and cardinals imposed feudalism and religious conformity. Hegemonic rule stultified innovation, science, and social advancement. Perhaps most important of all, ordinary people owned few assets, had few freedoms, and little stake in the economy, and long-run per capita growth was close to zero (North et al., 2009).

BOX 2.5 EARLY DEMOCRACY IN ATHENS AND ROME

Athens is regarded as the birthplace of democracy, which derives from the Greek '*demos*' (people) and '*kratos*' (power). In the 5th century BC, some 30% of Athenian adult males (who owned land and were not slaves) were classified as 'citizens' rather than subjects. They participated directly in the Assembly (but women, and men that did not meet the franchise, did not participate). Decisions were made by a show of hands or by using black-and-white stones, with a quorum of 6 000 being required for some votes. Most office bearers were selected by lottery each year, not by election, an idea that bears inspection in community conservation.

The Roman Republic emerged in 509 BC with the overthrow of the Roman Kingdom. Rome developed a unique system of government with checks and balances, elected representatives, a mixed (and largely unwritten) constitution, and three arms of government: legislative assemblies, the Senate, and executive magistrates. The legislative assemblies involved all Roman citizens, who met in large numbers to pass laws and elect executive officials. This was balanced by a senate, or assembly of elders, of some 300–500 senators, mostly from rich noble families. The Senate issued decrees, which 'advised' executive magistrates on day-to-day civil administration. Executive magistrates were elected officials at a number of levels, the most senior of whom were the two elected consuls who shared power in civil and military matters for a year, playing a role somewhat like a modern American president. Complex checks and balances reduced corruption and encouraged performance.

The Roman Republic (509–27 BC) was successful, and conquered or allied with the whole of Italy, North Africa, Spain, much of France, Greece, and the eastern Mediterranean to become the Roman Empire. The 500-year Republic ended when, following destabilising civil wars and conflict, the Roman Empire was established in 27 BC. The Roman Senate granted enormous power to Augustus in this year, who replaced the Republic as a dictator. For 200 years, a sufficient proportion of the emperors were competent, defended the land rights of the ordinary soldiers and legionnaires who fought for them, and provided peace, prosperity, and expansion. The Western Roman Empire was undermined by power struggles amongst the dictators, civil wars, and abuses of power. Rome was sacked in AD 476 by Visigoths and Vandals from the north, heralding the European Dark Ages. The Eastern Roman Empire endured for another thousand years as the Byzantium Empire until 1453.

The Glorious Revolution

Medieval Europe (from the 5th to the 15th century) was built around the doctrine that a monarch derives the right to rule directly from the will of God and is therefore subject to no earthly authority. Monarchy, together with the hegemony of the church and church doctrine, provided the political foundation for aristocracy and feudalism, a political system in which land was held by a few elites in exchange for service and labour. The arbitrary rule, unfairness, warfare, famine, and great suffering of this period, in which life was 'nasty, brutish and short' (Hobbes), is forcefully conveyed in novels like Ken Follett's *Pillars of the Earth*.

In the 1500s, gradually the compelling concept of individual rights, and that all men are equal, emerged, as Europeans began to challenge the divine right of kings. The turning point away from medieval misery is often considered to be the English Civil War (1642–1651) and the Glorious Revolution (1688). The Glorious Revolution marks the transition from the personal rule of the Stuart kings to the rule of the Protestant William of Orange as a constitutional monarch without absolute authority. It is a watershed between the feudal paradigm of governance based on the 'divine right of kings' (extractive governance) and kings ruling with the 'consent of the governed' (inclusive governance), as described by the English philosopher John Locke in *Two Treatises of Government* (1689). In 1689, the English Bill of Rights limited the powers of the monarch, set out the rights of Parliament and free speech, and brought about an end to feudalism. It was a defining moment in the recognition of the rights of individuals, and these ideas echo through the United States Bill of Rights, the French Declaration of the Rights of Man and of the Citizen (1789), and the United Nations Universal Declaration of Human Rights (1948).

The doctrine of equality gave birth to today's liberal democracies. At the same time the purpose of property, including land, was also transformed. In 1628, Sir Edward Coke introduced into English law the dictum that 'an Englishman's house is his castle'. By 1763, this phrase become very popular, quoted even by William Pitt the Elder, later prime minister of England, in a famous court case. Cadastral property was no longer a mechanism for the elite to control and tax farmers − who were subordinate, insecure, and not in democratic control of their own destiny or property (Bethell, 1998). Englishmen were free in their own right. These new ideologies began to transform the economic system by empowering citizens with secure rights that could not be taken from them without fair and consensual remuneration. As Tom Bethell observes:

> the Industrial Revolution came to England first because the rule of law came there first, and in 'democratizing' the security of property the laws of England stimulated the creation of wealth … [giving] to every man [the security] that he shall enjoy the fruits of his own labour.
>
> *(1998, p 89)*

He then quotes Adam Smith in *The Wealth of Nations* to emphasise the tremendous change in human rights and security of property that had taken place by the 18th century. In the present state of Europe, the proprietor of a single acre of land is as perfectly secure of his possession as the proprietor of a hundred thousand.

Secure property rights and inclusive governance reinforce each other (Acemoglu and Robinson, 2012). Thus, 'restricting control [through the devolution of property rights] enlarged liberty' (Bethell, 1998), setting off a virtuous cycle of scientific, industrial, and social revolutions. The Age of Enlightenment (18th century) consolidated emerging new norms such as individual liberty, religious tolerance, and the separation of church and state. This coincided with the Scientific Revolution (1543–1687) and the Industrial Revolution (1760–1840) as freedom and innovation came together in revolutionary new forms of institutions, social contracts, shareholder companies, and production systems such as manufacturing and factories. Before the Industrial Revolution, expanding human populations rapidly absorbed any minor improvements in agriculture or technology, and the quality of life hardly changed for many centuries. The Industrial Revolution unlocked this Malthusian trap. Rapid improvements in technologies, manufacturing processes, and the harnessing of new forms of energy (including water, but especially coal-fired steam power) resulted in massive and sustained economic growth, technological advancement, and the improvement of human conditions, including nutrition, human stature, medicine, and life expectancy.

Europe's population exploded from 100 million (1700) to 400 million (1900). The later stages of the Industrial Revolution were accompanied by a massive widening of democracy and inclusive governance. This period also marks the zenith of European colonialism and plunder, and the governing of much of the planet by Europeans or through European norms, including the Americas, Australasia, Africa, and parts of Asia.

European growth was not uniform. North (2005, pp. 132–135) sheds considerable light on the preconditions for prosperity by comparing the inclusive strategies adopted by England and Holland from the 16th century, to the extractive strategies that led to the decline of Spain's powerful empire. Warfare dominated Europe in these centuries, and the costs of this warfare escalated as military technology improved – cannons and ships-of-the-line were much more expensive than pike men and archers. There were three ways to pay for war: borrow money, mainly from Florentine bankers, whose loans dried up when war loans could not be repaid; extract money from citizens through confiscation and taxation; or build a larger economy as people gained property rights and created wealth through competitive markets. Spain chose the second course. To fund its wars, the Spanish throne confiscated the silver plate of the aristocracy, milked the treasury, and taxed peasants so heavily that it damaged the agricultural economy. This caused economic decline, which severely reduced the power and influence of the Spanish and Hapsburg Empires, despite the wealth plundered from the New World. England, by contrast, and not always purposefully, evolved a modern inclusive state of shopkeepers and factories, independent farmers, and a highly efficient navy and civil service. Consequently, it was England (which had previously been an irritant to the much larger more powerful Spanish Empire) that become the primary global power. The difference between extractive (Spain) and inclusive (England) forms of government (Box 2.6) are a strong predictor of socio-economic performance.

BOX 2.6 EXTRACTIVE AND INCLUSIVE FORMS OF GOVERNANCE

The distinction between extractive and inclusive institutions was developed and popularised by Acemoglu and Robinson (2012), and overlaps considerably with North et al.'s (2009) comparison of personalised (rule by man) and impersonal (rule of law) economies.

Extractive governance (Big Man rule)

In 'extractive' institutions, a small group of individuals (the 'elite') exploit the rest of the population. The rationale of the centre is to extract wealth from its subjects to enrich those in control of the state, and to perpetuate its rule. Policies serve the (selfish) interests of the ruling class. Economic decisions and benefits are often personalised (according to who you know), so people in charge of profitable firms are often the direct relatives of the ruler. Thus, elites enrich themselves by acquiring property rights, often denying these same rights to the poor even when these resources include the land they live on or the forests they live off.

Inclusive governance (rule of law)

'Inclusive' institutions include the majority of society in benefiting from and governing the economy. Political and economic power is widely distributed through a combination of democracy, property rights, and an inclusive market economy. Economic decisions are impersonal and follow rules that seek the greatest good for the greatest number. Both the economic and political systems are based on individual rights and property rights. The role of the state is to protect these rights, and the state in turn is held to account democratically. This leads to a virtuous circle whereby inclusive economic institutions support, and are supported by, inclusive political institutions.

Sources: Acemoglu and Robinson (2012); North et al. (2009).

The role of the state

Beginning with philosophers Thomas Hobbes and John Locke, a new conceptualisation of the role of the state emerged. Previously, the state had been a mechanism to extract wealth from its subjects. From the 17th century, its role was transformed to protect the rights of person and property through legal systems, courts, police, and the military. Consequently, the modern state has a monopoly on the legitimate use of force to protect people's rights against plunder – by foreign powers, criminals, gangsters, monopolies, charlatans, and the like. The state protects the rights of person, property, and exchange through both law (courts) and force (police). Providing this security avoids the need for people to violently protect what is theirs – Hobbes's 'warre of every man against every man'. Thus, an irony of the decentralised free market is that it requires a strong state to design institutions, rights, and markets, and to protect them, so that the conditions for free competition and exchange are met. This is the basis of Max Weber's rational–legal nation state: the competent and powerful centre that seeks the greatest good for the greatest number.

Bureaucracy

Large-scale human endeavour requires effective administration and a governing bureaucracy. The history of bureaucracy is ancient. China is the origin of the meritocratic bureaucracy with examination systems established by the Song dynasty (960–1279) lasting a thousand years until 1905 (Fukuyama, 2011). Throughout this history, regimes have resorted to powerful measures to counter the temptations for officials to use public office for private gain, or even to take over as rulers of society. Compared to using eunuchs and slave-administrators to prevent state capture by civilisations as diverse as Ancient Egypt, the Chinese Imperial service, and the Ottoman Empire (Fukuyama, 2011), modern democracy is effective, if less brutal.

For centuries, Europe suffered from the temptations and powers of public office characterised by buying offices in order to extract a surplus (tax farming). By today's standards, the system was corrupt and inefficient. Learning from ancient meritocratic Chinese bureaucracy, and from their exposure to the Imperial Indian Civil Service, Her Majesty's Civil Service of the United Kingdom was modernised by the Northcote–Trevelyan Report in 1854. Meritocratic performance replaced the practice of buying offices. Recruitment and promotion came to be guided by exams and merit, rather than preferment, patronage, purchase, or length of service. This led to the emergence of the modern state and the idea of efficient and rational bureaucracies. Max Weber, an influential

German sociologist (1864–1920), suggested that bureaucracy is essential for the functioning of modern societies and large-scale organisation. He set forth the highly influential bureaucratic (or legal–rational) model based on well-defined and hierarchical offices and tasks, meritocracy, responsibility, accountability, and the idea of the professional career civil servant and of primary loyalty to the organisation. Similar reforms in Germany, France, the USA, and elsewhere led to a period of highly effective, centralised, and professional civil service that spread globally on the wings of colonialism, globalisation, and the Industrial Revolution. We see the effects of the dedicated public servants of the progressive era in the initial forming of conservation norms and practices in the 1900s (Chapter 7), and also in the revision of these rules and norms in southern Africa in the 1960s (Chapter 8). However, we forget that the nation state and bureaucracy are remarkably recent phenomena in the evolution of human society (Box 2.7).

BOX 2.7 THE NEWNESS OF NATION STATES AND DEMOCRACY

Democracy, nation states, and good governance are remarkably recent phenomena that we take for granted. The idea of a nation state is often associated with the Treaty of Westphalia in 1648 which established a new system of order in central Europe and introduced the concept of sovereign states. Nation states are a new idea: the United States (1776), Germany (1871), Italy (1815–1871), and African countries (1884–1885) being less than 250 and even 150 years old.

Democracy is also new. England enjoyed a limited franchise from 1688, but women first acquired the right to vote on the Isle of Man (1881) and then in New Zealand (1893). However, women only received full democratic rights in the United States in 1920 and in the United Kingdom in 1928. Under highly restrictive franchise rules, free black men could vote in the USA and even pre-apartheid South Africa, but genuine democratic rights only followed the US Voting Rights Act of 1965. African Independence began in the late 1950s, and ended with democratic elections in South Africa in 1994, but African democracy only began to take hold at the end of the Cold War and the fall of the Berlin Wall in 1989.

A model for prosperity – and CBNRM

This brief history suggests that the current age of prosperity is rooted in the transformation of the role of property and the rights of man which, by protecting person and property, allow individual people to fulfil their individual potentials (Figure 2.1). Thus, prosperity depends on the free-market system (left side of figure), which in turn depends on the design of these markets. Markets that are free from nefarious influence, in turn, need to be designed and protected by a strong state and rule of law. The risk of the state becoming extractive and self-serving is controlled by inclusive (democratic) politics, including freedom of information and accountable officials (right side of the figure). As noted above, both inclusive governance and free markets are rooted in the protection of the rights of citizens and their property.

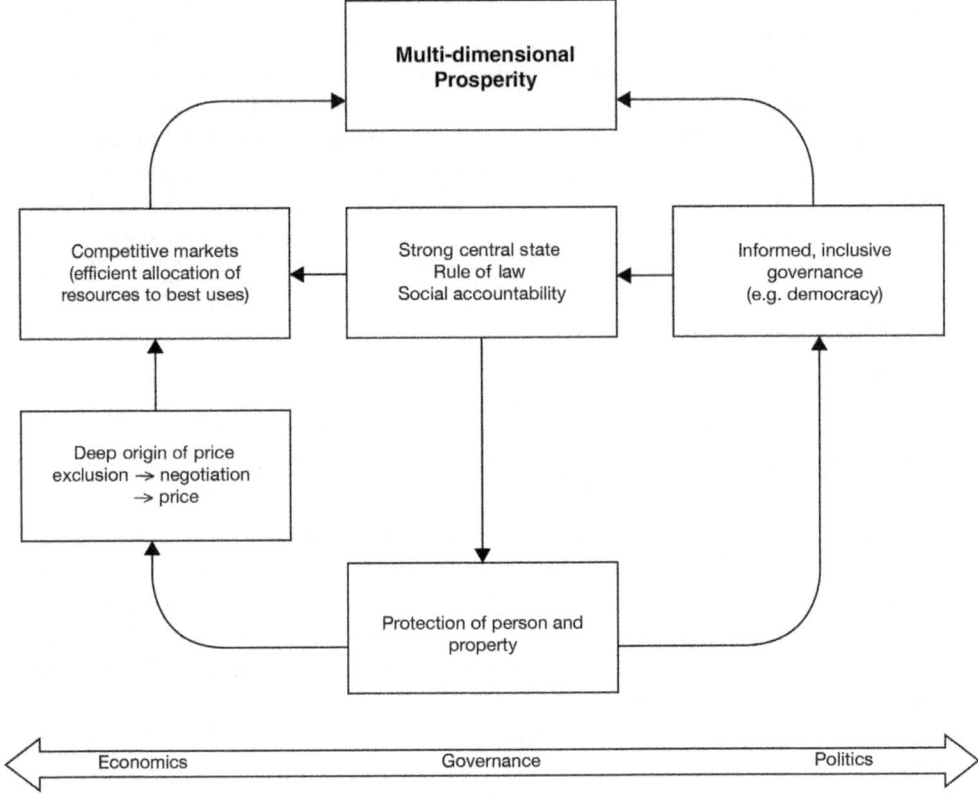

FIGURE 2.1 A model for prosperity and sustainability

Is there any reason not to apply this model to people and wild resources, including the forests and drylands of the global south? Do we need to devolve and entrust the rights to wild resources to communities through systems of governance that are inclusive at the national and global level (right side of Figure 2.1), but also at the level of the community where effective governance depends on all people who are affected by the rules being able to participate in making them (Chapter 14)? Do we also need to include the true value of wild resources and ecosystem services in the economic calculations made by people living on the land (left side of Figure 2.1)?

Lessons for community conservation

For much of human civilisation, Big Men, with some level of bureaucracy and administration, have ruled through extractive or personalised regimes, as we see with the pharaohs of Ancient Egypt, early Chinese Emperors and dynasties, and feudal Europe (Fukuyama, 2011). The well being of humankind has increased slowly at best, with many reversals.

Global prosperity emerged only after the 'invention' of inclusive regimes, where people have the right to retain the fruits of their labour and intellect, together with political systems that ensure that rule making continues in their interest (Figure 2.1). Consequently, the vast majority of countries with a GDP per capita exceeding $20 000 are liberal democracies (North et al., 2009), and highpoints in human society tend to be associated with broader systems of rights, such as humanity flowering briefly in Athens, then Rome, separated from our current prosperity by nearly 1 000 years of the Dark Ages.

What has this all got to do with the sustainable governance of wild life and community conservation? History emphasises the centrality of institutions and that societal outcomes depend on the nature and quality of rules and their implementation. Long-term economic growth and the absence of physical conflict is strongly associated with democracy (North et al., 2009) which has its foundation in individual security – to both life and property (Bethell, 1998; Hayek, 1944).

These are the very features that are lacking in the drylands, savannas, wetlands, and mountains of the world, inviting the hypothesis that insecurity of person and property are the root cause of the overlapping problems found in these lands – loss of biodiversity, land degradation, poverty, and even violence. I have called these 'ungoverned spaces', not because of violence and AK47s, but because they lack the institutions needed to protected people, property, and wild life. It follows that these ungoverned spaces need their own Glorious Revolution. Inclusive governance, community rights strongly applied (to land and wild resources), and markets (carefully designed) are the ingredients for a virtuous cycle of economic and political development necessary to break the current plunge into ecosystem collapse.

Ungoverned spaces are plundered by well-placed people who profit by not paying the full costs of their actions, and are a hiding place for industrial-age externalities like poverty and waste. New elites grow wealthy by claiming the very same modern property rights that they deny to rural citizens (Menard and Shirley, 2011), arguing that local people neither need nor deserve title to their own land or wildlife (often on the basis of history or custom). This logic maintains a frontier economy, with few incentives for sustainability, but many opportunities for rent seeking.

Wildlands, forests, fisheries, and wildlife are especially vulnerable to this phenomenon. They were nationalised by far-sighted officials seeking to protect them during the Progressive Era (Chapter 7). But nationalisation broke down systems of local control, replacing them with long accountability feedback loops and openings for private gains from public resources (this is the formal definition of corruption). This centralisation is often reinforced (unthinkingly?) by the global organisations that many of us work for.

Thus, in the 21st century, we still rely on public governance of wild resources which, as we will see in Chapter 5, are no longer public goods. We seem not to have noticed that low-value domestic species are replacing wildlife species on a global scale. Neither have we analysed why – because they enjoy individual and cadastral property rights, whereas wild species remain under national or even global governance systems. Moreover, the global or top-down planning approach that we apply to biodiversity conservation reduces local people to a servile status (Bethell, 1998) and social misalignment where the governing few (the bureaucratic-technical elite)[1] are at odds with the inclinations and aspirations of the people who live with wildlife.

Indeed, we should raise a red flag as soon as environmentalists depict themselves as heroes in the war to conserve wild life, especially if it is war against local people. My own research, and that of many of my students, shows that most local people like wildlife and parks (Mulindahabi, 2017). They value nature, for material and spiritual reasons and because they understand the benefits of ecosystem services. Why would this surprise anyone? These people live in and from nature, every day, and depend on it far more than we do. But when we pigeonhole them as the enemies of conservation, it is our own perception that may be in error. If local people are damaging the nature upon which their lives depend, we need to look critically for misalignments of the rights and rules governing wild species and spaces.

This leads to the central purpose of this book. We need new rules for a crowded planet, especially for ungoverned spaces and marginalised people. It has been my good fortune to see the transformational power of new rules first hand in the rewilding of private and community conservation in southern Africa, and in making wildlife meaningful to rural people. Although these rules are the polar opposite of the status quo – the public management of wild resources – they are highly complementary, because they apply to wild species on the vast areas of land outside public protected areas. My goal is to share my experience, using it to develop theory and practice for rewilding drylands and forests. Emphasising institutions (especially new and appropriate forms of individual and collective property rights), plus the reinsertion of wild life, writ large, into the global system of exchange, is controversial because it runs contrary to environmental norms. However, current approaches are not working, and I am confident that we have enough knowledge to design new institutions for governing wild resources in the context of private and community land with a high probability of success. I hope this book will provide guidance for how to do this.

Even if we understand what to do, doing it will not be easy. There will be considerable resistance to change, especially from people in positions of power who are benefiting from the current situation. This includes most of us who are sharing this book, whether we are government officials in charge of wildlife, forests, and fisheries, or are participating in international organisations and meetings.

Assumptions, and the way we think

Institutions, or the rules and norms that guide human behaviour, are deeply embedded in culture, including the way we think (Williamson, 2000). And, in this, environmentalists are seriously divided. Thus, Martin (2012, p 49) argues that 'the "old" science-based approach to conservation – dispassionate, data-driven, focused on habitats and suites of species rather than on narratives that anthropomorphize animals – is under dire threat'. Martin traces this to the Born Free generation which anthropomorphised wildlife, later manifested in the Kenyan-driven bans on hunting and the ivory trade, and the shift from 'broad-based ecological approaches to an obsession with individual animals'. Anthropomophisation has opened a major split in the culture and politics of conservation. Naming individual animals has shifted the focus of conservation away from understanding and managing the system pragmatically to idealising micro-components of the system – such as Cecil the Lion. Good intentions from afar, unfortunately, have seriously undermined conservation on the ground. They have drawn good people into nice offices, and distracted us from critical issues like local people and the economic reality of justifying and paying for land for conservation. If we want to conserve 'cute critters', we also need to conserve the intact ecosystem they live in (Martin, 2012) and to find a way to pay for this.

Affluent urban people are gaining ever more power to influence conservation. But, just as passion and good intentions are inadequate to build a rocket without technical know-how, so are they inadequate for managing complex social–ecological systems. Thinking fast and emotionally is not serving conservation well, and nor is the assumption, especially by urban people, that they can impose their moral certainties onto other people, who live in completely different circumstances and cultures. There has long been a cold war in which local people are depicted as being anti-conservation, or in which the presence of people is seen as antithetical to conservation (Terborgh, 1999). With illegal wildlife trade, we may even be entering a hot war in which the security of local people is disrupted not only by organised crime, but also by the shoot-to-kill policies of the conservation state and green militarisation (Cooney et al., 2017; Duffy, 2016). The current anti-hunting lobby, too, is self-defeating. It ignores the inconvenient facts that some 60–80% of the wildlife recovery in southern Africa has been paid for by high-value hunting, and that hunting has played a disproportionate role in the recovery of wildlife in America and Europe (Bond, 1994; Donnall, 2010; Naidoo et al., 2016; Rubino and Pienaar, 2018b). There is a need, too, to consider where fashionable approaches like 'demand reduction' will lead. If demand reduction is successful, wildlife will have no value to the people who live with it. This pulls in exactly the opposite directions from other popular ideas like payments for ecosystem services. I am concerned that, economically speaking, conservation policy is often incoherent. Too often, we rush with great passion on new crusades to 'save' another threatened species or ecosystem. Our knee-jerk reaction is to blame environmental destruction at the foot of human greed, population growth, trade, or poaching. Yet perhaps our own motivations are impure. Despite our passion and good intentions, we dramatise issues that will fill the financial coffers of our own organisations, becoming masters of storytelling and spin yet not holding ourselves to account for delivering measurable conservation. So the problems persist. And worsen.

Rather than thinking fast, perhaps we need to sit back and think slowly – ask the big questions about why our current solutions – business as usual – are failing. In this and subsequent chapters, we will trace the history of mankind and our relationship to nature, especially the rules that we use to govern it. My hypothesis is that it is in the rules that govern nature, and the changing of these rules – the Glorious Revolution mentioned above – that we will find the real solutions to enabling man to live in harmony with nature. This chapter is the start of a journey to frame conservation as being much more about institutions, governance, and economics than about conservation biology, and to think slowly and carefully (and in systems) about how to transform these rules.

Recommended readings and exercises

I highly recommend Glenn Martin's *Game Changer* for a beautifully written and insightful appreciation of the changing norms of wildlife management. Francis Fukuyama (2011) provides a comprehensive overview of the emergence of human civilization, with a far less Eurocentric view than I have provided, while Harari's (2014) *Sapiens* is highly readable and fascinating. Any scholar of institutions needs to read Douglass North (1990) from cover to cover, while Tom Bethell (1998) provides a partisan but strong argument about the relationship between property and prosperity over the ages.

Note

1 This is a term often used by Professor Marshall Murphree (in conversation *c.*1996) to refer to the over-sight of wildlife by environmentalists and experts at the global or national level, with little cognisance of the rights or knowledge of local people.

3

A BRIEF HISTORY OF MAN'S IMPACT ON THE PLANET

LEARNING OBJECTIVES

This chapter:

1. Suggests that the recovery of the environment in post-industrial society reflects strong institutions that are better at including the costs of nature in decision-making.
2. Suggests that environmental decline occurs in 'ungoverned spaces' that underprice the environment, so that waste is dumped and resources are extracted at below cost.
3. Describes how low numbers of hunter-gathers wiped out large mammals, except in Africa.
4. Describes agriculture and the role of farming in society, noting that high-input agriculture is important for food production, but that deforestation and degradation is driven mainly by poverty and low-input agriculture in a commons tragedy of global proportions.
5. Suggests that low-input farming may be ecologically mismatched to forests and drylands, and may be less viable than uses based on intact habitats.
6. Explains that the massive and increasing impact of farming on ungoverned forests and drylands may be an outcome of the institutions (property, markets) that favour farming and disadvantage wild species, rather than any ecological or economic advantages.
7. Introduces the concepts of total economic value and ecosystem services, and the important differences between economic and financial prices.
8. Uses simple models to suggest that natural systems may have an economic comparative advantage and to illustrate how policy reform can transform economic values into land use outcomes.
9. Closes by suggesting that 'new economic institutions' are essential for governing wild resources on a crowded planet, especially in ungoverned spaces.

Man and his impact on the planet

Ever since *Homo sapiens* expanded out of Africa, man's power to impact on the planet has multiplied, as a hunter-gatherer, farmer, industrialist, and urban citizen. Even low populations of mobile hunter-gatherers, probably less than 10 million people in total, devastated wildlife in their pathway as they gradually populated all corners of the planet. Man then settled down in sedentary, agricultural societies during the warmer, wetter Holocene, beginning some 10 000 years ago. This resulted in local environmental impacts, but human populations remained quite low, with perhaps 300 million people at the time of Christ, when even Europe remained a wild place. Human populations grew only slowly, stunted by warfare, disease, and famine, reaching 1 billion people only in the early 1800s, with the majority living in the fertile farming areas of China, India, Europe, the Niger river in West Africa, the East African highlands, and Meso-America (Goldewijk et al., 2010). As late as 1700 or 1800, the tropical forests and drylands of the world were still sparsely inhabited, mainly by hunter-gatherers.

Man's power as an ecosystem engineer was magnified a thousand-fold by the Industrial Revolution, colonisation, and globalisation. Initially, temperate forests were harvested for timber and replaced by farmland. Following the Great Acceleration from the 1950s, man began to have planetary impacts, affecting the atmosphere (ozone depletion, acid rain, global warming), the cycling of key nutrients (nitrogen, phosphorous), pollution (through the disposal of waste products such as plastics, chemicals, and even medicines), and losses of marine fisheries, tropical forest, wildlife, and land productivity (Steffen et al., 2015). Our ecological footprint is huge – man now consumes some 40% of global photosynthesis. There is no question that man will continue to have a massive impact on the planet. The challenge is to understand the forces that increase or reduce these impacts, and to manage them better, especially in tropical drylands and forests.

Governed and ungoverned spaces

Does the recovery of environmental quality in the latter stages of economic development, following a U-shaped environmental Kuznets curve (Lomborg, 2001; Dinda, 2004), mean that we can live prosperously and sustainably on the planet? The Living Planet Index shows a slight recovery of biodiversity in rich countries since 1970 (by 7%), offset by much larger losses in middle income countries (31% decline) and low-income countries (60% decline) (WWF et al., 2012) (Figure 3.1). With humans appropriating nearly 40% of net primary production (Vitousek et al., 1986), and rich people using far more energy and resources than poor people, is this a genuine phenomenon, or does it merely reflect the deflection of the environmental footprint of affluent society across the planet?

Much of the 'recovery' of the environment in affluent nations can be attributed to strong institutions. Thus, robust property rights prevent excessive extraction of resources, or dumping of waste, because users must pay the full costs of these actions. Markets for most economic activities, and improving markets for environmental effects, improve resource allocation and reduce free-riding – including cap and trade systems and taxes for pollution, individual tradeable quotas for fish, payments for ecosystem services, environmental offsets, and so on. The political system is sensitive to the demand by affluent people for better environmental conditions, resulting in the clean-up of rivers and smog, and the protection

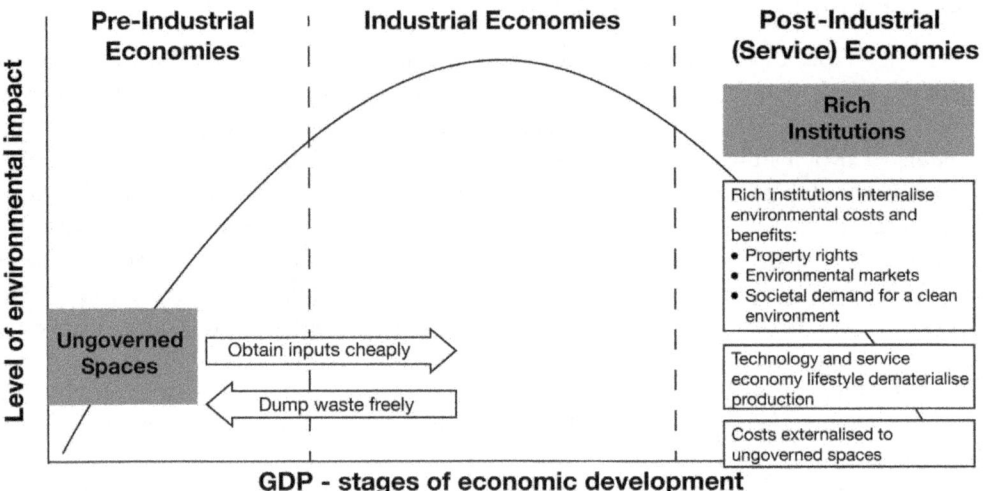

FIGURE 3.1   Drivers of the environmental Kuznets curve

of green space and endangered species. New technology and service-economy lifestyles may also partially dematerialise consumption, as urbanisation and high-input agriculture allows farmland to revert to forest (although the environmental costs of modern agriculture in terms of lost populations of birds and insects is currently being scrutinised). Affluent societies are managing their environments better as markets and political systems increasingly internalise the costs of environmental effects. However, the question is how much of their considerable environmental footprint is externalised into ungoverned spaces where its full costs are not accounted for – dumping plastics and hydrocarbons into the ocean and atmospheric commons, and sucking in huge amounts of environmental resources from elsewhere.

Where are we exceeding the safe operating space for humanity?

Scientists claim that we have transitioned from the Holocene to the Anthropocene, a new age in which earth's processes are dominated by man (Ellis, 2011). The world economy is depicted by classical economists as a closed system (left side of Figure 3.2), but it is not, so the assumption that it can expand continuously are wrong (Vatn, 2015). The globalised economy is an open system that demands inputs and raw materials from nature, and dumps waste back into the environment, but seldom pays the full costs of over-extraction or the dumping of waste and even poor people. This occurs differentially into 'ungoverned spaces'. This does not refer to violence, but to places that are ungoverned because they lack institutions and rules, so that costs and benefits are neither internalised nor paid for properly, as shown on the right side of Figure 3.2. It is no coincidence that the places where we are exceeding planetary boundaries are associated with ungoverned space (i.e. where costs and benefits are not internalised) in three forms (Figure 3.3):

- Genuine public goods, such as the atmosphere and oceans, which are non-subtractable and non-excludable (Randall, 1983; Ostrom and Hess, 2007).
- Processes where the attribution of cause and effect, and costs and benefits, is difficult (such as biogeochemical flows).

CLASSICAL CLOSED ECONOMY
Goods and Services
Money (payments for goods and services)
PRODUCT MARKET
FACTOR MARKET
FIRMS
HOUSEHOLD
Money (wage rent and interest)
Factors of production (capital, land, labour)
Waste
Poor people
Raw materials
including biodiversity
UNGOVERNED SPACES
Local Commons
Global Commons
Biogeochemical cycles (attribution problems)

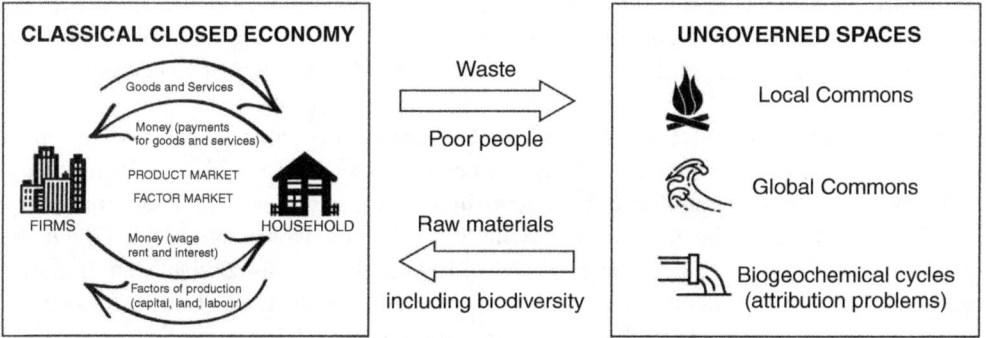

FIGURE 3.2 The 'closed' economy and ungoverned spaces
Source: Child, B. (this volume)

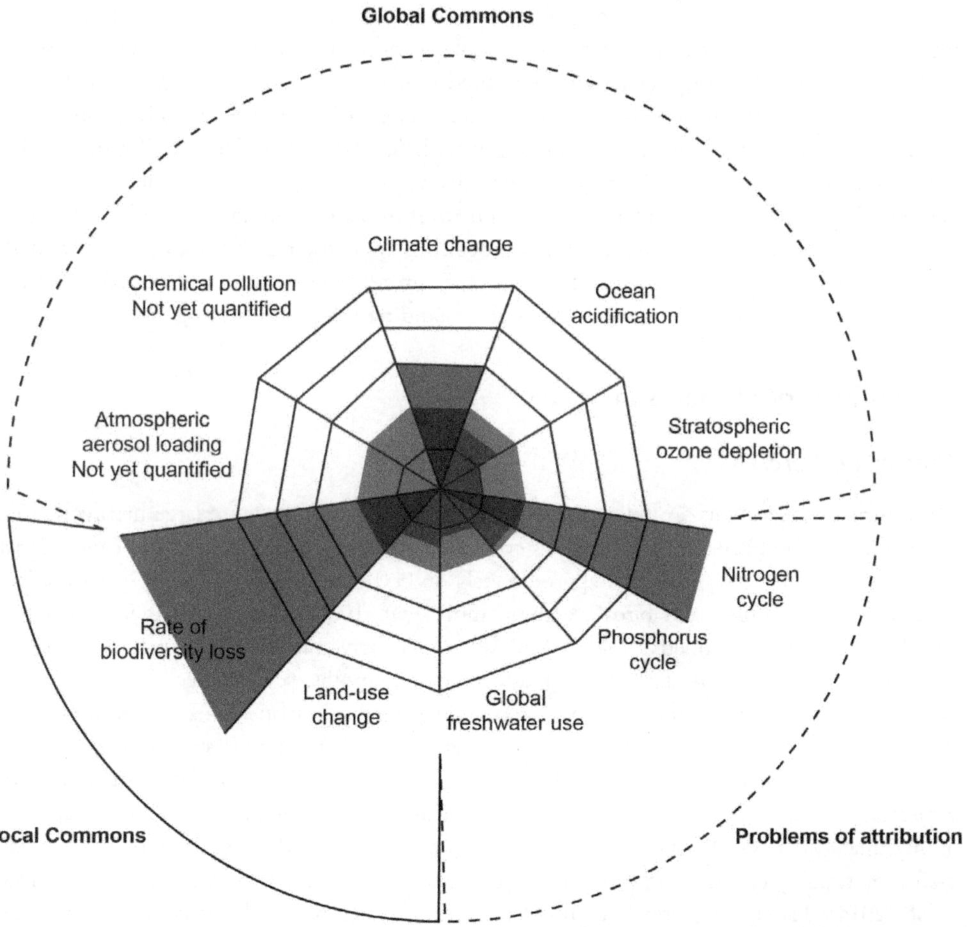

Global Commons
Climate change
Ocean acidification
Stratospheric ozone depletion
Nitrogen cycle
Phosphorus cycle
Global freshwater use
Land-use change
Rate of biodiversity loss
Atmospheric aerosol loading Not yet quantified
Chemical pollution Not yet quantified
Local Commons
Problems of attribution

FIGURE 3.3 Exceeding the safe operating space for humanity in ungoverned spaces
Source: Modified from Rockström et al., (2009).

- Places, especially drylands and forests, where institutions are outdated, crude, and mismatched to the characteristics of the resources in question, and fail to properly internalise the costs and benefits of land use and biodiversity into the economy.

Man fails to internalise the full costs and benefits of the economy for two reasons – incomplete or missing property rights, and difficulties of attribution. This results in (to use the dry language of economists) an under-supply of healthy oceans, atmosphere, and biodiversity. Most of the nine earth systems identified by Rockström et al. (2009) are public or near-public goods (being non-excludable and non-subtractable) and are categorised correctly as global commons. However, neither biodiversity nor land use change are global commons, as they are both subtractable and excludable (Chapter 5). Solutions do not therefore lie in global governance, but in empowering local commons for global benefit.

Land and biodiversity are subtractable (i.e. they get used up), and they only lie in ungoverned spaces (i.e. are non-excludable) for historical and political reasons. These ungoverned spaces, where marginal agriculture, biodiversity conservation, protected area management, and rural poverty overlap, are the subject of this book. Our quest is to create new and appropriate institutions that address poverty and restore complex ecosystems and their considerable values.

Wealth flows from ungoverned to governed spaces, and waste flows the opposite way. We need to address the dualism where affluent elites and societies, usually protected by strong property rights, prosper by extracting wealth (i.e. resources) from, and dumping their excesses (waste, poverty) into, ungoverned spaces, without paying the full costs of their actions. This is seldom a problem where deep institutions (including strong property rights, information, environmental regulation, innovations in market-based approaches), and the blessings of strong civil society and democracy, protect ecosystems and people. Without these institutions, however, tropical ecosystems (and people) are suffering.

A brief history of man's impact

Hunter-gatherers

Only Africa, especially its savannas, retains its full spectrum of 15 or more large herbivores. The occupation of the planet by *Homo sapiens* coincides with the extinction of many large animals. Australia's megafauna went extinct approximately 46 000 years ago including all species of large land mammals, reptiles, and birds weighing more than 100 kg and six of the seven genera weighing 45–100 kg (Roberts et al., 2001). When man arrived in North America about 12 000 years ago, there were cheetahs, lions, mastodons, mammoths, several species of deer, bison, sloths, tapirs, horses, and more. The fossils record suggests that as many as 48 species went extinct in a period between 8 000 and 11 000 years ago, and a further four species more recently, including woolly mammoths as recently as 2 000 years ago. There is a similar story in South America where ancient tigers, bears, wolves, sloths, horses, llamas, and rhinos disappeared approximately 10 000–12 000 years ago. The common factor in these extinctions is the sudden arrival of a highly effective big game predator who was also adept with fire – humans (Sandom et al., 2014). Large, slow breeding (K-selected) species clearly could not cope with this new predator on the landscape and were wiped out almost everywhere across the globe. Perhaps they were unaware of the dangers inherent in man, as we see with David Attenborough sitting amongst birds and reptiles in the Galapagos unperturbed by his presence. In Africa, herbivores co-evolved with hominids and were fully aware of the dangers of man the hunter. A full spectrum of wildlife species survived alongside low densities of hunter-gathers for thousands of years, and wildlife was certainly abundant and spectacular when Winston Churchill journeyed through East Africa (Churchill, 1908), although it is now under severe pressure (Craigie et al., 2010; Ogutu et al., 2016). This suggests that man has a significant impact on wildlife even at low densities, but also that the relationship between man and wildlife is not simple. Wildlife certainly survived in large numbers in some places, including bison in America and wildlife in some parts of Asia. Vast herds of wildlife in southern Africa were only exterminated recently by the arrival of white people, mostly by farmers and fences but also by hunters.

Farming and food

Farming has a massive environmental footprint, and it is difficult to discuss wild species and spaces without some understanding of agriculture. The strength and weakness of agriculture is that it greatly simplifies ecological systems. It replaces multi-species ecosystems with crop monocultures or near monocultures, and ecological feedback loops with man-made controls through tilling, fertilisers, pesticides, and irrigation. Thus farming simplifies ecosystems so that it can maximise and concentrate the conversion of light, water, and nutrients into a narrow range of food, fibre, fuels, and other raw materials that are useful to man.

Ecological simplification may work in the agricultural sweet spot – the fertile and well-watered prairies, deltas, and steppes that have been converted wholesale into farms. However, it is not suited to the ecological complexity of dryland and forest ecosystems, especially given the much lower yields here. Farming in the agricultural sweet spot is a wise use of land – we have to produce food, and we should do it where agriculture is economically more productive than other land uses. Interestingly, we often justify conservation to protect the original biodiversity for insurance and option values. Yet the agricultural landscapes where many of these important species presumably preside contain few protected areas and are seldom incorporated into maps of biodiversity hotspots (Myers et al., 2000).

Implementing community wildlife management requires a good understanding of the complicated relationship between people, farming, and nature, especially in the 21st century when we need to feed over 7 billion people, with the expansion of agriculture the primary threat to intact drylands and forests. The relationship between people, farming, and food is complex and emotional, and differs greatly on different continents, with no easy answers.

About 11 700 years ago, man emerged from the Ice Age (the 2.5-million-year Pleistocene) into the warmer wetter Holocene and began to domesticate plants and animals in the mid-to-late Stone Age. Most early civilisations occupied the 'agricultural sweet spot' – the Nile Valley and Fertile Crescent, the Indus and Ganges Valleys, the Yellow River Valley – where there was sufficient (and sufficiently reliable) water and soil fertility to grow crops consistently.

Farming has also allowed profound changes in society, not always for the better. The pastoral vision of farming as a comfortable lifestyle is far from the truth. Historically, farmers may have delivered a food surplus, but were often worse off. Farmers worked harder than hunter-gatherers, their diets narrowed towards grains and carbohydrates, disease was more prevalent, and farmers became shorter, less healthy, perhaps less happy than egalitarian bands of hunter-gatherers, and trapped at the bottom of hierarchical societies (Diamond, 2002; Fukuyama, 2011). Even today, small-scale farmers are often trapped in a life of drudgery, with livelihoods that are marginal and hard (Harari, 2014), and prone to food insecurity and disease. Even now, the term 'agricultural development' remains an oxymoron. Farming is important because it provides food and is a major activity for the poor. These are not the same things. Most of the planet's food comes from high-input farming, and the low-input farming that is doing so much environmental harm is hardly able to feed the people that do it. Historically, peasant farmers seldom prosper or gain political freedoms, and it is rare that countries have advanced to modernity by tilling the soil – consider the challenges faced by the peasants in feudal Europe and the vulnerability of contemporary farmers in the tropics.

The environmental footprint of farming

Farming is placing severe pressures on land and wild resources, utilising nearly 40% of available land by the end of the 20th century (Foley et al., 2011). Croplands comprised a third of this area, and the global area of cropland doubled every century after 1500 (Goldewijk et al., 2010). A further third of the planet comprises hot and cold deserts. Despite the increasing scarcity of farming land, land degradation affects 38% of the world's croplands, or 2 billion hectares, and is particularly serious in the drylands in poor countries (Rekacewicz, 2005). Not surprisingly, the major threat to nature is the plough and the cow, as the space for nature is squeezed. Using species-area curves, Wilson (2016) calculates that today's protected areas, which cover 15% of the planet, will conserve roughly 62% of species, but we need half the planet to save 84% of species.

Croplands expanded by 35–80% between 1961 and 2013 (FAO, 2009; Deepak, 2016), mostly at the expense of forests and drylands in poor countries including Africa. In Africa, farming yields have remained stagnant at about 1 tonne per hectare (Sanchez, 2015), so food production depends on clearing more land; croplands have expanded by 50% since 1960. This forces farming beyond the prime agricultural zone and into forests and drylands where it becomes a major threat to surviving wildlife and habitats, a situation that will worsen, even to the point of catastrophe, as the African population approaches 2 billion people in the next few decades. African farms and the environment are already not coping (IPBES, 2018), yet 54% of global population growth will be in Africa (and 82% by 2100[1]). Apart from Africa, most population growth will be in South Asia and Central America, where agricultural expansion is also a significant threat to wild habitats (20% expansion). By contrast, croplands are retracting in North America, the European Union, and Eastern Europe (FAO, 2009).

We are dealing simultaneously with several problems – feeding the world, rural poverty, and space for wild animals and places. These challenges deserve far more intellectual attention. Most of the world's food comes from high yield and industrial agriculture, and we have the technical ability to feed the world. The agricultural expansion that is so damaging to forests and drylands is mostly through low-input farming by extremely poor people who, paradoxically, are often hungry. Low-input farming in agriculturally marginal areas is not sustainable and will have severe human and environmental effects, not least in projections of 50–700 million environmental refugees (IPBES, 2018).

High input agriculture

Globally, food production is easily outpacing population growth. Food is getting cheaper and farming is becoming less and less profitable. Although human populations have quadrupled since 1900, the average inflation-adjusted price of crops (corn, wheat, soybeans) has declined fourfold since 1900. Improved technology is similarly lowering the relative price of many primary commodities such as wood and minerals (Krutilla, 1967). We can easily feed the world because crop yields increased as much as twelvefold in developed countries like the USA after World War II (Fuglie et al., 2007). The invention of the Haber-Bosch process revolutionised farming by converting atmospheric nitrogen into fertiliser. Further technological improvements include the sophisticated use of fertilisers (nitrogen, phosphorous, potassium) and micronutrients, scientific agronomy, mechanisation, plant breeding, genetic engineering, and irrigation. However, these miracles of modern agriculture are not without cost. Modern farming requires extensive inputs, especially water, energy, fertilisers, and pesticides. Furthermore, agriculture externalises many of its costs. Farming uses 70% of all water used by man, with the virtual water costs of a kilogramme of beef, cereal, and fruit being 15 000, 1 500, and 1 000 litres respectively (Rekacewicz, 2005). In addition to 2 billion hectares of degraded land, excess nitrogen and other farming inputs have serious environmental costs, including groundwater pollution, eutrophication, and coastal dead zones, and possibly contribute to the 82% reduction in the biomass of insects in Germany (Vogel, 2017). Industrial farming (especially factory farming) also faces ethical, distributional, and ecological challenges. It depends on remarkably few species – with as few as 12 types of grain and 23 species of vegetables providing the majority of food – and it is questionable whether we are doing enough to protect native progenitors.

Farming is heavily subsidised, especially in rich countries, and although subsidies have declined from 30–35% of farm gate prices in 2000 to 'only' 15–17% by 2015, this still represents some $672 billion in the 50 largest food-producing countries (OECD, 2016),[2] which is nearly five hundred times the budget of the Global Environmental Facility. Despite all this, the proportion of people in affluent societies who farm has declined from about 70% in 1900 to less than 2% today, with family farming becoming economically difficult. Diseases of affluence – diabetes, obesity, heart disease, and so on – are at least partly attributable to the poor quality of industrial-age diets. In addition, factory-farming poses at least as many ethical questions as trophy hunting, while the interaction between farming and environmental health is a huge issue.

BOX 3.1 FARMING MODELS

Farming and food will always be important, especially for poor people, and the 1–3–5–10 model described by Sanchez (2015) is useful for framing this conversation.

One tonne yields, or unimproved agriculture, is practised widely in Africa, with high environmental costs because food production relies on expanding the area of cultivation, rather than on improving yields.

Yields can be improved to 3 tonne yields by introducing simple agrarian practices (planting dates, spacing, weeding frequency, management of the soil surface), improved seeds that match the length of the rainy season, and fertilisation (manure, composts, manufactured fertiliser). These well-known farming practices have been rebranded as conservation agriculture, climate-smart agriculture, or sustainable agriculture. They complement CBNRM by allowing short-term gains in farming while freeing up land for wildlife and forests.

The next step up, to 5 tonnes (as in China), requires improvements to the whole system including research and extension, inputs and financing, farm production, processing, and storage, and marketing and institutions.

Ten-tonne yields are linked to high-tech modern industrial farming.

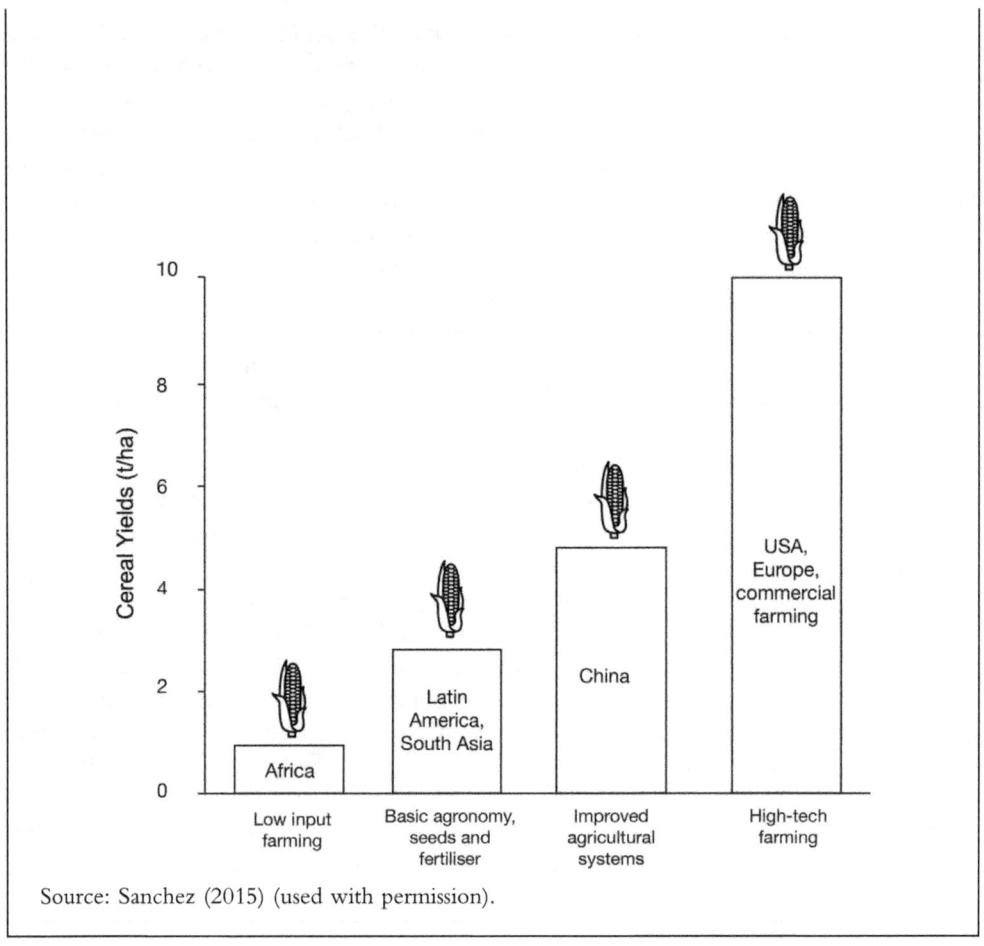

Source: Sanchez (2015) (used with permission).

Low input, small-holder subsistence farming

At the other end of the spectrum, hundreds of millions of people live on modest farms (less than 2 hectares) growing maize and beans with cattle, goats, chickens, squash, and pulses on the better soils and in the more reliable rainfall zones of the tropics and sub-tropics. These small farms are generally sustainable and economically sensible.

The same cannot be said of dryland and forest farming systems (Box 3.1), which are associated with extreme poverty (White et al., 2002). Hundreds of millions of people on the cusp of survival are forced to rely on low-yield agriculture with high environmental costs. Bowles and Choi (2013) suggest that, even in the Holocene, farming did not emerge because it was economically superior to hunting and gathering, but because crops, dwellings, and domestic animals (unlike wildlands and wildlife) could be unambiguously demarcated and defended. This raises the intriguing idea that the expansion of marginal agriculture reflects property rights rather than the economic superiority of farming.

Especially in Africa, poor people grow more food by expanding land area rather than by improving yields. The environmental costs of low-yield farming and open-access property regimes in forests and drylands are seldom internalised, but include soil degradation, deforestation, desertification, slash-and-burn farming, and the over-exploitation of wild resources such as bushmeat (Poulsen et al., 2009).

There is no dispute that low-input agriculture has a high costs in terms of deforestation statistics and loss of wildlife. An important question, then, is how well are the people doing? The short answer is: not well, and that rural economies are often severely distorted. In many drylands in southern Africa, people continue to clear and burn land for farming which usually provides less than 10% of livelihoods; rural people depend on employment and poverty grants for as much as 90% of their living. These villages look agrarian on the surface but, economically speaking, they are places to live rather than places of production, with heavy reliance (90% plus) on off-farm income (jobs, remittances, government grants) (Eguren and Sprague, 2014). In western Zambia, where agricultural conditions are slightly better (600–1 000 mm of rainfall) and people have fewer options in the modern economy (only 15–20% of livelihoods are off-farm), on-farm livelihoods include natural resources (29–40% of total livelihood), farming (13–16%), and livestock (22–39%), depending on rainfall conditions. However, the telling statistic is that up to 80% of these households experience hunger at the worst time of year, and even at harvest time 10% of families are food insecure (Muyengwa et al., 2014).

This situation is not helping anyone. People are wiping out nature without climbing out of poverty, and natural resources are being wasted.

These systems contain all the elements of collapse, and they are extensive. Communal lands, where people have few rights, cover as much as 65% of the global land area and up to 2.5 billion people (Vitousek et al., 1986; Pearce, 2016), while some 22% of the world's land is held by indigenous peoples, often with few rights (Sobrevila, 2008; Nakashima et al., 2012). Some 70% of the poorest people in the world live in rural areas, often in forests and drylands, and have so few alternatives that they migrate to towns in search of employment despite the squalid conditions.

This represents a tragedy of the commons on a grand scale, especially in Africa. In communal areas, people get 'free' land and grazing, and 'free' materials to build their houses. 'Free' wild-harvested food, fibre, and medicine contributes about 20% of their livelihood, even in South Africa (Shackleton et al., 2007). They also get free services, including water, health, and education (even if these are of low quality), and in some countries this is supplemented by social grants and food aid. Despite these social and environmental subsidies, people remain locked in poverty. Ungoverned spaces are also a politically convenient place to dump millions of poor people at relatively low expense, where they are less likely to riot and more easily influenced to vote for the 'right' party.

Communal systems of low-input farming are grossly inefficient. This leads to the hypothesis that productivity could be increased many times by better institutions (even before environmental costs and benefits are factored in). Child et al. (2012) show that communal lands with weak institutions in South Africa earn R232/hectare (mainly from firewood and livestock grazing) compared to R2 000 (and as much as R6 000) from ecologically similar private land (mainly from wildlife) that has strong institutions (see also Chapter 7). Conceptually, institutional reform can transform these systems from an unproductive to a productive basis of attraction (Figure 3.4). Economically, CBNRM stimulates this transformation by replacing a tragedy of the commons with accountable

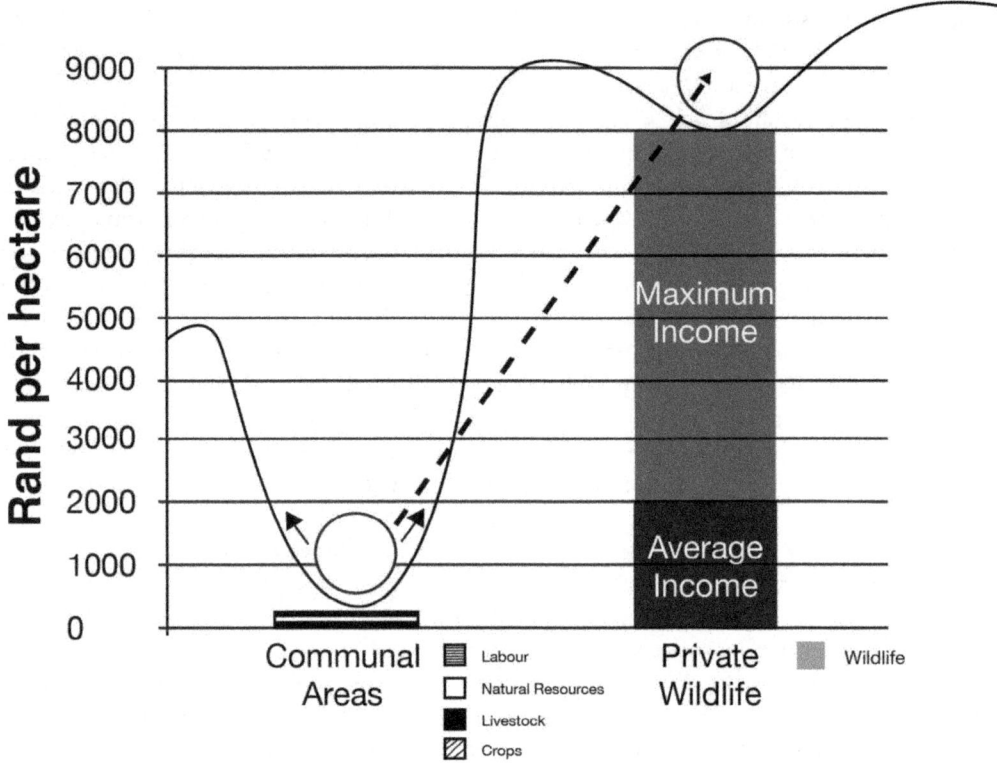

FIGURE 3.4   Comparison of returns from land with weak and strong institutions

institutions by (1) internalising the full costs and benefits of current practices such as low-input farming, (2) incorporating the value of wild resources into economic decision-making, and (3) reducing over-exploitation by closing the boundaries. CBNRM and collective action such as land use planning and village clustering also lower the costs of services like education, health care, and the internet, while freeing up land for high-value land uses, including wildlife. These will never be high-production zones, but they can be ten times more viable than they are today.

Property and pricing failures in ungoverned spaces

Ecological mismatches

Figure 3.5 traces the relationship between agricultural yields and rainfall, with yields peaking in the sweet spot where rainfall is sufficient and reliable for growing domestic crops. Farming may be particularly mismatched to ecosystem processes outside the agricultural sweet spot, where farming has expanded only recently.

In drylands, farming is unreliable, and the plants are harvested indirectly using animals like cattle and sheep. Productivity declines steeply because it now depends on secondary (not primary) production in the trophic pyramid.

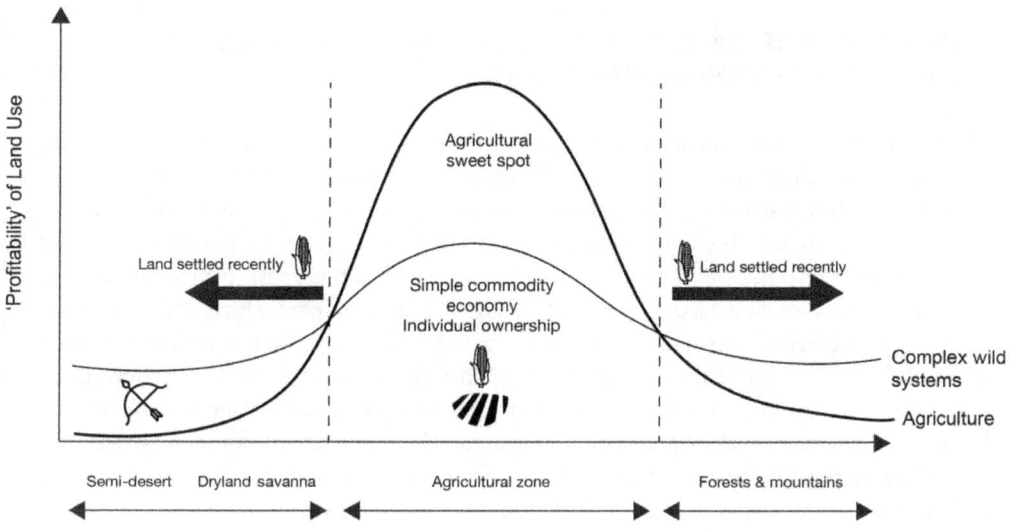

FIGURE 3.5   Expanding out of the agricultural sweet spot into drylands and forests

Forests, too, are complex and not suited to most domestic plants and animals. Productivity is low, and people harvest a wide variety of plants and animals (rather than simple crop monocultures). Ecosystem services such as carbon sequestration, water production, and biodiversity are valuable, but invariably excluded from the economic equation.

Hunter-gatherers and nomads have practised complex and mobile lifestyles in drylands and forests for thousands of years. We have already shown that farming does not adequately address poverty in these regions, while simple, sedentary agricultural systems are ecologically mismatched and harmful to these complex ecosystems.

Dryland ecology and land use

The ecological limiting factor in drylands, by definition, is the amount, seasonality, and variability of rainfall. Consequently, savannas are patchy and heterogeneous, and the diversity in the form and behaviour of Africa's 92 species of large mammals is an evolutionary response to this temporal and spatial variability. Giraffes, elephants, zebras, and antelopes fill a wide range of ecological niches, reflected in morphological differences, but also in seasonal movements and migrations, feeding and anti-predation strategies, and even physiology (Cumming, 1982). Yet, in Africa, we have replaced this natural diversity of 92 species with five species of domestic herbivore, fences, and artificial water. We should hardly be surprised (Chapter 8) that livestock commodity production, which is so unsuited to the underlying ecological conditions, failed on private land in southern Africa. Importing simple agriculture and institutional systems (i.e. cadastral plots) has not worked, and savannas in particular pose significant institutional challenges (Box 3.2).

BOX 3.2 AFRICAN SAVANNAS: ECOLOGICAL COMPLEXITY AND INSTITUTIONAL CHALLENGES

Starting in the 1950s, ecologists and range scientists began to study dryland savannas, which were being 'ruined' by livestock (Bigalke, 1966). Clemsian models of plant succession – the 'balance of nature' – seemed not to apply: savannas are defined as having both grass and trees, which compete vigorously for water. Damaging the grass layer can 'tip' a savanna from a healthy state (perennial grassland with some trees) to an unhealthy one (infestation of inedible bushes, with little grass), and a permanently less productive state. Removing grazing pressure seldom allows a linear recovery. This 'state-and-transition model' led to a paradigm shift in ecology, introducing principles like disequilibrium, complexity, resilience, transformation, and adaptive management (Gunderson and Holling, 2002; Walker et al., 2004). Dryland savannas are nonlinear and complex because it is difficult to predict the outcomes of interactions between rainfall, fire, frost, and grazing. This makes management, and the attribution of costs and benefits, difficult.

Moreover, drylands are widely misunderstood. It's not the animals that are important, but the grass. Plant-available moisture is the limiting factor in drylands, and a healthy grass cover is the Holy Grail for managing soil water relationships and preventing desertification. Grass protects the soil surface and determines how much of the rain that falls infiltrates the soil. In reality, therefore, cattle ranchers in drylands are grass farmers. Savannas are difficult to manage, and producing animal biomass is especially problematic, because yield and environmental health are in direct opposition. Grass provides the food for herbivores, but when farmers have too many cattle, this damages the grass permanently by letting in too many woody plants – bushes are a symptom of mismanagement (i.e. ecological scar tissue) not a solution. The highly variable climate makes it difficult to detect these changes in ecosystem health before it is too late, resulting in widespread range degradation.

This sets up the challenge for designing institutions and management practices to sustain ecological productivity in environments that are naturally low yielding yet complex to manage. The attribution of costs and benefits is difficult, as is accounting for mobile or cryptic wildlife. Wildlife mobility is an evolutionary response to the patchiness and variability of these systems, but this adaptation is lost with sedentary livestock production. In savannas, economic institutions need to cope with two problems: the attribution of costs and benefits in a complex system, and the ecologies of scale necessary to manage the spatial and temporal variability of the system. These features are absent with small, individualised ownership. The default position of an open-access property regime is hardly better, with widespread decline in savanna productivity and wild species.

Forests and livelihoods

Forests, like drylands, are complex, but this time complexity takes the form of a multi-layered ecosystem occupied by a huge diversity of plants and animals. Forest peoples therefore use a wide range of products, including fish, bush meat, fruit, and other plant foods, across space and time. Like drylands, forests degrade as farmers, often from outside, replace traditional and complex forest lifestyles with low-value subsistence farming, or simplify ecosystems with large-scale plantations (e.g. palm oil, soy) or cattle pastures. Deforestation has high environmental costs in terms of biodiversity and ecosystem services (MEA, 2005), because forests store carbon and produce oxygen on a global scale, and are an important source of water. Like drylands, new institutions for governing complex livelihoods and ecosystem services in forests are rare, so people fall back on simple individualised cadastral systems.

Property mismatches in drylands and savannas

If simplified domestic production systems (farming and livestock) are ecologically unsuited to complex drylands and forests, and are economically questionable, why are they so dominant? The reason may be as simple as ownership. As far back as the Holocene, domestic species had replaced wild species largely because they were privately owned (Bowles and Choi, 2013). Simple, individualised ownership of land and animals evolved with sedentary farming, and is deeply ingrained. When people expanded out of the agricultural sweet spot, they took their simple systems of individual cadastral rights with them, leaving wild species (for which institutions have not evolved – see Chapters 5–8) at a considerable disadvantage. Domestic plants and animals have flooded these complex environments, not because they are more viable or ecologically suitable, but because we know how to own them, but have not developed effective ownership systems for wild plants and animals.

Aboriginal local institutions for complex resources, such as hunting or rubber trails in the Amazon or beaver management in North America, did exist (Stroup and Baden, 1983; Duchelle et al., 2014). However, these institutions were undermined and replaced by the centralised control of wild resources (Chapter 7). Public ownership is no longer suitable for managing wild resources (Chapter 5), but it has prevented the evolution of decentralised systems. Wild resources are replaced by cattle, soy, and low-value subsistence farming because the transformation from an empty to a full world has been too fast for new economic institutions for wild resources to evolve, resulting in the extensive loss and degradation of forests and drylands (Box 3.3).

BOX 3.3 THE EXTENT AND LOSSES OF TROPICAL FORESTS AND DRYLANDS

Forests cover 31% of the world's land surface, or just over 4 billion hectares (FAO, 2016). Half of the temperate forests in the industrial North were harvested for timber or replaced by farms during industrialisation, but are expanding again in post-agrarian societies. Rapid forest loss has now switched to the ungoverned spaces in the tropics, mainly because of agrarian expansion but also because of the global demand for raw materials. Less than half of the tropical forests remain (MEA, 2005). Although communities hold and use a considerable portion of forests, especially in the tropics, only a small fraction is officially recognised by statutory law, and even less is protected and securely held by indigenous peoples and communities. In 2013, communities held some legal rights to 511 million hectares of forest – about 15.5% of the world's forests, but less than 1% in Africa (Tchawa, 2009).

Drylands cover over 40% of the earth's surface, 66% of Africa, and 72% of developing countries (MEA, 2005, Chapter 22). They support 2.1 billion of the poorest and most marginalised people on earth, 90% of whom live in developing countries. Some 10–20% of drylands are degraded, and ecosystem degradation is sometimes the principal factor causing poverty (MEA, 2005; Safriel et al., 2005). According to the (United Nations, 2011), drylands support 28% of endangered species, are especially important for wildlife, and store 36% of terrestrial carbon. Only 8% of drylands are classified as protected areas, and they include eight of the world's 25 biodiversity hotspots (Davies et al., 2012).

The multiple values of nature

If the institutions for owning nature are faulty and outdated, what about the mechanisms for valorising nature? It is only recently that the multiple values of nature have been recognised, and we are still far from understanding how to turn these values into land use incentives. As late as the 1960s, the focus was on the quantity of primary commodities: tonnes of gold, kilogrammes of beef, bushels of corn, board feet of timber, and so on. The concept of multiple use values for environments only emerged in the 1960s, with the growing importance of recreational and amenity values. Counter-intuitively, economists noticed that the price of natural resource commodities was going down in a more demanding world (Goldewijk et al., 2010). Foresters, in particular, recognised that their future lay in multiple forest values, not commodities (Krutilla, 1967).

Krutilla (1967) introduced the concept of total economic value with which we are familiar today (see left side of Figure 10.1). This was at a time when the value of wood in America's protected forests was going down and national forests needed to be justified by a much wider range of values. In addition to direct use values and indirect values, Krutilla brought clarity to the preservation of future options (e.g. the contribution of plants to important drugs like cortisone, digitalis, and heparin), existence values, and bequest values. Almost five decades later, the ecosystem services framework adapted from the Millennium Ecosystem Assessment (MEA) makes the same point, classifying ecosystem services to include:

1. Provisioning services: food, fibre, fuel, and water.
2. Regulating services: benefits obtained from ecosystem processes that regulate our natural environment, such as the regulation of climate, floods, disease, wastes, and water quality.
3. Cultural services: such as recreation, aesthetic enjoyment, and tourism.
4. Supporting services: services that are necessary for the production of all other ecosystem services, such as soil formation, photosynthesis, and nutrient cycling.

The MEA may have reframed these values as ecosystem services, but people have understood this for thousands of years, with Plato linking forests and mountains to water supplies, and the thousand-year-old Arabian *hema* system protecting ecosystem services (Gari, 2006; Hanks, 2006).

Pricing failures in drylands and forests

Market-based conservation approaches hinge on the assumption that wild resources have an economic comparative advantage in some (or even many) drylands and forests, and that the total economic value of intact habitats exceeds that of habitats converted to farming and other uses. However, these economic values are not reflected in the financial prices that guide land use decisions. Wild resources are invariably underpriced, while domestic species are often subsidised, leading to the wrong land use outcomes (see Chapter 6). There is a wide gap between the theoretical economic values and financial (or market) prices for wild resources (defined in Box 6.3), and the operational challenge is to close this gap.

Figure 3.6 conceptualises the values of various ecosystem services across a gradient from arid deserts to tropical rainforests.

Provisioning services are split into two categories. Farming refers to domestic species in converted habitats (agriculture). Non-timber forest products include wild food and medicines, natural materials such as the poles, grass, and reeds used for building houses, and fuels from firewood and charcoal. People in drylands and forests make considerable use of wild foods, fibres, fuel, and medicines, so the curve tips up slightly towards the drier and wetter regions. We make no bones about the fact that farming monocultures are highly productive within the agricultural sweet spot, but elsewhere the true economic value of wild plants and animals may well exceed that of domestic ones.

The value of regulating services like water production and carbon sequestration is assumed to increase roughly with biomass, and can be large (Börner et al., 2010).

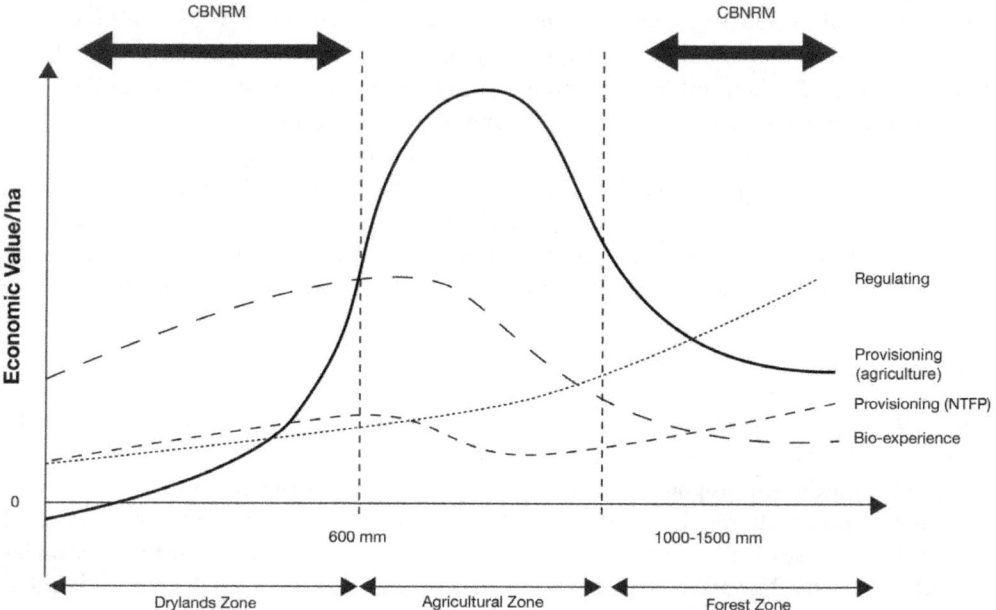

FIGURE 3.6   The value of ecosystem services relative to agriculture across a rainfall gradient

The estimate of the value of cultural services is limited to the monetised 'bio-experience' economy, with the high values of outdoor recreation and tourism peaking in semi-arid savannas because of the variety of wildlife, climate, and wide-open spaces.

While the exact shape and magnitude of these curves are hypothetical, they suggest that wild systems may have a comparative advantage over commodity agriculture outside the agricultural sweet spot. African wildlife is a case in point, and large areas of drylands in southern Africa have been rewilded because wildlife's theoretical comparative advantages (Bigalke, 1966) have been translated into real values and land use change (Chapter 9). Back of the envelope calculations suggest that this economic model can be extrapolated to East African savannas, and even the Congo rainforest, provided policy constraints (i.e. non-ownership, restrictions on high-value uses like hunting) are removed.

Is this comparative advantage limited to Africa wildlife, or does it apply more generally to wild species and intact habitats? In a much-quoted global study, Costanza et al. (1997) calculate that the services provided by nature are worth at least twice the total financial value of the global economy. The MEA concluded that the total value of intact forests and wetlands was often higher than converting these ecosystems to farming, intensive forestry, and other intensive uses, although they could only find a few studies to support this (MEA, 2005). Similarly, TEEB (2009) estimated that the total economic value of tropical forests was $6 120 per hectare per year – more than that of many agricultural options. However, only 8% of this value was reflected in market prices (i.e. food and raw materials). The highest values were not traded in the market – climate regulation (32%), water (22%), and erosion prevention (11%). Ding et al. (2016) found that security of forest tenure for indigenous and local communities significantly reduced deforestation rates in the Amazon, with high economic returns of $4 559–10 274 per hectare. In the Amazon, Steffen et al. (2015) suggested that the REDD (reducing emissions from deforestation and degradation) carbon price (R$2 703/ha) significantly exceeds the value of the land uses currently causing deforestation (R$1 433/ha). Carbon payments could reverse as much as 55% of forest loss, with slash-and-burn agriculture (smallholders) and land-extensive pastures (large land-holders) being especially inferior land uses. However, weak institutions, especially weak local tenure, were a major bottleneck for operationalising REDD payments effectively, preventing economic values from being reflected in decision-making.

The literature suggest that intact ecosystems may well have a comparative advantage in many places, but this still needs to be tested empirically by translating this advantage into financial prices. The institutional reforms that converted economic values into land use incentives for wildlife in southern Africa are therefore a model that may have much wider application.

Loss of wildlife

There is no doubt that intact natural systems and wildlife are in serious trouble. Man's impact on the planet is summarised by an article in *Science* in 2011 entitled 'A Global Perspective on the Anthropocene'. People and farming have expanded into most corners of the globe, mostly in the last 100 years (Goldewijk et al., 2010). Over 90% of total mammalian biomass is now made up of humans and domestic animals compared to under a tenth of 1% at the dawn of agriculture 10 000 years ago (Smil, 2002). This trend is accelerating. Since 1900, the zoomass of domestic herbivores has increased more than fourfold while that of wild herbivores has halved to only 4% (optimistically); some 50 000 cattle get added to this inventory each day (Ripple et al., 2015; Smil, 2011). The decline of Africa's long-surviving wildlife is catastrophic. We have lost well over half of Africa's wildlife since 1970 inside parks (Craigie et al., 2010) and a great deal more outside (Ogutu et al., 2016), and perhaps as little as 5% of these vast herds survive today.

Wildlife is being eliminated by three major drivers: the cow and the plough, over-utilisation (bush meat harvesting and the illegal wildlife trade), and conflict (Ripple et al., 2015). With the media sensation associated with poaching and trade bans, we seldom pause to note the deep economic inconsistencies in these outcomes. Wildlife is replaced by the cow and the plough because it has little or no value to landholders. On the other hand, wildlife is over-exploited through uses such as the bush meat trade and the illegal wildlife trade because it is too valuable. Which is it?

Failure to analyse the problem carefully leads to incoherent solutions based on erroneous assumptions, such as wildlife bans and demand reduction, which have almost totally failed. The no-value versus too valuable conundrum is a critical puzzle that we need to solve, and we will return to it later, noting in passing that solutions based on the secondary matter of price will remain futile until the primary issue – wildlife ownership – is resolved.

Getting prices right for wildlife

If the previous arguments hold, intact wildlife systems often have a comparative economic advantage in drylands and forests. This is illustrated by the solid (economic) lines in Figure 3.7. However, the property and pricing 'failures' illustrated by the black arrows favour domestic species and disadvantage wild species, so landholders produce the wrong things. The solid curves show how land should be allocated, and the dashed curves show how land is allocated after policy failures have distorted underlying prices. Therefore farming expands

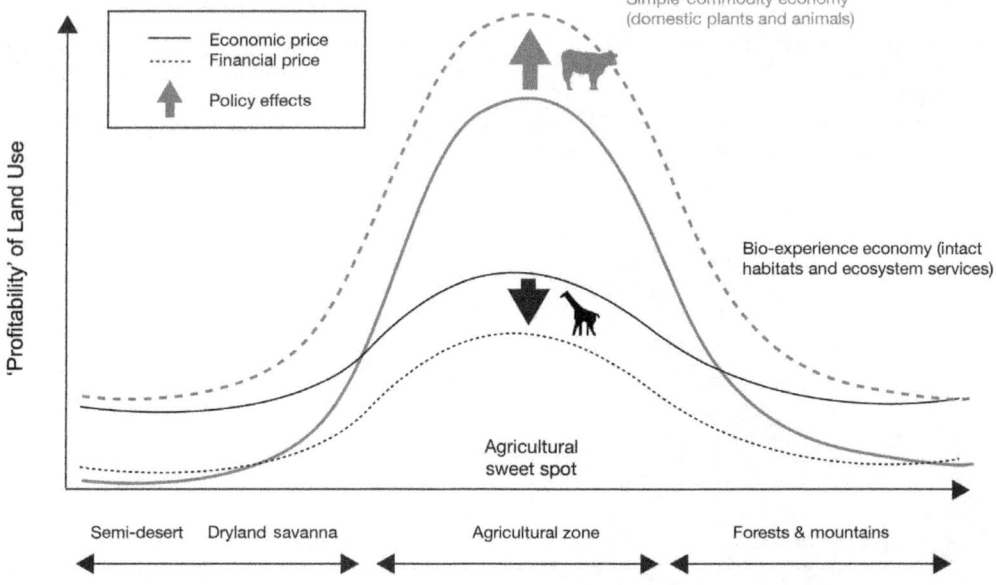

FIGURE 3.7 The financial and economic prices of wild and domestic production systems

even where it does not make economic sense, and replaces wildlife. This also exacerbates poverty, because it is economically less efficient.

The alternative future lies in getting these prices right so that land is allocated according to the relative positions of the economic (solid) curves. This pathway to sustainable poverty reduction requires local people having strong rights to utilise wild species and intact natural habitats as profitably as possible (rather than converting them to agriculture), maximising the value of wild species by trading them globally and correcting for missing markets such as ecosystem services. This is not a binary farming-versus-the-wild argument. Both are necessary, but we can improve the productivity and sustainability of drylands and forests by lifting the financial curve for wild resources upwards to better reflect total economic value, and ensuring that wild resources are competitive.

This is not a pipe dream. Models like the wildlife economy in southern Africa provide an empirical example of the proactive policy reforms needed to rewild domesticated landscapes (Chapter 8). Theoretically, wildlife landscapes have economic advantages, and terms of trade are moving against raw commodities and agriculture and towards the bio-experience economy so these advantages are increasing. However, leveraging the wildlife economy to address poverty and biodiversity losses will require bold institutional reform, based on new assumptions about the drivers of poverty and biodiversity. There will be no progress without effective local ownership of wild life and wild lands, effective community governance, and active promotion of new and global markets for wildlife. These new ideas began to emerge under the name of 'community-based' solutions in the 1970s as a response to the threats to forestry, fisheries, and wildlife in Africa, Asia, and America (Pomeroy, 1995; Arnold, 2001; Charnley and Poe, 2007). In the next chapter, we will discuss the historical reasons for ungoverned spaces, or de-institutionalisation, before embarking on an understanding of the principles and process of re-institutionalisation.

Notes

1 The spatial impact of future population growth (including on nature) is illustrated by a graphic presented in the EU Parliament in 2017. Available from: https://bayernistfrei.com/2017/04/25/epmigr/
2 These can be calculated from OECD Stat: www.oecd.org/tad/agricultural-policies/producerand consumersupportestimatesdatabase.htm#tables

4

INSTITUTIONS AND UNGOVERNED SPACES

LEARNING OBJECTIVES

The governance of wild resources has passed through five phases:

1. A pre-industrial local economy.
2. Over-exploitation on the frontier of the colonial period and the Industrial Revolution.
3. Strong public management as a response to over-exploitation in the late 19th century.
4. The biodiversity crisis of the late 20th century, when the flaws in the public management model resulted in wild species and spaces becoming, effectively, ungoverned.
5. Local proprietorship through the sustainable governance approach and community-based natural resource management (CBNRM).

In this chapter I attribute the crisis in wildlife and biodiversity to the deinstitutionalisation of wild resources through:

1. Loss of rights to land and wild resources.
2. Loss of rights to exchange wild resources in global markets.
3. Weakening of the social capital in communities through various forms of oppression and trauma.

Finally, the chapter:

1. Uses the historical example of the Glorious Revolution, and the transformation to the modern economy, to illustrate the importance of institutions.
2. Provides a model for institutional change.
3. Hypothesises that this model can be applied to ungoverned wild species and spaces through CBNRM and the sustainable governance approach.

Institutions and ungoverned spaces

Environmental outcomes reflect institutions more than the underlying ecology or even economics (Chapter 3). Thus, simple agricultural systems often replace wild species in drylands and forests because institutions favour domestic species, even where farming has serious economic and environmental weaknesses. This chapter explains why forests and drylands have become 'ungoverned spaces' and sets out the implications for wild resources and social outcomes. I contrast the security of person, property, and inclusive institutions that underlie modern prosperity (Chapter 2) with ungoverned spaces where people have few rights to land, few rights to wild resources, and social capital has been undermined by long-term insecurity, extractive forms of governance, and even violence.

The making of ungoverned spaces

There have been four major transitions in wildlife governance (Table 7.1). Aboriginal local regimes persisted for thousands of years. Wildlife was extirpated in frontier economies as 'new' lands were overwhelmed by colonisation and globalisation in the late 19th and 20th centuries. The response was centralised governance and public ownership of wildlife, forests, and fisheries. Effective public management led to the recovery of wildlife in rich countries with strong institutions. However, it further undermined local systems and proved ineffective where these institutional conditions (i.e. inclusive governance and effective public administration) were missing, leading to wide scale losses of biodiversity in the tropics. Bold rethinking of wildlife governance in southern Africa, and the emergence of CBNRM globally, now heralds the genesis of a new approach.

Contemporary crises in biodiversity, deforestation, and desertification are, to a significant extent, the outcome of failed governance. Effective answers may only follow a critical assessment of our underlying assumptions about wild life governance. This chapter reviews the doctrines of conservation governance historically, leading to the suggestion that the sustainable governance approach and CBNRM are new approaches that may be better aligned conceptually with contemporary realities.

Local commons

In the empty, pre-industrial world most institutions for governing wild life were simple and local – not much else was needed. Whether these were sustainable or not is hard to judge. Early hunter-gatherers wiped out many large animals as they colonised new continents (Ripple et al., 2015), but white colonialists also arrived to abundant wildlife. Examples of viable local commons range from the sustained use of bison and beavers by American Indians prior to white settlement (Stroup & Baden, 1983), to the ancient Arabian *hema* system instilled into Islamic Law (Grainger & Llewellyn, undated), to bushmeat hunting in the Congo rainforest (Mavah, 2015), to local forests in Germany, and so on. Many of these local commons collapsed where local people were unable to exclude outsiders from taking their resources, but enough examples have survived into modernity to show that they can work (Ostrom, 1990).

Frontier economy and public conservation

In Europe, the transition from feudalism to capitalism was accompanied by the enclosure of the commons (Box 4.4). Privatising land improved productivity, and excess farm labour was absorbed by the cotton looms of the Industrial Revolution and European colonization, which was a critical juncture in the governance of wild species and space. In the absence of rules to control the use of wildlife, fisheries, and forests, Europeans rapaciously over-exploited the 'frontier economy' of newly colonised lands, including bison in North America and elephants and other wildlife in southern Africa in the 19th century. As we will discuss in Chapter 7, far-sighted leaders like Theodore Roosevelt took action. They perceived the commons as a primitive pre-industrial form of governance, and commoners as slightly inferior. Consequently, they defined wildlife as a public good, giving the newly centralised state primary responsibility for the safekeeping of animals and habitats. The Europeans emulated these measures in their far flung colonies through the London Conventions for the Preservation of Flora and Fauna in 1900 and 1933, with Roosevelt being an honorary member of the Zoological Society of London (Fitter & Scott, 1978).

In this era of progressive government, forests, fisheries, drylands, and wildlife were characterised as public goods and managed as public assets through state agencies (i.e. budget-funded institutions, Box 4.1). This established a governing philosophy for wild life with global hegemony: the idea of state-protected areas, the banning of commercial use, and the centralised control of wild resources by powerful state bureaucracies.

BOX 4.1 A BRIEF INTRODUCTION TO 'BUDGET-FUNDED INSTITUTIONS'

The management guru, Peter Drucker (1973, Chapters 12–14), calls government agencies 'service institutions' and provides a valuable synopsis of their strengths and weaknesses. Service institutions include both government agencies and the service institutions within the private sector, such as human resource departments. There are three justifications for service institutions:

- Where the state has a natural monopoly, including protecting the rights of person and property through laws, courts, defence, policing, and the auditor general.
- Where it is necessary to manage natural monopolies and public goods (Chapter 5). However, the features of public goods change over time with advances in technology and other features. Thus, the state once provided telephones or railways, whereas today this is done by the private sector, because these services are no longer natural monopolies.
- Where institutions that provide education and health supply a need rather than a want and are considered to be a public right or in the public interest.

By calling service institutions 'budget-funded institutions' (or spending organisations), Drucker gets to the nub of why service institutions often do not perform well. Ordinary businesses are paid for satisfying a customer, and are severely disciplined by the market-place. By contrast, service institutions get budgets by persuading a single payer (often the general public) that they are worthwhile. Consequently, they are directed to satisfy the whole marketplace (rather than just a market share) without alienating people. To keep hunters and animal rightists happy, for example, a wildlife agency falls back on motherhood statements and good intentions, and its precision of purpose suffers; this would be fatal for a business. Budget funding affects the way service institutions function:

- They tend to be 'misdirected by budget', with the classic bureaucracy seeking more power, a bigger budget, or bigger staffing, rather than the provision of better services.
- Without the discipline of the market, service institutions can avoid focusing their effort on a small number of priorities that add value. They also defend redundant functions 'because this is the way we have always done things'.
- They often have coercive monopoly power.
- They are paid for good intentions and for not alienating important constituents (rather than satisfying one group, or achieving measurable results).

Drucker found that service institutions perform best where they define a clear purpose that adds genuine value, and are held accountable for the delivery of this value. This led him to develop ideas about performance management:

- Deciding, clearly, what is our business, and what should it be.
- Prioritising this in objectives and goals.
- Measuring performance.
- Creating accountability for performance – including public accountability. The performance of public institutions suffers in the absence of effective information and civil society mechanisms.

These ideas, and the influence of people such as Margaret Thatcher, led to the field of 'new public management' in the 1970s and 1980s. This sought to make public services more businesslike, focusing on the quality of services provided to the customers of the public sector, through mechanisms such as decentralised or outsourced service delivery models. The North American Model for wildlife conservation (Chapter 7) is, arguably, a reasonably effective service institution, with high levels of civic benefit (visitors to national parks, hunters) and scrutiny. However, many wildlife, forestry, and fisheries agencies, and many of the NGOs that have filled in gaps, suffer from the weaknesses listed above, including confusing mandates, a focus on intentions rather than results, and weak accountability for performance.

New game laws prohibited market hunting by the middle or lower classes, keeping big game hunting open for elites (often cheap or free). Colonial administrators turned a blind eye to subsistence hunting by local people, who had lived with these resources since time immemorial (Parker, 2004). In both North America and the British Empire, sportsmen were at the forefront of efforts to preserve game animals (Donnall, 2010). Non-sportsmen struggled to understand how people who killed wildlife for pleasure could claim to be 'real lovers of nature' (Fitter & Scott, 1978), a divide which continues to today. Roosevelt's solutions (taking wildlife out of the marketplace; replacing local institutions with public management) was a workable response in the colonial period when the primary threat to wild life was colonial over-exploitation. By the 1960s, however, the world was a very different place. The primary threat to wild life was now the cow and the plough. Without appropriate institutions to manage this threat, and an over-emphasis on outmoded responses (e.g. trade bans), wildlife has been decimated, compounded by the declining effectiveness of state agencies and public management (Box 4.2). In Chapter 7 we will flesh out this story, leading to new ideas for addressing them (Chapters 9 and 10).

BOX 4.2 THE ROLE OF STATE AGENCIES IN CONSERVATION[1]

State agencies have played a central, and often remarkable, role in the conservation of wild resources in many countries (including forests, wildlife, drylands, and fisheries). However, conservation requirements (Chapter 5), the role of the state, and the capacity of state agencies are changing rapidly. The fulcrum of governance is also shifting away from a singular reliance on state agency leadership towards polyvalent approaches (or chaos in the absence of some form of coordinated leadership).

In an insightful chapter, Grindle and Thomas (1991) explain how political appointees are replacing professional administrators in many state agencies, leading to the loss of independent technical skills and a divergence of agency capacity from the public good. Compounded by declining funding and professional independence, the technical capacities of once highly regarded state agencies for forests, fisheries, national parks, and wildlife are declining. Senior bureaucrats in wild life agencies, including international agencies, increasingly lack the field experience and technical judgement of their predecessors.

Moreover, in the absence of strong mechanisms of social accountability, natural resource agencies and the nationalised resources they manage are vulnerable to state capture and predatory and/or extractive behaviours.

The design of many wildlife or forestry agencies has also entrenched conflicts of interests. They fulfil contradictory roles as the primary regulating agency for the conservation estate and sector, but are also major economic players in them. This is a high risk situation, where public agencies are able to use their legislative powers to protect their monopoly position (Drucker, 1973). Thus, struggling park agencies often act against the public interest by blocking the emergence of competition from the private sector or communities.

As noted in Chapter 2, strong and accountable state agencies play a critical role in crafting effective rules. Confident bureaucrats are much more likely to trade influence over a booming sector for power over a declining one, and to obtain the new skills required for decentralised governance. We see examples of this in the following chapters, where wildlife administrators 'gave up power' but greatly expanded their credibility and influence as they rode the success of the wildlife economy and CBNRM.

Weak state agencies are often major impediments to CBNRM because, paradoxically, weak agencies resist innovation. In fighting to retain authority over wild resources, even when they patently have limited capacity to fulfil their responsibilities, they become impediments to change, blocking the devolution of rights to communities and landholders. Practising short-sighted economics, they instinctively grab the money produced by parks and communities, undermining an expanding wildlife economy that would pay back this money many times over in the form of taxes, jobs, and a growing economy.

Deinstitutionalisation: the making of the tragedy of ungoverned spaces

As state agencies took over responsibility for governing wild resources, they divested local people of their rights to these resources, causing local systems of governance to wither (Ostrom, 2000). However, state agencies are now struggling with the complexity of local resource management in a global world, leaving a huge gap. With neither communities nor states governing wild areas, these spaces are becoming, essentially, ungoverned.

Rapid human population growth since the 1800s has carved vast areas of wilderness into farms. In North America, hundreds of millions of acres of wilderness were opened to individual claim and settlement in the 1860s following the depopulation of native peoples and the Homestead Acts. The same phenomenon spread to Africa and Latin America in the mid-20th century, this time to forests and drylands filled with poor farmers, often in environments unsuited to farming, and without the underlying institutions of property rights and inclusive government. This led to the widespread emergence of communally held land and publically owned wildlife, ideologies that are defended with considerable tenacity despite paving a pathway to poverty and decline. What we forget is that these norms are the outcome of slavery, conquest and colonialism, violence, oppression, and abuse. Yet people defend and perpetuate these systems, disregarding the evidence that they entrench inefficiencies, inequities, and even human rights abuses. Change is more likely if we shine a light into the dark recesses of history to show how these institutions came about, sometimes for fair reason but more often for foul. While the institutional reforms associated with the Glorious Revolution in England heralded a period of unprecedented freedom and prosperity, history has burdened rural people in forests and drylands with institutions that are in many ways the exact opposite of these. I will group this loss of rights and freedoms under the term 'deinstitutionalisation', listing them briefly before elaborating them below. In terms of rights, communities do not have the rights:

- To the land on which they live.
- To use, manage, benefit from, and protect the wild resources they have lived with for thousands of years.
- To exchange their wild resources to the best advantage in the market economy.

In terms of freedoms:

- Communities have been severely traumatised by slavery, violence, and dispossession, which has devastated their social capital.
- 'Traditional' rule is often autocratic, tainted by a history of slavery (and even brutality) where people were traded for trinkets and muskets (Reader, 1999), and perpetuated by

indirect colonial rule and socialist or one-party states. Even today, top-down local governance is unthinkingly favoured by development and conservation agencies who should know better. The profound differences between representational or top-down governance and inclusive governance (or deep democracy) is dissected at length in the second half of the book.

Loss of rights to land

Most rural people in forests and drylands have insecure rights to land. Especially in Africa, the perpetuation of so-called communal land is a tragedy, holding people in neo-feudal conditions, where the land is owned 'on their behalf' by the state, and they are the 'subjects' of traditional leaders who allocate land at the local level and to whom they are beholden for their homes and farms. Such 'traditional' systems are supported by politicians and officials, despite their problematic social and economic consequences, and are wide open to abuse, corruption, and resource extraction.

Pre-colonial land ownership in Africa was fluid, with few examples of formalised individual or group tenure, the primary designation being tribal territories and kingdoms. Low populations and old and infertile soils meant Africans moved to new places once land was used up. Neither was land worth fighting over, with Reader (1999) contrasting the ancient graveyards of Africa that contain few signs of violence with the history of Europe which is brimful of the castles, battles, and skirmishes as kings, dukes, and lords protected their high-value land. Indeed, Reader speculates that conflict in Africa was quite rare until the European and Eastern slave trade set people against each other.

In the early colonial period, Africans were often consigned to 'native reserves' or 'tribal trust lands'. A few were created to protect local people from colonial exploitation, but most served to subjugate Africans and, to this day, facilitate the extraction of wealth by the modern economy and elites without proper recompense. An example from South Africa, although a particularly egregious situation, illustrates the nature of the political forces and economic consequences at play in the perpetuation of communal lands (Box 4.3). Communal lands were an instrument of oppression and extraction, but they remain ubiquitous in Africa (and beyond), holding in place a quasi-feudal system of governance.

BOX 4.3 THE PERPETUATION OF COMMUNAL LANDS IN SOUTH AFRICA

In the 1880s, individual property rights began to emerge in native reserves in what is now the impoverished Eastern Cape of South Africa. Local Africans began to acquire property rights, invest in agriculture, and compete quite favourably with white-owned farms.

Three sets of interest conspired to reverse the process of land entitlement. White farmers did not like competition from black farmers. Powerful mining companies, who needed labour to work in harsh employment conditions, did not want people making a good living from the land. Traditional chiefs did not want their 'subjects' to own land, because this greatly reduced their power as feudal lords. Consequently, land formalisation and titling was blocked and reversed, and the consequent absence of property rights, especially rights of exclusion, enabled the rich modern/urban sector to extract resources (including labour) without paying the full price. Without property rights, economic development was limited, allowing the South Africa regime to maintain people in ignorance and poverty. Land 'owned' by traditional leaders ('devolved despots') on behalf of the state, provided a mechanism of indirect top-down rule.

The outcome is today's communal lands, and a dual economy – a modern, urban, and commercial sector (that is institutionally rich with property rights, banks, etc.) that sits side-by-side with a communal sector that is unproductive, degraded, and associated with poverty and lacks security of property and other institutions.

Source: Acemoglu & Robinson (2012).

Unfortunately, non-ownership of land was re-enforced in the post-independence socialist period, while in many countries control over wild resources was further centralised. In the Republic of Congo, for example, age-old local systems of local forest management and communal territories ended abruptly in 1963 when all land was nationalised under the guise of scientific socialism (Mavah, 2015). Even today, only 10% of rural land in Africa is registered, leaving undocumented and informally registered land susceptible to land grabbing, expropriation, and corruption (Byamugisha, 2013). Over 99% of forests (Tchawa, 2009) and a similar proportion of drylands remain 'communal'.

Good traditional management does exist, and some local leaders are exceptional. The system may even have been reasonably effective when populations were low and land was plentiful.[2] However, the opportunity costs of communal lands increase exponentially with human population growth and globalisation. Communal lands are usually associated with hierarchical and personalised governance, and a dualism in which the elites have property rights but the poor do not (Menard & Shirley, 2014). Modern title trumps local ownership rights, opening the door to corrupt elites and even outsiders to grab and sell off land that does not really belong to them (Murombedzi, 2014). The poem about the Goose and the Commons (Box 4.4) illustrates that these traits are not unique to Africa, Latin America, or Asia.

BOX 4.4 17TH-CENTURY ENGLISH RHYME THAT STANDS THE TEST OF TIME: 'THE GOOSE AND THE COMMONS'

The law locks up the man or woman
Who steals the goose off the common
But leaves the greater villain loose
Who steals the common from the goose.

The law demands that we atone
When we take things we do not own
But leaves the lords and ladies fine
Who take things that are yours and mine.

The poor and wretched don't escape
If they conspire the law to break;

> This must be so but they endure
> Those who conspire to make the law.
>
> The law locks up the man or woman
> Who steals the goose from off the common
> And geese will still a common lack
> Till they go and steal it back.

The outdated institution of communal lands exacerbates poverty, environmental degradation, and the scourge of land grabbing (Murombedzi, 2014). Owned by no one, they exemplify the persistence of bad institutions (North, 1990). Although they have not adapted to the rapidly changing circumstances of the 20th century, replacing them is controversial. There is a strong attachment to the 'traditional' concept of non-ownership of land, especially by people in power who benefit from the system. Poverty remains, despite the economic consensus that persistent rural poverty can only be addressed if the fundamental failures in property are rectified, including titling individual farmers in productive areas, or a more sophisticated combination of individual and village-based tenure in forests and drylands.

Loss of rights to wild resources

If land rights are weak, local rights to forests, fisheries, wildlife, and other wild resources are even weaker. People have few rights to wildlife or forests, so they use domestic crops and animals (which they do own) to exploit unowned land and grazing. These diametrically opposite systems for governing domestic and wild species have massive consequences, including on the managerial norms of state agencies. Thus, agricultural ministries encourage local people to improve their farming. Wildlife agencies do the opposite. Claiming ownership of forests and wildlife, they seek to prevent communities from using them, sometimes even criminalising age-old hunting-and-gathering livelihoods. They also claim most or all of the income streams when these resources are exploited commercially (usually by outsiders), resisting the return of benefit to the communities that live with these wild resources. Even when wildlife benefits do get to communities, officials expect these to be used for public amenities like schools or clinics, whereas agricultural profits are perceived as private income, benefit people directly, and are seldom taxed in this way. The economic playing field is tilted dramatically against wild species. This results in a highly dysfunctional political economy for wild species. Communities do not own wild resources and disengage from solving problems of sustainability. Forest communities in Congo, for example, were fully aware that overuse of bushmeat was threatening a major part of their livelihoods, and were remarkably prescient about the causes of these problems. However, when asked why they were not acting to save their own livelihood options, they simply said 'this wildlife is not ours', leaving all responsibility for fixing the problem to the government despite knowing full well that the government lacked this capacity (Mavah, 2015). Even where local people have the rights to pursue 'subsistence' uses, they are trapped in low-value commodity production (i.e. meat) because they are prohibited from undertaking active management or participating in the exchange processes that can add so much value to wild resources.

Local people are consistent in saying that it is 'the government's wildlife, and therefore it is the government's responsibility to look after it'. However, the effectiveness of public management relies on voluntary observance of rules of use because it is prohibitively expensive to impose and enforce rules through policing and punishment (Wade, 1987). However, in the tropics, conservation rules lack legitimacy. Local people continue to be excluded from the process of making the rules, and perceive conservation rules as privileging national or global users over local ones. Westerners sometimes find it difficult to understand that *de jure* rules with no *de facto* legitimacy will simply be ignored, but in many disenfranchised communities the norm is for everyone (including criminals) to exploit wild resources before someone else can, with use also exemplifying everyday forms of resistance and protest and 'the weapons of the weak' (Scott, 1985). Officials and attendees at lofty meetings may fantasise that they are making rules, but to all intents and purposes wild resources remain ungoverned. Without recognised property rights and a genuine stake in the forests, wildlife, or fish that they live with, and participation in the making of conservation policy, these policies lack local social legitimacy and economic integrity, and will continue to fail. This contrasts starkly with the North American public model, where people have long participated in the process of making conservation rules, which are aligned with the public good, and adequately funded (Shelhaus, 2001).

Loss of markets for wild resources

Local people have clearly lost their rights to use wild resources. However, they have also lost their rights to valorise these resources in legal markets, reflecting the anti-market norms established by Roosevelt and colleagues in the 1900s and perpetuated by special interest groups today (through global institutions such as CITES). Taking rights away from people without their participation and consent, and without compensation, undermines their livelihoods and might even be called theft. These high-handed decisions cause considerable anger and resentment, illustrated vividly in interviews with Scottish fishermen in the 1990s when Africans proposed listing the North Sea herring on CITES to make exactly this point. As with the prohibition of alcohol and drugs in the USA, and bans on trade in 17th-century Europe, the over-riding effect of bans is to shift use into the shadows, and into the hands of criminals, stoking violence and criminality with serious social consequences for vulnerable people (Sobrevila, 2008). Removing the legal value of wild resources to local people takes away any incentive to reinvest in them and is a major factor in the widespread loss of wild species across the planet. Moreover, trade bans impose considerable opportunity costs on producer communities, given that the illegal trade in bushmeat, forest products, and high value items like ivory, rhino horn, and abalone (excluding fish and forest products) is worth more than $8–10 billion.

The unthinking reflex of shutting down wildlife markets requires critical examination. Taking away rights from the world's poorest people without compensation is legally and morally problematic. By preventing the full economic value of the wild resources from being included in land use decision-making, it undermines the opportunity for local people to play a positive role in conservation (Murphree, 1989), while it inflicts criminality on vulnerable people, including criminalising age old livelihood practices.

Social trauma and loss of social capital

In my early career working with African villages, I, like many others, assumed that these small communities had high levels of social capital. Surveys of livelihoods and governance in dryland communities in at least five countries in southern Africa suggests the opposite. Levels of trust within communities are low, and trust in leaders is low, especially with money. Associational capital is also weak, and largely authoritarian and top-down, with chiefs and churches predominating (Figure 15.3). There is some engagement in cultural activities, schools, and burial societies, but communities seldom work together collectively in resource management or business (Eguren & Sprague, 2014; Mavah, 2015; Muyengwa et al., 2014). Scattered settlement patterns also hint at an impulse to hide away from authority and even violence, forgoing the benefits of small-scale artisanal specialisation and diversification, and the provision of services associated with economic clustering.

This suggests a further source of deinstitutionalisation – severe social trauma and authoritarianism. Elites enriched and empowered themselves by selling people for muskets and trinkets in the inhumane slave trade, shattering communities that were also devastated by European diseases (Reader, 1999). Conquest and colonialism destroyed the rights of local people, who were ruled indirectly, presumably often by the same elites that had gained so much from the slave trade. The destruction of social systems in the colonial period (Hochschild, 1999) was rarely corrected in the post-colonial period, where centralising and authoritarian tendencies were exacerbated by the Cold War and opportunities for enrichment. A more benign example is the further centralisation and politicisation of the control of natural resources in Zambia (Gibson, 1999; HURID, 2002), and community resources everywhere continue to be appropriated and looted (Nelson, 2010).

Long histories of colonialism, racialism and tribalism, and extractive governance models, have left a legacy of divisiveness and powerlessness in which ordinary people are easily exploited by new elites with anti-democratic norms. We most often associate these features with troubled or failed states (Chabal & Daloz, 1999). However, community leaders and leadership structures display the very same attributes of oppression and extraction, leading to what Musavengane and Simatele (2016) call 'self-oppression', where local people have lost faith in leadership, collective action, and themselves.

Freedom, the philosopher John Locke argued, comes most seriously under threat when those in authority use their power to attack the property of their own citizens, increasing their own riches and power by taking what they want from the people (Dillon, 2006). By property, Locke did not just mean rights to land, but rights to 'Lives, Liberties and Estates' (Dillon, 2006). Just as we have prospered as Locke's freedoms and rights have been normalised in some societies, so has the opposite happened in many rural communities, who are denied the rights to the land they live on and the wild resources they live with. These rights, to this day, continue to be taken away. We can and should point accusatory fingers to the mechanisms of state capture and resource extraction that personally enrich Big Men and their cronies. However, we conservationists are also guilty when we disenfranchised local people with the best intentions in forums such as CITES, where decisions on local resources are often made without a single local person in the room.

The wild life tragedy

The shining example of effective conservation is Roosevelt's top-down 'North American Model' which has resulted in wonderful national parks and much more wildlife than a hundred years ago (Mahoney & Jackson, 2015). In a country built around capitalism, ironically, public management and central planning (perhaps even a socialist model) provides considerable benefits to citizens, including cheap access to hunting and outdoor recreation (Shelhaus, 2001). In terms of the alignment of property characteristics, rights, and ownership, the North American Model is somewhat misaligned (see Chapters 7 and 8), but it is held together by professional wildlife agencies, reliable funding mechanisms such as the Roberson-Pitman Act, federal and state budgets, by numerous democratic checks and balances provided by civil society of municipal, state, and federal agencies, and by the investments in over 500 universities with schools training wildlife and forest managers.

However, it is a mistake to project this model to places that lack similar levels of inclusive and accountable governance, reliable financing, and technical capacity in universities and state agencies. Yet, this is exactly what we have done, especially outside protected areas, resulting in extensive losses of biodiversity through over-exploitation (bushmeat, illegal wildlife trade) and replacement (the cow and the plough). The response to these dysfunctional systems is often dysfunctional, calling for more of the same, which is why we need to challenge the underlying assumptions of current approaches.

Responses to the wildlife crises, for example, are invariably emotional – thinking in ways that are 'fast', unconscious, automatic, and error-prone (Kahneman, 2011). Perhaps recalling Teddy Roosevelt's abhorrence of market hunting over one hundred years ago, the response to the illegal wildlife trade has not challenged the assumptions that markets are bad, or that public management is good. The fixation with demand reduction (despite the failure of this approach[3]), and the trend for ever more centralisation at the global level (rather than localisation), has established a vicious circle whereby centralised conservation models vulnerable to democratic failure are reinforced by special interests that benefit from the status quo (Olsen, 2000). In some instances this gets truly bizarre, depicting local people as being anti-conservation with only outsiders being good enough to take on this responsibility (Terborgh, 1999). We have gone too far and we need to turn away from uncritical ideology and special interests towards a pragmatic, thoughtful search for the greatest good for the greatest number. Solving multiple threats in complex social ecological systems, like building a rocket, will not emerge from good intentions. It requires pragmatic, considered, and tenacious application, and thinking slowly so as to understand whole systems. Economic and institutional history shines a new light on many of the challenges of conservation, and we now return to this discussion.

The greatest transformation

Extractive and inclusive institutions

In Chapter 2, we suggested that the prosperity of society is associated with individual freedoms and rights to own and freely exchange goods, services, and ideas, and to protect and enhance these rights through processes of inclusive governance (North et al., 2009). At a deeper level, freedoms, free markets, and inclusive governance depend on rights to both life and property (Bethell, 1998; Hayek, 1944), and the capacity of an accountable state to reduce the costs of defending these rights through policing and courts (North, 2005). Rural people in the drylands, savannas, wetlands, and mountains do not have these rights, including personal security (Allison et al., 2012). They inhabit extractive economies, and personalised, patrimonial societies where Big Men – kings and presidents, lords and chiefs – run the economy in ways that are self-serving, benefiting from monopolies, differential access to rights, graft, and corruption. Big Men make decisions, and it is who you know that counts. In the absence of the rule of law, ordinary people are vulnerable to elite predation and even violence. Hard work and innovation goes unrewarded. It is difficult for ordinary men or women to engage in the political process of improving the rules, which locks these inequitable and unproductive economies into a non-virtuous cycle. For millennia this was the natural way of things (North et al., 2009).

The tipping point between extractive and inclusive institutions was England's Glorious Revolution (Chapter 2), which transformed the mechanisms of authority and property. After 10 000 years of extractive regimes associated with the divine right of kings and feudalism, the Catholic King James II of England was replaced by William of Orange, the English Parliament, a constitutional monarchy, and a Bill of Rights that transformed 'subjects' into 'citizens' and laws that increasingly protected democratic and economic rights (Chapter 2). This virtuous cycle unleashed that talents of ordinary people and fuelled the European Enlightenment and the Scientific and Industrial Revolutions.

Liberal democracies, characterised by the rule of law, human rights, and property rights, are significantly more prosperous and equitable than other societies (North et al., 2009). They are 'inclusive' because political and economic institutions serve everyone, and they are 'impersonal' because they follow the rule of law and are not easily manipulated by powerful people (Acemoglu & Robinson, 2012). Rules enable people to protect the fruits of their property, labour, and intellect; and also to protect and enhance the rules themselves, thus establishing a virtuous cycle. At the scale of the local village, this is also the transformation that CBNRM seeks – unlocking the potential of ordinary people as inclusive governance and the rule of law replaces Big Men.

The greatest transformation

The last few centuries have seen the most rapid transformation of human society in history (North, 2003a). The misery of an agrarian economy based on feudal ownership was replaced by the current era of prosperity and personal freedoms that so many of us are fortunate to enjoy. This unparalleled wealth is rooted in fundamental changes in institutions, with critical lessons for the governance of CBNRM.

Two hundred years ago, most of us lived in villages and small towns. We made things ourselves and exchanged them with people we knew. People who deal with each other repeatedly in small-scale political and economic transaction develop high levels of trust, cooperation, and reciprocity (North, 2003b); if someone cheated, the whole village would soon know. Transaction costs were low because people dealt with each other repeatedly; but this world of small-scale artisanal production (tinker, tailor, baker, blacksmith) lacks economies of scale and can only take prosperity so far.

Adam Smith published *An Inquiry into the Nature and Causes of the Wealth of Nations* in 1776 amidst the economic excitement of the unfolding Industrial Revolution. He recognised that wealth is created by converting raw materials into useful products and through specialisation (Beinhocker, 2006) which in turn depends on exchange – the highly effective pin-maker of Adam Smith's story needs to trade his pins for food. Extreme specialisation and global exchange is the source of the enormous wealth that we enjoy today, allowing a dramatic transition from artisanal production to a global economy (North, 2003b). This is not automatic. Institutional economists like Douglass North have shown that high quality rules or institutions make this transformation possible. We no longer live in villages where we deal with each other repeatedly and can prevent cheating. In this massive globalised economy, we deal all over the world, every day, with people we do not know and with whom we will never deal again. For this system to work, the personalised checks and balances of the artisanal village are replaced by impersonal rules to prevent shirking, cheating, and defecting, including property rights, markets, contracts, banks, police, courts, information, Amazon reviews, standards, and so on.

Adam Smith's second insight is that people are best able to judge their own happiness. Therefore, he argued, the most just mechanism for allocating resources is for self-interested people to make their own choices (rather than have governments make these choices for them). This is also the most economically efficient mechanism, and we will return to Adam Smith's invisible hand in Chapter 6. Smith defined the free market as competition based on price and quality alone, free of monopoly and other distortions. This is not the same as laissez-faire economics, because a philosophy of non-interference implies that the most powerful will be able to make rules that advantage themselves. Reagan's statement 'the government is the problem' is incorrect; markets need professional, independent bureaucrats to design good rules.

The shift from personalised regimes to impersonal exchange has created vast wealth, and shared this wealth reasonably well. North (2003a) showed that this transition requires three things:

1. New economic institutions (property rights, exchange, contracts) to encourage low cost exchange.
2. Political institutions to design and enact rules.
3. Third-party enforcement of the rules.

We have sufficient knowledge about institutions to design them proficiently to get the economic outcomes we want (North, 2003a), and we will introduce some of this knowledge later in the context of sustainable governance and CBNRM. Far more challenging are the political institutions that form and shepherd these rules, and the mechanisms to enforce them. This includes the rules on wild life ownership (Chapter 5) and trade (Chapter 6), where permitting subsistence use, but preventing global trade, traps communities and wild resources in a pre-industrial economy that does not allow them to fulfil their comparative economic advantages (Box 6.1).

Institutional change as a complex system

So far, we have discussed the nature of economic rules. However, new institutional economists like Williamson (2000) and North (1995) provide a compelling four-level model to explain how institutions change. Thus, economic rules (L2) occur within the bounds of the cultural norms of the society in question (L1), leading society to organise itself in different ways (L3) and with different outcomes (L4) (Box 4.5).

As we saw in Chapter 2, for example, the European Enlightenment challenged the divine right of kings as the primary societal authority and prompted ideas like individual rights and choice, constitutional governance, the separation of powers, and science (rather than religious dogma) (L1). Consequently, age-old, personalised, neo-patrimonial practices like feudalism, serfdom, and peasantry, gave way to liberal democratic societies that valued individual rights, democratic processes, and the 'rule of law' (L2). Thus, by 1700, an 'Englishman's home was his castle' and could no longer be violated at the whim of kings. These new rules led to new forms or organisation (L3), including elected parliaments, political parties, impersonal courts, and public limited companies, allowing the economic game (L4) to be played in different and far more productive ways.

With four interacting layers, the economy is a complex system, suggesting that we can get to the same end point in different ways. For example, community conservation in much of Latin America (Chapter 13) arose out of social movements encouraged by the Roman Catholic church (i.e. L1), leading to stronger rights for local communities (L2) and, in some cases, better forest management (L4) (Schmink & Wood, 1992). By contrast, community wildlife management in southern Africa was conceptualised by enlightened bureaucrats (L2), who then used successful pilot projects (L4) to encourage greater social acceptance of community-based approaches (L1) (Chapters 8 and 11). Policy change is assumed to be hierarchical, with sensible discussions at high levels resulting in new rules (L2). However, the low returns from donor investments in 'policy reform' suggest that hierarchical policy reform is often disappointing. A more proactive mechanism is the 'long-hook short-hook' approach.[4] In essence, champions establish pilot projects (L4), often working beyond the bounds of the current rules. If these pilot projects are successful, they can be used to change overlying cultural attitudes (L1) by demonstrating empirical success, making it easier to consolidate gains by transforming the rules (L2).

BOX 4.5 FOUR LEVELS OF SOCIAL AND ECONOMIC ANALYSIS

The process of institutional reform is described by Nobel Laureates Oliver Williamson and Douglass North (North, 2003b, North, 1990, Williamson, 2000) and is summarised in the figure below. At L1, the national 'culture', including norms and religious beliefs, frames which types of institutions (L2 – rules and norms) are socially and culturally possible, and which will be resisted. Thus, post-socialist societies (and conservationists) (L1) often have difficulties understanding or accepting decentralised or free-market approaches and invariably resist devolution.

In turn, different configurations of rules (L2) result in different forms of organisation (L3 – the players). Thus, highly centralised and personalised regimes favour monopolistic and extractive forms of organisation (such as feudal lords, high priests, and chiefs), whereas liberal democracies are characterised by inclusive markets, competitive processes, and individualism.

This affects the way the daily game of give-and-take and supply-and-demand (L4 – the game) plays out, with the outcomes – for instance, of communist regimes, feudal systems and monopolistic guilds, and modern liberal democracies – being very different.

Governance is the formation and stewardship of the rules (Hyden et al., 2003) across these levels. Rules are designed, fought over, and enforced through the complex political interplay between forces promoting and opposing changes (as illustrated in the figure). System outcomes are emergent properties of complex interactions, varying from highly personalised and centralised systems on the one hand (Rule by Man), to democratic, impersonal, and inclusive societies on the other (Rule of Law).

The figure notes the challenging issue of 'third party enforcement of the rules'. Theoretically, enforcing rules through compliance monitoring, policing, and the judiciary is a fundamental role and rationale for the nation state. This gives rise to significant challenges where the state is not playing this role.

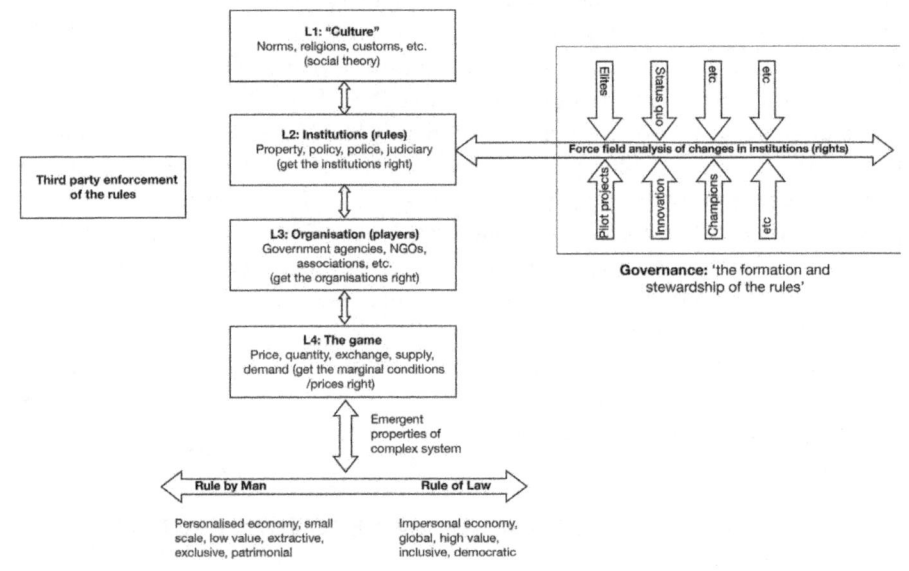

A model of the process of institutional change

Source: adapted from Williamson (2000).

The rules for free markets are not spontaneous: they are designed

Good economic institutions and policies are necessary to improve human livelihoods and the state of the planet. However, good institutions don't evolve; they are designed, and herein lies the challenge. Institutions such as democratic processes, property rights, and markets are created by people competing in political marketplaces, where outcomes are influenced by power, precedence, special interest, and enlightened technocrats. Even so–called 'free markets' are designed by people. The new institutions for good conservation outcomes will not evolve spontaneously. It is far more likely that old and failing institutions, and the vested interests behind them, will maintain themselves. If we want new institutions, we are going to have to conceptualise and design them, fight for them through the four-level economic process described above, and develop mechanisms to protect and enforce these (enlightened) rules.

Are we in the 'Dark Ages' of conservation?

The greatest transformation, and its inclusive institutions, have bypassed ungoverned forests and drylands. We are living in the 'Dark Ages' of wild life conservation. The (feudal) few dictate to the many who have little say over their own resources, and policy is often shaped more by emotion and ideology than dispassionate evidence. Communities living in, and depending on, forests, drylands, wildlife, and other wild resources are reminiscent of the Dark Ages of Europe. The rights to wild resources are controlled in the courts of feudal Big Men, and 'poachers' (and even criminals) often take on the anti-establishment personae of Robin Hood and his Merrie Men of Sherwood Forest – robbing the rich to help the poor.

About half of the planet is comprised of such communal lands, including most tropical forests and drylands. With weak institutions and a low capacity for collective action, these ungoverned spaces are locked in poverty and environmental degradation. Exactly as described by Garrett Hardin, people harvest rangelands, wetlands, forests, and wildlife on a first come first served basis (Hardin, 1968), while externalising the costs of over-use to the community, society, and the planet. This catastrophe is avoidable. The previous chapter showed that the output of land with rich institutions exceeded that with weak institutions by a factor of 12 or more. The institutional dividend of secure tenure and strong institutions is potentially huge. This is supported by the impressive gains demonstrated by new models of conservation, including private conservation (Chapter 8) and CBNRM in southern Africa (NACSO, 2015; Taylor, 2009). In the Amazon, too, partial reinstitutionalisation provides for better outcomes. Where local and indigenous communities have regained territorial rights to their forests (but not always full rights to utilise these efficiently), the loss of forests is two times lower than where forest tenure is less secure (Nepstad et al., 2006), and the application of Ostrom's principles to community harvesting of the giant pirarucu fish in the Amazon produced excellent outcomes (Koziell & Inoue, 2006).

CBNRM and the Glorious Revolution

Like the Glorious Revolution, CBNRM is an institutional approach to conservation and development. CBNRM has a similar cultural starting point (Level 1) – ownership of resources by the citizenry, and a switch from a personalised to impersonal governance, entrusting rural communities with legitimate rights to use, manage, benefit from, and protect their wild resources. This will require new rules, including the ownership of wild resources, markets for these resources, and the application of the principles of participatory self-governance (Level 2). There is a lot of damage to repair, including establishing community rights to land and wild resources, developing markets for legal products, rebuilding social capital, and restoring environmental damages. CBNRM, writ large, is the process of reinstitutionalising ungoverned spaces as local commons through private-community ownership. New rules will give rise to new forms of organisations, such as well-governed villages that own and manage wild resources (Level 3). If the prices of wild resources reflect their full values, the outcome will be a more sustainable bio-economy (Level 4).

With communities having been in the dark for so long, CBNRM is highly unlikely to emerge spontaneously. CBNRM combines the devolution of rights with the development of participatory forms of governance. The next few chapters explain why the devolution of rights is a precondition for successful CBNRM. Chapters 11 to 16 discuss the importance of participatory governance and democracy, because this is often forgotten and will seldom emerge spontaneously; rural people have rarely experienced inclusive governance and have limited capacity to aspire to it, as well as limited knowledge of its workings. Moreover, elites are benefiting from the status quo at multiple levels, and will naturally resist the transformation from extractive to inclusive governance. Over the last 10 000 years, participatory forms of governance emerged rarely and grudgingly (Tilley, 2007). Thus, one of the ironies of CBNRM is that participatory governance needs to be 'imposed' (albeit on willing supplicants), and then protected. This requires a clear understanding of institutional design, considerable effort and persistence, and strong third-party monitoring and enforcement because even where the gains of inclusive governance are recognised by communities, there is a strong tendency for elites to re-exert themselves (Muyengwa & Child, 2017).

Conclusion

As conservationists, we rush with great passion in new crusades to 'save' the next in a long line of threatened species or ecosystems. We lay the blame for environmental destruction at the foot of human greed, population growth, trade, or poaching, and see its salvation in our good intentions, often raising more money to do more of the same. But what we rarely do is sit back, think slowly, and ask the big question: What is going on when we are wiping out things that are so obviously important? If we change our assumptions about the causes of environmental decline, fixing them will be more likely.

Wild resources are, to all intents and purposes, ungoverned and over-exploited. The rules established in 1900 are no longer working. The crowded planet of the 21st century deserves new economic rules. We are sitting with a lowest common denominator wildlife economy in which wild resources are vastly over-exploited, but where this exploitation is so inefficient that it does little do countermand poverty. The poor live in ungoverned spaces with few rights to protect their resources which are prone to elite predation (Menard & Shirley, 2014). The rich benefit from resources they can extract from these ungoverned spaces. So what is the incentive to change?

Institutions are highly path dependent, and institutional reform is rare, difficult, and sometimes slow. One such critical juncture was the Glorious Revolution in England and the global prosperity that emerged from it (North, 2005). Another, which is closer to our cause, is the transformation of the wildlife sector in southern Africa on both private and communal land (Chapters 8, 11, and 12). Within these two examples, I suggest, lie the technical concepts and guidelines for the global transformation that we need. But providing the technical solutions is easy. Changing the system is not. How likely is it that the techno-bureaucratic elite will let uneducated local people preside over the fate of forests and fisheries, wildlife and wilderness?

Yet, unless the local communities that live with wildlife see it as an essential part of their livelihoods, they will continue to acquire and raise domestic plants and animals because these are the only resources that they can own legally. Ungoverned spaces are locked in a negative sum economy that over-utilises the environment, while under-delivering economically, in a way that is so inefficient that it is quite possible to achieve greater levels of economic development from lower levels of environmental use. The cause of this inefficiency is institutional failure. In the next chapters, we will begin to develop new institutional models that address these weaknesses. The 'sustainable governance approach' is about 'getting the economics right' by getting the institutions right – returning the proprietorship of land and wild life to landholders and communities and removing barriers to exchange and trade. CBNRM is the subset of this approach concerned with the 'getting the governance right' within the community. Undoing so much negative history and experience is challenging, and CBNRM will require tenacity and considerable recapitalisation of land, resources, communities, and institutions. However, when done properly, the results can be rapid and remarkable.

Notes

1 For readers interested in the current and future functions of state conservation agencies, Drucker (1973) describes the nature of budget-funded institutions, Grindle and Thomas (1991) describe the changing nature of state agencies in post-colonial societies, and Reed (2002) assesses the public and private nature of the goods provided by park agencies, and suggests new ways for organising and financing park agencies to provide these goods.
2 Good traditional leaders can play an important role uplifting their people, but are few are far between. Unfortunately, traditional leaders are much less accountable to their people than of old. They conflate a multitude of roles – from businessman, to politician, to local patriarch – often using their position for personal enrichment rather than the good of their people. I have witnessed more than a few cases where traditional leaders have facilitated outsiders to get title to prime areas of community land, often in exchange for an old vehicle or even as little as a few bottles of brandy. In these same places, I note with a sense of both irony and exasperation, how government officials and politicians cannot see their way to (and argue strongly against) giving the people living on the land the same rights to what is rightly their own land, through some combination of individual and village title.
3 For example, rhino owners and managers do not support the rhino ban (Rubino & Pienaar, 2018). Despite the rhetoric about demand reduction, African states participating in the Global Environment Facility (GEF)'s Global Wildlife Program have allocated no more than 1% of their grants to demand reduction.
4 I am using the term shared with me by Nik Sekhram of UNDP in the design of the UNDP GEF Grasslands Project in South Africa. In essence, this project established multi-stakeholder forums, with the practical task of solving on-the-ground problems, including mining and biodiversity in Mupumalanga Province in South Africa, and commercial forestry, farming, and biodiversity in Kwazulu Natal. The project was very well managed. Its core strength was to facilitate stakeholders, many of whom were in conflict with each other, to come together through pilot projects, such as mining and biodiversity. These experiences were then written up as guidelines, and through the participation of senior people from ministries, industry bodies, and universities in these pilot projects, the guidelines were often adopted at much high levels, in essence setting policies.

5

PROPRIETORSHIP

LEARNING OBJECTIVES

This chapter introduces the theory of property in the context of CBNRM and wild resources, and covers:

1. The functions of property in the context of CBNRM, including the efficient allocation of resources, equity, security, and capital.
2. The four types of property ownership – private, public, common, and open-access.
3. The concept of exclusion and the effectiveness of property in internalising the costs and benefits of resource use, to guide the economic allocation of resources.
4. Property and proprietorship as a bundle of rights (access, withdrawal, management, exclusion, alienation), including the effects of liability and regulation on proprietorship.
5. The features of wild resources – fugitiveness and complexity – that make internalising costs and benefits difficult, and the challenge of matching jurisdictional scale to the fugitiveness of resources.
6. A framework for categorising resources as being private, public, club, or common pool, using the characteristics of subtractability and/or excludability.
7. The characteristics of wildlife in relation to this framework as a moving target, suggesting that wild resources are miscast as a public good.
8. The differences between resource attributes, rights, and owners, suggesting that wild resources are increasingly suited to private and common pool regimes rather than over-simplified systems of public management.
9. Where and how collective action responds to the challenges of managing fugitive and complex wild resources.
10. The use and misuse of regulation in managing these challenges.
11. Devolved CBNRM proprietorship in the context of decentralisation and public administration.

Understanding property in the context of CBNRM

In the previous chapter I suggested that the problems afflicting resource-dependent communities in forests and drylands originate from insecurity of person and property. This chapter describes the institution of property in the context of the reinstitutionalisation of ungoverned spaces through CBNRM, where property has four main purposes:

1. The efficient allocation of resources as widely understood by economists.
2. The equitable ownership of resources and the empowerment of the people who live with wild resources and bear their costs through entitling.
3. Security of person, as the foundation for rebuilding social capital, inclusive governance, and a sense of pride in one's own resources.
4. Clarifying the ownership of assets to improve opportunities for use and investment.

Property rights are central to economic efficiency, economic justice and, I would argue, environmental and social sustainability. While the characteristics of assets are given, the rights over them are socially constructed. In order to design new configurations of property to match the changing economic characteristics of wild resources, we therefore need a better understanding of property in the context of wild resources and rural communities. Indeed, the concept of proprietorship is a central component of the sustainable governance approach and CBNRM, along with price (see next chapter). Proprietorship and price are suitors in the budding romance between economic principles and conservation.

CBNRM is a rights-based approach, with rights providing the foundation for effective economic and political governance (Chapter 2). Ostrom and her colleagues have made immense contributions to the understanding of rights in the context of common pool resources and collective action, and this chapter outlines these theories in the context of CBNRM. I only sketch out the big lines of these fascinating ideas. Readers are referred to the deeper conceptual analyses about property as a bundle of rights (Schlager & Ostrom, 1992), definitions of private, common-pool, and public goods (Randall, 1983; Hess & Ostrom, 2007), and the confusion between resources, rights, and owners (McKean, 2000; Hess & Ostrom, 2007).

Policy-makers tend to focus either on private goods and the market, or on the public regulation of goods not amenable to market allocation, and little in between. For goods in between, Ostrom showed collective management was not primitive, but remarkably sophisticated and outperformed private or public solutions under certain conditions (Ostrom, 1990). Practitioners already knew this intuitively, with Ostrom and others learning from long-lasting commons in the Swiss Alps, mountain villages in Japan, underground water management in California, medieval irrigation systems in Spain, irrigation systems in Sri Lanka and the Philippines, Turkish fisheries, and others (Wade, 1987; Ostrom, 1990; McKean, 2000; Hess & Ostrom, 2007). The brief background on property rights provided below underpins an understanding of the governance and economics of wild resources.

The purpose of property

Property and the efficient allocation of resources

Private property is the basis of economic efficiency (Coase, 1960). The classical economic argument (Figure 5.1) is that clearly delineated property rights internalise the costs and

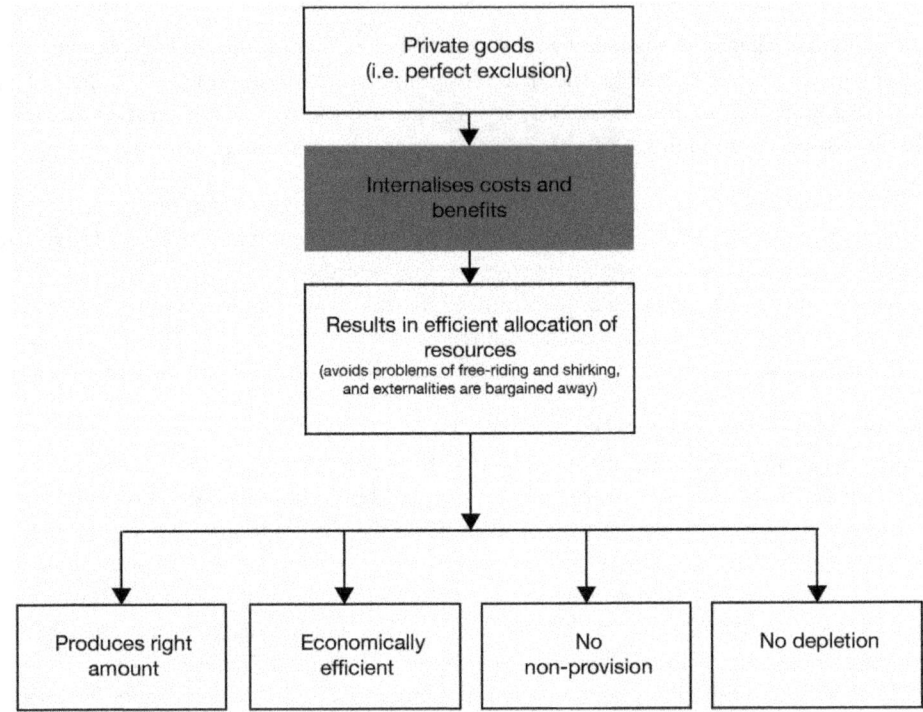

FIGURE 5.1   Why private goods have non-problematic characteristics for economic allocation

benefits of land use to the owner, creating full accountability for action which (combined with open, competitive, free markets) allocates resources to their highest valued uses. Harold Demsetz (1967) argued that 'a primary function of property rights is that of guiding incentives to achieve a greater internalization of externalities' and that such property rights were largely concerned with holding the owner fully accountable for his or her actions by controlling externalities. Where property rights are clearly delineated, the costs and benefits of resource use are internalised to the decision-maker, resulting in the optimal allocation of resources (Coase, 1960).

An important attribute of property is the right of exclusion. This empowers the owner to exclude others from extracting resources without permission or payment, and prevents people who have not contributed from free-riding and reaping the rewards of the owner's efforts.

In the imperfect real world, we nonetheless need to design new forms of property rights for fugitive and complex natural resources that internalise costs and benefits, and exclude free-riding to replace the current institutions that are crude, entrenched, and inadequate (Chapter 4).

Incomplete property rights and the tragedy of the commons

The importance of property for resource allocation is best illustrated by observing what happens where property rights are absent or incomplete. The classic example is Garrett

Hardin's tragedy of the commons, where the individual cattle herder reaps the benefits of his livestock but externalises the costs of over-grazing to the rest of society:

> the rational herdsman concludes that the only sensible course for him to pursue is to add another animal to his herd. And another; and another … But this is the conclusion reached by each and every rational herdsman sharing a commons. Therein is the tragedy. Each man is locked into a system that compels him to increase his herd without limit – in a world that is limited. Ruin is the destination toward which all men rush, each pursuing his own best interest in a society that believes in the freedom of the commons. Freedom in a commons brings ruin to all.
>
> *(Hardin, 1971, p 1244)*

The economic logic of Hardin's tragedy of the commons applies widely to the problem of ungoverned spaces (Chapter 4). Each rational individual maximises his or her private gain by adding one more animal, cutting one more tree, or hunting more bushmeat. The same conclusion is reached by all the people sharing the land, who privatise benefits while ignoring the social costs of their actions, inexorably degrading resources to the detriment of all.

A similar argument applies to land use trade-offs between wild and domestic species. If a person owns his livestock, but does not own wildlife (which is the property of all), he will increase his livestock until the wildlife has been completely replaced. The differential ownership of wild and domestic species means that the decisions made by the individual diverge from the best interests of society. These are market failures, caused by incomplete or differential property regimes. In these examples, individuals privatise the benefits of the cattle they own, but externalise the costs of unowned resources to society, be these over-grazed rangelands or the replacement of wildlife.

Economic efficiency is most likely where private goods are bought and sold in free markets, and where complete property rights internalise all costs and benefits. In imperfect markets, however, prices are distorted by a number of factors, with incomplete property rights, externalities, asymmetric or incomplete information, non-competitive or rigged markets, and other factors resulting in resources being allocated inefficiently. Fugitive, complex wild resources are particularly vulnerable to these imperfections, so private action does not contribute to the social good in the way Adam Smith anticipated, resulting in widespread over-use or under-supply of environmental goods. We have already described the over-use of bushmeat or grazing lands in open-access property regimes. Under-supply is associated with public goods like clean air, national radio, or national parks, which are consumed without contributing to their costs, with 'free-riding' resulting in less of these goods than we would ideally want. Partial or asymmetric property rights are also highly problematic. For example, weak institutions enabled colonial and later regimes to plunder raw materials, obtain cheap labour, and maintain political control cheaply through indirect rule (Acemoglu & Robinson, 2012). Likewise, rural areas and wild resources are still exploited by the modern urban or global economy because weak property rights allow the latter to acquire these resources (i.e. privatise them) cheaply, with few requirements for reinvestment. This allows for excessive rents, profits, or power that encourages the winners to maintain the system as it is.

Property and equity

Economic efficiency results from clearly defined property rights and exchange in free markets, and is independent of who owns the resources. Nonetheless, Coase (1960) and others acknowledge that the question of who owns the resources is also important. Ignoring these distributional issues gives free market economics a bad name, especially as property regimes and markets are often defined by the rich and powerful to benefit themselves (Stiglitz, 2002). Thus, CBNRM seeks stronger property rights (1) to create wealth through the efficient allocation of wild resources, but also (2) to ensure that wildlife and the wealth it generates are owned by the communities who live with it. Property rights also empower communities by providing legal protection against resource extraction and give communities the same incentives to protect 'their' wild resources as they have for domestic crops and animals.

Property, security, and inclusive governance

Property rights are usually discussed in the context of economic efficiency, where there is wide consensus about their importance. However, an equally fundamental (and political) purpose of property is its guaranty of freedom (Hayek, 1944). Security of person and property are a necessary condition for inclusive governance, allowing ordinary people to engage politically without fear of retribution. In feudal systems, it is difficult for ordinary people, or 'subjects', to challenge the power of lords or chiefs without putting at risk their property or freedom. Property rights are particularly important for the poor and disenfranchised (Bethell, 1998) and, in the context of CBNRM, proprietorship promotes economic justice by empowering marginalised communities with real choices about their own lives.

As Murphree describes in Masoka Village (Jones & Murphree, 2001; Taylor & Murphree, 2007), the psychological or attitudinal component of proprietorship is powerful – 'this resource is ours' awakened a sense of nurturing for wild resources that had long lain dormant under a centralised wildlife regime. 'We see now,' said one elder, 'that these buffalo are our cattle' (Murphree, 1991). Thus, proprietorship enhances social and collective processes of reciprocity, fairness, and control, beyond the cold individualistic calculus of cost, benefit, and maximisation. We should also not presume that these advantages are linked only to individual private property (Hess & Ostrom, 2007) because in certain circumstances collective property has significant advantages. Individuals gain pride and identity from being members of sports teams, clubs, and even public limited companies. Why would they not also gain pride and identify as members of private-community management regimes in CBNRM?

Property as capital

In his thought-provoking book, *The Mystery of Capital*, Hernando de Soto (2000) suggested that poor people own large amounts of capital, but that this is dead capital because it cannot be put to good use because of the legal haziness surrounding informal property rights. Formalising informal property rights, he argues, allows poor people to turn their dead assets into capital and to increase their wealth. Simply knowing who owns the resource is important, because it is difficult to do business where property rights are unclear or contested. Legal clarity is the foundation for community-private partnerships and the recapitalisation of many community wildlife areas in southern Africa (NACSO, 2015). De Soto also emphasised the importance of property as equity, for instance in developing wildlife businesses with private sector partners. However, outright mortgaging of land rights is risky, and CBNRM can be highly effective with non-transferable property rights and title.

In transitional economies, the definition of property rights is further weakened by the ambiguous co-existence of modern and traditional systems. Thus legislated *de jure* rights, granted and enforced by governments, overlap and compete with, and have ultimately displaced, de facto local rights (Pomeroy, 1995; Ostrom, 2000; Arnold, 2001) that have evolved over many generations and have considerable local legitimacy even where they are not entrenched in the formal system of laws and legislation. It is difficult to do business where there is confusion over rights and responsibilities. Moreover, mechanisms of exclusion and control break down in ill-defined dual systems so that wild resources are over exploited as non-property.

Understanding property

Types of ownership

Most people are familiar with the four types of property rights defined in Table 5.1 (though, later, we will suggest that the boundaries between these rights are not as clear as they seem):

- Individual private property (*res privatae*): Private or freehold property is when an individual, or an individual entity such as a firm, owns the resources. Theoretically, the owners are fully accountable for the costs and benefits of actions, leading to economically optimal solutions.

TABLE 5.1 The four categories of right owners

Property rights regime	Owner	Economic attributes
Private *Res privatae*	Private individuals or corporates	Exclusion is uncomplicated, costs and benefits are internalised, and economic allocation is unproblematic.
Common *Res communes*	Owned by a clearly defined group of people that has legal or traditional rights to exclude others (not to be confused with open access)	Highly variable, depending on strength of the group and its ability to exclude others.
Public *Res publicus*	Owned by the public or government	Highly variable, depending on the accountability of officials for costs and benefits of their actions, and their capacity for exclusion.
Open access *Res nullius*	Owned by no-one: ownerless property where anyone can take the resource on a first-come basis	Highly problematic, with well-known problems of inefficient economic allocation and unsustainable resource use. These are the conditions described in Hardin's 'tragedy of the commons'.

- State property (*res publicus*): Where the rights of ownership and management of land and wild resources are vested in the state. Where states are publicly accountable for costs and benefits this can provide economically sound outcomes, but this is often not the case for reasons we will explain.
- Common property (*res communes*): Common property is often confused with open-access property regimes. However, common property is where rights are held and exercised in common by a group with clearly demarcated membership, with the legal right and capacity to exclude non-members from using the resource without permission or payment (Ostrom & Hess, 2007).Personally I avoid the term 'common property' because outside of a small number of academics it is widely misinterpreted as common to all, rather than common to a small, defined group. Academics defining common property as rights 'vested in some form of collective' (Randall, 1983) or as being held by 'an identifiable community of interdependent users' (Feeny et al., 1990) leave open the scale, character, and function of the collective. These are critical issues, which we will come back to in later chapters. Moreover, until recently, common property was seen as primitive or archaic, counter-economic, and a hangover from less developed times, especially the period before the enclosure of the commons in 18th-century Europe.
- Open access (*res nullius*) or non-property: These regimes are where rights are not well defined so that resources are open for anyone to use, resulting in over-exploitation (Hess & Ostrom, 2007). This is ownerless property, but is free to be owned and belongs to the first taker.

Exclusion

These four categories of property rights are not always useful for predicting resource outcomes. A much stronger predictor of outcomes is the concept of exclusion. Functionally, this suggests two extremes of property: property where there is effective exclusion (by the state, individual, or communities) and non-property where exclusion is ineffective. Non-property includes weakly implemented state ownership, weakened forms of common or community property, and *res nullius* and open access property regimes.

Exclusion is a critical concept and implies that anyone wanting to use the resource has to bargain with the owner (see Chapter 6) – who can be a state or a non-state actor. Well-protected national parks are a good example of exclusion achieved through effective public management. However, poorly protected parks, where the state claims ownership without the capacity to enforce exclusion, quickly degrade towards de facto open-access or non-property. The decline of wild resources is correlated with weak exclusion. Public ownership of wildlife may have been crudely adequate in an empty world, but often degenerates into non-ownership in a full world where the value of resources increases without a concomitant improvement in the capacity for exclusion.

Exclusion is commonly associated with private property applied by individuals, groups, companies, and corporate bodies. An important, but neglected, form of ownership is 'private-community property'. It is a fallacy that wild resources cannot be owned as a community's private property. Indeed, the core purpose of CBNRM is to create community-private resource regimes with strong rights and the capacity for exclusion to overcome the problem of open-access or non-property. Some academics would argue that the term 'common property' implies private-community property, but in the real world the term 'common property' is too ambiguous and too often confused with non-property, concealing the importance of private community ownership in CBNRM.

I prefer explicit terms like 'private-community ownership' and see the goal of CBNRM as incorporating a Village Company. I use the term 'Village Company' to emphasise the need to give communities legal rights and personality, and a structure akin to business and shareholder ownership. This structure protects the rights of members because they are legal shareholders of the CBNRM business. It also defines the roles and rights of members and officers, and the procedures to be followed for making decisions (especially about money). Moreover, it differentiates CBNRM communities from the rather nebulous administrative committees of project interventions and the problems commonly associated with them. Internally, members of a Village Company have legal status as shareholders and can hold their leaders to account through commercial legal mechanisms when local social processes fail. Externally, giving communities legal personality as a Village Company provides for more protection than administrative or policy arrangements.

Defining proprietorship as a bundle of rights

Proprietorship is a complex concept, concerning the acquisition by an individual or organisation of a bundle of rights and responsibilities for resources. In the context of CBNRM, Murphree (1991) defines proprietorship as:

> a sanctioned use right, including the right to decide whether to use the resources at all, the right to determine the mode and extent of their use, and the right to benefit fully from their exploitation in the way they choose … [and to] determine the distribution of such benefits and determine rules of access.
>
> *(Murphree, 1991)*

Murphree's definition is operationalised in the CAMPFIRE cases study (Chapter 11) as the right to use, sell, benefit from, and manage wildlife. A more widely used definition is that of Schlager and Ostrom (1992), who define property as including rights of access and withdrawal, management, exclusion, and alienation (Table 5.2). However, rights are associated with responsibilities and liabilities, and I have expanded Table 5.2 to include these for reasons explained below.

Thus, CBNRM requires the devolution to communities of the rights to access and use resources and to manage them, to exclude others from using them, and to sell them (or at least the annual production from them). CBNRM is predisposed to fail without strong rights, and preferably title, across this range of entitlements (Table 5.3). We can classify community conservation according to how many rights are devolved, and to which level (see Table 11.3), with CBNRM corresponding to fully devolved citizen control (or privatisation) in Arnstein's ladder of participation. Where CBNRM initiatives disappoint, the cause often lies in the failed or aborted devolution of this full range of rights (the other primary cause of failure is weak systems of micro-governance and elite capture). There is some debate about full alienation. In Mexico, communities have title to their land and, following recent legislative changes, can sell their land following agreement by a super-majority of the community (Bray et al., 2005). This seldom happens, except where communities become included in major expanding urban conurbations. In the main,

TABLE 5.2 Defining rights as a bundle of entitlements and obligations

Formal definition of rights

1. **Access**: The right to enter a defined physical property.
2. **Withdrawal (benefit sharing)**: The right to obtain the 'products' of a resource (e.g. catch fish, appropriate water).
3. **Management**: The right to regulate internal use patterns and transform the resource by making improvements.
4. **Exclusion**: The right to determine who will have an access right and how that right may be transferred. This includes the right to bargain and exchange (i.e. sell) the products of the resource to best advantage (my addition).
5. **Alienation**: The right to sell or lease either or both of the above collective choice rights.
6. **Liability**: The responsibility to prevent the resource from causing damage and/or the requirement to pay compensation in cases where damage is caused.
7. **Regulation**: The responsibility for ensuring that rights are managed and for structuring rights to ensure the greatest good for the greatest number by preventing private action from imposing social costs on society.

Source: Expanded from Schlager and Ostrom (1992).

TABLE 5.3 Defining proprietorship in terms of use rights

Rights and responsibility	Definition of rights	Owner	Proprietor (CBNRM)	Claimant	Authorised user
1. Access and withdrawal	The right to enter a defined property and the right to obtain the 'products' of that resource	X	X	X	X
2. Management/ participation	The right to regulate internal patterns of use and to transform the resources	X	X	X	
3. Exclusion	The right to determine who will have access to the resource and to protect the resource	X	X		
4. Alienate (sell) flows	The right to sell products	X	X (flows)		
5. Alienate (sell) stocks	The right to sell use-rights, after which the former rights holder has no rights	X (stocks)			
6. Liability	The responsibility to prevent the resource from causing damage and/or the requirement to pay compensation in cases where damage is caused	X	X		
7. Regulation (self-regulation)	The responsibility for organising and controlling the resources and managing externalities	X (especially collective action)	X		

Source: modified from Schlager and Ostrom (1992) and (Murphree, 1991).

CBNRM requires only non-transferable community title, which reduces the risk of exploitation and landlessness.

Liability and regulation

I have expanded on Schlager and Ostrom's (1992) definition of rights, because issues of liability and regulation act on rights and can change them significantly. I will use the fascinating example of Zimbabwe to illustrate how this principle is often ignored when a country nationalises wildlife. Like much of the colonial world, early Zimbabwean wildlife legislation was built around the principle of the 'King's Game' (see Chapters 7–8). Wildlife was legally defined as a public asset, owned by the state on behalf of the people. However, contradictions inevitably arose with the management of a public asset on private land. In the mid-1970s, the government prosecuted a farmer for shooting wildlife, including royal game. Claiming he was protecting his cattle, land, and grazing resource, the farmer flipped the argument around. He sued the government on the basis that if the state claimed ownership of wildlife, it was also liable for the costs of that wildlife to his ranch, including grazing and damages to crops and livestock. The court agreed with the farmer (Beadle & Macdonald, 1969) making conventional wildlife laws unworkable, especially as the legal definition of wild life[1] could be interpreted to include pest species such as quelea and locusts. This resulted in an ingenious legal innovation: in the 1975 Parks and Wild Life Act, wild life was *res nullius*, but proprietorship was devolved to landholders (Chapter 8).

As a practical matter, I have added regulation to Table 5.2 because regulatory restrictions on property rights can significantly change the nature of rights and can radically shift the distribution of costs and benefits between different constituencies. The Convention for the International Trade in Endangered Species (CITES), for example, removed the right of landholders to sell rhino horn or elephant leather without their consent and without paying for it. These actions might be considered normal for wildlife, but would be called theft if applied to private goods like cars or cows, and could easily be reversed in the courts, as recently happened for the internal sale of rhino horn in South Africa. Even eminent domain, or the right of a government to expropriate private property for public use, usually requires fair compensation. Conservation relies heavily on regulation and somehow gets away with not paying the full costs of appropriation because of the history of public ownership of wildlife and the weak rights of communities that live with it. Despite ineffective results, there is an over-reliance on negative incentives and wildlife regulations that often do more harm than good. In South Africa, for example, less regulated species have recovered much faster than heavily regulated ones (Dry, 2010). This is because regulators seldom assess the effect of regulations on the all-important economic and financial curves (Figure 3.7), and differential regulation and taxation reduces the competitiveness of wild species. The efficacy of regulations in wildlife conservation needs critical review, with far greater consideration of positive incentives (Murphree, 1999).

The question of who the regulator should be also deserves more attention. While the state is the primary regulator, decentralised self-regulation through associations and localities is highly effective. Centralised regulation is costly, and there are many advantages in decentralised or devolved self-regulation. This usually requires (a) empowering participants to self-regulate through associations, landholder communities (Chapter 8), standards, and certification mechanisms and (b) inspecting the quality of this regulation on a periodic basis.

Internalising costs and benefits, and the problem of fugitiveness and attribution

Property regimes fulfil the critical economic function of getting prices right by internalising the costs and benefits so that the user is fully accountable for his or her actions. This is relatively easy for private goods like cars and houses, but is more difficult for wild resources for a number of reasons (Table 5.4). Wild resources are inherently fugitive in space and time, so their effects spill over the boundaries of simple cadastral property regimes. Being associated with complex ecological systems and multiple values, it can be difficult to attribute costs and benefits to individual actions and investments. This is similar to the problem of marking – it is easy to attribute the damage caused by a cow that is owned and branded, but who do we blame for the damage caused by a wild buffalo? Finally, the costs of marking and drawing boundaries around wild resources are relatively high, especially compared to the relatively low productivity of wild systems (Ostrom, 2009). Historically, rather than taking on the challenge of designing property institutions suited to wild resources, policy-makers have defined wild resources as public assets, an approach that now needs to be reconsidered.

TABLE 5.4 Challenges to using simple cadastral property regimes for wild resources

Source of difficulty	Challenges to the attribution of costs and benefits
Fugitiveness	Wild resources are economically fugitive in that they are mobile (in space and time), making it difficult to draw boundaries around their effects.
Attribution	Complexity – ecosystem services and wild resources are complex. They provide a multitude of complementary uses and values at different levels in society. This makes the attribution of costs and benefits difficult.
Marking wild resources	Lack of distinctive markings – unlike domestic crops and animals, wild resources usually lack the distinctive markings that are important for attributing costs and benefits or, indeed, property rights, as we do for instance with the branding of cattle (Ostrom, 2007).
Low-value per unit area	Wild resources often occur in agriculturally marginal zones, and have a low value per unit area. The combination of low returns and the higher costs of defining and managing property rights means that the return on investment in defining property rights can be low or even negative. However, the counter-argument is that the low value of wild resources is an artefact of the absence of suitable systems of property rights, which fail to reflect the true (and high) economic value of wild resources (Chapter 3).
Historical treatment of wild resources as public assets	The governance of wild resources as public assets has become a social norm with strong resistance to change or re-imagining better alternatives (despite the mismatch between characteristics of resource systems and the property regimes for managing them).

Scale, and the challenge of managing fugitive resources

In the real world, institutional design needs to account for scale. Economically speaking, the goal of jurisdictional boundaries is to internalise costs and benefits as fully as possible. This introduces the practical challenge that the mobile resources such as elephants and rivers are fugitive in space, and we can only internalise their costs and benefits at the scale of, say, the village, district, or region. Soil degradation, with its multi-generational effects, is fugitive in time, creating additional challenges. Added to this is the difficulty of attributing the costs and benefits of ecosystem services, with the provision of clean water, carbon sequestration, or ecosystem health and productivity, emerging from complex, non-linear, and multi-scalar processes.

The relationship between the fugitiveness of a resource and jurisdictional scale is illustrated in Figure 5.2. In finding the right balance between scale and accountability, we need to consider two factors. The higher we move up the jurisdictional scale, the more we are able to internalise the full suite of costs and benefits. However, accountability feedback loops weaken rapidly as the decision-maker moves further from the resource.

This raises two central questions about scale which will recur throughout the book: At which level do we draw the boundaries? Through which process do we get there?

The level of proprietorship should match the scale at which the externalities occur, and should be as low as possible because this maximises transparency, accountability, and the efficacy

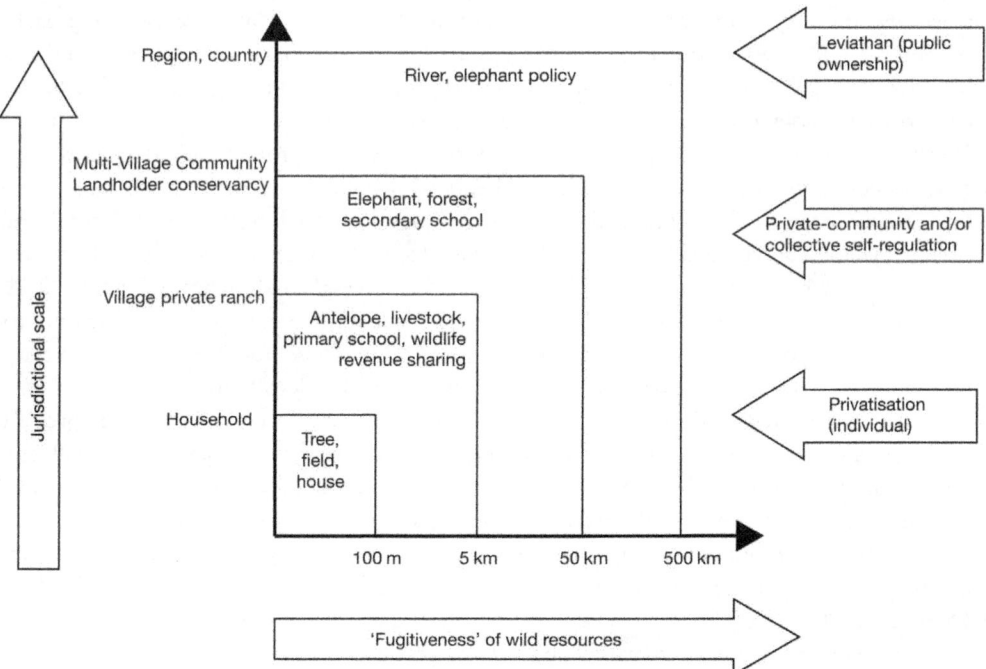

FIGURE 5.2 Internalising costs and benefits by matching resource fugitiveness and jurisdictional scale

Source: this figure illustrates concepts discussed by Murphree (2000) and Martin (2003).

of feedback loops. Moreover, small groups work better, and social cohesion and cooperation is maximised by acting at the lowest scale possible (Dunbar, 1993; Murphree, 1994a).

We can get to the optimal situation in two ways. We can design it centrally. Alternatively, we can let it evolve from the bottom. The ubiquitous top-down approach appropriates rights and powers far more centrally than is optimal, and has democratic deficiencies. Bottom-up approaches are rare; rights are devolved to the lowest level possible, and are then scaled up parsimoniously from the bottom up through a process of delegated aggregation (Murphree, 2000) – thus maintaining ultimate authority in the citizenry, as when Dutch farmers aggregate to manage dykes. In taking on the challenge of reconfiguring the governance of, say, wildlife, the starting point would be the devolution of the resource proprietorship to individuals or villages (Figure 5.2). It is sensible to manage sedentary assets like fields and homes at the level of individuals where costs and benefits are internalised. Using the same logic, local schools, grazing, village taxes, forests, water points, and most wildlife would be managed at the level of a village to match the fugitiveness of the resource to the scale at which it is managed. Resources that are more fugitive, or are associated with economies of scale, might need to be managed by a group of villages (elephant, secondary school), and a few at even higher levels. We will return to these issues in later chapters.

This suggests that the binary approach of locating property rights either with individuals (privatisation) or with states (nationalisation) is too simple for wild resources. In the late 1800s when many wildlife laws were formed, collective approaches like the English Commons were seen as primitive, while the experience of frontier economies, and especially the near-elimination of the American bison, discredited private approaches (Chapter 7). This led to the assumption that environmental allocation problems could only be solved by governments with major coercive powers, and the hegemony of Leviathan authority in natural resource management (Ostrom, 1990, p 8–9). However, neither Leviathan nor free market approaches based on individual property (Anderson & Leal, 1997) adequately account for the costs and benefits of wild resources. Leviathan acts at the top of the system, and accountability feedback loops are too long to be effective, while central planning is vulnerable to the self-serving decisions of officials and special interests. Private ownership and the free market, conversely, have the advantage of acting at the bottom of the system, but need to deal with the challenge that land units are sometimes too small and decision-making too short-term to fully internalise the costs and benefits of utilising wild resources, especially when dealing with small-holder farmers.

Ostrom suggested options that acted in the middle scales, notably common property management (i.e. private-collective ownership). With empirical evidence that collective action can be effective (Hess & Ostrom, 2007), she challenged condescension towards 'primitive' commons and the theory that 'rational, self-interested individuals will not act to achieve their common or group interests' (Olson, 1965), suggesting common property institutions could be effective if they incorporated a set of design principles (Wade, 1987; Ostrom, 1990). We therefore need to broaden the governance possibilities for wild resources beyond public ownership and simple privatisation. Ostrom discusses collective management largely in the context of individuals sharing and using a common resource under common rules. CBNRM, however, also includes communities acting collectively as a unitary business. Before we discuss these, however, we need to develop a much more concise economic definition of wild resources.

Defining public, private, and common-pool resources

Randall (1983), Polski and Ostrom (1999), and others provided conceptual clarity to the definition of public, private, and common goods by introducing and defining the concepts of 'exclusion' and the awkward term 'subtractability' (which has a second, equally unattractive, label, 'non-rivalry'):

'Excludability' is the extent to which people can be excluded (or not) from using a resource through physical barriers, technical means or legal instruments. High excludability implies that consumers will have difficulty consuming the good or service without contributing to its cost; low excludability implies that consumers may be able to 'free-ride', consuming the good or service without contributing to the cost of provision or production (Polski & Ostrom, 1999, p 10).

'Subtractability' is the extent to which the use of a resource by one person reduces the supply to others. High subtractability implies individual consumption; low subtractability implies that more than one person will consume the good or service at the same time (Polski & Ostrom, 1999). Thus, an impala is subtractable because hunting one impala means less are available for other hunters, whereas viewing mountain scenery does not subtract from its availability for other people, who are therefore not rivals in its use.

This provides the well-known framework for classifying goods (Figure 5.3) according to whether they get used up or not (i.e. are subtractable, or rivalrous) and whether it is easy to exclude others from taking the resource. Economic efficiency is achieved only for private goods in competitive markets – all other goods are subject to market failure in some way.

Pure private goods are excludable (i.e. its owners can exert private property rights, preventing others from using it without paying for it) and subtractable because consumption by one person prevents consumption by another. Free markets allocate private goods well. Clubs' goods are similar, except that they come in large lumps (e.g. golf clubs).

		Excludability (is it easy to exclude other people from using the resources?)	
		Easily	**With great difficulty**
Rivalrous/Subtractable (do resources get used up?)	Subtractable	**Private Goods** (cars, houses, phones, etc.)	**Common Pool Resources** (forest, fish, etc.)
	Non-subtractable	**Club Goods** (golf club, private neighbourhood, etc.)	**Public Goods** (legal systems, air, views, etc.)

FIGURE 5.3 Definition matrix for private, common pool, club, and public goods
Source: based on Randall (1983) and Polski & Ostrom (1999).

Public goods, by contrast, are non-rivalrous (use by one person of a legal system does not prevent use by someone else) and non-excludable (it is not possible to exclude people from fresh air or street lighting).

Common pool resources fall in the middle. They are, to some extent, subtractable and can be over-exploited, and it is difficult (but not impossible) to exclude potential beneficiaries from using them. The combination of high subtractability but low excludability often leads to over-use or under-supply by, for example, overfishing or under-investment in managing the fishery.

The miscasting of wild life as a public good

Analysing wild life (including forests, drylands, fish, and wildlife) with these criteria suggests that it is not a public good. Forests and wildlife act much more like private goods, with some common-pool properties, while a few attributes such as existence values might be public goods. Wild resources share with private goods the attribute that one person's use subtracts from another ('subtractability'), but they also face challenges of excludability, with both of these attributes being dynamic in a world of increasing scarcity and technology (Figure 5.4).

Moreover, any simple categorisation of wild life (forests, fisheries, drylands, and wildlife) hides a lot of granularity. Wild animals cover a spectrum from private to common pool goods because they are clearly subtractable, while the degree to which they are excludable varies. Wildlife can simultaneously display properties of private, common pool, and public goods (and even a club good in the form of a private reserve or national park). As arrow 1 in Figure 5.4 shows, the costs and benefits of a small, territorial antelope like a duiker can be internalised at a very local level as a private good, whereas elephants and grazing are common pool resources, while medium sized antelopes (such as kudu) fall somewhere in between. Existence and option values might be considered public goods.

The properties of individual resources can also be disaggregated in this way. Elephants are an important resource for many communities in Africa in both positive and negative ways (Bond, 1994). The income from an elephant, for instance, is best managed as a private good (at the level of a community). The elephant itself may be managed collectively by the several villages across which it roams, but the costs it imposes when it tramples fields are private. Further, elephant policy is national, and elephants have global existence values.

The point is that crude institutions lose sight of this granularity. 'Solving' this complexity through a lowest-common-denominator approach by lumping them all as public goods has significant costs in terms of our goals – which is to internalise the costs and benefits of wild resources in ways that encourage full accountability.

A further wrinkle is that subtractability and excludability are moving targets (arrow 2 in Figure 5.4). Not many decades ago, wildlife and forests were perceived as so abundant that they could not be exhausted (non-subtractable), but they are now clearly subtractable, as is the capacity of the atmosphere and oceans to absorb pollution and plastics. Similarly, technology is providing new methods of exclusion, while the costs of these methods is declining relative to the value of the resources, changing the economics in favour of exclusion. For example, rhinos are now so valuable that we can afford to use satellite radio-collars, micro-chips, drones, smart-fences, automated night-vision equipment, and so on to protect them. The combined effect of increasing subtractability and excludability is that wild resources are being transformed from the public goods of the 19th century into common pool resources and, quite often, private goods, at a significantly faster rate than governance regimes are able to adapt to.

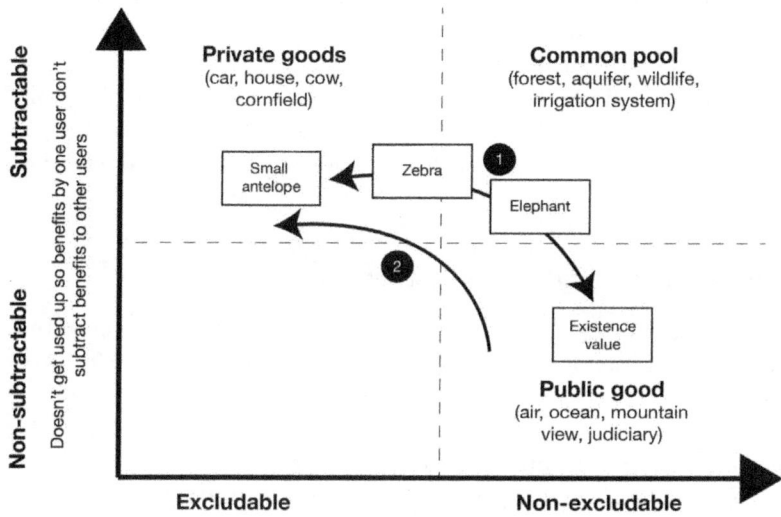

FIGURE 5.4 Wild resources are private or common pool goods and becoming more so

This brings us to the critical issue. The governance regimes for wild resources no longer match their economic characteristics. Wildlife, forests, local fisheries, grazing, and communal lands are publicly owned and managed, but are no longer public goods, being both subtractable and, to a significant extent, excludable. This implies that reconfiguring their governing institutions, especially proprietary rights, can significantly improve their allocation (and conservation). African wildlife, for instance, is subtractable and excludable (as it is quite possible to protect wildlife using boundaries and law enforcement), so that private management is theoretically likely to be as or more effective than public management.

Ostrom recognised these complexities with her statement 'beyond markets and states'. Thus, redefining the jurisdictional authority for wildlife at the level of a community or a private ranch, converts it into a private resource. This has the considerable advantage of enabling us to entrust the allocation of wildlife far more to processes of pricing, markets, and individual choice, where it will look after itself economically, provided it is valuable and proprietors have sufficient powers of exclusion. However, some costs and benefits extend beyond the boundaries of the individual wildlife properties (i.e. externalities) and are complex, and it is here that Ostrom recognised the power of social processes (e.g. peer-to-peer accountability) and collective action to strengthen mechanisms of accountability, thereby complementing privatisation (see Chapter 8). Collective action often strengthens environmental accountability, but it can also add value by expanding the scale at which environments are managed (Figure 5.6).

Treating wild species and spaces as public or global goods also results in a highly inequitable allocation of costs and benefits (Table 5.5). Benefits are global, but costs are local (Wells, 1992). The evidence suggests that this is not sustainable, politically, socially, or economically.

Confusion between goods, rights, and owners

Confusion in designing effective governance regimes for wild resources stems from using the words 'public' and 'private' for three different things (McKean, 2000; Hess & Ostrom, 2007):

- The natural attributes of the wild resource in question.
- The rights applied to this good in the form of human-constructed institutions or rules.
- The owners who acquire these rights (Table 5.6).

Therefore, in addition to defining wild resource more accurately, it is important to differentiate (1) the type of goods from (2) the different forms of property rights applied to these goods and (3) the different kinds of owners that can acquire them (Table 5.6). The use of similar words to mean different things across these different dimensions of governance leads to further confusion.

Matches and mismatches between resources, rights, and owners

Figure 5.5 brings together a complexity of factors that need to be accounted for when designing governance systems for wild resources: the fugitiveness of the resource, and how different combinations of owners and property regimes affect the rights and capacity for exclusion, and therefore whether landowners are accountable for wildlife or not. The large arrow at the top of the figure reminds us that wild resources are fugitive at different scales. The y-axis illustrates the four main categories of owners. The scope of the property regimes associated with these owners (illustrated by the length of the arrow in relation to the x-axis) depends on the nature of the resource and the capability of the owner to internalise costs and benefits and enforce exclusion.

Exclusion is illustrated on the x-axis, ranging from effective exclusion that internalises costs and benefits (left side of axis), to weak exclusion that results in market failure and inefficient use of resources. Thus, private ownership is generally good at accounting for all costs and benefits (though elephants may stray beyond the bounds of a private land unit), non-ownership is not, and public ownership is highly variable. The line arrows labelled a, b, and c show that the system is dynamic and the effectiveness of ownership regimes changes.

The central message portrayed by Figure 5.5 is that private, community, and public management can all be effective (i.e. through strong exclusion) under the right circumstances, but that community and public property regimes are vulnerable to decaying towards open access.

TABLE 5.5 The unhealthy political economy of wildlife

Governance regime	Costs	Benefits	Consequences
International/ global	• Few (limited willingness to pay)	• Existence and option values • Ecosystem services • Central financing	**Free-riders** – the incentive is to protect benefits by using regulations to impose conservation (and its costs) locally. Free-riding results in under-investment in wildlife, and is unsustainable.
National	• Funding of protected area agencies (usually at inadequate levels)	• Status • Donor funding • Tourism economy • Ecosystem services	Incentive is to play two games – tell a positive conservation story (to access financing) while not protecting the resource at home (so as not to alienate the rural political constituency) (Gibson, 1999).
Landholder/ community	• Loss of life, crops and livestock • High opportunity costs – reduced farming, fear of walking at night because of lions, etc.	• A little meat, usually obtained illegally	**Losers:** Costs outweigh benefits, so wildlife disappears, passively or actively.

Private ownership of many wild resources is sensible, especially if properties are large enough to internalise most of the costs and benefits associated with wild resources. However, combining private ownership (including community-private) with collective self-regulation is an even better solution because it accounts for both the private and common-pool attributes (i.e. externalities) of these resources.

Unfortunately, there has been widespread decay of community property regimes towards an open-access situation (illustrated by arrow 'a' in Figure 5.5), as colonial and post-colonial regimes eroded the rights, capacities, and confidence of local communities to manage forests, wildlife, and fisheries.

Wild resources, and especially those owned by communities, were often nationalised (arrow 'b'), with public ownership becoming the norm. However, many public management regimes have decayed towards open-access regimes and ungoverned spaces (arrow 'c'), especially when attempting to manage public wild resources on private or community land under modern, global pressures.

TABLE 5.6 The confusing use of the terms 'public' and 'private' in natural resource governance

Good	Rights	Owners
Private	Private *res privatae*	Individuals Corporations
Common pool	Common property *res communis*	Groups (defined legally or traditionally)
Club		
Public	Public (state property) *res publicus*	Public or government Not for profit
	Open-access/non-property *res nullius*	Ownerless

Good. This is an attribute of the good, based on (1) how easy it is to exclude others from using the good, and (2) how much the use of the good by one user subtracts from the use by others.

Rights. These human-constructed rules and norms (i.e. an institution) refer to the clarity, security, and exclusivity of a right enjoyed by the right-holder.

Owners. The agent, agency, or player that acquires rights. This ranges from a public body, which represents the public, to a private body that represents only itself.

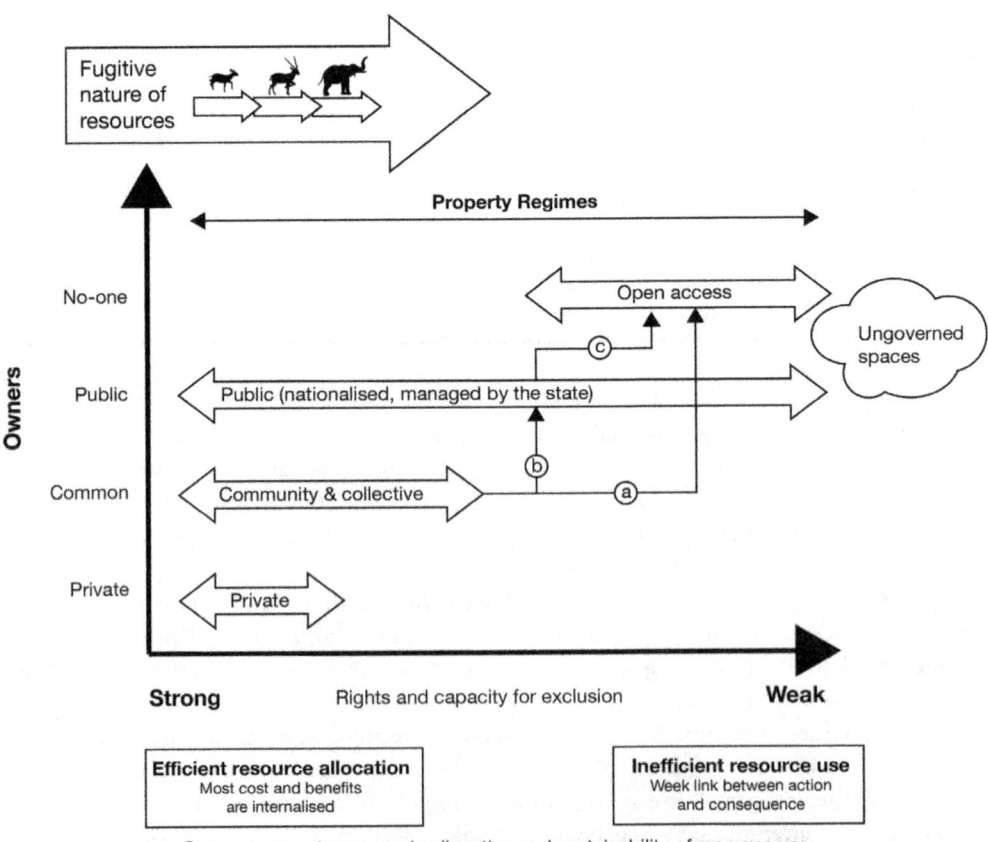

FIGURE 5.5 The interaction between resource fugitiveness, rights, owners, and exclusion.
Source: building on Hess and Ostrom (2007) and McKean (2000).

In some circumstances, nonetheless, public ownership and management of wild resources is highly effective (Chapter 7). National Parks are a good example, provided they are socially legitimate, provide benefits to society, and are capacitated to manage the resources and exclude free riders, poachers, and land encroachment. Park boundaries, incidentally, are a cost effective mechanism of exclusion, and can be remarkably effective even with relatively low investments in patrolling and exclusion. However, when under-resourced or lacking social legitimacy, many parks rapidly become paper parks (arrow 'c') unable to exclude illegal hunting or settlement.

In places where states fail to manage the relatively straightforward task of protecting national parks, we need to be even less confident of the efficacy of public management of wild life and biodiversity under much more challenging circumstances outside parks. Yet this pattern of nationalised wild resources, inadequate state capacity for exclusion, and alienated communities is the norm in post-colonial states in Africa, Latin America, and much of Asia (i.e. arrows a, b, and c), and explains the magnitude of losses of wild resources.

To reverse the trend towards ungoverned spaces, we need to move to the left of Figure 5.5 where mechanisms for internalisation costs and benefits and exclusion are strong, through various combinations of private, community, and public governance. Public governance is historically dominant, but is overstretched and underperforming. This points us towards private and collective forms of management. Private property regimes are simple to administer and can be very effective. We are most familiar with individual private property – such as farms or, more recently, private wildlife reserves and forests. However, the vast extent of ungoverned spaces occur where communities and biodiversity priorities overlap, suggesting that we have most to gain on a global scale by developing private-community resource management and production units – the subject of CBNRM.

Forms of collective action

Collective action takes several forms, with different purposes and benefits (see Figure 5.6). Ostrom mainly discusses collective self-regulation by the users of a shared resource. My focus is different. I am primarily concerned with the internal governance of communities as private-community production units, or Village Companies. 'Collective production' is the topic of the second half of the book. Before this, Chapter 8 introduces 'collective self-regulation'. There I discuss peer-based self-regulation by groups of landholders sharing Zimbabwean landscapes, where collective action and social interactions are invoked to manage environmental effects that are too complex for the market, such as social controls over soil erosion and deforestation (Child & Child, 2015). 'Collective management' refers to the action of bringing villages, or even private ranches, together to manage much larger landscapes (ecologies of scale), or because of economies of scale linked to brands, markets, supply-chains, and so on.

Regulations and property rights

Policies, rules, and regulations have a tremendous impact on the performance of society even though they are largely invisible (Chapter 3). By acting on property rights and on markets, regulations and regulators significantly affect the distribution of costs and benefits,

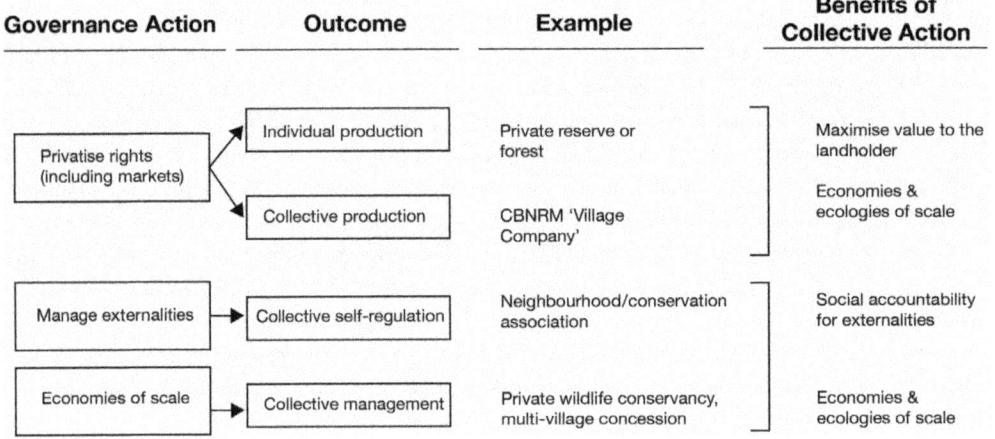

FIGURE 5.6 The different purposes of collective action in natural resource management

and can enhance or retard devolved, market-based approaches. An unfortunate consequence of the highly centralised governance of wild resources is that regulations are also thrown around like confetti (by bureaucrats and at international conventions) with far too little accountability given for the fact that these actions are far from costless. Wild resources are invariably over-regulated, with little thought as to how regulations move the cost curves (Figure 3.7) or undermine the credibility of the regulatory agency – regulations are worse than worthless unless they can be enforced and funded, and over-dependence on them is undermining this approach.

Legislation, policy, and regulations are best applied sequentially. In the context of the argument in this book, the first priority of policy and legal reform is to get prices right for wild resources through appropriate combinations of proprietorship and markets (Table 5.7). To shift the profit curves for wild species upwards (Figure 3.7), we need to reverse historical norms such as public ownership, market closure, and the imposition of transaction costs (e.g. red tape), fees (e.g. licences), and limitations on management options.

Once the pillars of price and proprietorship are in place, markets and regulations need to be carefully crafted to get the right outcomes. For example, we might need to design the market for rhino trade to drive landscape conservation and avoid pen-fed rhino production. In places where we are promoting private-community ownership, regulations and mechanisms are necessary to control the standard of devolved governance (see Chapters 11–16). Finally, taking a market-based approach does not mean neglecting regulations for externalities that cannot be controlled through privatisation and local self-regulation, or by ignoring the need to subsidise biodiversity that is valuable but that cannot finance itself.

Finally, the process by which regulations are made is important. It should involve people in regulating their own affairs, and regulations should be parsimonious, and as close to the action as possible.

Regulations are also political, with major distributional implications, and we can never ignore the political question of 'who decides'. Regulations work best in inclusive societies, such as North America where hunters and recreationists benefit significantly from the

TABLE 5.7 Checklist to illustrate a strategic approach for regulating the governance of wild species and spaces

Regulatory strategy and goal	Example
Establishing devolved market-led private or private–community conservation	
Proprietorship. Shift proprietorship of wild resources towards private and collective management, and away from open access.	Change legislation or administrative procedures to devolve proprietorship of rhinos (and other wildlife) to landholders and communities. Ensure 100% of revenues are retained by landholder. Protect property rights (e.g. take legal action to challenge the removal of property rights without fair compensation).
Price. Use markets to drive cost curves to favour conservation and rewilding, where this is sensible.	Promote legal wildlife trade and product development (e.g. tourism, hunting, trade) to replace illegal markets. Remove bureaucratic costs and regulations.
Designing governance and markets	
Design the market to achieve specific conservation goals, e.g. use the high value of rhinos to rewild large areas of land, while avoiding factory farming of rhinos.	Allow unrestricted trade where rhinos originate from large landscapes, but increase regulatory restrictions or taxes for rhinos in small areas or even pens.
Set the technical and democratic standards for devolved self-regulation by communities.	Rules and principles for community self-governance (see Chapters 11–16), or for industry bodies (e.g. a private forest association).
Non-market values and externalities.	Provide subsidies for important habitats, or use regulations to control externalities that cannot be controlled locally.

public management of wildlife and wild spaces. Box 5.1 provides a perspective on why southern Africans are so frustrated by how rhinos (and other wildlife) are regulated with imperial high-handedness and economic dysfunction.

BOX 5.1 A SOUTHERN AFRICA PERSPECTIVE ON RHINO REGULATION AND POLICY

Nearly 6 000 rhinos live on private land in southern Africa. Rhinos in South Africa have recovered from less than 60 animals in the 1950s to some 20 000 today. This is an amazing achievement, but the park managers and landholders who have done so much to bring white rhino back from the brink of extinction view CITES with anger, frustration, or resigned cynicism.

To them, the way CITES has regulated the rhino trade is anti-democratic, ignorant, and counter-productive. They believe that CITES has been captured by, and benefits, special interest groups. Having seen a new way to the rewilding of land, landholders view CITES's decision as extremely retrogressive, imposing an old fashioned, unworkable, and foreign ideology that wildlife is a public or global asset, which is disrespectful in the extreme regarding their commitment, success, and knowledge in bringing rhinos back from the brink of extinction through private ownership and commercial utilisation. There is wide acceptance and growing evidence that species that are owned and utilised by land-holders are increasing far faster than species that are specially protected (Dry, 2010). Even national and state park agencies rely heavily on selling live rhinos to cover their costs. Yet this progress is almost entirely ignored, which is intensely frustrating to South African rhino owners. Indeed, they are unanimous in their opinion that sustainable use has driven wide-scale rewilding of the land, and that rhino horn should be traded to pay for rhinos, the land they live on, and the high cost of rhino protection (Rubino & Pienaar, 2018).

However, CITES, in its wisdom, has turned a deaf ear to this evidence, and has trodden all over the democratic process and the rights of rhino owners. Rhino owners argue that CITES has removed their property rights to rhino without compensation, affecting a potential income of some $250 million annually. CITES has not replaced this income with alternative funding models to cover the burgeoning costs of rhino protec-tion, and nor have they compensated rhino owners for their loss of rights and income. It is true that philanthropy is financing several hundred pop-up rhino NGOs, but land-holders are curious about how much money is raised and what happens to it.

Rhino owners do not need charity. A rhino's horn grows about 10 cm per year, with females producing 0.75 kg and males 1.5 kg annually. This can be harvested every 2 to 2.5 years (or approximately ten times in a rhino's lifespan), before the horns become too attractive to poachers (John Hume, personal communication). Rhinos are truly the goose that keeps on laying the golden egg. Removing the horn is not too different to fleecing a sheep. However, horn from a single rhino is worth about $50,000 each per year, and allowing landholders to retain this value would transform vast areas of southern African landscapes back to wildlife. Some 20% of South African farmland has already been rewilded, paid for by the sustainable use of wildlife, mainly hunting and the sale of meat and live animals.

The CITES rhino decisions provide a poignant example of the persistence of bad regulations that are not working. These decisions originate in ideologies that define rhinos as a public or global resource. They persist because they benefit the status quo, through status, employment, and even money. Yet, like colonialism, the parties most affected by the rules have little say in writing them. Rhino producers have been stripped of a valuable asset without consent, and now bear the huge costs associated with criminality, itself partially a result of the rhino prohibition. The southern African rhino population that was recovering nicely has now been put quite firmly on a pathway towards extinction by CITES. Rather than supporting an approach that was clearly working, CITES has imposed a public goods approach that is unworkable because it is not supported by a reliable financing model. This is irresponsible beyond belief. Yet the special interest groups who overruled the wisdom, economic rights, and democratic wishes of rhino producers will bear no personal accountability for this decision. A combination of special interest and highly centralised governance systems is toxic to the common good as explained by Mancur Olson (2000).

CBNRM, public administration, and decentralisation

Decentralisation

Common property management, including CBNRM, overlaps significantly with the practice and theory of public administration, and especially decentralisation. Decentralisation, or the process of redistributing functions and powers away from a centralised authority, has received a lot of attention in recent literature on CBNRM and community conservation (Ribot, 2008; Ribot et al., 2010).

Decentralisation is complex and takes a number of forms (Box 5.2). Although definitions vary between scholars, I consider 'deconcentration' to merely shift central government functions to district level officials. 'Delegation' empowers entities that are independent or semi-independent of government, such as community producer associations or local park management committees. CBNRM is about 'devolution', economic, political, and administrative.

BOX 5.2 DEFINING TERMS USED WITH DECENTRALISATION

Decentralisation: the redistribution of functions away from a central locality or authority.

Administrative decentralisation:

- Deconcentration: passing the responsibility for regulation or service provision to a central government institution that is located outside the capital (e.g. district administrators, field offices of government agencies);
- Delegation: shifting the regulation or provision of goods from government to public corporations, associations, or publicly regulated private enterprises (e.g. trade unions, farmers associations, regional planning authorities);
- Devolution: empowerment of autonomous and independent local bodies (e.g. local municipalities, indigenous and community forests);
- Privatisation: transferring responsibilities or functions and allowing them to be performed by private businesses;
- Deregulation: the reduction or removal of state regulations, often to free up market allocation of goods.

Political decentralisation: measures that give citizens more power including democratisation and subsidiarity.

Source: adapted from Rondinelli (2003).

CBNRM often involves three different but simultaneous forms of decentralisation (Figure 5.7). Economic decentralisation, including privatisation and deregulation, is central to CBNRM. This empowers communities to act as economic production units with discretionary choice. It shifts the allocation of goods from the centrally planned state to the landholders, communities, and markets through the devolution of rights to access, use, manage, sell, and exclude others from a resource. In terms of community self-governance, CBNRM also includes elements of political devolution. Administratively, CBNRM communities are empowered to plan and zone natural resource use, and are often called upon to provide their own social services using income from wild resources

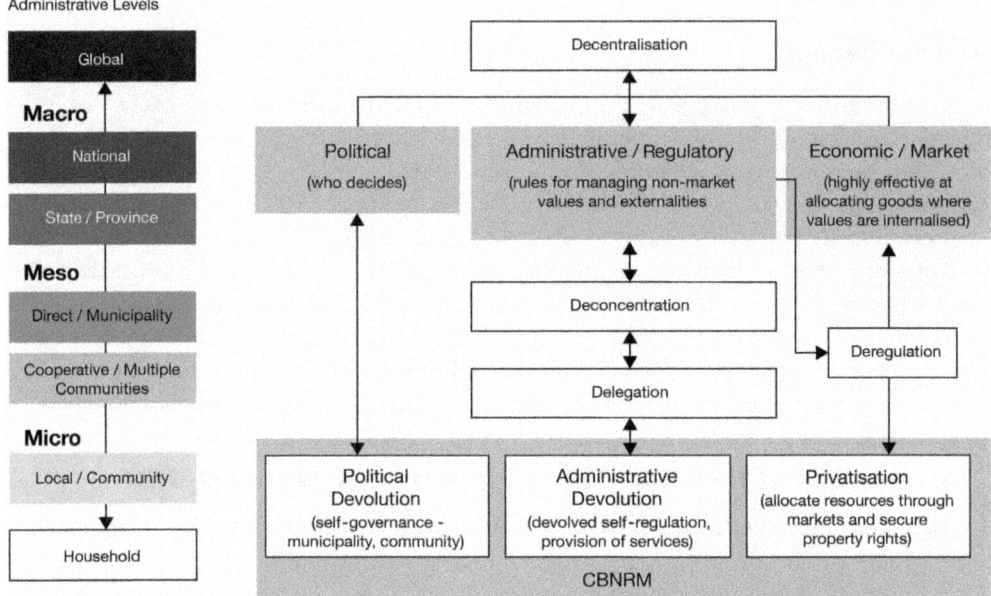

FIGURE 5.7 CBNRM and decentralisation
Source: Grenville Barnes provided the initial idea for describing administrative levels.

(e.g. building schools, health points, or providing water), especially where government is remiss and unable to provide these.

In CBNRM, these distinctions between deconcentration, delegation, and devolution are not always made clear (Ribot et al., 2010). CBNRM should not be seen as deconcentration, converting community bodies into mini-natural resource agencies (forestry, wildlife, fisheries) focused on regulatory responsibilities rather than production, as sometimes happens when it is confounded with district administration. Rather, CBNRM is analogous to private natural resource management, except that the production unit is a community rather than an individual.

The economic and political implications of property rights

A discussion of tenure is incomplete without highlighting the massive impact it has on the political economy and governance of wild life. The locus of property rights determines the function of macro-institutions and their prominence relative to productive activities. Defining wild life as a public good means that most decisions are made centrally in the political market place (Figure 5.8). By contrast, property rights pull the locus of power and decision-making towards the local level and the discretionary choices of a market-based system (Barnes & Child, 2014).

I used the examples of rhinos, and the bizarre bazaar that is CITES (Box 5.1), to illustrate the weaknesses, dangers, and dysfunctions of the centralised systems in which we (mis)placed so much responsibility. Certainly at a global level they epitomise democratic

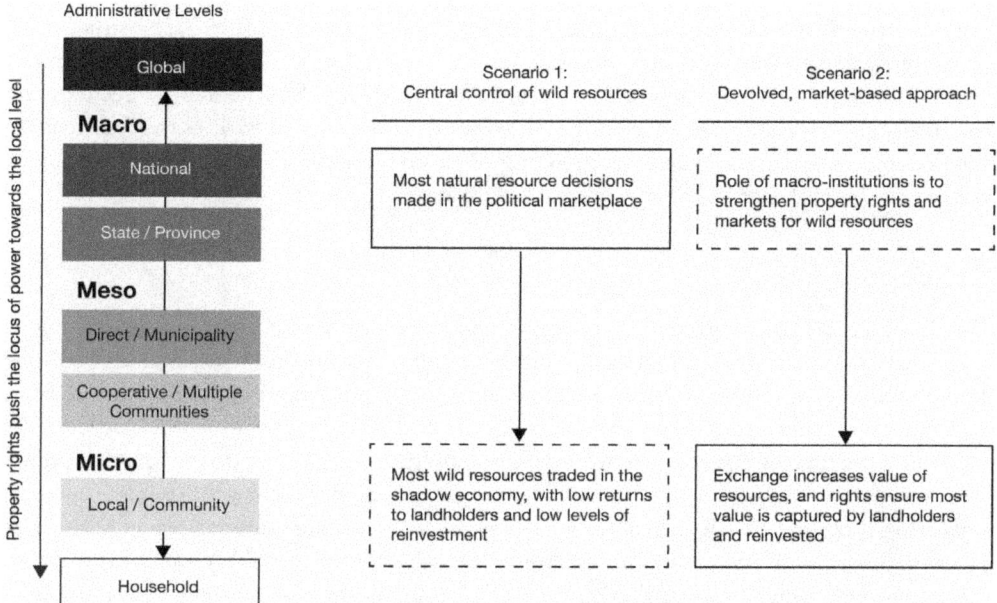

FIGURE 5.8 Rights and the governance of resources in political or economic market places
Source: Barnes & Child (2014).

failure, are leapt upon by special interest (Olsen, 2000), write rules they are unwilling to pay for, and have a near-criminal tendency to impose opportunity costs on others without accountability. I would go so far as to say that this dysfunction is the central cause of a market failure that prevents many people who are willing and want to pay for environ-mental goods from dipping into their pockets, because they simply do not trust these mechanisms and the players feeding off it. But make no mistake, these are still market places, where people trade power, treasure, and influence, with limited accountability for performance in terms of measurable conservation or poverty reduction. This is not to blame the many wonderful and committed people who are doing their best to save the planet, but to highlight how ineffective and frustrated we will all continue to be unless we evaluate, challenge, and change the system.

Not only is the devolution of rights the right thing to do in terms of community rights and the wildlife economy, but it will radically alter the political structure of the system, shifting most decisions into the economic market place. With privately owned livestock, for example, most cattlemen live on the farm, exchange takes place between individual livestock owners in the economic market place, and there is no need for international meetings to 'fix the problem'. Despite so few international conferences for the conservation of goats or the international trade in cattle, domestic species are increasing rapidly. By contrast, landholders and communities do not have strong rights for wildlife, and the plethora of global conferences about wildlife do not appear to be doing much to stop it from disappearing so fast. In a centralised system, exchange blossoms in two places, neither of which results in serious investment in the resource base: the political market place of national and global bureaucracy, central financing and philanthropy, and special interest, and the economic shadows where exchange is informal and often criminalised. The structure of property rights also has significant implications for the distribution of talent. Skilful farmers stay on the farm, whereas talented conservationists are rewarded for attending meetings and spinning stories that raise money. One of the tragedies of the current times is the disincentives for dedicated and capable conservationists to work deep in the field where they are most needed, at the coal-face of effectiveness and innovation.

Conclusions

This chapter has introduced key concepts about proprietorship, emphasising that it is a foundational concept that underpins land husbandry, free markets, sustainable economic growth, equity, and even inclusive governance. It affects how the politics of wild life plays out, and even where conservationists work.

After a century of governing wild resources publicly, we need to adapt to an emerging realisation that wild species are essentially private goods with common-pool properties, becoming progressively more excludable and subtractable on the crowded planet of the 21st century. There is an increasing mismatch between the nature of wild resources (private or common-pool resources) and their governance as public goods. These institutional mis-alignments create many of the problems faced by wild resources, including replacement by domestic species, over-exploitation, under-investment, and the inequitable distribution of costs (local) and benefits (national or global). On a crowded planet, public wild life is unlikely to thrive on private land. Unless we change direction, wild species will be steadily relegated to the 15% of the planet that is public land. We need to strengthen the world's protected areas, but outside them we need new institutions for wild species and spaces, with private-community proprietorship and CBNRM being especially important given the overlap between wild species and indigenous and local communities.

Note

1 In Zimbabwe's Parks and Wild Life Act (1975) 'wild life' means all forms of animal life, vertebrate and invertebrate, which are indigenous to Zimbabwe, and the eggs or young thereof.

6

ECONOMIC PRINCIPLES FOR WILDLIFE GOVERNANCE

LEARNING OBJECTIVES

This chapter introduces:

1. The application of economic concepts to wildlife conservation and CBNRM.
2. The exclusion of wild resources from the market economy by environmentalists and economists, which renders them priceless but worthless.
3. The importance of property rights for internalising the costs and benefits of wild resources.
4. The challenge that the economy does not pay the full costs of extracting resources, or dumping waste, resulting in an under-supply of nature.
5. The evolution of *Homo sapiens* as economic man and the creation of wealth through man's unique abilities to specialise and trade within a set of rules.
6. A simple definition of sustainability: more, from less, for more, forever.
7. A model for analysing wildlife conservation as an economic problem, including the difference between financial and economic prices.
8. Property rights as necessary for price formation and allocating resources efficiently under certain conditions.
9. Classical economic theory as it applies to wild resources, including: the creation of wealth through specialisation and exchange; Adam Smith's invisible hand and the equilibrium theory of supply and demand; a definition of free markets as competing on price and quality alone (not laissez-faire) and the need for institutions to protect these conditions; and how market failures, including differential taxation and regulation, affect the economic and financial price of wildlife.
10. Issues to consider in market-led conservation, including new forms of organisation that combine privatisation, collective action, and public governance in ways that are matched to the economic characteristics of wild resources.

Priceless but worthless

Wild life and wild resources are priceless but worthless because environmentalists and economists have excluded them from the mainstream economy. Conservationists abhor the dollarisation of wildlife, while economists treat the economy as a closed system, ignoring the true costs of the waste they dump into the oceans, rivers, and the atmosphere, and of the raw materials and services they get from nature. Prices guide the allocation of resources in a global world, and we can no longer afford to leave wildlife beyond the economic side line.

Taking wildlife out of the marketplace

John Muir, the father of the 'preservation' movement, sired the ideology of keeping nature separate from economics and 'saving the American soul from total surrender to materialism'. In a famous argument over the the conservation of Hetch-Hethy Valley in Yosemite National Park, John Muir set his preservationist ideals against the utilitarian economic argument to dam the river to provide water to Californian cities, remonstrating lyrically against the desire to dollerise everything (Muir, 1912). By contrast, Theodore Roosevelt promoted the ideal of 'conservation' as wise use, setting in place the divide between conservation and preservation that continues to this day. Nevertheless, Roosevelt responded to the wildlife crisis of the late 18th century by removing it from the marketplace, outlawing market hunting and preventing inter-state commerce in wild plants and animals (but not timber production or fisheries) (Chapter 7). In essence, wild life and other environmental resources became 'public goods' outside the market, implying no rivalry in use and no need for mechanisms of exclusion (Vatn, 2015, p 52).

The mainstream economy and the environment

Theories of neo-classical economics share the viewpoint that nature is not part of the economy. Most standard economics textbooks depict the economy as a closed system of exchange between firms and individuals (left side of Figure 3.2), but ecological economists have pointed out that the economy is not a closed system and therefore cannot simply grow itself out of trouble as is too often assumed (Ropke, 2004; Daly, 2005; Vatn, 2015). The real world economy extracts vast quantities of raw materials from the environment, and dumps huge amounts of waste into it, where it also hides millions of poor people (see right side of Figure 3.2). The economy also depends heavily on a living planet for pollination, agriculture, flood protection, and so on (Costanza et al., 1998). The economy, however, fails to pay many of these costs, externalising many of them, predominantly into ungoverned spaces – the global commons of atmosphere and ocean, and the local commons where poor people live in forests and drylands, often alongside wildlife and national parks.

Ungoverned spaces

The over-extraction of resources and dumping of waste (and poor people) occurs differentially into 'ungoverned spaces' which, for our purposes, are defined as the places where costs and benefits are not fully internalised and we can get away – for the meantime – with not paying the true costs of these actions. As noted in Chapter 3, we are exceeding planetary boundaries disproportionately in ungoverned spaces, including the atmosphere and oceans which are genuine global goods and, for our purposes, the 'local commons', where the institutions for land and wild species are outdated, crude, and mismatched to the characteristics of the resources in question. We know that the differential performance of social systems and economies is closely related to the quality of institutions (North, 1990), and our argument is to apply this very same institutional logic to wild resources and communities. In Chapter 3, we illustrated this argument with data comparing the economic productivity of land with rich and weak institutions in drylands in South Africa (Figure 3.4). Trends in the Living Planet Index suggest this pattern is global. Biodiversity is recovering slightly in countries with rich institutions despite a high per capita environmental footprint, whereas middle or low income countries with weaker institutions are losing species rapidly despite low per capita consumption (Figure 6.1). In the early stages of economic growth today's rich countries also had weak institutions and over-exploited their forests and environments. However, as they developed more sophisticated and polyvalent institutions, and began to internalise the costs and benefits of the environment, there was some ecological recovery, following public actions and investments in protected areas and endangered species and, not least, new rules that began to establish markets for nature and pollution.

Rich countries, perhaps nearing the limits of what public management and regulation can achieve, are experimenting with new kinds of property rights, markets, and governance arrangements for environmental effects that were unpriced until recently. Examples include pollution taxes, cap-and-trade markets, individual tradeable quotas for fishing, multiple stakeholder approaches, as well as payments for environmental services, biodiversity offsets, certification and standards, and so on. Some of these market based solutions have been quite successful,

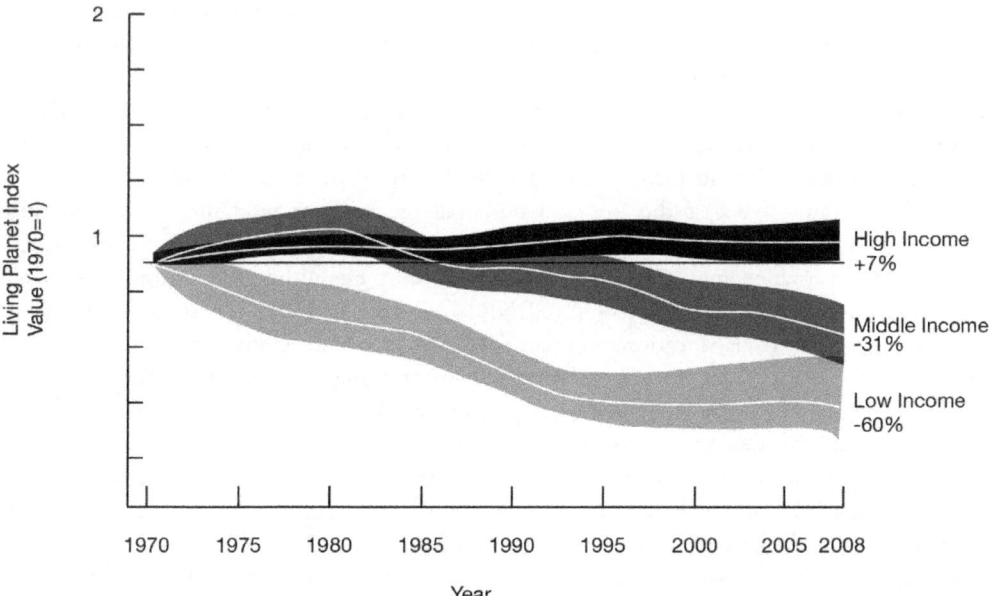

FIGURE 6.1   Biodiversity loss in high, middle, and low-income countries over time
Source: adapted from WWF et al. (2012) WWF Living Planet Index.

including the protection of ozone through the 1987 Montreal Protocol, cap-and-trade systems incorporated into the Clean Air Act in the USA in 1990, and the idea of individual tradeable quotas for fisheries (Mckean, 2000; Tietenberg, 2006). Other economic mechanisms for ecosystems services are beginning to emerge in the form of payments, certification, and biodiversity offsets, and their effectiveness is still developing (Wunder et al., 2010).

As I explained in Chapter 3, the combination of property rights, markets, and social action has controlled or internalised many of the costs of waste and resource extraction in prosperous nations. The effect of these 'rich' institutions has been to partially price ecosystem services and to prevent environmental harm, leading to some environmental recovery. For example, strong property rights prevent people from dumping or extracting resources in an uncontrolled manner, even before we take into account the development of 'markets' for pollution, tradeable quotas, and environmental offsets. Even then, it is quite likely that much of the environmental footprint of high-consumption societies is being exported into ungoverned spaces where the costs and benefits of resource extraction and the dumping of waste materials are not fully internalised. Examples are numerous, and include plastics in the sea, the migration of dirty manufacturing to poorer countries, and climate change where poorer people, especially Africans, will pay a much higher cost than the societies that contributed the most to this problem. Similarly, wealthy, urban societies tend to free ride on the services provided by forests and wildlife, which they expect poor people and poor countries to provide as a global public good.

Ungoverned spaces and their genesis

The environmental recovery of modern urban society stems, to a significant extent, from the underlying strength of institutions such as secure property rights, the rule of law, and inclusive governance. Similarly, we can blame much of the loss of ecosystem health and diversity in the ungoverned spaces of low-income countries on institutional failure. The causes of this are described under the heading 'deinstitutionalisation' in Chapter 4, pointing to the loss of rights by local communities over land and wild resources, and to the devastation of social capital and local systems of governance through a history that includes slavery, colonialism, conquest, diseases, one-party dictatorship, and, more recently, the continued erosion of rights through the globalisation of environmental policy.

In this chapter, we cast the loss and potential recovery of wild life as an economic problem that can, in many places, be resolved by reinstitutionalising wild resources (i.e. the sustainable governance approach, Chapters 7–10) and by establishing communities as viable natural resource management units (i.e. CBNRM, Chapters 11–16). These new approaches respond to the mis-classification and continued public management of wild resources outside public lands, when they are neither non-subtractable and non-excludable (Randall, 1983; Ostrom & Hess, 2007). They also respond to weaknesses in public capacity (especially exclusion) and the degradation of already misaligned and vulnerable public governance systems into de facto open-access regimes, so that wild resources and wild spaces have become, essentially, ungoverned and highly susceptible to replacement and overexploitation (Figure 5.5). As the son of a man who established protected areas in three countries, I am a strong proponent of national parks. However, like my father, I recognise that we need new approaches to conserve wild species on the 85% of the planet that is not in parks and protected areas, and that we need to use scarce conservation dollars as efficiently as possible by enabling wildlife to pay its way wherever it can.

Economic man and sustainability

Human success and prosperity is rooted in a 'cognitive revolution'. Biologically, we are economic man, with unique abilities to cooperate, specialise, exchange, and control these processes through institutions and technology (Bethell, 1998; Reader, 1999; Harari, 2014) – features that lie at the heart of the subject we call economics. Despite our success as economic man, our governing myths and ideologies have excluded nature from the economy. Priceless nature is unpriced or under-priced, resulting, in the dry words of economists, in a serious problem of under-supply. Our natural planet is in jeopardy. We need to question the underlying assumptions and ideologies about how we govern nature, and to factor the sustainability and value of nature into the economic future of the planet.

A practical definition of the somewhat elusive concept of sustainability is provided by the ditty: 'more (well-being) from less (resources), for more (people), forever'. Sustainability is deeply economic and includes three components. 'More from less' is about economic efficiency. Essentially this is about 'getting prices right' to allocate resources to the highest valued uses possible. It is the subject of this chapter. Social equity, the second issue, is about who benefits from this efficiency, and is a political question about environmental justice and who owns resources in the first place. For many reasons, the people who live with wild life should own it, and get a fair price for their assets. The third issue is about ecosystem sustainability and maintaining the capital in our ecological bank account. We might under-stand the wonder of these natural systems through the biological and ecological sciences, but establishing the incentives for maintaining these environmental stocks lies firmly in the realm of economics and governance. This defines three circles of sustainability – economic viability, socio-political acceptability, and ecological sustainability (Child, 1995).

Wildlife conservation as an economic problem

The proposition that we can save nature by incorporating it into the economy is bold, controversial, and certainly not mainstream. I make this case using examples from southern Africa in Chapters 8 and 9 to provide proof of concept, and to illustrate important principles and policies. But before analysing the economic history of wildlife in southern Africa against the North American counter-factual (Chapter 7), this chapter introduces some useful economic ideas as briefly as possible.

The starting point is the simple economic model (Figure 6.2) that underpins the sustainable use approach (SASUSG, 1996). In drylands, wildlife has an economic comparative advantage (the upper dashed curve in Figure 6.2), because the priced and unpriced values of wildlife exceed that of other land uses. However, wildlife's economic comparative advantage (defined in Box 6.1) is not reflected in financial prices (the lower dashed curve) because of market or policy failures. Many conservationists rely on public funding or rich benefactors to subsidise it, but to get wildlife back administrators in southern Africa identified and addressed these market failures to enable wildlife to pay for itself.

Wildlife administrators, however, have less power to influence the profit curves of domestic species, which are over-priced (or subsidised) for a number of reasons. For some 5 000 years, simple cadastral tenure (Bowles & Choi, 2013) has allowed farmers to own, invest in, and benefit from domestic species, but not from wild species. Moreover, farming is

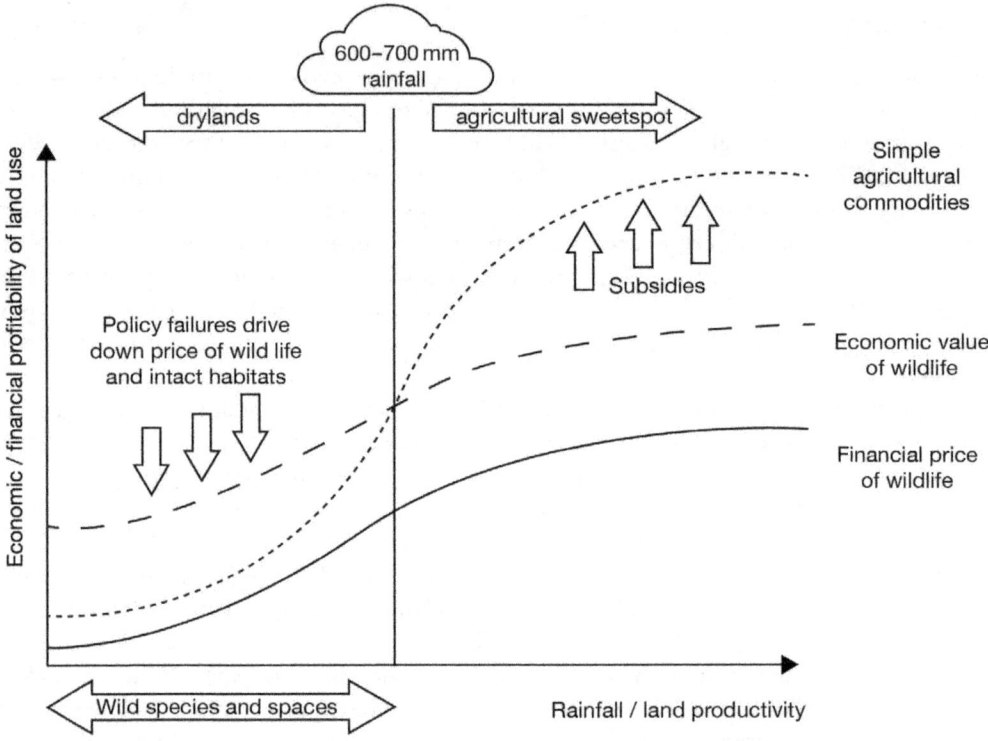

FIGURE 6.2 Economic model of the wildlife economy and the sustainable use approach

Source: modified from a power point developed by Chris Barnes and Greg Stuart-Hills (personal communication) (with permission)

invariably subsidised, directly through price support, indirectly through research, agricultural services, and infrastructure, and by not paying the full environmental costs of production. Together, these factors (i.e. differentially favourable institutions, subsidies, taxes, and regulations) have driven down the profit curve for wild species, driven up the profit curve for domestic species, and given us the wrong outcomes. On a global scale, high-value wild resources are being replaced by low-value domestic crops and animals (Anon, 2011; Smil, 2011).

Wildlife recovered in southern Africa because wildlife officials, especially in Zimbabwe and Namibia, incorporated these profit curves as a cornerstone in policy-making. Thus, to quote my father, the director of the wildlife agency in Zimbabwe:

> Wildlife is a renewable resource, which, like other resources, must be conserved and used wisely. Its continued existence outside protected areas will depend on its competitive ability in terms of landholder benefits.
>
> *(Child, 1995, p 70)*

However, correcting the economic curves requires institutional reform, and we see the importance of the concepts of proprietorship and price emerging. Thus:

Wildlife can compete with other land uses because it has an economic comparative advantage, but it will be conserved only if this advantage is reflected in market prices and landholders receive a sufficient share of the benefits.

(Child, 1995, p 70)

Knowing that prices coordinate the allocation of land, wildlife administrators sought to 'get prices right' for wildlife through policy reforms that moved the financial curve upwards towards the economic curve. Where prices were missing (as for wildlife), or subsidised (as for livestock), the pricing mechanism could not be trusted to allocate land to the best uses, or to maintain the capital stock of the environment. Wildlife officials inherited policies of not dollarising wildlife. However, they learned that without a price, wildlife on productive land was replaced by commodities that do have a price. The costs of wildlife, such as crop damage or livestock losses, were measured in dollars, but the benefits of wildlife were not measured in this way, so it was categorised as a nuisance. Indeed, whole departments – the game departments of colonial Africa – were established to 'control' wildlife because of its negative value. Politically, too, wild land was perceived as 'under-utilised' and open to ever more agricultural commodity production because the value of natural systems were not dollarised.

To 'get prices right', wildlife administrators set out to maximise the value of wildlife but also to internalise the full costs of land use, including deforestation and soil erosion. With Zimbabwean administrators being obsessed by healthy soils and soil erosion, conservation became a quest for wise and efficient land use and 'economic efficiency', rather than to promote any single land use (like wildlife). Instead of a singular focus on wildlife, officials began to prioritise the rights and well-being of landholders and communities. Good farming practices were encouraged in the agricultural sweetspot, because growing more food, better, would reduce the demand to farm drylands more suited to the wildlife economy. Wise land use and economic efficiency are laudable goals, and the vehicle to get there is institutions that internalise the costs and benefits of different land uses as completely as possible (see Chapter 8), together with the concept of comparative advantage (Box 6.1).

BOX 6.1 THE CONCEPT OF COMPARATIVE ADVANTAGE

Comparative advantage is a critical concept in economics. Its message is that we are better off if we specialise in producing the goods and services that we are best at (relatively speaking) and use the profits from these to buy the other goods and services that we need.

For example, if Bill Gates can type at 120 words a minute and his personal assistant types at 80 words a minute, it still pays Bill Gates to delegate the typing. He has an absolute advantage in typing, but he has an even greater gift for software development.

Most people will agree with this, yet they take the opposing viewpoint when you suggest people in drylands should achieve food self-sufficiency by producing wildlife and buying food. Wildlife has a comparative advantage in some drylands, including southern Africa, and earns as much as two to four times that of conventional beef production (Child et al., 2012). Like Bill Gates, the sensible economic choice is for farmers and communities to produce wildlife and sell it to buy food. Counter-intuitively, producing wildlife reduces their vulnerability to hunger, whereas expecting them to grow their own food in areas where the climate is unreliable, increases food vulnerability.

Resource allocation

The central challenge of economics is how to best allocate scarce resources (Pearce & Turner, 1990) with a debate over whether to achieve this through central planning or a decentralised market place. Centralised planning seems orderly and assumes that politicians, bureaucrats, or technical experts are best placed to make choices about how resources are allocated. The former USSR epitomised central economic planning, as do the public management and regulatory approaches that are a strong feature of natural resource management. In the free-market system, by comparison, the allocation of resources is guided by prices and exchange between many thousands of individuals making their own choices. The question is whether we have placed too much confidence in central planning and the public management of biodiversity. Have we got the balance wrong, and should we strengthen tenure, pricing, and market mechanisms to improve the allocation and sustainability of wild resources outside protected areas?

Prices

Economists would call this 'getting prices right'. But, what are prices, how are they formed, how do they work, and why should we trust them to allocate precious wild resources? Introductory economics courses explain that prices provide the signals that guide resource allocation, and that prices emerge through the interaction of supply and demand, or what is called equilibrium theory. However, less recognised is the reality that prices cannot form in the absence of ownership and exchange. This gives us two theories of price. We will introduce equilibrium theory quickly, but belabour what I have clumsily called the 'property theory of price' because this is crucial for developing countries and natural resources.

Internalising costs and benefits: the Coase theorem

Central to economic thinking is the concept that if the costs and benefits of an action are fully internalised, individuals will make the right decision for society after weighing these up. Ronald Coase (1960) made exactly this point in 'The Problem of Social Cost', cited by the Nobel committee in 1991.[1] Coase set up the problem of a cattle rancher living next to a farmer, where cattle ate the crops unless constrained. With elegant reasoning, he showed how the two parties negotiated an optimal solution depending on the price of crops, cattle, or fencing, provided that property rights were clearly delineated (and in the absence of transaction costs). The nub of Coase's argument is that people bargain their way to the most efficient use of resources, provided costs and benefits are fully internalised. Thus, clearly defined property rights are a necessary condition for establishing prices and for efficient resource trade-offs. This captivating argument underpins the free market economy, but these conditions are piercingly missing in ungoverned spaces.

Property rights and exclusion as a basis for price

Conventional economic theory assumes that property rights are widespread, well-defined, and operational. However, this is not the case for wild resources and ungoverned spaces, and it is surprising that the importance of property rights and exchange is so widely ignored in economic debates about wild resources. In a discussion with the distinguished economist, Mancur Olson, Tom Bethell writes:

> Starting with Adam Smith, all the leading economists came from just those countries where the essential legal preconditions for real economic advance did exist. So they took them for granted. This was a 'tremendous oversight,' Olson admitted. 'Economics developed, loosely speaking, in a particular type of society, namely democratic societies with secure rights and independent judiciaries, so people haven't bothered to think about these things very much in economics'.
>
> *(Bethell, 1998)*

The 'property theory of price' that I will now describe is more foundational than the theories of supply and demand and general equilibrium. It is not new, just so obvious that it is often forgotten.

The property theory of price is illustrated by a simple example. If you live in a society without property rights, I can simply take your pen when I want it – it has no price, except the cost of brute force. This was the nasty and brutish condition of mankind described by Thomas Hobbes before rights of property and exclusion were entrenched and protected. By contrast, in a society which gives you clear rights to your pen, and protects them, I cannot take your pen by force because it is, indeed, your pen. If I want it, I will need to negotiate with you for it. Being stronger, or having more weapons or personal connections than you have, does not allow me to take it. Through this negotiation, the pen gets a price, and this price regulates the production and consumption of pens. We create wealth by exchanging things between us, and we do this peacefully.

This theory of price rests on the following sequence: ownership, exclusion, negotiation, and price. You own your pen. This gives you the right to exclude me from simply taking the pen. Therefore, if I want it, I have to negotiate a trade. This negotiation sets the price of pens, and the price of pens guides the economy to produce exactly the right number of pens. Economists call this 'getting prices right'.

However, to get prices right, we also need to design and protect the institutions governing property rights and exchange (North, 2003). Protecting order and the rights of person and property is the foundational purpose of the modern liberal state. In these circumstances, so familiar to the citizens of liberal democracies that they are almost forgotten, goods change hands peacefully through mutual agreement and negotiation – the free market. We forget to our peril that many people, especially in forests and drylands, are not living in such circumstances. Without rights, and low-cost, peaceful, exclusion, they live in a world far more akin to Hobbes's description. Economic systems as we know them are dysfunctional, including for wild resources. Resources are used wastefully and inefficiently, in circumstances leaving local communities subject to violence and criminality, land and resource grabbing, and landlessness (Murombedzi, 2014).

In the absence of both proprietorship and exchange, nature is locked in a low-value grab-all economy. By now it should be obvious that the wild life economy cannot function in the absence of property rights and low cost exclusion. What is less obvious is that it also cannot operate in the absence of exchange and trade, and we turn now to this issue.

Exchange and specialisation as the basis of wealth

Published in 1776, Adam Smith wrote the *Wealth of Nations*. Smith showed that wealth is created by converting raw materials, often from the environment, into things that people want. The secret to wealth creation was the productivity of labour, which he illustrated using the famous example of the pin factory. Through specialisation (the 'division of labour') and cooperation, a pin factory produced 48 000 pins per man per day, compared to about 20 by a man working on his own (Beinhocker, 2006). But specialisation requires trade. Pin makers couldn't eat pins, but needed to trade them for bread, fish, and everything else they needed. Trading what you're good at, for what you need, is essential for wealth creation, and the only species that has developed the cognitive ability to trade – man – is also the wealthiest. If you give one orangutan two oranges, and the other orangutan two bananas, they will hang on to what they have for dear life, never trading to a better situation. If you give one human two cups of tea, and the other two cream scones, they will bargain and trade so that they are both better off, eating a scone with their tea. The process of consensual exchange, by definition, make both parties better off. It creates wealth.

Efficiency and the invisible hand

Having addressed the first great question of economics (How is wealth created?), Smith tackled the second: What is an efficient, and just, allocation of resources, and how does this occur (Beinhocker, 2006)? Smith was writing at the end of the Enlightenment, when the question of authority and legitimacy was being resolved in favour of individuals rather than the king, the state, or the church. Smith presented the attractive argument that people – not authorities – are the best judges of their own choices. As a frugal Scotsman, he also believed that resources should be allocated efficiently, not wastefully, to maximise the total wealth of society. Smith's argument was captivating. The best way to organise the economy was, he said, for people to trade freely, with self-interest and prices guiding them to provide the goods and services they needed. Competitive markets were, Smith concluded, the most morally just mechanism for allocating society's resources. People were free to make their own choices. Free markets gave people more of what they wanted and allocated society's resources more efficiently. In an insight that lies at the heart of free market economics, Smith (1776) suggested:

> [The merchant] intends only his own gain, and he is in this, as in many other cases, led by an invisible hand to promote an end which was no part of his intention ... By pursuing his own interest he frequently promotes that of society more effectually than when he really intends to promote it.
>
> *(Smith, 1776, Book IV, Chapter II, para. ix)*

The invisible hand was, of course, the mechanism of supply and demand operating in competitive markets. In 1862, the English mathematician and economist, William Stanley Jevons, published *A General Mathematical Theory of Political Economy*, while the French economist, Leon Walrus, published *Elements of Pure Economics* in 1854, believing that turning economics into a mathematical science was a good thing. Smith's ideas evolved into a theory of general equilibrium, strengthened in the 1960s by a series of famous economists including Marshall, Samuelson, and Arrow who established that prices act like a nervous system, causing all the markets in the economy to automatically coordinate themselves, thus allocating all the resources perfectly to their best uses[2] (Beinhocker, 2006).

This concept is captivating, and is the essence of free-market ideology. However, it is also a simplified and mathematical abstraction, because it depends on numerous assumptions that are only approximated in the real world – perfect information, homogeneous products, costless transactions, the absence of externalities (i.e. perfect property rights) and economies of scale, and so on. Real markets range from close to perfect (where these conditions are near to being met) to greatly imperfect (where they are not, as is the case for wild resources). We will revisit the problem of imperfect markets in the next section.

Supply, demand, and free markets

Markets are places that bring producers and consumers together to trade, with price being the point at which supply meets demand. This is micro-economics, with its ubiquitous supply and demand curves. In a perfectly competitive market, more firms will enter the market to supply a good as its price increases. Conversely, as the price increases, fewer customers will buy the good. The equilibrium point is where the quantity of the good demanded by consumers (at that price) exactly equals the supply of that good (at that price). Resources are allocated to their highest valued uses and no resources are wasted. This is expressed mathematically using the supply (S) and demand (D) curves of economic text books. Prices, however, are not static. If, for instance, fashions change and the demand for tourism increases, the equilibrium point and price for tourism also increase. Prices are not a capitalist trick. They reflect the socially negotiated value of resources and provide the mechanism for making resource trade-offs and allocating land to its highest value uses. This is why the pricing, and mis-pricing, of wild resources is so important. Prices only become a capitalist trick when the underlying structures – property rights, markets, and rules – are incomplete, or capitalists are able to manipulate them.

Who designs free markets?

There is some confusion about the definition of 'free markets'. One interpretation, that of laissez-faire, is the ideology that markets should be free from the interference of governments (Beinhocker, 2006). This was not Smith's interpretation, and nor is it ours. For Smith, supply and demand should be determined by price and quality alone, free from the distortions caused by monopolies, price distortions, government price-setting, and so on. This raises the question of how the rules that frame free markets arise. They will not just emerge in perfect form. In a laissez-faire situation, rules will be 'written' by the rich and powerful who benefit most from markets, to the disadvantage of the many. Alternatively, we might hope that highly professional public administrators in inclusive democracies write the fairest rules possible (and we should certainly strengthen natural resource agencies for this purpose). Indeed, one of the primary justifications for the rational-legal state is the need for a collective authority to set in place and protect impersonal rules that seek Smith's conditions (see Figure 2.1). However, making good rules in the real world is difficult. Non-ideal situations persist because they benefit the power groups that maintain them. Thus, ungoverned spaces persist because they allow rent seeking. Centralised environmental governance, likewise, benefits those in charge. This creates resistance to change. For example, most environmentalists are viscerally opposed to a wildlife economy and its central tenants – local ownership, the utilisation of wildlife for profit, and global markets. If they allow wild life to re-enter the global processes of local choices and exchange on an even economic playing field, their role as the Lords of Conservation will surely shrink (Figure 5.8).

Local commons and global benefits

Are wild species so important for the global good that we cannot afford to allow local people to own them? I will make the opposing case: That local ownership, and the re-creation of the local commons, are an essential mechanism for global benefits. Without local ownership, wild resources are, essentially, free and unprotected, and users can take as much of them as they want without contributing to their upkeep. This is business as usual. Devolving the ownership of wild resources to landholders and communities, by contrast, will lead to proprietorial protection and to the prices that allocate them to the best use.

I will illustrate these concepts in the context of the seemingly unsolvable and rapid decline of forest wildlife through the bushmeat trade. I will propose the unthinkable, and challenge current norms, by giving to communities the rights to use, manage, and benefit from bushmeat and the capacity to exclude others from simply taking it, together with considerable encouragement to trade it to make as much money as possible. Under these new conditions, consumers will need to negotiate with the producer community to buy bushmeat, because they now own their wildlife and can exclude other people from taking it without paying. The price of bushmeat emerges through these negotiations – for example, four bags of maize for one duiker carcass – and now incorporates the costs of producing wildlife, not merely the costs of hunting it in an open-access situation. These prices, in turn, guide resource allocation. The relative price of duikers and goats, for instance, enables the community to decide if they should use their land to produce duikers or if they would be better off by replacing the duikers with goats. The more people want bushmeat, the higher goes the price of wildlife, and the greater is the incentive for managing wildlife – not goats – sustainably. Once bushmeat becomes a profitable community business, it is in the interest of the community to protect the forest to produce more wild animals. In this way, the rising demand for bushmeat is converted into incentives for the community to conserve wildlife and its habitats.

Should we encourage trade in wild life?

Proprietorship is a necessary condition for the 'true' price of duikers, or buffalo for that matter, to emerge, but it is not a sufficient condition. The second question is whether we should encourage exchange and even global trade in wild life. The pin makers can't eat pins. Likewise, without trade, forest communities are highly unlikely to get the best price for their wildlife or therefore to specialise in wildlife production. Banning trade is a sacred cow of conservation ideology, but this does not mean it is right.

Exchange creates wealth, and the more global are the markets, the more value is created. We can illustrate this in a striking way by comparing the value of a buffalo in a no-trade scenario (subsistence use only) with a scenario where exchange is actively encouraged. When communities hunt down a buffalo, it is worth, at most, $500, or the value of the carcass. However, trading on the global market, they can expect American or European trophy hunters to pay about $7 500 for a buffalo (plus a similar amount in outfitting fees) and still keep the meat. Now, the buffalo earns the community $8 000, or nearly 20 times the no-trade, subsistence value. In addition, we need to factor in the economic impact of an additional $7 500 in outfitting fees, wages, tips, travel, and significant ancillary expenditure and multipliers on the local economy. This example shows that the exchange process alone increased the value of the buffalo some 20–40 times, with the corollary that restricting markets destroys wealth by the same amount.

This has major land use implications. A $500 buffalo probably cannot compete with cows for land, but an $8 000 buffalo can, provided communities formally own the wildlife and get 100% of these payments, and can prevent other people in the value chain from grabbing this money. Correcting these policy failures requires two reforms. First, we need to maximise the price of wildlife through global markets by removing trade restrictions and developing markets and products (price). Second, we need to ensure that this money gets to the community (proprietorship). With buffalo now worth $8 000, compared to cows at $500, there is a strong incentive for communities to rewild their economies. Wildlife has thrived under these conditions in southern Africa, but not where policies sound good but governments only allow communities to retain, say, 20% of the value of the buffalo.

Banning trade might have worked in an empty world, but in a global and full world the biggest threat to wildlife is habitat replacement. We cannot fight this if forest communities are locked into using wild resources in traditional, unspecialised, low-value ways (Box 6.2). Exchange and trade are necessary to multiply up the value of wild species. However, trade has a positive impact only when it is twinned with proprietorship, because in the absence of proprietorship it will accelerate unsustainable resource extraction.

BOX 6.2 THE FALSE ECONOMY OF SUBSISTENCE USE AND ALTER-NATIVE LIVELIHOODS

To avoid the discomfort of banning the age-old livelihoods of hunter-gathering communities, a compromise is often reached where subsistence uses of wild species are acceptable, but commercial exchange is not. Subsistence hunting, like low-input agriculture, is a low-value and wasteful use of resources. Large numbers of animals are harvested for low gain, because wealth creation is reduced by an order of magnitude in the absence of free exchange. With buffalo, for instance, we can support the same livelihoods from two trophy buffalo as we can by harvesting 40–50 for bushmeat. Consequently, trophy hunting is highly sustainable biologically and economically, but wild meat production is not. Trapping local communities in low-value subsistence uses does little to address poverty, and even less to enable wildlife to compete economically for land with domestic resources like crops and livestock.

Another motherhood strategy associated with attempts to conserve wildlife is the somewhat empty idea of 'alternative livelihoods'. If alternative livelihoods were so easy to generate, people would have found them. Moreover, the alternative to wild life-based livelihoods is to replace wild life with domestic species. What problem does this solve? Counter-intuitive as it may seem, the best 'alternative livelihood' to low-value bushmeat and the illegal wildlife trade is often the high-value, legal use of wildlife.

What would happen if, rather than limiting forest communities to the subsistence use of bushmeat, sophisticated markets allowed people all over the world to buy healthy, forest-grown meat, rather than meat-like substances produced in factory farms? First, the price of bushmeat would skyrocket. Provided communities owned the wildlife and the land it lived on, and could exclude others from taking them, this would increase incentives for allocating land to wildlife and protecting wildlife. Forest communities could then specialise in managing forests and bushmeat sustainably, rather than replacing wild meat with domestic plants and animals. It would pay them to act like livestock owners the world over – maintain the stock of animals and sell the surplus. As with private livestock, the most likely problem is too many animals on the range, not too few.

To conclude this argument by introducing some economic terminology, forest communities have a comparative advantage in producing bushmeat and forest products. Increasing the price of these through well-managed exchange dematerialises the economy, shifts land to higher-value uses (wildlife), and enables these communities to make more (livelihood) from less (environmental raw materials) compared to low-value commodities. Moreover, selling high-value bushmeat through legal channels, rather than badly prepared meat in the shadows, might well reduce criminality. This transformation depends on institutions – the property rights and markets that convert low-ordered raw materials into more highly ordered products and services (Beinhocker, 2006).

Free markets won't solve all conservation problems, but there are important opportunities to use them to improve the management of wild spaces and wildlife outside protected areas. An effective market-based approach has several requirements, namely that:

- Most of all, proprietorship is devolved to the people living with wildlife and forests.
- Wildlife is traded, wildlife products and missing markets are developed, and ideologically imposed restrictions on uses are removed.
- Costs and benefits are fully internalised through a combination of property rights, collective self-regulation, and regulation (in that order).
- People and wild resources compete on the basis of price and quality, not power, personal connections, or brute force.
- Wild resources are not disadvantaged compared to domestic resources by differential taxes or regulations (including trade bans).

The problem of differential taxation and regulation

We have already discussed the first four issues. However, the negative effects of differential taxation and regulation on wild resources is seldom recognised. Differential taxation is where, for example, a proportion of the income from wild resources is extracted by administrators, such as national agencies or district councils, when this is not done for domestic species. 'Taxing' crops or livestock by retaining 50%, 80%, or even 100% of the sales price would cause a public outcry, and even riots, yet it is common practice for wild resources. Similarly, governments and NGOs often insist that communities use wildlife income to provide social services and infrastructure, yet the dollars from domestic species go directly into people's pockets. They forget that this undercuts the price of wild products compared to domestic ones in the eyes of the people living on the land and making land use decisions. Moreover, any situation where some resources are taxed but others are not causes economic distortions and the misallocation of resources.

The same logical applies to regulations. In parts of South Africa, a farmer needs 17 permits to sell the carcass of a springbok but only one to sell a sheep. This is a typical example of differential regulation. The extra checks and balances put in place by environmentalists have the perverse effect of reducing, in the case of South Africa, the ability of springboks to compete for land with sheep.

Differential taxation and regulation are an anachronism that arises from the historical treatment of wild life as a public good. They shift the cost curves (Figures 3.7 and 6.2) very considerably to the disadvantage of wildlife, which is why more regulated species often recover more slowly than less regulated species in South Africa (Dry, 2010). Regulators are irresponsible if they ignore the effects of fees and regulations on the economic competitiveness of wild species, and they should check themselves by asking if they would apply the same taxes, fees, or regulations to domestic corn or cows.

Market failure and the difference between economic and financial prices

The free market approach, with its elegant mathematics and the democratisation of choices, is ideologically captivating. However, the assumptions that underpin the mathematical perfection of supply and demand break down in the real world, so that what is good for the individual is not always good for society. This is called 'market failure'. The difference between an economic price and a financial price (Box 6.3) is enormously important, but seldom recognised by conservationists. An economic (or social) analysis is holistic. It is undertaken from the perspective of society, and uses the real value of wild life, priced and unpriced. By contrast, a financial analysis is undertaken from the perspective of an individual or firm and uses the actual prices prevailing in the market. Wildlife is subject to substantial market failure, which is why it is often priceless to society (economic price) but worthless to the people who live with it (financial price). In economic jargon, the true economic value of wild life is not translated into the financial prices by which land use decisions are made.

BOX 6.3 ECONOMIC VERSUS FINANCIAL ANALYSIS

There are two components to the differences between economic and financial prices, and these important distinctions run through our discussion.

Economic price (or value) refers to the total economic value of an ecosystem to society. It includes unpriced values plus market prices, corrected to reflect their true value. Thus economic analysis uses true prices judged from the perspective of society, and reflects prices as they should be.

By contrast, a financial analysis uses accounting prices from the perspective of the individual (not society), and reflects prices as they are (not as they should be) – the real-world prices by which individuals make decisions.

	Whose perspective does it take?	What prices does it use?
Financial price or analysis	Evaluated from the perspective of the individual or firm making day-to-day decisions. Reflects what is good for an individual or firm (rather than society).	Reflects the price prevailing in the current marketplace – so-called accounting or financial prices.

(Continued)

(Cont.)

	Whose perspective does it take?	*What prices does it use?*
Economic price or analysis	Is evaluated from the perspective of society – the greatest good for the greatest number.	Reflects the true value, including priced and unpriced values, and 'corrects' prices to reflect their real value or opportunity cost to society.

> Market failure is the difference between private and economic prices, and results in private actions not optimising the allocation of resources for society.
>
> Wild resources are prone to market failures, leading to over-exploitation and under-investment in them. This includes over-exploitation where, in the absence of property rights, the individual internalises the benefits to himself but externalises the costs to society. However, we should not ignore the problem of shirking, where people benefit from the values of forests or elephants without paying for these values.

Markets 'fail' when they violate the assumptions required for perfect markets, including well-defined property-rights, a large number of buyers and sellers, perfect information, perfect factor mobility, rational decision-making, and no transaction costs. Market failure is especially problematic for wild resources for reasons discussed above: the public management of private resources, open-access and weak property rights, bureaucratic limitations on uses and markets, missing markets especially for ecosystem services, and differential taxation and regulation.

Can we use markets for conservation?

Arild Vatn (2015) suggests that the market has weaknesses in allocating wild resources for three reasons. First, Coase (1960) makes the case that property rights allocated resources efficiently in the absence of transaction costs; however, the latter are a very large part of getting anything done in the real world, and would be high for complex and fugitive wild resources. Second, it is theoretically true that if all resources are owned, and costs and benefits are fully internalised, the price mechanism alone can allocate resources efficiently. However, splitting up all aspects of the environment into pieces, and then trading according to prices, is simply too many transactions to contemplate as being manageable by the free market alone (Vatn, 2015, p 222). Third, economic efficiency is only part of the problem – we also need to consider who owns the rights in the first place (Vatn, 2015, p 350).

Vatn is right in saying that natural resource trade-offs are too complex, numerous, and subtle for the market to cope with. But so is the global economy, which has solved this problem by creating organisations. In the first of his papers quoted by the Nobel committee, Coase suggested that firms have evolved for this very reason – to reduce the costs of millions of transactions (Coase, 1937). Coase agrees that, theoretically, all transactions can be made through the price mechanism. However, firms exist because it is more efficient to bring people together in organisations in order to make many of these transactions through social processes.

The same logic applies to wild resources. So far we have focused mainly on better rules and institutions – economic level 2 (Box 4.5). To internalise the costs and benefits of complex wild resources, however, these rules should also encourage new forms of local organisations (level 3) that fully internalise costs and benefits. Because wild resources combine the attributes of private goods and common-pool goods (Figure 5.4), there is considerable merit in combining (1) private ownership (by landholders and communities) and market exchange with (2) local, collective self-regulation, such as neighbourhood associations, to manage numerous environmental effects and externalities. This combines the strengths of private production units with strong individual property rights for wild resources, with social mechanisms of reciprocity and collective action to manage the many costs and benefits of nature that are not captured privately.

An example of the second – ingenious – level of organisation are the conservation catchment communities described in Chapter 8. Groups of landholders living in the same catchment meet democratically to set rules to govern themselves, using peer pressure and collective action to internalise costs and benefits and reduce externalities associated with soil erosion, forest management, or mobile wildlife. This combination of private property rights and collective self-regulation will not perfectly internalise all costs and benefits, but it is a massive improvement on either Leviathan or simple privatisation (Child & Child, 2015).

Local collective organisations occasionally emerge spontaneously, but this process can be accelerated by a well-designed legal framework and technical support. Communities of practice and stakeholder groups also add value. However, care must be taken to avoid the capture of such a process by special interests under the guide of stakeholders, if they are free riders and unaccountable for their power. Thus, in designing local collectives, it is important to distinguish landholders, who have decision-making powers and are accountable, from stakeholders.

Free markets versus extractive capitalism

Private property and markets are a powerful mechanism for conservation, especially where they are combined with local social controls by resident landholders and communities. Environmental markets can work well if combined with carefully crafted institutions and ancillary organisations as just described. Nonetheless, environmentalists and social activists are right to be fearful of extractive capitalism, and should address these weaknesses, rather than completely shying away from an approach based on just property rights and free markets.

In the real world, markets are seldom as free as Adam Smith would have liked. Even in democratic countries such as the United States, markets are shaped by the powerful for self-enrichment – Big Oil, Big Pharma, Big Banks, and so on. In developing countries, similarly, elites are protected by property rights, at the same time that they argue against giving these very same rights to the poor, leaving them vulnerable to exploitation (Menard & Shirley, 2011).

After the fall of the Berlin Wall, free-market economists rushed to privatise state assets in post-communist Russia with too much faith in markets and too little consideration of institutions and equity (Stiglitz, 2002, Chapter 5). This created wealth, but this wealth was captured by oligarchs and led Russia down a path towards a mafia state. With wild resources, therefore, we are not advocating simply for property rights, markets, and economic efficiency, but for community property rights that incorporate social equity to ensure that the communities that live with wildlife are the rightful beneficiaries. Likewise, family owned enterprises with a sense of neighbourliness and a sense of place are likely to factor unpriced environmental costs and benefits into economic equations (especially if these processes are enhanced by education and collective action). By contrast, corporates, with absentee management, and a focus on scale or short term profit metrics, are not.

Conclusions

The common view is that wild resources and ecosystem services are too important or too complicated to be governed through the market system. This is not supported by the facts. Both environmentalists and economists have mistakenly accepted (and even promoted) the separation of nature and economics. Wild resources protected from the 'selfishness and greed' of man have usually fared far less well than resources that have been incorporated into the global market system through ownership and trade (e.g. agriculture, forestry, livestock, and urban property). Wild species and ecosystem services may theoretically be worth several times formal global GDP (Costanza et al., 1997), but in practice they are worth very little, especially to the people who live with them. In this chapter I have made the case that wild resources are disappearing because they are not part of the market system that guides so much human activity. There is merit in reinserting wild resources into the market economy, provided institutional conditions are well designed.

Private exploitation of wildlife was discredited by the damage it caused under frontier conditions, but private exploitation of a public resource differs fundamentally from utilisation through private ownership (Stroup & Baden, 1983). When enforceable property rights are combined with trade, 'selfish' private actions can lead to the common good and rapid wildlife recovery (Carruthers, 2008). However, private action and ownership alone is a very crude instrument. To properly account for environmental effects, it needs to be combined with mechanisms of social accountability, such as local self-regulation, which is a far superior mechanism for controlling externalities than governments with coercive and regulatory powers (Ophuls, 1973 and Hardin, 1978, quoted in Ostrom, 1990, pp 8–9). With their long feedback loops, centralised systems struggle to internalise the costs and benefits of land use properly. Their accountability often lies elsewhere; and neither is central control immune to self-serving officials and special interest. As Ostrom suggests, 'there is no reason to believe that bureaucrats and politicians, no matter how well meaning, are better at solving problems than the people on the spot, who have the strongest incentives to get the solution right'.

There is no one-size-fits-all approach to the challenge of natural resource governance (Ostrom, 2009). Improved governance of wild life requires combining and sequencing privatisation, collective action, and public ownership in ways that are much better aligned to wild resources:

- First, we put as much as possible in the private box through individual freehold or private-community ownership, getting prices right at the lowest possible level. This requires maximising the value of wild resources through specialisation, exchange, and market development, and ensuring that this value gets to landholders. Clear proprietary boundaries and strong property rights are necessary to internalise costs and benefits.

- Second, we use collective action, democratic peer pressure, and local regulations to manage externalities and complexity as far as possible, and as low as possible. People who live and work on the land are better at this than absent landlords.
- Finally, the centre should frame the rules to support devolution and devolved regulation. It also has the responsibility to ensure that outcomes are monitored to inform adaptive policy reform. We fall back on central regulations only when all other measures fail.

Thus, wild resources are privatised to landholders and communities so that incentives favour wild species over domestic ones. Second-order problems of externalities and attribution are managed through collective action. This significantly changes the role of the state, which needs more capacity to design and protect economic institutions, and less for the day-to-day management and policing of wild resources. The state retains the residual responsibility for effects that are not internalised through privatisation and collective action, with the challenge being to remain effective without undermining the devolved approach by being high-handed. The politics of these changes will not be easy. However, we now understand the economics and institutional economics of wild resources sufficiently well that we can at least propose technical solutions with considerable certainty that they can work if applied correctly.

Notes

1   It is interesting how many Nobel economists are central to the arguments in these pages: Frederick Hayek (1974), Ronald Coase (1991), Douglass North (1993), Danial Kahneman (2002), Elinor Ostrom (2009), and Oliver Williamson (2009). Others referred to include Paul Samuelson (1970), Simon Kuznets (1971), Amartya Sen (1998), and Joseph Stiglitz (2001).
2   In persuading my students, many of whom are social or natural scientists, that the economics that they need for conservation is manageable, I refer them to Beinhocker's excellent book. Chapter 2 outlines Smith's theories of wealth creation and describes the emergence of economics as a mathematical science. Chapter 3 contrasts the precision of economic mathematics and its assumptions with the real world, but nevertheless suggests that these problems do not prevent economics from providing a sound framework for understanding how the real world works.

7

THE INSTITUTIONAL HISTORY OF WILDLIFE AND ITS GOVERNANCE

LEARNING OBJECTIVES

This chapter describes the history of three of the four stages in the political economy of wildlife, including:

1. The pre-modern wildlife economy.
2. The destruction of wildlife on the frontier of the Industrial Revolution.
3. The public management and nationalisation of wildlife and wildlands.

It then describes:

4. The Yellowstone Model for National Parks.
5. The North American Model for Wildlife Management.
6. The London Convention and the nationalisation of wildlife in Africa.
7. The public management of wildlife in Africa.
8. Parallels with the nationalisation of community lands and forests.

It concludes by suggesting:

9. That institutions established in 1900 for wildlife result in a misfit between the public governance and private nature of wildlife and wildlands in the tropics under contemporary current circumstances.

Four stages in the political economy of wildlife

To analyse wildlife governance, it is useful to know where the current policies came from. This chapter traces the economic and institutional history of wildlife policy, especially in

North America and Africa. This history of wildlife governance has similarities with forestry and fisheries, which were also nationalised, but the exclusion of wildlife from private ownership and global markets is perhaps more extreme. The four stages in the political economy of wildlife (Table 7.1) reflect changes in its scarcity. Contemporary conservation approaches emerged at a critical juncture in the early 20th century, when the primary threat to wildlife in an empty world was commercial over-exploitation on the colonial frontier. However today's full planet faces the additional (and greater) threat of habitat and species replacement, which demands a fresh approach and fresh thinking.

TABLE 7.1 The changing political ecology of wildlife

Phase	Wildlife scarcity and threats	Institutional response
Pre-modern	Wildlife abundant relative to low human populations	Limiting factor was the ability to harvest wildlife with primitive technology, so institutions focused on sharing the spoils of the hunt, not limiting use
Frontier economy (1850–1930)	Over-exploitation of wildlife as value increases in the absence of institutions to control use, leading to the rapid depletion of bison, passenger pigeons, elephants, forests, etc.	• Few rules or norms to control use on the industrial frontier; • Costs of harvesting greatly reduced by technology (guns, railways), while globalisation increases market access, demand, and prices.
Public management (1900–present)	Rules are a response to over-exploitation in the previous frontier era, rather than emerging threats – the expansion of agriculture after World War II. Wildlife recovers in some rich urban societies with strong institutions	Public management: • Control of wildlife centralised in state wildlife, fish, forestry agencies; • Commercial use greatly restricted or banned; • National Parks established. The urbanisation of Western society changes locus of power over wildlife policy from rural to urban
Ungoverned spaces (where public management fails)	Wildlife declines rapidly in the absence of these conditions in developing countries	More of the same
Sustainable governance approach (1960–present in southern Africa)	Major threat to wildlife in a full world is habitat and species replacement, and also over-exploitation	Policies on wildlife ownership and use reversed in response to new threats: • Proprietorship of wildlife devolved to landholders (and, later, communities); • Commercial uses encouraged.

Phase 1: the pre-modern economy

The extinction of the Quaternary mega-fauna is attributed to the colonisation of the planet by hunter–gathering societies (Ripple et al., 2015). Nonetheless, when white people arrived in the New World, wildlife was plentiful. While there are few written records about pre-modern institutions for managing wildlife, they seem to follow a predictable economic logic (Stroup & Baden, 1983). Wildlife was abundant, and the risk and effort of killing large animals was high, so rules focused on sharing the spoils of the kill, rather than controlling offtake levels. This is reflected in some African traditions today where, for instance, certain parts of the elephant carcass (e.g. the front leg that is closest to the ground) belong to the traditional leader. Nonetheless, some African communities recall restrictions on hunting of pregnant animals or hunting at biologically unsuitable times of the year. The Plains Indians of North America used every part of the bison they killed with great frugality, until they obtained horses and firearms, and shared in the wholesale slaughter and trade of bison alongside white hunters. Reflecting Hardin's tragedy of the commons, individuals bene-fitted from killing ever more buffalo, sometimes for just their tongues, while the costs were shared by all (Stroup & Baden, 1983).

However, environmental tragedy is not the inevitable outcome. The Montagnais Indians on the Labrador Peninsular harvested beavers sustainably because of the low pressure on the resource. When French fur trading markets increased demand, a tragedy of the commons did not automatically follow, unlike the bison. The Montagnais 'enclosed the commons', which is easier for sedentary beavers than far-roaming bison. Individual hunters became responsible for individual beaver trapping areas, excluded others, and managed beavers sustainably. Much later, these local institutions were overwhelmed by white trappers in the 19th century (Stroup & Baden, 1983).

Phase 2: the destruction of wildlife on the frontier of the Industrial Revolution

The European Industrial Revolution, and European colonisation of Africa and the Amer-icas, radically altered the balance between wildlife and people. In the absence of rules to restrict offtake, market hunters decimated North American wildlife. American Bison were reduced from some 30 to 60 million bison to a few hundred remnant animals in the five decades to 1900. Explosives were used to harvest egret feathers in the swamps of Florida and hunters killed 25 000 passenger pigeons a day in Michigan alone (Brinkley, 2009). This, too, has an economic explanation. Hunting wildlife became profitable, as new technologies (e.g. rifles) lowered the cost of harvesting, and railroads and urbanisation opened up new and lucrative markets. Africa, too, lost some 2 million game animals during the frontier period. Ivory hunting decimated elephant numbers in southern Africa although, in retro-spect, the greatest threat to wildlife was the expansion of farming. Modern examples of frontier (or open-access) economies include many fisheries, and the illegal trade or poaching of timber, bushmeat, high-value elephants, rhinos, and seafood where the rights of, or capacity for, exclusion have broken down.

Modern wildlife policy is a response to the excesses of the 'frontier economy' defined as open-access property regimes where the profits from utilising wildlife are high, in the absence of rules for limiting use. Profiteering from wildlife in a frontier economy is often conflated with the privatisation of wildlife. This is a significant misinterpretation because privatising the harvesting of wild animals with no constraints is very different from managing a resource privately.

This chapter tells the story of the extirpation of wildlife, which is vivid and bloody, but the lessons can also be applied to forests, rangelands, and fisheries, which are affected by similar economic processes and governance responses.

Phase 3: the public management and nationalisation of wildlife and wildlands

The institutions that govern wildlife and wildlands today emerged to counter the over-exploitation of wild resources on the European colonial frontier in the late 19th century. They radically altered the relationship between people and wildlife for the next 100 years, and institutionalised three norms that permeate conservation to this day:

1. The establishment of state-protected areas.
2. The prohibition of the commercial use of wildlife including bans and restrictions on trade.
3. Public management and 'ownership' of wildlife through powerful central forest, fish, and wildlife agencies.

In this discussion about the origins of modern wildlife conservation philosophy, I have included national parks because they have dominated the way we think about conservation. However, our primary concern is the governance of wildlife and wild resources outside parks and the mismatch between the public persona of wildlife and the economic realities of land set aside for production.

The North American Model

Roosevelt and the Boone and Crocket Club

Sportsmen, hunters, and fishermen in North America raised the alarm about the catastrophic loss of wildlife and wildlands on the industrial frontier. With a sense of emergency, farsighted leaders, invariably 'sportsmen' like Theodore Roosevelt, George Bird Grimmel, and Richard Onslow (Fifth Earl of Onslow, chair of the London Convention), took action. Roosevelt was a remarkable man who changed the course of global conservation, becoming the most active US president in the history of American conservation (Krausman & Mahoney, 2015). Roosevelt alone set aside over 30 million acres of protected areas and established bureaucratic agencies to manage and control forests, fish, and wildlife – the US Departments of Forestry and of National Parks, together with state fish and wildlife agencies. As an avid hunter who travelled to all corners of the country, Roosevelt was keenly aware of the rapid demise of wildlands, big game, and other species. In 1887, he called highly influential citizens to his home and formed the Boone and Crocket Club. Those present included: George Bird Grinnell, an anthropologist and naturalist who edited the highly influential *Field and Stream* and founded the Audubon Society and the New York Zoological Society; Gifford Pinchot, first head of the United States Forest Service; and Stephen T. Mather, the first director of National Parks Services. They promoted the concept of protected areas, leading to the powerful United States Forest Service under the highly influential Pinchot in 1901, soon followed by the National Parks Service in 1916 under Mather.[1] Roosevelt blamed market hunting – largely by European poorer and middle classes – rather than open access property regimes, for the demise of bison, passenger pigeons, egrets (for their feathers), and all manner of wildlife (Brinkley, 2009). This led to the Lacey Act of 1900, which prohibited any commercial use of wildlife. Roosevelt believed that public access to hunting would bring many benefits to society, and, through his inspiration, wildlife became recognised as a Public Trust Resource. Although Roosevelt may have diagnosed the symptoms rather than the causes of the problem (as over-exploitation rather than weak property rights), his banning of market hunting and the commercial use of wildlife established a precedent that continues to today. Public management was effective, at least in North America, but it dealt a deathblow to institutions like local commons and markets for wild life. Thus, we have the ingredients of the modern conservation movement: the development of protected areas, the prohibition of the commercial use of wildlife, the management of wildlife by all-powerful state bureaucracies, and the public trust doctrine. This combination of norms is known as the North American Model of Wildlife Conservation.

The Yellowstone Model for National Parks

Parks often dominate the way we think about wildlife conservation, forgetting that most wildlife lives outside parks. The stereotypical American protected area is highly preservationist, unconcerned about economics, allows no hunting and permanent settlements within their boundaries, and is managed by the state. This model was adopted by the International Union for Conservation of Nature (IUCN), which defines six categories of protected areas, of which the first four assume public ownership, while the important Category II protected areas (national parks) prohibit settlement and hunting of wildlife (although killing of fish appears to be quite acceptable). The global goal is to protect 17% of the planet, and most recent growth towards this target is in Category VI 'sustainable use areas' which include many community areas.

Murphree (2002) expresses concerns that a bio-centric and state-centric view of protected areas has 'set up false dichotomies in protected area policy: that protected areas are confined to state management, that they are about non-use rather than use, and that they are about exclusion rather than regulated access'. In a sophisticated analysis, he suggests that 'Protected Areas are commons in that they are sites and bundles of collective entitlement for their constituents which require protection through controls on their use'. He goes on to argue that 'their essence is collective and controlled access' and that they may be 'owned' by a wide range of constituencies, from local communities to nation states. The assumption of public ownership and management of parks been challenged only recently, with Grazia Borrini-Feyerabend et al. (2013) proposing four governance regimes – public, private, community, and co-management. Both Murphree and Borrini-Feyerabend challenge the association between protected areas and the state, and between protected areas and non-use policies. While agreeing that the objective of parks is to conserve nature, parks can be owned and managed by communities, the private sector, or the state, or by some combination of these. They are set aside for economic gain (in its broader sense, including unpriced values), provided the primary land use is conservation, and should provide benefits that are aligned with the needs of their governing constituencies.

Moreover, it is a misconception that the 'Yellowstone Model' is uninterested in economic value (as defined in Chapter 6). Viewing parks as economic black holes is harmful, especially in developing countries. Even in the United States, the social and economic values of parks are important. Parks were established only on 'useless land' with low opportunity costs in terms of agriculture or mining (Runte, 1979). The strong public and political support of US parks originates in the US Parks Service's excellence in providing public value and ensuring that Americans, and the politicians that represent them, recognise these values. As Shelhas (2001) argues so lucidly, America's parks were an instrument of economics and public values from the beginning, with the National Parks Services seeking to make itself socially relevant and valuable to ordinary middle class Americans by providing them with quality and affordable access. Even the development of Yellowstone was heavily influenced by commercial interests, with railway magnates exerting political pressures to provide new destinations to fill the trains on America's rapidly expanding railway network (Chase, 1987). American park managers are keenly aware that public pressure is necessary to encourage politicians to allocate them workable budgets. Recently, they have begun to make this case economically, using methods developed by the late Daniel Stynes (2005), to suggest that parks are a high-value use of land in terms of public access, jobs, and economic impact, especially in rural areas. Thus:

> In 2013, the National Park System received over 273 million recreation visits. NPS visitors spent $14.6 billion in local gateway regions (defined as communities within 60 miles of a park). The contribution of this spending to the national economy was 238 thousand jobs, $9.2 billion in labor income, $15.6 billion in value added, and $26.5 billion in output. The lodging sector saw the highest direct contributions with 38 thousand jobs and $4.4 billion in output directly contributed to local gateway economies nationally. The sector with the next greatest direct contributions was restaurants and bars, with 50 thousand jobs and $2.9 billion in output directly contributed to local gateway economies nationally.
>
> *(Thomas et al., 2014)*

Despite this, US Parks are facing a current maintenance backlog of $11 billion, suggesting that the link between citizen value and public funding is fallible (Gee, 2018).

Interestingly, biodiversity conservation is a latecomer in the wide set of values provided by parks, with the first park explicitly set aside for what we now call biodiversity in the United States, namely the Florida Everglades, as recently as 1947. Americans today are comfortable with the public ownership and management of parks, but also with outsourcing management and commercial functions.

The Yellowstone Model has been exported globally, very often in a form that is more purist than Yellowstone itself. Where mechanisms of economic and social accountability are weak, however, the colonial model of state-run parks systems is under-performing. Parks are invariably under-funded, with disappointing results in terms of biodiversity conservation, economic impact, and community well-being. As we will show (Figure 9.7) parks can create significant economic value but are mostly badly under-funded. I have tested this claim for a number of years with graduate students in my class on parks. Invariably, case studies of private, community, and public parks in all continents show that parks provide substantial economic benefits in terms of ecosystems services, biodiversity, and tourism income and multipliers, yet, almost without exception, parks are under-financed relative to these values. This suggests that there is little understanding of the high economic return on investment in parks, or of the difference between financial and economic performance. In much of Africa, for instance, parks are perceived as catering largely to rich, white foreigners and the tourism businesses that service them. Consequently, there is inadequate investment in park infrastructure and protection, putting at risk the substantial economy that depends on parks. Chapter 9 makes a strong economic case for parks and nature, using data from South Luangwa in Zambia. The upside-down pyramid in Figure 9.7 shows that a small investment in funding parks (i.e. the apex of the pyramid) supports a much larger economic pyramid, while livelihoods near the park gate are twice as good as in counter-factual communities further way. These data show that many parks can and should be managed as 'rural economic engines', with the deliberate intention of providing jobs and developing the community around them (Child, 2004). Unfortunately, the economic structure of parks is poorly understood by policy-makers. They perceive parks as cash cows and rake in fees, neglecting the reinvestment necessary for park protection, management, and infrastructure. This has large opportunity costs in terms of biodiversity, but equally large costs to the economy, neighbouring communities, jobs, and taxes, an argument that we seldom hear.

Parks are a good example of the potential of the wildlife economy, and of the associated economies of scale. Without expounding too much about protected areas, parks are one of the few engines for economic growth available in remote rural areas, and provide a beachhead for a bio-experience economy in which communities can participate. Parks are often closely tied to CBNRM and often the only substantial economic force in remote rural areas. However, achieving these potentials requires a much more nuanced understanding of park economics, greater investment in competent park management,[2] and an acknowledgement that the governance of wildlife outside parks needs to be very different from the public management of the parks themselves.

Public wildlife on private land: the North American Model for wildlife management

Environmentalists tend to view wildlife outside protected areas through the lens of the North American Model, managed as a public asset. In North America, state 'Fish and Wildlife' agencies manage fish and wildlife that is resident on, or which traverses, private lands (endangered species and trade are administered federally). This system works quite well in the unique context of North America but papers over the mismatch between the resource and its ownership and governance regimes (see Chapter 5).

As with national parks, wildlife and hunting provide wide societal and economic benefits. Roosevelt's legacy was to treat hunting and fishing as a public good as he wanted all Americans to have the right to pursue a manly outdoor activity. Low hunting fees encouraged wide participation and political support from millions of middle-class hunters. Some 101.6 million Americans (40% of the US population aged 16 and above) participated in wildlife-related activities such as hunting, fishing, and wildlife viewing in 2016 (USFWS, 2017). Fish and Wildlife agencies ensure that there is a broad understanding of the economic value of hunting, which generates nearly $1 billion in annual fees, while its economic contribution in terms of employment and added value exceeds this by far. Thus Arnett and Southwick (2015) state that:

> 13.7 million American hunters and 2.1 million Canadian nature-recreationists spent more than $38.3 billion and $1.8 billion, respectively, on non-commercial hunting-related expenses each year ... [supporting] an estimated 680,937 jobs and $26.5 billion in salaries and wages [with] estimated state and local taxes [of] $5.4 billion, and federal taxes [of] $6.4 billion for 2011.
>
> *(Arnett & Southwick, 2015)*

North Americans recognise that the economic values of hunting far exceed the financial price of hunting tags and have established reliable funding mechanisms to reinvest in wildlife management. The famous Federal Aid in Wildlife Restoration Act of 1937 (the Pittman-Robertson Act) taxes all hunting and fishing equipment and uses this money to provide matching funds to state Fish and Wildlife agencies of about $300–500 million annually (Arnett & Southwick, 2015). Parallel to this, a large civic sector of hunting and conservation clubs and societies invests a further $1.6 billion annually into wildlife and land management, rehabilitation and purchase. Arnett and Southwick (2015) conclude that 'such economic numbers cannot be ignored by decision makers'. Hunters also argue that their substantial contribution to the conservation of wildlife and wild spaces is far in excess of vocal urban environmentalists that oppose hunting, and may be as much as 70% of conservation dollars (Donnall, 2010).

One of the ironies of capitalist America is the effectiveness of its well-managed public (centrally planned and socialist) system for conserving and valorising wildlife. Wildlife populations in North America have recovered. Wildlife is sufficiently important to large numbers of Americans that they hold public agencies to account; and wildlife benefits from a financing model that recognises the economic (not only the financial) value of wildlife.

These conditions do not hold in many tropical countries. The public gets few benefits from wildlife, and has few incentives or mechanisms to hold wildlife officials to account. Moreover, officials do not yet have the economic training to make the economic case for parks, hunting, or tourism, so parks and wildlife are invariably underfunded. Passionate environmentalists often neglect the importance of reliable financing models. Thus, North America's public trust model works because economic values are reflected in conservation agency budgets. Southern Africa's sustainable use model works (Chapters 8 and 9) because landholders own the wildlife and fund it by using it profitably. What does not work is a public park model that is neither adequately resourced nor staffed. Even less sustainable is a public model that centralises the control and benefits of wildlife on private land, while still expecting local landholders to bear the costs.

Hunting and fishing

In recent years, hunting has become controversial, reflecting the growing divide between urban and rural people in empathy and understanding.[3] Urban people, somewhat disconnected from nature, forget that the livelihoods, cultures, and norms of many rural communities have been inseparable from hunting and gathering for hundreds of generations. Moreover, starting with Theodore Roosevelt, the role of hunters and fishermen in recovering wildlife is undisputed, especially in North America (Donnall, 2010). Hunting has also played a disproportionate role in rewilding much of southern Africa, including communities. It remains a critical mechanism for financing conservation, especially in situations where wildlife is depleted; this is discussed briefly because of its disproportionate value to communities embarking on CBNRM.

As societies settle down as farmers and urban dwellers, hunting often transitions from a subsistence activity to one that is primarily symbolic and cultural. Arnett and Southwick (2015) describe this transition in Ancient Greece where, like North America, hunting was largely democratic. They suggest that hunting has considerable social importance, maintaining social networks (often through the sharing of meat), preserving cross-generational bonds and stories, providing mental and physical challenges, connecting people to the outdoors, conserving ecological knowledge, and encouraging understanding of and reinvestment in wildlife. The inclusion of so many values is why the total economic value of sport hunting usually exceeds that of meat commodity production (e.g. bushmeat), sometimes by an order of magnitude.

However, hunting is often not so democratic. Hunting rights are often controlled by wealthy or noble classes, as recorded in ancient Persia and Egypt (Arnett & Southwick, 2015). When the Normans conquered England under King William I, they introduced game laws which prevented Saxon commoners from hunting, entrenching the elitist concept of the 'King's Game' (and the Robin Hood Era) that was in many ways exported to European colonies. Consequently, in much of the world, wildlife is synonymous with the exclusion of ordinary people by the ruling elite, and is destroyed as much through social protest (Heffelfinger et al., 2015) as through ungoverned over-use.

Recreational or sport hunters had a major influence on the North American Model and the Public Trust Doctrine. With the growth in wealth and leisure after the Great Depression and World War II, the number of hunters in North America increased rapidly after 1950, providing a demand for hunting which funded wild lands. Hunters ventured into Africa and Asia where, as we will see in Chapter 8, the high fees paid by hunters provide a vital impetus to wildlife, including: its recovery in southern Africa; ungulates and snow leopards in Tajikistan (Kachel et al., 2016); and Bharal sheep and Himalayan tahr in Nepal (Aryal et al., 2015). Hunting is an invaluable tool for financing wildlife conservation because it generates high values from limited offtakes and can be practised in many areas that are unsuitable for tourists carrying cameras.

Yet there is considerable controversy about hunting for 'sport'. Roosevelt abhorred market hunting by ordinary people (today's 'commoners') and differentiated it from 'True sportsmen … [who] do no harm whatever to game'. Today, the urban elite promote the exact opposite. It is 'politically correct to accept subsistence hunting while abhorring frivolous killing of animals just for "sport"' (Heffelfinger et al., 2015). In the furore over 'Cecil the Lion' I compiled comments from *New York Times* articles to describe the complexities of hunting and anti-hunting sentiments. Some consider trophy hunting to be pathetic, repulsive, and unsportsmanlike. Others expressed their disgust that yet another wealthy man did as he pleased without concern for the consequences. Some saw Cecil as a road sign in the wanton destruction of the planet by 7 billion humans, crying not only for lions, but for the planet, injustice, the voiceless, and the loss of beauty. A much smaller group made the case for legal and responsible hunting, and against the growing anthropomorphisation of wildlife. A lonely African voice (Goodwell Nzou, 4 August 2015) in a sister article said he was quite happy that there were no longer lions in his rural village, and also on the US campus where he was doing his doctorate.

Hunting is a powerful conservation tool because it generates high values from low offtakes. Some 60–80% of wild land outside protected areas in southern Africa is financed by hunting. Indeed, the net effect of hunting on conservation has little to do with offtake, or even biology. What matters is the governance of hunting, and what happens to the money. Proprietorship of wildlife is key, and measured by the return of the income from hunting to landholders or 'producer communities', where a 100% return represents full proprietorship. The current strident anti-hunting chorus endangers conservation as much or more than the illegal wildlife trade.

Economics and governance of the North American Model

Roosevelt framed wildlife conservation in the public sphere, resulting in the Yellowstone Model for National Parks and the North American Model of wildlife management outside parks. These models are important, because of their influence on global conservation approaches. In North America, state and federal agencies are highly accountable to citizens, who benefit by using parks and hunting opportunities at very low cost to themselves. America's combination of vibrant democratic culture and economic education ensures that taxes are reinvested in parks and wildlife, and that management is held to high standards. Wildlife also gets substantial support from civic organisations. North American wildlife is a massive industry that benefits large segments of society through a multitude of values. It is resulting in widespread recovery of wildlife and habitats through sophisticated institutions including taxes and well-publicised economic surveys that promote the case for funding the wildlife economy. This includes a significant investment in professional wildlife management agencies, and more than 500 universities with specialised degrees in aspects of wildlife management (Mahoney & Jackson, 2015).

While the value of wildlife is transferable to many other countries, what is less certain is the transferability of the North American governance model, which is predicated on social accountability, public benefit, and competent state agencies. With this background, we turn to the story of wildlife conservation in Africa.

The London Convention and the nationalisation of wildlife in Africa

By the end of the 19th century, European colonial powers were becoming extremely concerned about the preservation of African fauna (Heijnsbergen, 1997). Some 20 million wild animals were eliminated in southern Africa between 1780 and 1880 resulting in the extinction of the bluebuck (*Hippotragus leucophaes*), cape lion (*Panthera leo melanochaita*), and quagga (*Equus quagga quagga*). Hunters and adventurers, such as the well-published Courtney Selous, set out in wagons to bring back ivory from the interior, and many farmers also survived off game meat and selling skins (Carruthers, 1989). Hunting no doubt killed a lot of wildlife, but it is always likely that the expansion of livestock, farming, and fences was the real cause of the loss of wildlife.

The European Powers implemented norms, very similar to the American models, in their African colonies through the London Conventions of 1900 and 1933[4] (Heijnsbergen, 1997). Mirroring the ideas of Roosevelt and the Americans,[5] the text of the London Convention of 1933 emphasises the same things: national parks without hunting or settlements; the nationalisation of wildlife and public management; and considerable restrictions on the trade and commercial use of wildlife. The Convention specifically seeks not to prejudice the hunting and other rights of native chiefs or tribes. Today's conservation policies (except for native rights) have hardly evolved.

The public management of parks in Africa

African countries began to set aside national parks – managed by Parks departments – from as early as 1898, many decades before other continents, and this accelerated after World War II (UNEP-WCMC, 2012). Separate 'game departments' were set up to manage wildlife outside parks, with the primary role of 'controlling' human wildlife conflict by, for instance, killing elephants or exterminating wildlife to reduce the risk of tsetse flies or diseases to domestic animals. Many of these agencies were amalgamated after the 1960s as the contemporary Parks and Wildlife Department.

The administrative and economic relationships between parks and rural communities has been changing rapidly. The game rangers that staffed early parks and wildlife agencies were always white and very often military. Living in remote areas, their companions in patrols and hunting were invariably black assistants, and they worked much more sensitively with local people than is the modern perception. For example, Ian Parker sought synergies between the elephant hunting culture of the Walunguli tribe and Tsavo National Park in Kenya (Parker, 2004), and Norman Carr promoted community benefits from tourism in the Luangwa. Some game reserves in Botswana – Moremi and Central Kalahari Game Reserve – were established with considerable sensitivity to the needs of the local bushmen (Child, 2009). Early game rangers and administrators believed that the major threat to wildlife was 'big white hunters', not the local people who had lived with wildlife since time immemorial (Parker, 2004), and even the London Convention explicitly recognised 'hunting or other rights already possessed by native chiefs of tribes'.

National Parks are a positive force in the African landscape, especially when managed well. However, parks provide a much larger suite of benefits than proscribed by biologists, and the potential of parks as economic engines in remote rural areas, and synergies between parks and the livelihoods of rural people, has not yet been realised.

There is considerable controversy about the relationships between parks and local people. Anthropologists and political scientists often cast parks in a negative light, following the narrative that parks displaced poor black people and are associated with human rights abuses and loss of land and access to resources (Brockington & Igoe, 2006). Environmentalists sing the opposite tune, perceiving communities as the enemy, so they get caught in the cross-fire over illegal wildlife trade and green militarisation (Duffy, 2016; Cooney et al., 2017).

The real picture is more nuanced. As in North America, many of Africa's parks were established between 1920 and 1960 on 'useless land' (Carruthers, 1995), or in critical watersheds such as mountain catchment forests. In their heyday, African wildlife departments were tiny, with a few dozen dedicated field rangers setting up game parks with limited resources and considerable ingenuity. Africa's population was one-tenth of today, and the number of people moved to make way for parks is miniscule compared to those now affected by these parks. Moreover, the narrative of park administrators running roughshod over local people may not be the whole picture. Certainly in Botswana and Zambia, park formation was sensitive to local people and followed years of negotiation with them (Campbell & Child, 1971; Astle, 1999), and citizen hunting was allowed for many years because wildlife hunting was integral to local livelihoods (Astle, 1999). Of course, the formation of parks, even on useless land, affected some local people. Even in America, Roosevelt evicted white farmers and miners to establish Grand Canyon National Park, but it would be hard to look back on Grand Canyon and say that Roosevelt's vision does not provide far more livelihoods and jobs today than if the area had remained fragmented into ranches and mines. In Africa, clumsy laws and high-handed actions in the name of parks and wildlife still causes long-term resentment and hostility to conservation, especially where traditional livelihoods were arbitrarily made illegal, sometimes directly, but more often by defining traditional hunting methods as 'primitive' and illegal.

Rather than getting stuck with a narrative and counter-narrative about whether parks are good or bad for people, it is more productive to ask how the changing administrative realities of parks have affected the relationships between parks and neighbouring communities, and where this can be improved. Parks in developing countries are a low economic priority, with shrinking budgets but swollen staffing levels. Consequently, park agencies become office-bound bureaucracies, gradually losing contact with and sensitivity to local people. In neo-patrimonial economies, state agencies, including park agencies, are politicised, lose technical proficiency, and even public accountability (Grindle & Thomas, 1991; Hyden, 2006). Officials are distrustful of the private sector which they don't understand (Gibson, 1999; Hyden, 2006), especially in post-socialist states, so opportunities to develop parks as economic engines are lost. Underpaid park managers can become distanced from, and distrustful of, local people, and when this is combined with pressures to stop poaching and the illegal wildlife trade, human rights abuses are not uncommon. In these circumstances, frustrated professionals then leave government service to work for NGOs. The growing gap in national capacities is filled by international NGOs and short funding cycles, which are often even more distant from local realities. In the absence of holistic long-term thinking and investment, parks lose their potential to conserve wildlife and promote local economic growth. The scale of these opportunity costs in terms of lost wildlife and lost income can be large (Lindsey et al., 2014).

There is also a wrong perception that local people don't like parks. Many people living around parks in Africa appreciate existence values ('I want my children to see wildlife'), ecosystem services ('the park brings rain'), and economic opportunities ('they bring jobs and development') (Mulindahabi, 2017). People might not like apes and elephants that raid their crops, especially when these are not managed effectively by park managers. But what they resent the most is the way they are treated by some park managers, and the unfairness of parks and wildlife policies. Rather than building on the underlying desire of most local people to see the park succeed, policies alienate them. They are excluded from the wildlife economy on their own land. Through 'green militarisation' communities are treated like criminals and made vulnerable to organised crime in a way very similar to blighted urban communities and the illicit drug trade. Consequently, even if local people like parks, they do not necessarily like park managers or wildlife policies. This situation can be reversed by managing parks specifically to create local economic growth, by providing local people with the safety and security they need to resist organised crime, and by involving them in the wildlife economy through CBNRM.

When parks are well managed and publicly accountable, they contribute significantly to the local rural economy, as we see around many parks in North America and some in

Africa. Indeed, most viable examples of CBNRM are in park buffer zones, where they can feed off the wildlife economy generated by the park, as well as excess animals. Proactively managing parks as engines for economic growth is therefore an important, if indirect, way of encouraging rural development, including CBNRM. However, parks are seldom geared up for this challenge. The goals of economic growth and social upliftment (to which they are well-suited) are seldom included alongside metrics like biodiversity and tourism in a park's key performance areas. Moreover, it is rare that parks' staff are trained in these capacities – namely the economics and governance of the wildlife economy and CBNRM. Nonetheless, the return on investment in an effective park–community relationship and a wildlife economy are usually large. Even small, thoughtful actions significantly improve relationships, such as the provision of medical and veterinary assistance to Bedouin in Egyptian parks, or micro-projects around parks in Rwanda (Mulindahabi, 2017). Thus, strategies for integrating park and communities as more sustainable social-ecological systems include (1) respectful park outreach, where nothing is more effective than accessible and respectful park managers; (2) revenue sharing and other support; (3) tourism and purchasing policies that deliberately include and benefit local communities; (4) the provision of security to prevent the disruption of local society by criminality linked to wildlife; and (5) devolving proprietorship so that communities can enter the wildlife economy as equal partners, with the goal of extending the wildlife economy and conservation landscape well beyond the park (i.e. CBNRM). There is a considerable need to build these principles into protected area management.[6]

The public management of wildlife outside parks

The institutions of parks and protected areas left to us by Roosevelt and his peers are broadly a good thing, disregarding current problems with management. Unfortunately, we cannot say the same about the institutions for wildlife management outside protected areas, where both centralised control and removal from the market threaten the survival of wildlife.

Living in the 'progressive era', policy-makers like Roosevelt and the European powers entrusted wildlife to government agencies. They did not trust the private sector with wildlife – indeed, they blamed market hunting for its demise. Moreover, even if they knew about common property, they associated it as a primitive form of economic organisation from which society was rapidly emerging, sometimes referring to ordinary people and non-elites in a slightly derogatory tone as 'commoners'. Thus, in establishing new institutions to tackle the massive threats to wildlife and wildlands, policy-makers only considered a narrow set of options. They nationalised wildlife, and often forests and fisheries too, with wildlife becoming subject to the doctrine of public ownership.

This is the historical root of the contemporary mismatch between the economic characteristic of wildlife and the structure of the systems governing it outside parks (see Chapter 5). Wildlife and wildlands get used up – they are clearly subtractable. With the possible exception of highly migratory species, we now have the economic incentives and technology for exclusion, especially if managed at the right local scale. As noted (Chapter 5), this defines wildlife and wildlands as private goods with some common-pool characteristics, not public goods. The norms of public ownership and management (i.e. nationalisation of wildlife) established in the early 20th century are human constructed, and can be changed.

Indeed, as the threats to wildlife increase, and as states and people recognise their rights, these norms are being contested. This debate plays out vividly and acrimoniously at international conferences such as the Convention on International Trade in Endangered Species (CITES), where some range states contest the rights of non-range states to dictate the use of their wildlife, and to remove rights without adequate compensation. Similarly, indigenous communities, especially those in Canada and Latin America, have been partially successful in reclaiming their ancient rights to and sovereignty over land, forests, wildlife, and fish. Where local people have less recourse to the political system, they simply ignore the rules and overuse or replace wildlife. Since this trend is likely to continue in a democratising world, it would be sensible to reconsider the governance of wildlife before we lose it all.

Too often, we confound the public conservation of wildlife on public land with the public conservation of wildlife on community or private land. The economic forces are very different. It is often difficult for Westerners who have grown up with deep norms about the public nature of wildlife to understand this distinction, which is key to good wildlife policy-making, as we will see in the next chapter. In initiating the transformation of the wildlife sector in southern Africa, Graham Child (1995) identified why internal contradictions in the 'old game laws' can only lead to failure. They entrench the privileges of the ruling class to hunt and use wildlife at much less than its true costs, while also marginalising the people who actually live with wildlife, especially rural communities. There is often an 'arrogant lack of sensitivity to the needs of people', 'an inability to understand that people and wild animals share the same habitats', and 'too much faith in the effectiveness of law, irrespective of whether it is acceptable, equitable, enforceable or even sensible'.

Taking wildlife out of the market place

The third Rooseveltian norm (after the formation of parks and the nationalisation of wildlife) was the removal of wildlife from the marketplace, a livelihood activity conducted mostly by commoners. An unintended consequence of cheap public hunting and outdoor recreation is that wildlife is under-priced. Establishing low administrative prices for wildlife reduces the incentives for the private sector and communities to produce it, and limits the wildlife economy to public lands (Child, 1995). Moreover, placing the management of wildlife in the public realm alienates local people who should have a genuine stake in its future, but increases the influence of special interest, media, and people without a genuine stake, with unintended negative consequences (Box 7.1).

BOX 7.1 THINKING BEYOND MORAL CERTAINTIES AND SIMPLISTIC SOLUTIONS

Hunting is a polarising topic. Campaigns to ban hunting are one of the most serious threats to wildlife conservation outside parks in Africa. Because moral arguments are so culturally specific, they are a bad foundation for global policy. We quickly lose our way when we allow special interests from one culture to impose their values on other cultures, especially the poor. While affluent urban special interest seek to deliberately block hunting and its benefits to rural communities, they often live in societies that condone factory farming, have a huge environmental footprint in terms of the use of the world's energy and atmosphere, have paved over their own biodiversity, and drink expensive wines and lattes in a world of poverty.

Moreover, a New Yorker might find a wealthy man killing an elephant for 'pleasure' abhorrent, whereas to an African $50 000 justifies the continued existence of a dangerous animal. Morals also ignore systems thinking. Killing an elephant is unattractive, but can be done purposefully to finance the conservation of elephants.

While some morals are widespread (such as the humane treatment of animals), it is better to govern wildlife pragmatically so that we have more wildlife, not less, rather than getting bogged down in unsolvable moral arguments. Advocating for stronger community rights (rather than weakening them) and for better wildlife governance (i.e. ensuring that the benefits from wildlife, including hunting, get to the people who live with animals) may be less emotive than opposing killing wildlife, but it might actually work, empowering poor people and saving wildlife.

The nationalisation of community lands and forests

I have focused on the economic history of wildlife because I am familiar with it, but similar drivers affect the forests and drylands that support so many local people. Like wildlife, forests and drylands were nationalised by the colonial state and renationalised by post-colonial socialist states. Communities who live in forests and drylands usually have no legal personality, and there is seldom formal recognition of their rights to the land and forests they live in (Table 7.2; see also Chapter 4). Local people usually cannot exclude others from external claims on their resources (Hatcher & Bailey, 2009), leading to a typical tragedy of the commons. They are also limited to using their resources in low-value subsistence uses and prevented from managing and trading their resources to best advantage (Chapter 6). Although inhabited by indigenous and local communities, some two-thirds of tropical forests remain in public hands, exceeding 98% in Africa (Tchawa, 2009). This has not led to good forest outcomes, leading to calls for tenure reform (Alden Wily, 2009).

Recently, there has been a significant transfer of forest ownership to communities in the Amazon (Hatcher & Bailey, 2009). Social movements in the late 1970s and 1980s reclaimed indigenous and community lands, responding to threats to local rights from frontier settlement and conservation and national development plans that excluded local people. With charismatic leadership (e.g. Chico Mendez), the support of the Catholic Church, and alliances with international organisations, these grassroots movements gained some control

TABLE 7.2 Tropical forest tenure

	Latin America	Asia	Africa
Administered by government	36	68	98
Designated for use by communities and indigenous peoples	25	3	1.6
Owned by communities and indigenous people	32	25	
Owned by individuals and firms	7	4	

Source: adapted from Hatcher & Bailey (2009).

and influence over the large tracts of forests in which they lived (Schmink & Wood, 1992; Cronkleton et al., 2008). Rights were formalised through designations such as indigenous territories and extractive and indigenous reserves. However, these rights were also significantly truncated through restrictions and even bans on the commercial use of forests and the wildlife in them. Nonetheless, evidence is emerging that indigenous and local people in indigenous and extractive reserves, respectively, are conserving forests better than alternative arrangements, including national parks (Nepstad et al., 2006). We also see this in the case of Mexican *ejidos* (Box 13.1).

Implications for wildlife and wildlands in the tropics

The nationalisation of wildlife has weakened or replaced local institutions and has also prevented them from evolving to respond to rapid changes and threats (Chapter 4). Especially since World War II, the pressures of the modern world have exposed the failings of the institutional misfit between the public ownership and governance regimes for wildlife, and the private and common-pool characteristics of these resources, especially in developing countries and the tropics. One consequence is a strong geographic association between:

- Communities with weak rights and institutions for land and wild life.
- Communities with weak social capital.
- Desertification, deforestation, land degradation, and loss of biodiversity.
- Protected areas and poverty.

This is also associated with policies that are economically incoherent. On the one hand, high-value wild resources are legally worthless, so they are replaced by low-value domestic resources. On the other hand, this same wildlife is over-exploited in the informal and criminal economy because it is 'too valuable'. These contradictions were papered over for a short time by public management, which may have been adequate to control access to wild resources in an earlier era with such abundance that wildlife 'could not finish'. However, the weaknesses in the public wildlife model are evident in the global disappearance of wildlife.

Across the world, state capacity to exclude non-authorised use has declined for many reasons. States face wide pressures and under-prioritise protecting wild resources; the jurisdictional reach of the state has always far exceeded its implementational grasp. This leaves us to rely with blind faith on environmental laws even where they lack social legitimacy. Local people never participated in drawing up the rules that prevented them from legally using, owning, and protecting their resources. Nor did they support policy decisions that, at the stroke of a pen, rendered age-old hunting and gathering livelihoods illegal. They are still not genuinely participating in global forums and deciding the future of the resources they live with. The only surprise should be that wildlife is not disappearing faster.

On the positive side, many wild landscapes and species are valuable. They are also potentially excludable. This suggests that there is more than one response pathway to the crisis of tropical biodiversity, ranging from further centralisation to full devolution. To inform this choice, we can draw on the competition for European supremacy between

Spain and England in the 1400s and 1500s, an example used by North (2005) to make a similar point. At this time, Europeans were consumed by wars that were becoming more and more expensive. There were three ways to fund these wars: by borrowing money, extracting wealth from the citizenry, or through economic growth. The Spanish chose to finance their wars by extracting the wealth of their people, plus treasure from their new colonies in Latin America, triggering the decline of the Spanish Empire. The English, partly in desperation, muddled through along alternative pathways, as much by accident as through contemplation. The break from the extractive governance and the divine right of kings was a critical juncture in the evolution of the Western World (Murrell, 2017). The gradual strengthening of the political and economic rights of Englishmen was a global tipping point, and is correlated with the Scientific and Industrial Revolutions and modern freedom and prosperity. English economic supremacy was built on inclusive governance, individual freedom, and a market economy of shippers and shopkeepers. This replaced Spanish hegemony, and the English-speaking people established a global empire.

The intriguing question is whether today's environmentalists face a similar choice. Do they follow the Spanish route of external funding and stronger central control, or take the English pathway of Locke's social freedoms and the protection of individual rights and property together with Smith's invisible hand? At the international and national level, conservation is following the Spanish option – more regulation, more centralisation, and more public financing. However, as North's history lessons suggest, perhaps the enlightened choice is to let go, carefully, and with well-crafted solutions. At the margins, we see glimpses of these processes beginning (Chapter 13), with evidence that locally owned forests are performing better (Nepstad et al., 2006), even in the absence of carefully designed policy requirements. In the next chapter, we turn to southern Africa where enlightened administrators deliberately transformed the wildlife economy with the result that the region is one of a very few places on the planet where wildlife has recovered at a substantial pace in the past half-century.

Notes

1   As president, Roosevelt declared over 230 million acres of new protected area including five new national parks, 18 new US national monuments by signing the 1906 Antiquities Act, 51 bird reserves, four game preserves, and 150 national forests. Brinkley (2009).
2   There is considerable need to unpack the differences between private, common pool, and public goods provided by parks (as we did for wild resources in Chapter 5) to provide additional insights about their governance and economic management. This is discussed in detail in an excellent paper prepared for the World Bank by Reed (2002).
3   To understand the growing divide between pragmatic conservationists and contemporary, more anthropomorphic conservation, I highly recommend Glenn Martin's *Game Changer*. This book is beautifully written, and partisan, but makes an insightful and valuable contribution when it traces the bifurcation in conservation back 60 years, through the persons of Ian Parker and George Adamson, both in the Kenyan Game Department. Parker is the quintessential pragmatist, with a deep understanding of African wildlife and people, and the need to balance human needs and wildlife management through legal, rational exploitation. Adamson was the dreamer, introducing the world to Elsa the Lion and *Born Free*, and bringing anthropomorphic conservation to the TV screens of much of the world, even if his joint efforts with his wife, Joy, disrupted natural systems with the introduction of semi-domesticated lions and had little or no real effects on habitat and wildlife conservation. See Martin (2012).

4   There were two London Conventions: the Convention Designed to Ensure the Conservation of Various Species of Wild Animals in Africa Which Are Useful to Man or Inoffensive, London, 1900, attended by the United Kingdom, Germany, Spain, Belgium, France, Italy, and Portugal; and the Convention Relative to the Preservation of Fauna and Flora in their Natural State (London Convention 1933), presided over by Richard Onslow, Fifth Earl of Onslow, and concluded by Belgium, Egypt, France, Italy, Portugal, South Africa, Spain, Sudan, and the United Kingdom.

5   According to their brief history of the Society for the Preservation of the Wild Fauna of the Empire, now Flora and Fauna International; Roosevelt was a member of this society. See Fitter & Scott (1978).

6   The Global Environmental Facility recognises that parks have two major outputs: biodiversity conservation and socio-economic impact. This is seldom measured with any consistency. Consequently, GEF's Scientific and Technical Advisory Panel is currently finalising a set of tools for monitoring socio-economic impact which includes (1) a simple assessment of ecosystem services; (2) top-down and bottom-up methods for assessing the total economic value of tourism including economic and employment multipliers; (3) simple ratios for assessing the financial sustainability and efficiency of park management; (4) the Social Assessment for Protected Areas tool which uses quantitative and qualitative methods to assess the positive and negative effects of a park on communities; and (5) livelihood surveys to understand local household economies and the impact of parks on them. See Child et al. (in prep.).

8

CHANGING THE GAME

LEARNING OBJECTIVES

This chapter describes:

1. How a critical juncture of circumstances and personalities led southern Africa to adopt a radically different governance regime for wildlife after 1960.
2. The emergence of the cow and the plough (not over-exploitation) as the primary threat to wildlife after the 1960s.
3. The hypothesis that wildlife had economic advantages in drylands.
4. Policy changes in Zimbabwe, including the devolution of proprietorship to land-holders, the removal of differential regulation and fees, and the development of collective self-regulation at both local and national levels.
5. How local self-regulation of natural resources in Zimbabwe matches Ostrom's (1990) principles for long enduring common property regimes.
6. Graham Child's role in rethinking and implementing conservation inside and out-side parks.
7. The Africa Special Project and the Arusha Conference, which laid down the principles of sound park management and a 'use it or lose it' philosophy outside parks.
8. The importance of communities of practice in driving change.
9. Comparative data from Kenya and southern Africa showing how wildlife numbers increased in response to institutional changes in southern Africa, but declined under outdated colonial modes of wildlife governance.
10. How wildlife recovery in southern Africa was a result of new institutions (privatisation and collective action) that better matched the nature of wildlife (subtractable, excludable), got prices right, and that had features similar to classic liberalism (rights of person, property, self-determination, and markets) that led Europe out of the Dark Ages.

A critical juncture in wildlife governance in Africa

Institutions are persistent and path dependent and rarely change, although critical junctures of circumstances, personalities, and processes, such as the Glorious Revolution of England, do occasionally result in radical transformation. Southern Africa is the only place where the governance of wildlife has undergone such a transformation, and it is also the only place in the Global South, and one of the few places in the world, where wildlife populations are expanding rapidly.[1] This chapter describes the history and thinking behind this transformation, and demonstrates an alternative governance regime that may be better suited to 'ungoverned spaces' than the public or North American model.

Africa's famous national parks

With its spectacular wildlife, Africa was an early adopter of the concept of national parks and game reserves. Famous parks such as Serengeti, Tsavo, Etosha, and Kruger are now the cornerstone of valuable tourism industries. At least up until World War II, however, much of Africa's wildlife still roamed outside parks, often in enormous 'controlled hunting areas' where big game hunting occurred alongside small populations of local villages. After World War II, however, an eight-fold increase in human population in countries like Kenya and Zimbabwe began to threaten these wildlife landscapes and wildernesses. Most conservation efforts focused on a few important parks in the mould of the Yellowstone Model. Management outside parks also followed the centralised model introduced by the London Conventions of 1900 and 1933, with game departments focusing largely on crop-raiding animals.

The cow and the plough

Africa's post-World War II colonial agenda emphasised 'development' and the clearing of wildlife as 'preservation of game must not be allowed to stand in the way of the urgent need for proper land usage' (Matheka, 2008). Returning war veterans were given farms and ranches in Zimbabwe, Kenya, and elsewhere. With farming expanding rapidly, landholders resented having to support the 'King's Game' on land they were struggling to develop for farming and ranching, driving wildlife outside parks into steep decline.

Zimbabwe invested heavily in agriculture. With grand visions of replicating the beef production on the Argentinian pampas, Southern Rhodesia subsidised the development of beef production (Child, 1988). Drylands were carved up into paddocks by fences, and stock numbers rocketed. Zimbabwe's leading cattle ranches had considerable political power with the white minority government. They made a case to the government that 'you cannot farm in a zoo'. In an agricultural state, farmers neither owned nor valued wildlife. Zebra and wildebeest were shot to reduce grazing competition with livestock. Giraffe were killed because they broke cattle fences. Buffalo were eradicated on the assumption that they transmitted diseases to livestock, especially foot-and-mouth disease. Lions, leopards, cheetahs, and elephants were actively eliminated by farmers, and game departments were paid to help. Massive areas of bush were cleared of wildlife across southern Africa to eliminate tsetse fly, on the assumption that removing game would remove the pest's food supply. When this did not work, water pans were bulldozed so game had no water to drink, and spectacular riverine trees were felled so that the flies had nowhere shady to rest (Child & Riney, 1987). Throughout the region, wildlife has been decimated on the massive game fences that now block seasonal movements and migration patterns. This environmental destruction was done in the name of progress and beef production, including to obtain preferential access to European beef markets as late as the early 1990s. The contemporary concern is with illegal wildlife trade, yet far more wildlife disappeared, and continues to disappear, through neglect and through competition with livestock which, by 1980, made up a staggering 90% of the large mammal biomass in most countries in southern Africa (Cumming & Bond, 1991), and perhaps as much as 96–98% of global large mammal biomass (Ripple et al., 2015). The primary threat to wildlife is the plough and, in particular, the cow.

The riddle of wildlife decline

This presented a riddle to the wildlife administrators of the 1960s: Why was a diversity of wild species, long-adapted to harsh and complex savanna ecosystems (Cumming, 1982; du Toit & Cumming, 1999), and intuitively of high value, being replaced wholescale by an exotic monoculture which did extensive ecological damage? The simple answer was that it had no value thanks to the institutions established through the London Conventions. Yet, replacing Africa's full spectrum of indigenous herbivores with an exotic monoculture of domestic animals seemed wrong. Surely multi-species systems of locally adapted wildlife were ecologically and economically competitive? Writing in 1966, Rudi Bigalke captures the essence of the times:

> In the 1950s several ecologists visiting East African countries began to urge a new philosophy … These men were struck by the large number and great variety of mammals living in harmony with their environment … [and] game areas were in good condition with little over-grazing and erosion … Land carrying domestic animals … was often overused and degraded. The inference was clear. The indigenous mammals had evolved in the country and were well-adapted to local conditions. Every available food niche was occupied … Domestic animals were ruining the country. Why not crop the game?
>
> *(Bigalke, 1966)*

Gamekeepers of Africa

Beginning in the 1960s, pragmatic, field-based wildlife conservationists in southern and East Africa adopted the phrase 'use it or lose it'. They saw that wildlife could only survive in the long term if the African people who lived with it also prospered by using it (IUCN, 1963; Parker, 2004). In the late 1950s, experimental game cropping schemes of hippo and elephant and other species were tested in countries ranging from Uganda to Zambia (IUCN, 1963). In East Africa and Zambia, charismatic game wardens like Ian Parker (Kenyan Game Department) and Norman Carr (southern Luangwa Valley in Zambia) were convinced that African communities must be involved in wildlife management and the sharing of revenues (Parker, 2004).

The winds of change blew through Africa, and these nascent ideas were never tested as Kenya gained independence in 1963 and Zambia in 1964. Changing direction, newly independent countries renationalised and centralised control of wild resources under the guise of African Socialism (Matheka, 2008), including fisheries, forestry, water, and wildlife as exemplified by Zambia (HURID, 2002). As Matheka (2008, p 633) emphasises:

> For the politician, wildlife was not only a national asset but also a source of patronage, while to the bureaucrat and the emergent rancher it was a source of easy wealth. Conversely, local communities gained little from conservation even after the establishment of game reserves in their names.

Wildlife in East Africa was renationalised. Kenya banned hunting in 1977, and has lost two-thirds of its wildlife since then (Ogutu et al., 2011).

Zimbabwe: crucible of change

The crucible for change lay further south, especially in Zimbabwe, where the trend towards wildlife utilisation through policies of devolution continued. We can describe the transformation of wildlife policy in Zimbabwe using Williamson's four-tier economic framework (Box 4.5). Institutional path dependency was unlocked by professional wildlife administrators, who had an alternative vision of the future of wildlife and who changed the rules to make this happen (Level 2). Pilot programmes (Level 4) provided short-term wins, and were then used to convince society (Level 1) that change was necessary, leading to further legal changes (Level 2), more field level results (Level 4), and eventually an entirely new cultural understanding of wildlife and society (Level 1). The wildlife economy is now widely accepted in southern Africa, whereas before 1960 wildlife was perceived as a public good and an impediment to the sound use of productive land.

Wildlife policy changes in Zimbabwe were driven by professional civil servants. Emerging from the British South Africa Company in 1923, Southern Rhodesia had a professional Victorian-style public service, with a strong emphasis on education, agriculture, research, and the environment. Zimbabweans had a reputation for pragmatism, for not running with the herd, and for letting facts, rather than opinions or conformity, shape outcomes. Moreover, decision-making was informed by science with agricultural research stations and a scientific journal (*Rhodesia Agricultural Journal*) being established within 13 years of white settlement. Under self-rule and settled by practical men rather than the aristocracy, the country was decentralised and democratic, at least as far as the small white population was concerned. By the 1930s, far-sighted officials in the water courts became concerned about soil and water management. Viscerally opposed to top-down solutions, they established a highly democratic, grassroots conservation system through the Natural Resources Act of 1941 (Box 8.1). Collective self-regulation was so effective for controlling soil erosion that it was later integrated into the wildlife act to control externalities associated with wildlife's mobility.

Compared to agriculture, wildlife was an afterthought, and the government did far more to harm wildlife than to help it. Several protected areas were developed from as early as 1902, but were only consolidated in 1975. However, public attitudes to wildlife began to change following 'Game Rescue', an audacious operation to save wildlife from the rising waters of the Kariba Dam in the late 1950s. The first professional wildlife officers were recruited in 1959 to complement the rugged field men that feature so prominently in early wildlife folklore. These professionals recognised the inevitable decimation of wildlife caused by the expansion of agriculture in the 1950s and 1960s, and pioneered a new philosophy – that of 'maximising the value of wildlife to the people on whose land it lives' – through bold policy and legislative reform (Child, 1995). A new way of doing conservation was invented by quiet, thoughtful, and determined men, who sought collaboration and results, rather than the limelight.

A senior wildlife administrator, Archie Frazer (appropriately nicknamed 'the Arch Phraser') rewrote the Wild Life Conservation Act of 1961 to open the door for landholders to crop and hunt game commercially through a permit system. His friend and colleague, Reay Smithers, the highly respected Curator of Museums, invited three Fulbright Scholars who were to have considerable impact on wildlife in Africa and beyond to Zimbabwe from 1958 to 1961 (Thane Riney, Ray Dasmann, and Archie Mossman).[2] With quiet official sanction before the new Act, Ray Dasmann and Archie Mossman initiated a game cropping experiment on the Doddieburn–Manyoli ranch owned by the Henderson brothers in southern Zimbabwe (Dasmann & Mossman, 1961). Their scientific assessment of game cropping strongly influenced southern Africa's conservation approach (Carruthers, 2008). Considerable experimenting with game cropping and wildlife production was also occurring in East Africa at the time (Talbot et al., 1961, 1965b). Building on the game cropping experiments, the wildlife industry began to grow. From a base of one in 1959, 169 game cropping permits were allocated to Zimbabwean landholders by 1974. Quite soon, wildlife officers realised that the administrative burden of counting wildlife and setting quotas was impossible, and unnecessary; ranchers were protecting wildlife, rather than eliminating it, as some suggested would happen (Mossman & Mossman, 1976). Realising that it had not gone far enough, wildlife administrators did away with permits and other requirements to use wildlife on private land.

The third Fulbright Scholar, Thane Riney, had great breadth of vision and influence. He introduced many innovative ideas about wildlife management that were well ahead of their time, including holistic ecological management and the sustainable use of wildlife (Riney, 1964, 1967, 1982). He discussed the new concept using wildlife with the Transvaal Department of Nature Conservation and the Natal Parks Board (led by Colonel Vincent) in South Africa, where they were taken up, and he also attempted to do so across the rest of Africa (Riney & Hill, 1967). Riney's ideas were eagerly adopted, including by young biologists in the Southern Rhodesian Game Department, especially Graham Child, who went on to turn many of these ideas into reality. Political attitudes toward wildlife in Southern Africa softened, owing much to the world-wide publicity generated by Operation Noah during the filling of Lake Kariba from 1958 to 1961 (Child, 1968).

By the early 1970s, Archie Frazer and Graham Child (the permanent secretary and former parks director, and the park director, respectively) concluded that 'most species are best protected by landholders and landholder communities that live with them' and that the legislation needed to be changed to reflect this. The evolution of wildlife legislation was hastened through legal action that highlighted contradictions in the public ownership of wildlife. In 1973, a cattle rancher (Morseby White) living near Victoria Falls shot a roan antelope, which was then Royal Game. Upon being arrested, he claimed that by eating grass on his land the wild animal was destroying his property and affecting his livelihood; if the state claimed ownership of wildlife then, surely, it was also legally liable for damage caused by 'its' wildlife. Chief Justice Hugh Beadle, ebullient, controversial, and an avid hunter, was sympathetic to this argument and ruled in favour of the landholder (Beadle & Macdonald, 1969). Wildlife officials were quietly supportive and used this ruling to launch the radical changes they were thinking about. They had already concluded that public ownership of wildlife was unworkable, except in the case of protected areas; it was only a matter of time before the state was taken to court for the costs imposed by wildlife, including use of water, grazing, loss of property, and life (G. Child, personal communication, 2015). With 'wild life' being legally defined as 'all forms of animal life, vertebrate and invertebrate, which are indigenous to Zimbabwe', including plague species such as quelea and locusts, this was a high risk to the state.

These ideas merged in the far-reaching Parks and Wildlife Act of 1975, which altered the core philosophy underpinning wildlife ownership. The technical task of designing policy to devolve custodianship of wildlife to landowners was relatively simple, but the main impediment was urban conservationists who opposed the idea that landholders could be trusted to manage wildlife responsibly if they were allowed to utilise it freely. Over the next five years, Frazer and Child shepherded a new act into law through numerous public hearings and protracted meetings with members of Parliament.

The Act was path breaking. It clarified the legal purpose of six categories of parks, sanctuaries, and safari areas (preceding the categorisation adopted by the International Union for Conservation of Nature (IUCN) in 1994, Phillips 2007) and, by 1975, Zimbabwe had a well-crafted and consolidated system of state-managed protected areas that secured 15% of the country (Child, 1995). It also established the primacy of holistic environmental management, including soil, even before the term 'biodiversity' was coined.

In 1967, Robert MacArthur and E. O. Wilson developed the theory of island biogeography, species area curves, and the possibility of parks becoming isolated islands (MacArthur & Wilson, 1967). Zimbabwe's park officials, conversant with the science, and aware that conserving 15% of the country was never going to be enough, set about conserving wildlife on land outside the protected area estate. A concise, and vital, section in the Act devolved to landholders the rights to use wildlife commercially with minimal interference from the government. The Act established landholders as the 'appropriate authority' for wildlife and gave them powers. To quote directly from Section 59, Subsection (4) of the Parks and Wildlife Act (1975):

Subject to this Act, the appropriate authority for any land may:

(a) Hunt any animal on the land; or
(b) Remove any animal or any part of an animal from the land or from one place to another on the land; or
(c) Issue a permit to any person allowing him or any other person or any class of persons to hunt any animal on the land or to remove any animal or any part of an animal from the land or from one place to another on the land.

Legally, wildlife became *res nullius* (owned by no-one), but, as 'appropriate authorities', landholders acquired full rights to use, manage, protect, and sell wildlife while it was on their land. The 'spirit of the Act' was to make wildlife as valuable as possible to the people living with it, and to use wildlife as profitably as possible provided use was humane.[3] To increase wildlife's financial competitiveness, all fees and most other regulations were removed – wildlife had to compete with livestock, and the Ministry of Agriculture did not set quotas for livestock, require permits, or charge fees.

Collective self-regulation

The biggest argument voiced against privatising wildlife was the fear of over-utilising species that moved freely between properties. This was solved more easily in practice than in theory, echoing Ostrom's (1990) confidence in local collective action. The far-sighted Natural Resources Act of 1941 (Box 8.1) provided a platform for the collective self-regulation of wildlife. Private landholders were already constituted as democratic communities (called Intensive Conservation Areas or ICAs) with the legal powers to regulate themselves through the Natural Resources Act of 1941. The system had proved remarkably effective at controlling and recovering from soil erosion and other environmental effects, was inexpensive, was hugely popular and respected by landholders, and was a near perfect match with the principles developed by Elinor Ostrom nearly fifty years later (Ostrom, 1990). Although these communities had strong legal powers to enforce destocking in cases of range degradation, to require soil erosion controls for roads or fields, or even to implement these and send the bill to the landholder, they seldom used these powers, relying instead on social pressures and peer accountability. At the heart of the system was informal self-monitoring, strengthened by education and extension to ensure that farmers were knowledgeable about the causes and consequences of soil erosion, burning, overgrazing, and so on. In well-designed, democratic systems, farmers placed checks on each other as humans have done since they evolved on the African savannas, and most problems were solved quietly and subtly. Rules for addressing complex environmental problems and externalities were effective because they were self-made, parsimonious, tailor-made to local conditions, and sometimes even ingenious (Ostrom, 2007, 2009c). It is particularly revealing that the system faltered as soon as it lost its essential democratic grassroots nature, following a decision by government to prioritise efficiency over democracy by consolidating natural resource management through rural and district councils (Child & Child, 2015). Nonetheless, landholders had learned the value of working together. They responded by consolidating wildlife properties at the level of a landscape through legal conservancy agreements. Thus, Zimbabwe used the Parks and Wildlife Act of 1975 to privatise wildlife and the Natural Resources Act of 1941 to manage its common-pool attributes collectively. This institutional approach provides a contrast with South Africa and Namibia where wildlife ownership rights are acquired through game fencing and a 'certificate of adequate enclosure', with the result that the landscape is highly fragmented by expensive game fences around individual properties, with negative financial and ecological consequences.

BOX 8.1 ZIMBABWE'S INTENSIVE CONSERVATION AREAS (ICAS) AND NATURAL RESOURCES BOARD (NRB)

Zimbabwe's natural resource governance system consisted of democratic catchment communities, with strong legal powers for self-governance, headed by the civic NRB (Child & Child, 2015). This system, legislated in 1941, is assessed against Ostrom's eight principles for governing a commons.

Boundaries clearly defined. ICAs were legally recognised associations of some 50–70 private landholders in ecological units, usually within the same drainage sub-catchments.

Those affected by the rules can participate in modifying the rules. ICAs were democratic, with regular community-wide meetings and field days, and wrote their own rules and regulations for natural resource management, including wildlife.

Match rules governing use of common goods to local needs and conditions. Rules were parsimonious, locally appropriate, enforceable, and effective. For wildlife, for example, the community would set quotas for high-value species, giving higher quotas to properties that had more wildlife, but didn't bother to set quotas for species that were numerous or low-value unless a specific problem arose. Game fencing was not necessary, because the use of wildlife that moved from one property to another was arbitrated by the community.

Monitoring by the community or by people accountable to the community. Most monitoring was informal, but the ICA partnered in an annual over-flight with the Lands Inspectorate to ensure erosion, deforestation, and so on were under control.

Use graduated sanctions for rule violators. Compliance was mainly through local social sanction and peer pressure, though the ICA had access to powerful legal measures if need be. If the ICA failed to act, government retained the ultimate authority to intervene. In reality, few cases went beyond the local ICA. In the first 16 years after the Parks and Wildlife Act came into force, the wildlife department had to deal with only one referral – which concerned the baiting of lions out of a national park by a cattle rancher.

Provide accessible, low-cost means for dispute resolution. Most actions were solved locally through peer pressure or legal action by the ICA. If a member disagreed with the ICA's ruling (e.g. to reduce cattle numbers or wildlife offtake), he could refer the matter to the elected national Natural Resources Board (which had 14 days to rule), with a further and final arbitration possible through the Natural Resources Court (which also had 14 days to rule). In the case of wildlife, either side could appeal to the Director of National Parks and Wildlife Management to consider appropriate action, again with a 14-day decision period.

The rule-making rights of community members are respected by high-level authorities. ICAs had significant legal powers of self-regulation. They could set quotas, ban hunting, and legally require a member to invest in environmental rehabilitation (e.g. erosion controls) or desist in activities (e.g. over-stocking, deforestation) that harmed their environment. Each ICA acquired the services of a government extension officer.

Build responsibility for governing the common resource in nested tiers from the lowest level up to the entire interconnected system. The authority for natural resource governance was entrusted to the civic Natural Resource Board, democratically accountable to a structure of some 200 landscape communities (ICAs). Excellent horizontal and vertical information sharing through these structures resulted in proactive and innovative responses to threats and opportunities involving natural resources.

Zimbabwe's wildlife legislation was path breaking and simple in its sophistication. The Parks and Wildlife Act of 1975 devolved wildlife proprietorship to landholders and set out to make wildlife as profitable to the landholder as possible by encouraging the commercial use of wildlife (provided it was humane), while removing permits and licence fees that reduced wildlife's competitiveness. The Natural Resources Act of 1941 then used Ostrom-like principles to control externalities and other features of complex systems that could only be managed through informed collective action. The principles of entrusting wildlife to landholders, making it as valuable as possible, and collective responsibility for landscapes, began to permeate the DNA of the park agency, the country, and the region.

Here I pay tribute to my late father, Graham Child, a field man and senior administrator who, as a deep observer of ecosystems and their interactions with people, worked diligently to developed a cogent narrative (or theory of change), allowing facts to shape his opinions, even if these did not follow those of the herd. Never afraid to provide strong leadership, this often reflected his belief that empowering people, be these park rangers, farmers, rural communities, or wildlife producer associations, would lead to the best results, combining healthy ecosystems with the greatest good for the greatest number. Then, when he found himself in the right position, and working with the right people, he transformed the wildlife sector by applying these principles. His legacy is summarised in Box 8.2, modified from a similar tribute in *Wildlife Ranching* in 2016. I am eternally grateful that my father fact-checked this tribute a week before he died. Ever humble, he said it 'made him blush'. The journey towards this new paradigm of wildlife conservation, developed intuitively through a highly collaborative processes between landholders, professional bureaucrats, and some scholars, is laid out in *Wildlife and People* (Child, 1995), a simple book of remarkable prescience to which I find myself constantly referring. These silverbacks knew what they were doing, even if it is left to our generation to frame these lessons in theory.

BOX 8.2 GRAHAM CHILD: GAME CHANGER[4]

Graham Child grew up on remote field stations in Zimbabwe, learning Shona and Ndebele, and absorbing the work ethic and culture of an effective civil service. His father, Harold Tamplin Child, survived two years on the Western Front in World War I to become a District Administrator and, later, to head African Agriculture in Southern Rhodesia. In the 1930s, 1940s, and 1950s, he implemented concepts that, today, we call CBNRM, including strengthening communities with a combination of individual (for fields and houses) and collective title (for common grazing areas), and ideas now called conservation agriculture, herding for health, and so on.

As a boarder at Plumtree School, way out in the bush on Zimbabwe's western border, Graham learnt that a man is measured by what he does, not by what he says. Serendipitously, he shared school dormitories and played sport against young men who were already passionate about a new role for wildlife in marginal agricultural systems. Bob Vaughan-Evans became a highly influential extension officer, encouraging Zimbabwe's Department of Conservation and Extension (CONEX) to view wildlife as an acceptable land use. The practical Peter Johnstone and the Style family pioneered game farming as early as 1959. Allan Savory was an early leader in wildlife utilisation before making a name for himself in holistic resource management.

Graham fell under the influence of extraordinary men. Reay H. N. Smithers, the 'grand old man of vertebrate research' and author of *Mammals of Southern Africa* (1996), developed the museum service of Southern Rhodesia as a centre of expertise in wildlife and history. Ever practical, Graham supervised the building of the Bulawayo Museum (which was opened by David Attenborough) and was keeper of mammals. He was seconded to the Southern Rhodesian Game Department as its first ecologist (under Archie Frazer), where he was one of a handful of rugged young men who lived in the wild to rescue stranded animals from the rising floodwaters of the Zambezi when Lake Kariba was being established. We still have tins of Mycota foot powder as a testament to months of jumping off boats to capture animals, ranging from snakes to rhinos. Awarded a prestigious Beit Research Fellowship, Graham turned this opportunity into a PhD, studying the effects of excessive numbers of wild animals on shrinking habitats. His examiner, and friend, was George Petrides, long-time editor of the *Journal of Wildlife Management*.

Graham always had a dog-eared copy of Aldo Leopold's *Sand County Almanac*, and we can follow a chain of events backwards from Graham to Aldo Leopold. Graham's inspiration and mentor was Thane Riney, an American and a prodigy of Starker Leopold (Aldo Leopold's son) at UC Berkeley and of Sir Frank Fraser Darling. Riney fine-tuned Graham's natural attributes as an observer of ecosystems and oral case histories, as well as his abilities to build 'a syndrome of evidence' (what we now called triangulation). Taking measurements and talking to local people, Graham was one of the first ecologists to document ecosystem thresholds, or today's 'tipping points' (Child, 1968; Parris & Child, 1973). Throughout his career, Graham understood the importance of whole ecosystems, starting from soils and grass, as well as adaptive management, as he assembled park agencies in Botswana, Zimbabwe, and Saudi Arabia. Riney was 'a breath of fresh air' who challenged orthodoxy, a characteristic shared by Graham, who was able to sum up the nub of the situation and initiate bold change.

With positions in the Museum and Game Department, Graham was involved in much of the early experimentation on game ranching in Zimbabwe. In the meantime, Riney came to head the United Nations Food and Agricultural Organization's (FAO) African Special Project, a continent-wide effort to conserve wildlife and develop a wildlife industry in Africa in the 1960s. Riney recruited Graham to survey and develop Chobe National Park, and later to work with Alec Campbell, a dedicated anthropologist, archaeologist, and naturalist, to establish Botswana's Department of Wildlife, National Parks and Tourism. With his good friend Alec Campbell, he recognised the importance of local people when developing Moremi Game Reserve and the Central Kalahari Game Reserve. To quote:

> A tremendously interesting and quite admirable scheme is now afoot to make this area available as a reserved home for Bushmen who do not wish to live in contact with the food-growing economy of the Bechuana Bantu peoples and prefer to live their own lives. The Bushmen, it is hoped, will be able to live their own lives as hunter-food gatherers without interference or encroachment.
>
> *(Wild, 1968)*

Graham would spend months in the field, accompanied only by his Bushmen[5] trackers and local assistants, researching animals, measuring vegetation, and collecting oral histories, returning only when he finished his two drums of fuel and drum of water – he allowed himself one cup of water per day for 'hygienic purposes'. On a field trip deep in the Kalahari, Graham and his Bushman trackers apprehended an elderly man trapping springbok on the edge of a saltpan – what many might call 'poaching'. My father applied his usual interrogation techniques – copious cups of tea. The old man berated Graham for interfering in his age-old lifestyle and scoffed when Graham pointed out conscientiously that wildlife money enabled the government to provide schools and water in the regional town, many days' walk away and of no benefit to him or his children. This was a transformative moment in Graham's life. He was struck by the absurdity of the system – that this man, practising his age-old livelihood, was considered a criminal in the eyes of the law, and that the benefits of wildlife accrued largely to urban areas.

At a young age, Graham had considerable influence on wildlife policy. From about 1967, he became vice chair of the highly influential Standing Committee for Nature Conservation, Wildlife Utilization and Management (MUNC) of the Southern African Regional Commission for the Conservation and Utilisation of the Soil (SARCCUS). By 1971, aged 34, he was recruited to director of the parks and wildlife agency in Zimbabwe, which was flailing after poor leadership.

Meeting for a week each year at MUNC, Graham and a pioneering coalition of wildlife directors and scientists recognised that received wisdom and inherited policies were detrimental to wildlife. Wildlife had ecological advantages in southern Africa's semi-arid ranch lands, but would have to pay its way to survive. Consequently, these men introduced radical new legislation in much of the region, devolving ownership of wildlife to farmers and, later, communities, and creating a sea change in wildlife conservation (Suich & Child, 2009). Graham often reminisced about his partnership with Bernabie de la Bat, the first director of Namibia's Department of Nature Conservation (and colleagues like Ted Riley who continues to do so much for conservation in Swaziland), and it is no coincidence that Zimbabwe and Namibia took the early lead in changing the way wildlife conservation is done in southern Africa.

In 1971, Graham was appointed to head the Department of National Parks and Wildlife Management. He recognised the power of devolved management before its time and, under his leadership, Zimbabwe became a global example of effective protected area management. Empowered field wardens held each other to high standards, building a world-class protected area system and tourism industry with limited resources, including decent housing and schooling for all 2 500 employees. Snippets of the extraordinary lives and adventures lived by these hardworking, dedicated, and sometimes wild men were collected and published by Mike Bromwich (Bromwich, 2014).

Growing up in a country passionate about soil erosion and healthy environments, Graham introduced a holistic approach to ecosystem conservation. He believed effective conservation is the outcome of 'ecological sustainability, economic viability, and socio-political acceptability'. Thus, he warned of the threats to biological integrity of excess numbers of elephants in Chobe National Park in the 1960s, but he also saw parks as an engine for sustainable economic development (Child, 1968). He took ecosystem health so seriously that he ordered the culling of over 30 000 elephants to protect habitats.

Graham was acutely aware that wildlife on private land was in trouble, having grown up with local farmers and having researched and practised game cropping. In his firm, pragmatic way, he 'recognised that most species are best protected by landholders and landholder communities' and the only way to reverse this negative trend was to 'maximise the value of wildlife to the landholder'. With Permanent Secretary Archie Frazer, he rewrote the landmark Parks and Wildlife Act (1975), boldly devolving to private landholders the right to use wildlife as profitably as they saw fit, providing use was, as he always emphasised, humane. He did away with most government controls and all government fees, believing them to be a hindrance to conservation, and instead placed his faith in local collective action to prevent environmental abuses. This belief extended to local communities. Well before the transformation to black majority rule, Graham persuaded Parliament to return money from hunting and culling wildlife directly back to communities. He attended Parliament in his khaki game ranger uniform to drive this unprecedented change in government financial management.

Two principles shine through Dr Graham Child's career: his belief in devolving power to ordinary farmers, communities, park managers, and industry associations to get work done effectively; and his confidence that the future of wildlife lay in making it more, not less, valuable, providing landholders and communities owned it. Graham played a significant role in developing a vibrant civic society for the wildlife sector by actively empowering associations of wildlife producers, safari operators, professional hunters, anglers, kapenta fishermen, ivory manufacturers, and falconers to take up responsibility for their own industries.

He created change by fostering in his department a culture of always finding ways to do things more cheaply and more beneficially for wildlife, farmers, and rural communities, even if it was radically different from the norm, and by doing things 'properly'. It is no accident that Zimbabwe pioneered private and community wildlife conservation, with the Communal Areas Management Programme for Indigenous Resources (CAMP-FIRE) becoming a globally iconic template in the early 1990s.

Graham was visionary, but he was effective because he was a strategic administrator, converting the emerging philosophies he promoted into succinct policy documents to guide the management of parks, crocodiles, rhinos, elephants, quelea, the wild bird trade, and others. These were gradually incorporated into departmental, national, and even regional 'culture'.

Graham's influence was almost always through a team-based approach and spread well beyond Zimbabwe. As one of two Regional Councillors for the IUCN, he worked to up-end what he called 'outdated colonial top-down wildlife legislation' (Sharp, 2017). The World Parks Congresses in Bali (1982) and Caracas (1992) were early catalysts in the transformation towards people-centred conservation. Graham convened major workshops at both congresses, emphasising that wildlife conservation could no longer be viewed as separate from people, and highlighting the importance of ownership and trade for transforming land back to wildlife. I still meet high-ranking people today who understand the importance of sustainable use because of Graham's influence at IUCN and CITES during these years. It is also no coincidence that IUCN's protected area categories, adopted in 1994, resemble those developed in Zimbabwe in 1975.

Graham encouraged people at all levels and of all races to strive for a new way of conserving wildlife, by owning it, making it as profitable as possible, and by ensuring that this profit was returned to the people who lived with the wildlife. His path-breaking ideas are captured in two books. The first, *Managing Protected Areas in the Tropics* (1986), which he co-authored, reflects the professionalism for which he strived in Zimbabwe's parks agency. The second, *Wildlife & People: The Zimbabwean Success* (1995), even today remains both pragmatic and conceptually visionary, encapsulating Graham's paradigm-shifting contribution to wildlife conservation in southern Africa.

Graham was no armchair philosopher, writing wistfully about how conservation should be done. He was always in the field, observing ecosystems, talking with rural people, and turning pragmatic knowledge into policy and administrative practices. Where conservation is etched with gloom, perhaps the true dismal science, Graham's pragmatism was combined with a determination and enthusiasm to demonstrate that, if done properly, the loss of wildlife could be reversed on a massive scale. His approach was grounded, conceptually informed, and used new institutions to empower the people who live with wildlife to benefit from and conserve it. To quote from an obituary written by Robin Sharpe:

> A colleague recalls 'I arrived in Zambia in 1969 when Graham was working with Thane Riney, Graham Caughley and Anton de Vos and others who were promoting some of those early projects to make conservation work for local people. Conservationists are still struggling to achieve that. I sometimes think that the concepts, approaches etc. at that time were ahead of where we are today. At that time conservationists spent their time in the field and had first-hand experience of the people and environments in which they were working. Today's conservationists spend their time at conferences and live in a grossly simplified world'.
>
> *(Sharp, 2017)*

The Africa Special Project: use it or lose it

With southern Africa attracting the limelight in global conservation, the IUCN and the FAO jointly sponsored the Africa Special Project to survey the wildlife situation in 20 countries in Africa in 1962 (Riney & Hill, 1967) and to develop it. Conservation is a small world. Riney was asked to lead the Africa Special Project, and was replaced as Chief Scientist at the IUCN by Ray Dasmann, another Fulbright Scholar who promoted renewable resource conservation in Africa and beyond.

By 1960, the IUCN concluded that the 'accelerated destruction of wild fauna, flora and habitat in Africa – without adequate regard to its value as a continuing economic and cultural resource – was the most urgent conservation problem of the present time' (IUCN, 1963). In 1961, Riney convened the pivotal Arusha Conference – the Conservation of Wildlife in Modern African States – which brought together leading conservationists from Africa, Europe, and America. The conference opened with the statement: 'Only by the planned utilization of wildlife as a renewable natural resource, either for protein or as a recreational attraction, can its conservation and development be economically justified in competition with agriculture, stock ranching and other forms of land use' (IUCN, 1963).

The Arusha conference made a critical distinction that is often forgotten – that completely different strategies are needed for managing parks, compared to the land outside them. It envisaged 'first, conservation of the national parks and faunal reserves'. However, in ways not done anywhere else in the world, it emphasised a 'use it or lose it' philosophy towards 'the management of wildlife stocks on lands outside the existing parks and reserves, especially on those lands not suited to agriculture' (IUCN, 1963). The Africa Special Project invested in developing parks and park agencies, experimented with game cropping (e.g. the Luangwa elephant cropping programme), and invested in training Africans by building the Mweka and Garoua wildlife training colleges for anglophone and francophone Africa respectively.

Recentralisation of wildlife

Many of the leading speakers and proponents of wildlife utilisation at the Arusha Conference were from East Africa. Although early ideas about game cropping (Talbot et al., 1965a) and even community involvement (Parker, 2004) were initiated in these great wildlife countries, including Kenya, Tanzania, and Zambia, they never took off as wildlife and forests were nationalised in the immediate aftermath of independence. Public conservation agencies were underfunded and lost capacity, while wildlife and forest products began to enter illegal markets, often with the connivance of high-ranking politicians (Parker, 2004; Milledge, 2007). Internationally funded NGOs, rather than landholders, began to dominate the conservation narrative, especially in Kenya (Kabiri, 2010). Both state agencies and NGOs benefited from centralisation, looking upwards to funding using non-use ideologies rather than downwards to the well-being of the people who live with wildlife. This perpetuated 'outdated colonial top-down wildlife legislation' (Sharp, 2017), which was occasionally challenged by effective champions and support coalitions, as we see in the decentralisation of the forestry in Tanzanian in the late 1990s (Nelson & Blomley, 2009).

Devolution of wildlife rights and communities of practice

By contrast, southern Africa charted a new course that can be described using Kotter's (1996) change management process (Figure 8.1). The rapid loss of wildlife created a sense of urgency or 'discomfort with the status quo'. Wildlife administrators were spurred to work together to develop a new vision, to enact bold new legislation, and to work with landholders to test game cropping and, more importantly, safari hunting. This created short-term wins, and maverick game ranchers like Peter Johnstone, mentioned in Box 8.2, showed that wildlife could pay (Johnstone, 1971). Based on these successes, the khaki-clad wildlife professionals, who were good friends, and who challenged and learned from each other, made further changes. Wildlife recovered very rapidly on private land, despite scepticism and considerable opposition from the agricultural sector. Within a decade or so, the vibrancy of the wildlife economy was indisputable as thousands of ranchers began to manage and benefit from wildlife.

In the late 1980s, this regional community of practice reconvened itself as the IUCN Southern African Sustainable Use Specialist Group (SASUSG), playing a major role in transferring the wildlife economy to rural communities (Suich & Child, 2009), with southern Africa emerging onto the world stage to champion community wildlife management and CBNRM (Borgerhoff Mulder & Coppolillo, 2005). This cross-regional learning was conceptualised as a number of statements and books (SASUSG, 1996, 2003; Martin, 2003, 2009a; Child, 2004; Spenceley, 2009; Suich & Child, 2009; Nelson, 2010) which are read by many practitioners, but seldom apparently by scholars.

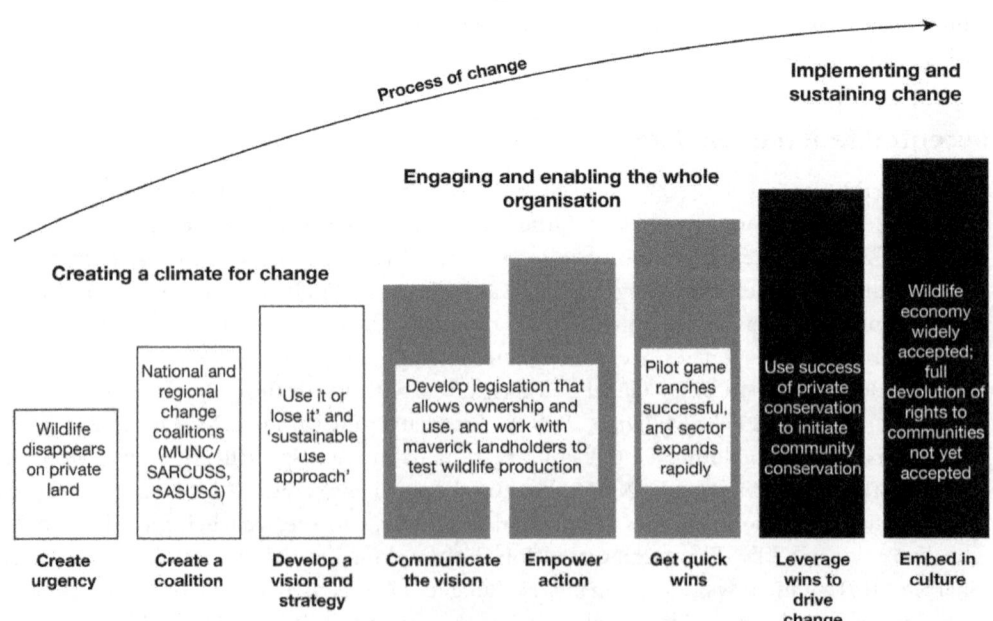

FIGURE 8.1 The process of change in southern Africa
Source: adapted from Kotter's eight-step change management process

This network also invested heavily in defending sustainable use principles, markets for wildlife, and the rights of local people to self-determination at forums like CITES and the IUCN (Hutton & Dickson, 2000; Oldfield, 2003), including through community theatre (Guhrs et al., 2006).

On the basis of effective park management, and pioneering progress in private and community conservation, southern Africa continued to influence global policy in forums such as the IUCN and CITES. Graham Child played a significant role in the Third World National Parks Conference in Bali in 1982 which, for the first time, 'links conservation with sustainable development and the rational use of the world's natural resources' in its final Declaration (Scriabine, 1983). By 1992, southern Africa's tenacious participation in CITES led to 'recognition of the benefits of trade in wildlife' in a resolution.[6] Many of the sustainable-use principles developed in southern Africa (SASUSG, 1996, 2003) find themselves repeated (if watered down) in the Convention for Biological Diversity's (CBD) Addis Ababa Principles and Guidelines for the Sustainable Use of Biodiversity (CBD, 2004).

Centralisation and the bureaucratic impulse

Marshall Murphree, a political scientist and spokesman for the poor, provided scholarly language to frame many of the empirical and intuitive lessons of the new wildlife movement in southern Africa. One of Murphree's best known 'laws' was that bureaucrats grasp power from above while resisting giving up power to those below them (Martin, 2009b). Yet why were wildlife administrators, especially in Zimbabwe and Namibia, doing the opposite? Like my father, I believe that prescient wildlife administrators devolved power because they recognised that in this direction lay the greatest good for the greatest number – including more of the wildlife they loved. These administrators were highly qualified scholar-practitioners. Often having PhDs, being observant and pragmatic, they were the sons of farmers, rural administrators, or extension officers and spent a great deal of time in the field. In these white minority regimes, administrators were respected, accountable politically to farmers and society, and operated in social environments where the metrics of performance were tangible progress, not conferences or statements, and certainly not the superficial celebrity leadership of today. Self-promotion was particularly frowned upon. This was especially so in Zimbabwe, where natural resource governance was scientific and deeply democratic in a well-educated sector that had long emphasised the importance of soil and environmental conservation (Beinart, 1984).

We also note that neither the private nor the public sector were solely responsible for innovation; they both were. The public sector invented new policies, and were better at seeing the big picture, while the private sector was entrepreneurial and single-minded in developing wildlife-based businesses. Interestingly, the scientists often focused on the wrong thing. They invested heavily in evaluating wildlife's ecological advantages over livestock (Talbot et al., 1965b; Taylor & Walker, 1978) and never predicted the key economic breakthrough, which was the shift from low-value meat (commodity) production to high-value safari hunting. Once they owned the wildlife, a few maverick farmers worked this out for themselves, transforming degraded cattle ranches into vibrant wildlife properties because hunting earned much higher profits at lower stocking rates. This experiential learning was only possible once farmers had the legal rights to use wildlife commercially, and explains why southern African's were so wary of the need for 'blue-print scientific certainty' in international forums, and preferred to place their confidence in devolution and adaptive management (Martin, 1999). Extensionists and scientists were important for analysing these lessons and spreading them across the sector, including into further policy reform.

Wildlife ownership as a policy experiment

When I began looking for wildlife properties to study for my PhD in the early 1980s they were rare, but within two decades at least 14 000 private landholders shifted partly or entirely to wildlife-based enterprises in South Africa, Namibia, Zimbabwe, Botswana, Zambia, and, more recently, in Mozambique (Chapter 9). The key to the recovery of wildlife in southern Africa was neither technical nor ecological, but lay in carefully crafted legal and institutional reforms that maximised the value of wildlife to the people who lived with it. If we consider these policy changes to be an experimental treatment, we can compare them to the control situation exemplified by public administration and the London Convention (Table 8.1). While maintaining the Rooseveltian approach for public protected areas, they deliberately and boldly reversed strategies outside parks, applying the concepts of proprietorship and price.

Wildlife generally increased in protected areas in southern Africa from 1970 to 2005, but declined dramatically in Eastern (48%) and Western Africa (80%) (Craigie et al., 2010; UNEP-WCMC, 2012) as illustrated by the data and emptying jars in Figure 8.2. Outside parks, these differences are even starker. Kenya, a premier wildlife country, and the only country that has consistently monitored its wildlife, has lost 67% of wildlife outside parks since the 1970s and much more of some species (Figure 8.3) (Ogutu et al., 2016). By

TABLE 8.1 The sustainable governance approach[7] as a policy experiment

Policy	Control (Business as usual)	Experimental Treatment (Policy reform)
	Public Approach (e.g. London Convention 1900, 1933; North American Model)	**Sustainable Governance Approach** (e.g. new wildlife legislation in southern Africa from 1960 onwards)
Protected Areas	Establish protected areas to provide public access and conserve fauna and flora and public enjoyment, while excluding 'consumptive' use and human settlement	As before, but with an increasing focus on self-funded parks that are engines for economic growth, providing jobs and economic growth in remote rural areas (Child, 2004)
Wildlife Ownership	Centralised in the state	PROPRIETORSHIP: Devolve rights to access, use, manage, and exclude (i.e. exclude others from taking their resources) to landholders and/or communities
Commercial Use of Wildlife	Restrict and/or ban (e.g. demand reduction)	PRICE: Devolve rights of sale to entrepreneurial landholders, and encourage the development of markets and products to make wildlife as valuable as possible (provided use is humane)

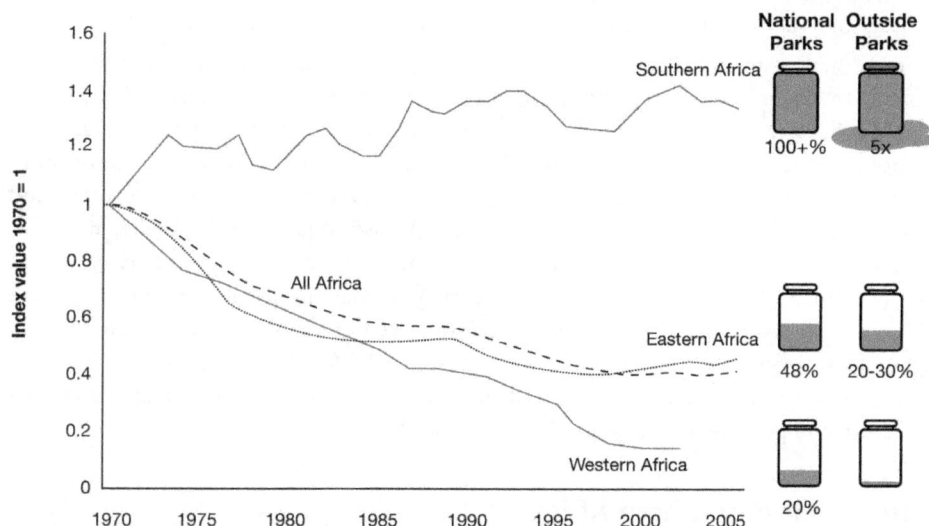

FIGURE 8.2 Trends in wildlife populations in Africa (1970–2005)
Source: adapted from Craigie et al., (2010) (with permission).

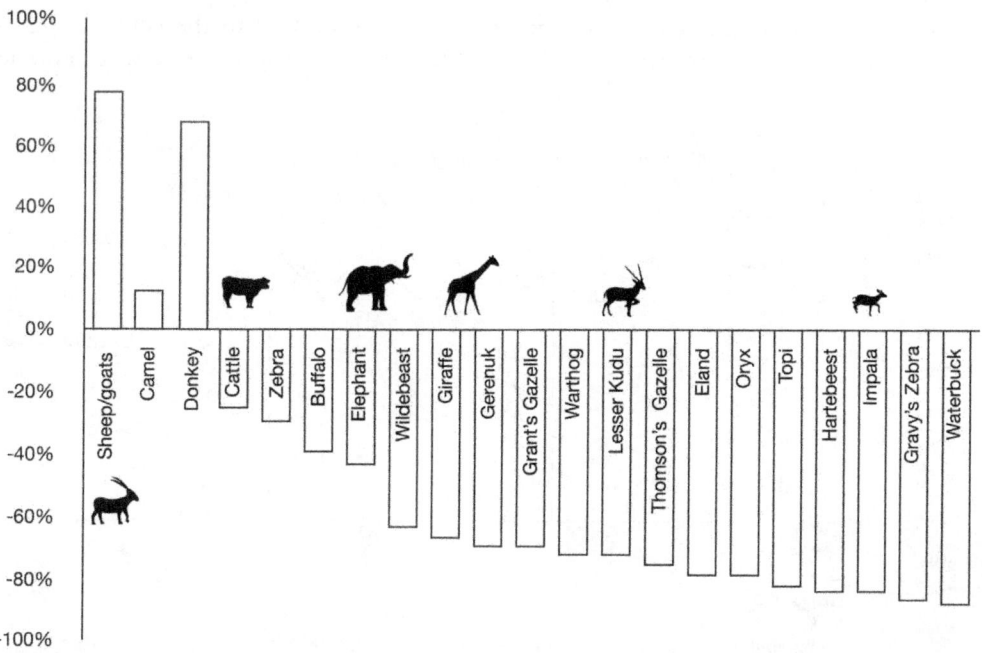

FIGURE 8.3 Trends in wildlife numbers in Kenya
Source: adapted from Ogutu et al. (2016) (with permission).

contrast, wildlife has increased five-fold in southern Africa (Chapter 9). While the under-lying national conditions differ, these changes nonetheless suggest that public models are less effective than new approaches based on devolution.

Ogutu et al.'s remarkable dataset is worthy of further inspection (Figure 8.3). Livestock biomass has doubled since the 1970s, especially hardy animals like sheep and goats (by 76%), camels (by 13%), and donkeys (by 7%), although cattle biomass declined by 25%, presumably reflecting range degradation. Despite all the publicity about poaching and the illegal ivory trade, the decline of elephants in Kenya (43%) was much lower than most other species. The decline was lowest for wildebeest (64%) and highest for Grevy's zebra and waterbuck (87% and 88%), while the decline of giraffe, gerenuk, Grant's gazelle, warthog, lesser kudu, Thomson's gazelle, eland, oryx, topi, hartebeest, and impala fall between these extremes (Ogutu et al., 2016). What is really worrying is that animals that are far less in the public eye, and far less threatened by organised crime, are in fact in much more danger than elephants. This suggests that the biggest threat to wildlife is a combination of non-ownership and non-use because, in terms of biomass, livestock exceeds that of wildlife by a factor of 8.1 compared to 3.5 previously (1977 to 1980).

Land-use trends in southern Africa

Data of this quality is not available in southern Africa. Nonetheless, compiling evidence from the literature (Table 9.7) suggests that wildlife has increased by a factor of more than six on private land, whereas livestock populations on rangelands have halved (Figure 8.4). In a stagnant economy, the average rate of growth in wildlife populations and the wildlife sector is about 6% per annum (Taylor et al., 2016).

Dry (2010) shows the remarkable contribution of private land to the conservation of threatened species in South Africa (Table 8.2). He suggests that species governed more in

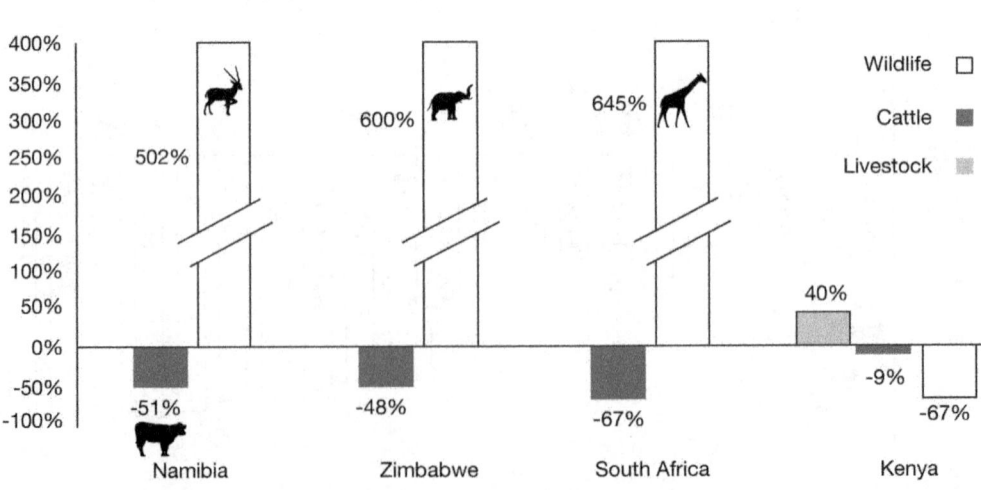

FIGURE 8.4 Comparing wildlife and livestock populations in southern Africa with the Kenyan counter-factual

TABLE 8.2 Trends in numbers of rare wildlife species in South Africa

	1950	*2015*		
	Total	*In parks*	*Private land*	*Total*
White rhino	30	12 000	5 000	17 000
Black rhino	30	1 510	450	1 960
Blesbok	2 000	25 000	225 000	250 000
Bontebok	19	1 000	7 000	8 000
Sable Antelope	450	500	15 000	15 500
Roan Antelope	150	381	4 500	4 881
Cape Mountain Zebra	80	1 925	865	2 790
Black Wildebeest	500	1 800	15 700	17 500
TOTAL	3 259	44 116	273 515	317 631

Source: Dry (2010) (with permission).

the character of old fashioned colonial protection laws (e.g. Cape Mountain zebra, bontebok) have recovered more slowly than species subjected to less bureaucratically encumbered utilisation such as sable and black wildebeest and, until recently, white rhino.

Most of the growth has been on private land because of a reluctance to give communities the same rights as their white counterparts, but where these rights have been devolved (as in Namibia) communities have performed almost as well as their private counterparts. Moreover, with 12 000 landholders continually experimenting with new wildlife products, the range and price of such products continues to increase, whereas the price of commodities such as beef and corn has been declining since the 1960s.

Referring to the price curves in Figure 3.7, southern African policies of devolving ownership and encouraging commercial use have pushed the wildlife profit curve above that of livestock, guiding landholders to devote more land to wildlife and less to livestock. By contrast, the potential of Kenya's wildlife is significantly higher than in southern Africa. However, Kenya has not devolved wildlife rights, despite significant pressure from landholders to do so over many years. This has restricted and banned markets for the meat and hunting which has financed 60–80% of wildlife in southern Africa where, lacking the scenery and scale of Africa's spectacular parks, ecotourism only provides about 5% of the income to private wildlife landholders.

BOX 8.3 WILDLIFE ECONOMICS AND THE INCOHERENCE OF 'DEMAND REDUCTION'

The economics of wildlife policies are strangely incoherent. Environmentalists promote payments for ecosystem services to boost the value of wild resources, yet simultaneously destroy the price of wild resources through 'demand reduction'. There are also serious conflicts between special interest groups that have closed the legal markets for rhino horn and landholders who look after half the world's rhinos and who want legal markets to pay the high costs of protecting them (Rubino & Pienaar, 2018).

How do we interpret these contradictions? First off, price is not the primary variable. Proprietorship is. Ownership (i.e. security of person and property) separates an economy of plunder from one of production, exchange, and investment. If communities own wildlife, a high price provides powerful incentives to protect the wildlife. If they do not own wildlife, this resembles a frontier economy, and high price combined with few controls on offtake are likely to lead to the decimation of wildlife. The key factor determining if price will increase or decimate a species is excludability and ownership (Chapter 5), suggesting that the global obsession with price and demand reduction is misplaced. We see this in the examples provided in this chapter. Wildlife is traded globally, but the effect of this trade depends on local policies about wildlife ownership. In the very same global markets, wildlife in southern Africa is increasing (where it is owned) while that in the rest of Africa is declining. This suggests that we would be far more successful if we promoted wildlife ownership, while recognising that markets are important, but very much secondary. For instance, would CITES become more successful if it adopted a policy of allowing trade wherever 100% of the profits from wildlife were returned to the landholder (state, private, or community), but questioning trade where these governance conditions did not apply?

Price is a second-order consideration, and a double-edged sword. High prices of ivory and rhino horn will decimate these species where policies shift this value away from landholders and into government coffers or criminal markets. By contrast, if these prices get to landholders, it is highly likely that they will protect their rhinos, as they have done so well for other species. Rhinos are like a sheep with a golden fleece. The horn regrows each year, and is literally worth more than gold – about $50 000 per rhino per year. If this value was incentivising rhino owners, the rapid expansion of rhinos on private land since the 1950s would accelerate. By contrast, with the current policy of demand reduction, landholders are divesting their rhinos because much lower benefits do not offset much higher costs and the dangers of facing off with armed criminals.

Conclusions

The shift in wildlife policies and threats described above is summarised in Figure 8.5. Over-exploitation in frontier economies led to public management of wildlife and parks (1). However, the expansion of domestic agriculture threatened the connectivity between parks (2). The southern Africans responded with a new paradigm for wildlife governance (3) – building sustainable wildlife landscapes by combining the privatisation of wildlife on private and community land, with a more aggressive role for parks as engines of economic growth in remote rural areas (Child, 2004).

The policy reaction to the late Victorian slaughter of wildlife occurred at a critical juncture when government administrative capacity was also emerging; this led to the presumption that an external Leviathan should assume responsibility for biodiversity conservation (Ostrom, 1990). For over 100 years, policy has stagnated, resulting in a mismatch between the private and common-pool nature of wildlife, and its public ownership and management. Paradoxes such as wildlife being priceless but worthless, and economic incoherencies like demand reduction for species that are being replaced because they cannot compete for land (Box 8.3), suggest that there is something radically wrong with the rules of the game.

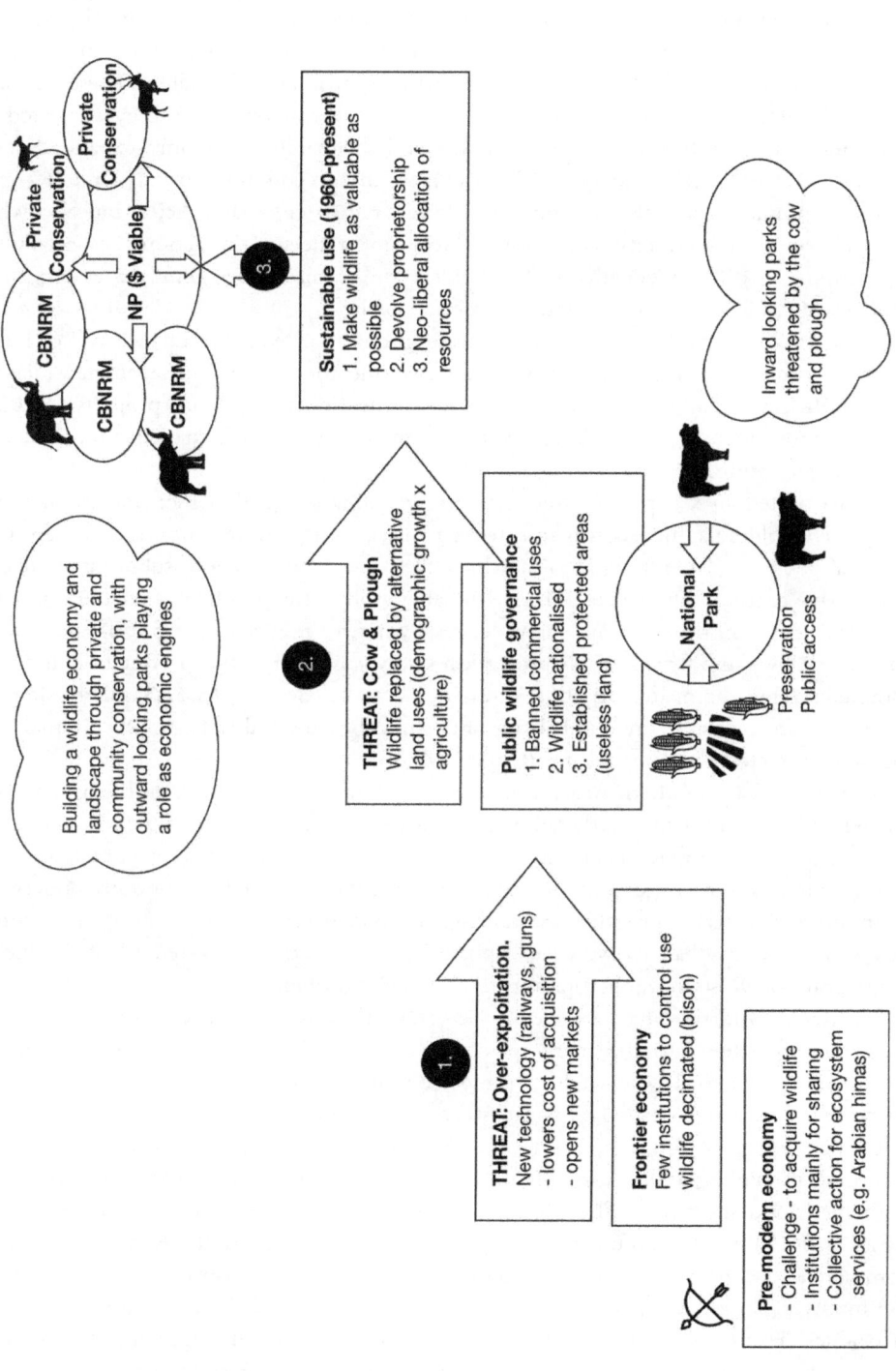

FIGURE 8.5 Changing wildlife policies and threats

Outside of North America, taking wildlife out of the marketplace has not been a successful strategy. By contrast, the experiences from southern Africa shows that reinserting wildlife into the economy can be effective, through new governance and property regimes that are much better aligned to the natural attributes of wildlife. These measures translated wildlife's comparative advantage into land use outcomes (i.e. they corrected the economic curves in Figure 3.7), and wildlife and the wildlife economy expanded at 6% annually. Where policies did not seek to correct these economic curves, wildlife's rapid decline continued, not only in countries that adhered to colonial policies, but also where the rhetoric of community conservation was not reflected in actions, especially the devolution of 100% of benefits to landholders and communities (Lindsey et al., 2014). The loss of wildlife is avoidable, rather than inevitable.

Southern Africa wildlife policy has far more in common with the English Bill of Rights of 1689 than with North America's public doctrine for wildlife management. Like the English Parliament in 1689, it legally recognises the rights of person, property, and self-determination, focusing on devolved resource ownership, wildlife markets, and collective local self-regulation.

As we noted in Chapter 6, free markets are designed, and design affects outcomes. Progressive wildlife administrators in southern Africa developed institutions that were good for rural people and the wildlife with which they lived. Nonetheless, subtle differences in design have profound influences. Thus, Zimbabwe solved the problem of wildlife's mobility with collective institutions and few fences, whereas the South African landscape is fragmented by game fences. High-protection status was less helpful to bontebok and Cape Mountain zebra, compared to more utilised species in South Africa. Similarly, leopards recovered on farmlands in Zimbabwe once they became valuable trophy animals, but specially protected cheetahs did not (Child, 1995).

The lesson from southern Africa is that the antidote to conservation problems may be to put wildlife back into the market, not to remove it. This requires a much better understanding of the economic characteristics (Chapter 5) and economics (Chapters 6 and 9) of wild resources and their governance. Unfortunately, these principles are poorly understood. Economically logical approaches are said to be controversial. Grossly simplified solutions may sound catchy ('ban ivory, save elephants'), but despite a track record of decline the momentum of Western public opinion remains behind them.

However, wild resources are clearly of great value. Reframing economic institutions to bring this value into the marketplace benefits wildlife and the people who live it. With some technical knowledge, these economic rules are fairly easy to design. However, the greater challenge by far is the political process of designing and protecting rules that work. Fortunately, a set of examples based on local control and benefits is emerging to highlight that this different (and contrarian) view of the future may have both merit and substance. The positive experience of wildlife conservation in southern Africa is not alone. Community forestry in the Himalayan foothills (Agrawal, 2005), the commercial use of vicuna in the Andes (Jacobsen, 2012), community management of the massive fish, piraracu, in the Amazon (Koziell & Inoue, 2006), the sustainable use of crocodiles (Hutton & Webb, 2003), amongst others, are small bright lights that should lead us to re-evaluate our scepticism, or even abhorrence, of rights based and market-smart solutions.

Notes

1   Wildlife is increasing in North America following the public model described in Chapter 7. It is also recovering in parts of northern Europe, often linked to a combination of hunting and forestry.

2   The genesis of wildlife conservation in southern Africa can be traced back to Aldo Leopold. All three Fulbright scholars studied under the legendary Starker Leopold at the University of California, Berkeley, and went on to have major impacts on global wildlife conservation. Thane Riney was chief scientist to IUCN, and headed the FAO Special Project. Ray Dasmann was also chief scientist at IUCN, played a significant role in the Stockholm Conference on the Human Environment, and initiated the Man and Biosphere Program at UNESCO. Archie Mossman specialised in game management and wildlife utilisation at Humboldt State University. They, in turn, mentored leaders like Graham Child, forming a strong link between policy experimentation in southern Africa and global conservation through Riney and several IUCN Director Generals who were closely connected to Africa – especially Lee Talbot (who did experimental game cropping in East Africa) and Martin Holdgate (who wrote the forward to Graham Child's *Wildlife and People*).

3   These intuitive policies were developed by innovators such as Thane Riney, Bernabie de la But (Director of Parks in Namibia), Archie Frazer, and others, and came to be known as 'the spirit of the Act'. As the intellectual leader of this new approach, Graham Child develops their underlying philosophy in his *chef-d'œuvre, Wildlife and People*. These men revised policy instinctively, and through observation, without the scholarship that we lean on today. In his later years, Graham Child was an avid student of the scholarship of wildlife economics, property rights, and common property theory, serving to validate and conceptualise the ideas developed in long discussions with colleagues and landholders in more theoretical language.

4   This box is modified from articles I wrote for *Wildlife Rancher* and *Pachyderm* that were published as obituaries to Graham Child.

5   I use the term 'Bushman' happily, having been told on several different occasions, and with some pride, that 'I am a Bushman', not a Khomani-San (or some such label).

6   Resolution Conf. 8.3 (Rev. CoP 13), https://cites.org/eng/res/08/08-03R13.php

7   In southern Africa, this is often called the 'sustainable use approach', and concerns wildlife ownership, markets, and the rights of landholders and communities. However, the term 'sustainable use' is also interpreted biologically, with an emphasis on the biology of offtakes, maximum sustainable yields, and so on. The mechanics of use, and whether the harvest of flow exceeds the ability of the capital stock to reproduce, is important, but is secondary to the incentives to manage use sustainably. Therefore, I have rebranded these concepts as 'the sustainable governance approach' because they focus on governance, with the sustainability of use being a secondary, and technical, issue.

9

ASSESSING THE ECONOMICS OF WILDLIFE

Tools and lessons

LEARNING OBJECTIVES

This chapter illustrates the practical use of a number of tools for assessing the economics of wildlife in dryland savannas in southern Africa. These include:

- Assessing the viability of cattle and wildlife enterprises using gross and net margin analysis, and return on investment.
- Historical case studies to understand a sector (e.g. beef) or the complexity of an enterprise where profits depend on the environment (e.g. wildlife and cattle production on Buffalo Range ranch).
- Policy analysis matrices to understand if wildlife has a comparative advantage, and to what extent policies tilt the economic playing field for or against wildlife.
- The use of multipliers to assess the total economic value of a protected area and the economic implications of this.

Together, the results of these analyses provide a strong case for rewilding drylands to create jobs, make profits, and expand the economy.

This chapter describes the economic methods I used to understand and build a case for wildlife in Zimbabwe and southern Africa. Together with economic ideas presented in Chapter 6, I hope it removes the conservationist's fear of economics and illustrates how relatively simple economic methods can be used to analyse a wild resource sector in some detail, and manage it better.

The white settler state of Zimbabwe is the background to this economic assessment of the relative merits of wildlife and livestock in the semi-arid rangelands of southern Africa. When hopes of vast mineral wealth never materialised, Zimbabwe (then

Southern Rhodesia) began to invest in agriculture, including agricultural research stations and journals, within a few years of the country being settled. The country's agriculture potential is high in the croplands on the 'Highveld' around Harare (Natural Region II) but much lower in the much more extensive drylands (Natural Regions IV and V) where variable rainfall of less than 600 mm annually limits production to livestock and wildlife (Box 9.1). Hoping to rival the beef industry of Argentina, the country invested heavily in livestock. White settlers acquired ranches large enough to carry 1 000 head of cattle (about 10 000 hectares), especially after World War II when military veterans were given farms. Tribal Trust Lands were established, initially to protect black people from land grabbing by the white settlers in their traditional areas of lighter soils, but these areas later became notorious for restricting the rapidly growing black population to lower value farmland.

BOX 9.1 A BRIEF HISTORY AND GEOGRAPHY OF ZIMBABWE

Zimbabwe, a small, landlocked country in central southern Africa, is the location of one of Africa's earliest civilisations (Reader, 1999). Around 2 000 years ago, the southern migration of the Bantu people displaced the original bushmen, developing first the Mupungubwe civilisation on the Limpopo River (c.1220) and, later, the substantial stone cities of Great Zimbabwe, Khami and Dhlodhlo and the Monomotapa civilisation (c.1450). These were founded on livestock, agriculture, gold mining, and trade with the East, especially gold and ivory.

In the early 1800s, Chaka Zulu's warfare disrupted South Africa. Fleeing northwards, Chief Mzilikazi settled the warlike amaNdebele at Bulawayo in western Zimbabwe with their livestock and annual raiding parties. In 1890, in pursuit of more gold, Cecil Rhodes's British South Africa Company financed a white settler state. With self-government in 1923, this 'development company' invested in a meritocratic civil service, agricultural research (the *Rhodesia Agricultural Journal* was first published in 1903), infrastructure, and education. These merits were over-shadowed by the dark cloud of racially biased policies. The white government declared a unilateral Declaration of Independence in 1965 and, following a civil war (1972–1980), black majority rule came to Zimbabwe in 1980 under Robert Mugabe. The first 15 years of independence were innovative and productive, but when Mugabe began to lose elections and referendums, a dynamic economy was replaced by an oppressive, personalised, and extractive state.

Land apportionment in Zimbabwe's Five Natural Regions

Zimbabwe is defined by two maps. In 1960, Vincent and Thomas (1961) categorised the country into five agro-ecological regions:

- Regions I and II (21%) on the well-watered Highveld were suited to intensive farming including maize and tobacco.
- Region III allowed a mixture of livestock and rain-fed agriculture.
- Regions IV and V (65% of the country) had low and unreliable rainfall suited to 'extensive' agriculture, a euphemism for livestock. This is where the wildlife economy blossomed.

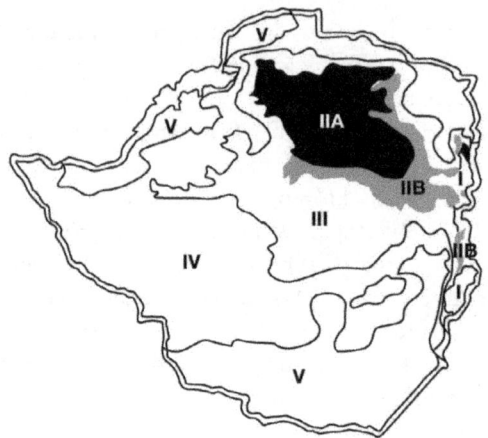

In 1930, the controversial Land Apportionment Act of 1930 divided the country through race-based tenure: free-hold white farms, Tribal Trust Lands, Native Purchase Areas, and Crown or National Land including most parks and reserves. Historically, the Shona lived mostly in an arc of easily tilled sandveld in Region III, and the Ndebele favoured the cattle range in Region IV and V. White farmers settled in the fertile and heavy soils of the Highveld, and also in 'good cattle country' in Regions IV and V. However, as technology (e.g. the introduction of steel ploughs) and populations changed, the ownership of the most 'productive' land by whites came under scrutiny and criticism.

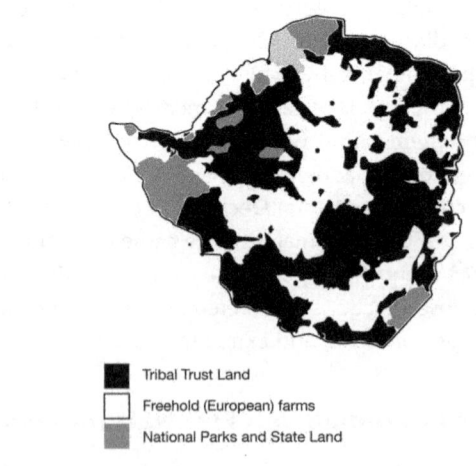

Studying game ranching and private conservation in Zimbabwe

In 1984, I was recruited by Zimbabwe's wildlife agency to support the private wildlife sector with extension and economic research. Having been involved in widespread rewilding of private lands, I can speak positively about the transformative role of economic research and economically informed policies. On my first field trip in my new job, I visited

Peter Seymour Smith's Iwaba game ranch. I still have the picture I took with an old Pentax government-issue camera (Figure 9.1), contrasting the abundant grass and healthy game ranch with serious overgrazing on the cattle ranch across the fence. This presented a land use paradox, a research question, and a career goal. Why were healthy wildlife systems being replaced on such a huge scale by domestic beef production, which seemed so unsustainable? Was this a market failure, and could we bring back wildlife if we got the prices right?

At this time, the politically powerful beef sector was scapegoating wildlife as it fought to survive in the face of declining environmental productivity and declining world beef prices, resisting any suggestion that wildlife was a legitimate land use and actively eliminating all buffalo on private land. While this chapter contrasts wildlife and livestock production, it is not intended to be anti-livestock or pro-wildlife – both have their place in drylands. I will provide methods and concepts for assessing the viability of a land use option to guide policy, although we cannot escape from the irony that, in the final analysis, wildlife saved many livestock ranchers from ecological and financial bankruptcy.

Understanding wildlife production ('game ranching')

The early research on utilising wildlife focused almost entirely on how much meat it could produce, the quality of lean game meat compared to marbled beef, whether wildlife was less harmful to the environment, and whether or not it had physiological or behavioural adaptations to Africa's harsh environment (Talbot et al., 1965; Mossman & Mossman, 1976; Taylor & Walker, 1978; Walker, 1979). What was missing was an understanding of the economics and economic history of drylands. With this in mind that I spent six years interviewing game and cattle ranchers, and delving into the archives to understand the political economy of these sectors.

Seymour Smith introduced me to a number of his neighbours, and I began to piece together the history and use of these drylands. At this time, all 22 ranchers (except Seymour Smith) farmed cattle, dividing their 10 000 hectare ranches into cattle paddocks. Before white settlement, this had been excellent big game country. However, by the time of my visit, the range was depleted and grazing species (sable, tssesebe, waterbuck, wildebeest, and zebra) had declined or disappeared. Big game and predators (elephants, buffalo, and lions)

FIGURE 9.1 Fence line photo comparing game and cattle ranches (1984)

had been eliminated by the late 1960s, including by government hunters. With cattle grazing causing the bush to thicken up, there were a small but significant number of browsing herbivores (eland, kudu, impala, bushbuck), plus some of the small mammals (steenbok, bush pig, warthog, klipspringer).

From the literature, I had expected to find that ranchers were selling game meat and skins. Instead, landholders were making money from mini-safaris, throwing most skins away, and selling meat only to recover costs. Ranchers were hosting American and German hunters, normally for a week, to shoot a small bag of animals, including one or two browsers with good trophy heads (kudu, bushbuck, eland), a grazer or two (zebra, wildebeest, reedbuck, etc.), plus a few smaller animals (warthog, duiker, etc.).

At the time, armchair opinion was that ranchers would eliminate wildlife if they were allowed to use it. This was far from the case. Ranchers were passionate about their wildlife, spending many hours proudly showing me how they were improving their wildlife populations and habitats. Fortunately, the chief extension officer for the Midlands (the highly energetic Bob Vaughan Evans, who I introduced in Chapter 8) conducted a baseline survey of wildlife populations in 1975 when the ground-breaking Parks and Wildlife Act gave farmers ownership of wildlife. Comparing my data with his (see Table 9.1) showed that browsers and common animals like duiker and warthog remained widespread, and were widely utilised, but that the range of large grazing animals was expanding rapidly (by 29% in a decade that included six years of civil war). Vaughan Evans's 1975 data did not include wildlife numbers, but farmers were unanimous in claiming that the increase in wildlife numbers was far greater than the expansion in range.

Most ranchers shared their audited accounts or enterprise budgets for cattle and wildlife. We analysed these together using gross and net margins of their enterprises (Table 9.2), and I made a point of returning results to ranchers regularly at farmer field days.

Of the 22 ranchers I visited in the Midlands, 21 were primarily cattle ranchers, but 12 of these were outfitting safaris themselves, and six sold trophy animals to outfitters. Although less than half of the offtake of 1 873 wild animals was by hunting clients, trophy hunting generated 85% of income compared to only 15% from meat and hide sales.

These were still cattle ranchers, and 80% of the total biomass was cattle. On average cattle earned $13.57 per hectare, but after deducting variable costs profits were low ($4.52/ha) (Figure 9.2). By comparison, wildlife earned $3.36/ha but most of this was profit ($2.93) because wildlife requires few inputs.

As an ecologist, I was aware that the limiting factor in drylands is grass fodder production. As an economist, the important question was if ranchers should use this grass for cattle or for wildlife. The data showed that cattle earned 7 cents per kilogramme of livemass compared to 17 cents from wildlife (livemass reflects how much grass is consumed). At the margin, therefore, Midlands ranches increased their profits by 10 cents for every kilogramme of cattle they switched to wildlife (Figure 9.2). Not surprisingly, wildlife expanded rapidly in the next decade, with several properties changing completely to wildlife and even introducing rhinos.

I then surveyed 15 properties (446 818 hectares) in the drier South East Lowveld, which was settled in the 1950s to promote cattle ranching. Nonetheless, by 1986 two-thirds of these ranchers were managing all or part of their ranches exclusively for trophy hunting. Wildlife had a financial and ecological comparative advantage. It generated 64% of profits from only 32% of the biomass, and was also allowing the land to recover from over-grazing. Hunting earned profits of 32 cents per kilogramme, but livestock only earned 7 cents,

TABLE 9.1 Changes in wildlife ranges in the Midlands (1975–1984)

Species on Midlands ranches	Occurrence in 1984 (% of properties)	Range expansion, 1975–1984 (%)	Wildlife utilisation (% of properties)	Comment
Impala	100	0%	91%	Some browsers survive cattle
Bushbuck	55	1	42	pressures to provide initial wild-
Kudu	100	0	91	life income through hunting
Eland	86	36	63	
Browsers total	*5*	*9*	*72*	
Wildebeest	45	17	100	Once they become valuable, the
Zebra	86	21	68	number and range of large
Waterbuck	91	41	50	grazing species expands to
Tsessebe	68	43	53	diversify wildlife and increase
Sable	73	21	75	value of hunting
Grazers total	*73*	*29*	*69*	
Duiker	100	0	36	These small animals survive in
Steenbok	100	0	27	moderate numbers with cattle
Bushpig	100	0	55	and provide initial wildlife
Warthog	100	0	68	income through hunting
Klipspringer	23	−13	60	
Reedbuck	100	12	55	
Small game total	*87*	*0*	*50*	
Leopard	91	33	20	Predators recover as landholders
Cheetah	100	75	0	switch to wildlife

Source: modified from Child (1988).

TABLE 9.2 Simple financial calculations for comparing the profitability of wildlife and livestock

Category	Amount
Gross income (GI)	100
• Sale of livestock	
• Sale of wildlife trophy animals	
Variable costs (VC)	40
• Fuels and oils	
• Vehicle repair	
• Cattle feed	
• Food/guiding of hunters	
Gross margin (profit) = GI–VC	60
Fixed costs (FC)	40
• Manager's salary	
• Depreciation on equipment	
Net margin (profit) = GI–VC–FC	20

Survey of 22 Ranchers, Zimbabwe Midlands, 1994

	Calculation	Units	Cattle	Wildlife
Turnover	Total value of sales	Z$/Ha	$13.57	$3.36
Profit	Sales – costs	Z$/Ha	$4.53	$2.93
Profit at the margin	Profit/kg biomass	Z$/Kg	7c	17c

Cattle versus Wildlife in the Lowveld in 1996

	Livemass (kg)		Turnover ($)		Profit ($)	
Cattle	10.1m	68%	$2.0m	60%	$0.7m	36%
Wildlife	4.7	32%	$1.4m	40%	$1.2m	64%

Profitability of Cattle and Wildlife Enterprises on 15 Lowveld Ranches (1986) 371 043 ha

	Z$/Ha	Z$/kg livemass
Wildlife	$4.47	32c
Cattle	$2.93	7c
Game meat production (on cattle ranches)	$1.16	11c (overheads costs carried by cattle)

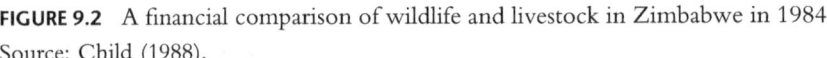

FIGURE 9.2 A financial comparison of wildlife and livestock in Zimbabwe in 1984
Source: Child (1988).

which forced cattle ranches to over-stock in desperation. Impala thrive on degraded cattle ranches, and ranchers often cropped excess wildlife for meat, and sold a few trophy animals to outfitters. However, these ranches only earned 11 cents from wildlife (if none of the overhead costs are attributed to wildlife), with the implication that meat commodity production was not viable, regardless of whether wildlife or livestock was involved. The money lay not in commodity production, but in the bio-experience economy of high-value animals and the provision of outfitting or guiding services.

The devastating 1992 drought was a tipping point, with high cattle mortality and feeding costs bankrupting most cattle ranchers. The rewilding of this land is summarised in an excellent documentary about the Savé Valley Conservancy (Taylor, 2002). In this film, a leading rancher (Clive Stockil) recounts the massive support given to cattle ranching since the 1950s, with overgrazing pushing land into environmental overdraft, and the elimination of wildlife to make way for livestock following the adage 'You can't farm in a zoo'. My data, supported by Pricewaterhouse (1994), concluded that wildlife would earn considerably more than cattle, provide more jobs, and allow environmental recovery. Meat production was not viable without damaging these drylands but hunting increased per-biomass earnings four-fold. Occasionally, where there was enough wildlife and scenery to attract tourists with cameras, tourism provided a cherry on top. By 1998, 27 ranchers (including many that I surveyed) combined their land as the 344 200 hectare Savé Conservancy, the largest private wildlife reserve on earth (Lindsey et al., 2009). They removed all their cattle, pulled down 6 000 km of internal cattle fences, and brought back much of the wildlife they had only recently eliminated, including 600 elephants, lions, buffalo, giraffe, rhinos, and other animals.

In Zimbabwe, hunting paid for close to 100% of the recovery of wildlife on some 1 000 properties, highlighting the misconception that ecotourism can flip rangeland economies back to wildlife – it is nice to think that tourism can pay for wildlife's recovery, but tourists only visit the very best wildlife areas. It may seem strange and contradictory to suggest that hunting allows wildlife populations to recover, so I need to explain this. When cattle ranchers turned to wildlife for financial survival, they profited by hunting a few large, high-value trophy males. The market for hunting is demanding, and if ranchers got greedy and shot more than 2–3% of the herd, they would soon find themselves hunting younger, non-trophy animals and ruining their reputation with hunting clients. Consequently, trophy quotas are usually set at 2–3% of a herd expanding at 20–30% annually, allowing ranchers to hunt profitably while their wildlife recovers rapidly. Without the fees from hunting (Box 9.2), as much as 80% of wild land in southern Africa would revert to livestock.

The progression from cattle ranching to large game reserves takes several decades and financial acuity unless large amounts of money are available. Initially, wildlife begins to recover on cattle ranches once they introduce mini-safaris and, with many ranchers doing this, the range and variety of species also increases, as we saw in the Midlands example above. Where they have the financial resources, some ranchers reintroduce game that had been eliminated in the name of cattle ranching – buffalo, elephant, lions, and rare or sensitive grazing species such as sable, rhino, and roan antelope. Similarly, in some landscapes, ranchers have combined their land as huge conservancies, which closely resemble national parks. Even then, it is quite difficult for tourism lodges and game viewing to pay, and they usually supplement hunting as the main breadwinner except in rare circumstances. Counter-intuitively, it is not the loss of wildlife that is the problem on private land, but over-abundance and range degradation, especially where there is tourism. Ranchers crop and sell excess animals like impala for meat, more for ecological reasons than for profit because, after 60 years, reliable markets for game meat have not yet developed and game meat cropping remains marginal. Excess elephants or lions pose more of a problem, because they impose large costs on the environment and other animals, are less profitable than other species because of these costs, yet are highly political.

BOX 9.2 HOW DOES SAFARI TROPHY HUNTING WORK?

Ranchers understand a great deal about the number and health of their wildlife, especially if they spend a lot of time hunting. Each year they calculate an offtake quota for trophy animals, usually 2% for large animals like kudu or buffalo, and 3% for smaller animals like impala. Theoretically these calculations can be based on wildlife numbers, but counting animals is difficult, and trends in trophy quality and field experience (i.e. it was easy/hard to find a big kudu last year) are usually more important. For marketing purposes, quotas are then split into 'bags', including one or two key animals (a buffalo, sable, or leopard), some larger plains game (wildebeest, zebra), and small animals (impala, warthog, etc.) to act as fillers.

In general, two fees are paid for hunting. The trophy fee is a payment for the animal, and is usually paid to the landholder to reinvest in wildlife conservation. The daily rate covers outfitting services such as tented accommodation, bush lodges, catering, the services of a qualified professional hunter (PH), 4x4 bush vehicles, and experienced trackers and skinners. On many ranches, the farmer is also an outfitter and keeps both the trophy fee and the daily rate.

Outfitters, many of whom are ranchers, travel to hunting shows in Europe and America to book clients, or book them through agents. Hunting clients purchase animals based on their hunting preferences, prices, the reputation of the hunter, and the reputation of the area in terms of quality of the trophy and the likelihood of getting it.

Hunts are marketed in many ways, with big-game and plains-game safaris being popular. A classical big game safari costs the client $2 000 or more per day for 18–21 days (regardless of how long he actually hunts). He pays a separate fee for each animal, including expensive 'big game' like elephants ($20 000 +), lions ($20 000+), leopards ($7 500 +), or buffalo ($7 500+). Most big game hunters will also take a bag of 'plains game', such as kudu, oryx, zebra, or wildebeest. By contrast, a plains game hunt is shorter (7–10 days) and much more affordable, costing $300–600 per day, plus trophy fees for one or two large animals (sable, kudu) and a few smaller ones (impala, warthog).

On the hunt, the client usually rises very early in the morning and loads up onto a hunting vehicle with the PH, trackers, and skinners. When they find tracks, they follow their quarry on foot, often stalking for many hours. For high value species, the team often tracks a number of animals over days or even weeks before they find a suitable animal to shoot, and they do not always succeed. With lower value plains game, hunters usually find their prey in a couple of days.

Once the animal is killed, it is loaded onto the vehicle and taken back to camp. The meat is doled out to staff or neighbouring communities. The skin and trophy is carefully cured and is then sent to a taxidermist for mounting. The PH records the date, location, sex, and trophy quality of the animals killed. These records are used for payments and to support biological management and quota setting.

Buffalo Range Ranch: using a case study to understand the interactions between environmental and economic factors on a game and cattle ranch

In highly variable non-linear savannas, it is notoriously difficult to incorporate the environmental costs and benefits of livestock and wildlife production into economic calculations without long-term records. I was fortunate that the Style family of Buffalo Range Ranch, who raised cattle and pioneered wildlife for over 30 years, kept fastidious records and were eager to share them with me.

George Style obtained a 20 000 hectare ranch on the Chiredzi River when he returned from World War II. The river provided the only perennial water in the landscape, and the land near the river had been degraded over the years. Style fenced off 8 000 hectares of this land for the wildlife he was so passionate about. He divided 12 000 hectares of healthy land away from the river into cattle paddocks and provided artificial water (Figure 9.3). George and his son, Clive, were keen cattlemen but also pioneered game production and safari hunting, building a hunting lodge on their property. They kept fastidious records of every cow or wild animal killed and sold for nearly 30 years, complemented by regular game counts. They also encouraged research on their ranch, including ecological surveys of the cattle and wildlife sections in 1973, 1986, and 1990 (Taylor & Walker, 1978; Child, 1988; Taylor & Child, 1991).

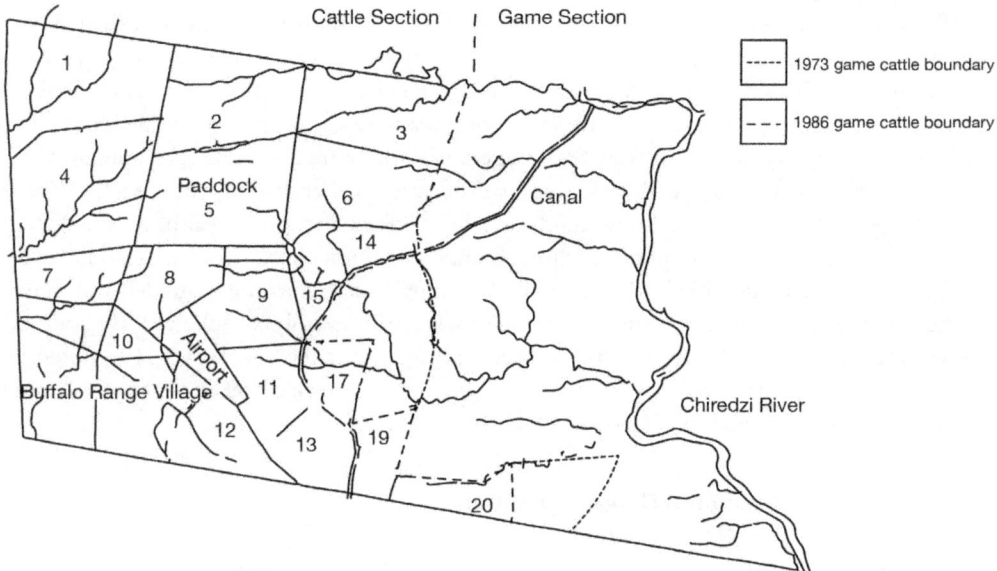

FIGURE 9.3 Map of Buffalo Range cattle and wildlife sections
Source: Child (1988).

In the early 1980s, the very notion of range degradation was being challenged by prominent British academics (Sandford, 1983; Schoones, 1994) who were being taken seriously by policy-makers. By contrast, Zimbabwean agriculturalists were obsessed with soil erosion and healthy veld (Beinart, 1984; Stocking, 1985). There was also a long-standing debate about whether wildlife could produce more meat than livestock, and whether multi-species systems were better for the environment.

TABLE 9.3 Meat production from cattle and wildlife reflects range condition on Buffalo Range

	Status of ecological capital	Beef meat yield (kg/ha/year)	Game meat yield (kg/ha/year)
1962–1972	Cattle section healthy and heavily stocked. Degraded habitats used for wildlife.	5.42	3.82
1971–1978	Good rainfall years, with cattle and wildlife sections in similar condition, but cattle range deteriorating through over-stocking, whereas wildlife section was recovering (Taylor & Walker, 1978)	8.32	6.14
1978–1982	Wildlife section in better condition (Child, 1988)	6.37	7.36
Overall	Yield per hectare	6.36	5.01
	Yield per kg livemass	0.128 kg	0.119 kg

Source: adapted from Child (1988).

   The question of whether wildlife or cattle produced the most meat was, as it turned out, the wrong question. Meat production reflects the condition of the range far more than the species using it (Table 9.3). Both cattle and game ranchers were, in reality, grass farmers. Cattle production thrived for 10 or 12 years with heavy stocking in the new paddocks, but then declined rapidly as grass tufts were damaged, precious rainfall ran off bare soils, and tree density increased from 3 000 to 12 000 stems per hectare, further shading out the grass. Cattle calving rates, which are a well-known indicator of livestock and range health, declined from 70% to less than 40%, mirroring the decline in range condition over a 25-year period (Taylor & Walker, 1978; Child, 1988) (Figure 9.4). Put simply, overstocking degraded the range, reduced livestock productivity, and pushed livestock profits steadily into the red. By contrast, game meat productivity improved steadily on a recovering range (Table 9.3). Promoted by the 1992 drought, many farmers in the Lowveld recognised that beef was not viable, and switched to wildlife which was (Taylor, 2002; Bond & Cumming, 2006).

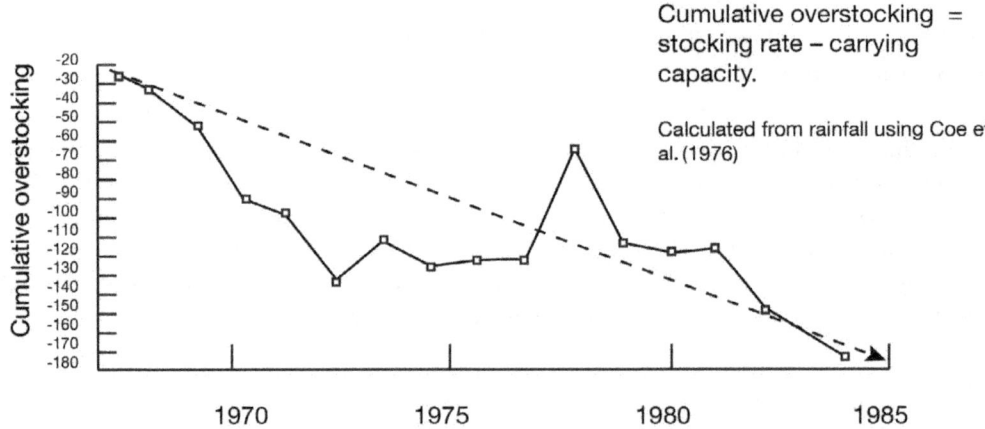

FIGURE 9.4 Comparing cattle productivity (calving rates) and range condition on Buffalo Range
Source: Child (1988).

Return on investment using financial and economic prices

Despite this accumulating evidence, agriculturalists and vets continued to block the wildlife sector. The wildlife sector consequently undertook a major study of ranch profitability (Jansen et al., 1992) using detailed financial records from 184 cattle and wildlife enterprises on 139 ranches (Table 9.4). This study calculated return on investment by dividing profit (i.e. sales minus costs) by the capital value of all the fences, water points, buildings, and vehicles supporting the business. Almost half (48%) of wildlife enterprises were profitable, with a return on investment greater than 10%, compared to only 5% of cattle businesses. Most wildlife enterprises were new and heavily dependent on sport hunting income, with a few minor tourism enterprises. The average return on investment from ranches with access to a wider spectrum of game, especially buffalo, was 17.3%, indicating the real potential of wildlife. Adjusting farm accounts using economic or shadow prices showed that price and exchange controls in the highly controlled Zimbabwean economy were undermining both wildlife and livestock production.

Cattle ranching

The preceding evidence gained little traction with the powerful agriculture and veterinary sector, who continued to deny the limitations of dryland beef production. Ignoring the growing potential for wildlife, they placed veterinary restrictions on game meat and investing in massive fencing projects and the eradication of buffalo (which halved wildlife profitability). This called for a deeper understanding of the history and economics of the livestock sector.

In the late 1890s, the rinderpest pandemic swept down through Africa and wiped out most of the cattle and cloven-hoofed wildlife in Zimbabwe. Comparing the recovery of the communal livestock with the white-owned commercial herd provides considerable insights into the livestock economy (Figure 9.5), with the fortunes of commercial livestock reflecting global market trends, but communal cattle telling us little about economics. Communal cattle provide multiple values (including as a store of wealth or for draught power), but are not good for gauging livestock economics because these values are privatised, whereas costs of open-access rangeland, free services, and so on are externalised to society.

TABLE 9.4 Return on investment from cattle and wildlife enterprises in Zimbabwe 1989/1990

Return on Investment	Wildlife Profitability: financial (economic)	Cattle Profitability: financial (economic)
Profitable (> 10%)	48% (79%)	5% (54%)
Marginal (< 10%)	38% (18%)	59% (35%)
Unprofitable (negative)	14% (3%)	36% (11%)

Financial value refers to the actual prices received by cattle and game ranchers.
The economic prices are based on assumptions about the shadow prices of imports, exports, and local factors in the heavily controlled Zimbabwean economy of the time.

Source: adapted from Jansen et al. (1992)

After the rinderpest, both herds recovered rapidly on grassland that was healthy and well rested. Although the white-farmer government dreamed that Zimbabwe would become a new Argentina, the commercial herd stagnated in the Great Recession of the 1930s. Economic recovery after World War II was reflected in world beef prices, and the herd grew steadily, with the government providing services and subsidies to establish new ranchers on the land, often veterans returning from the battlefields of Europe, North Africa, and Asia.

By this time, the communal lands were over-stocked, and concerned environmentalists enforced a highly unpopular programme of compulsory destocking. The growth of the communal herd from the early 1930s until the late 1960s was stilted by compulsory sales through a system of pricing that also cross-subsidised the commercial herd. Not only is one of the best ways to make money from cattle by buying thin cows which fatten up quickly, but the prices set by the national Cold Storage Commission also favoured 'A' grade beef rather than 'scrub' cattle from overstocked communal lands.

The massive growth in cattle populations in the 1970s (Figure 9.5) reflects technological change. Before this, most cattle were grass fed on the nutritious but sparse 'sweetveld' grass in drylands (Natural Regions IV and V). Although grass was abundant in the 'sourveld' of Region II, cows starved here because this rank grass lacked the necessary 4% nitrogen required for their rumens to digest cellulose properly. The invention of nitrogen supplements and grain-fed livestock radically changed production, with the spike in cattle numbers reflecting the swing towards grain-fed production and feedlots. Globally, feedlots increased efficiency so much that global beef prices fell by at least 25%, rendering dryland grass-fed beef marginal and even unprofitable.

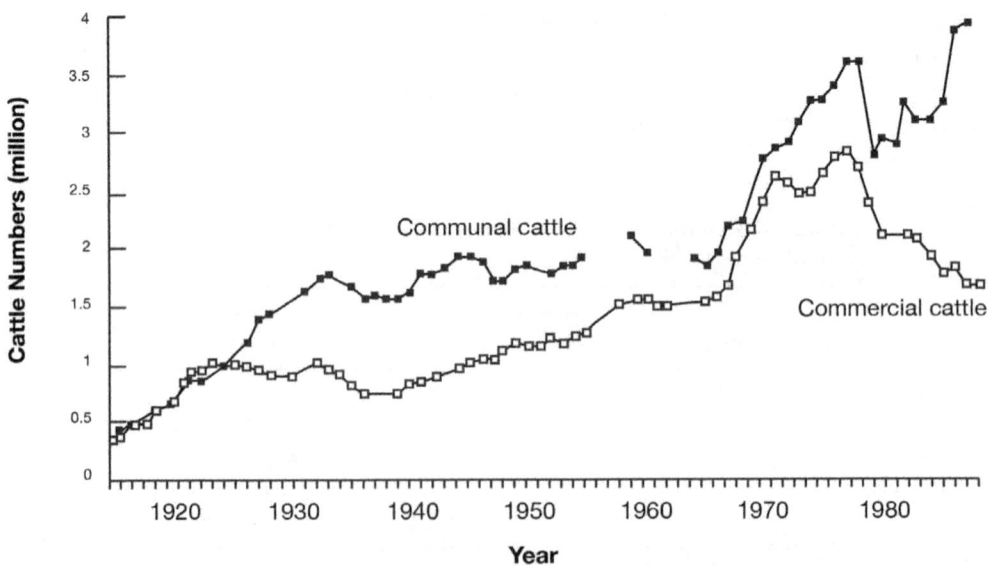

FIGURE 9.5 Commercial and communal cattle populations in Zimbabwe (1910–1988)
Source: Child (1988).

The collapse in both herds in 1975–1980 reflects the civil war. The commercial herd continued to decline after this despite massive government efforts to rebuild the beef sector. The cost of beef production included Z$62 million in livestock services, a direct price subsidy of Z$50 million through the Cold Storage Commission, and $74 million in farmer variable costs compared to beef sales of Z$149 million. With a net loss of $35 million, even ignoring the costs of capital, the beef sector was clearly not viable. Ignoring this analysis, the powerful beef sector panicked and took measures to seek preferential access to European beef markets, despite even higher costs in the form of new abattoirs, disease control, and further limitations on the wildlife sector.

Reading eight decades worth of government reports and parliamentary inquiries suggested that the livestock sector enjoyed a plethora of subsidies over the years including price support, transfer payments, and farm development grants for fencing and water provision (Phimister, 1978; Child, 1988). Reading historical inquiries into the beef sector was fascinating, but it was not good economics. It is much quicker and more reliable to compare the domestic producer prices of beef with world prices (Figure 9.6) to assess if beef producers were subsidised (or taxed). After the mid-1970s local beef prices were on average 40% higher than the world price, confirming that ranchers needed increasing subsidies to compete in global markets where technology was lowering the price of production (Rodriguez, 1985; Williams, 1993).

Comparative advantage and policy analysis matrices

The next step in the sophistication of economic analyses is the policy analysis matrix (PAM) (Table 9.5), which I include briefly for the sake of completeness. PAM is a theoretically

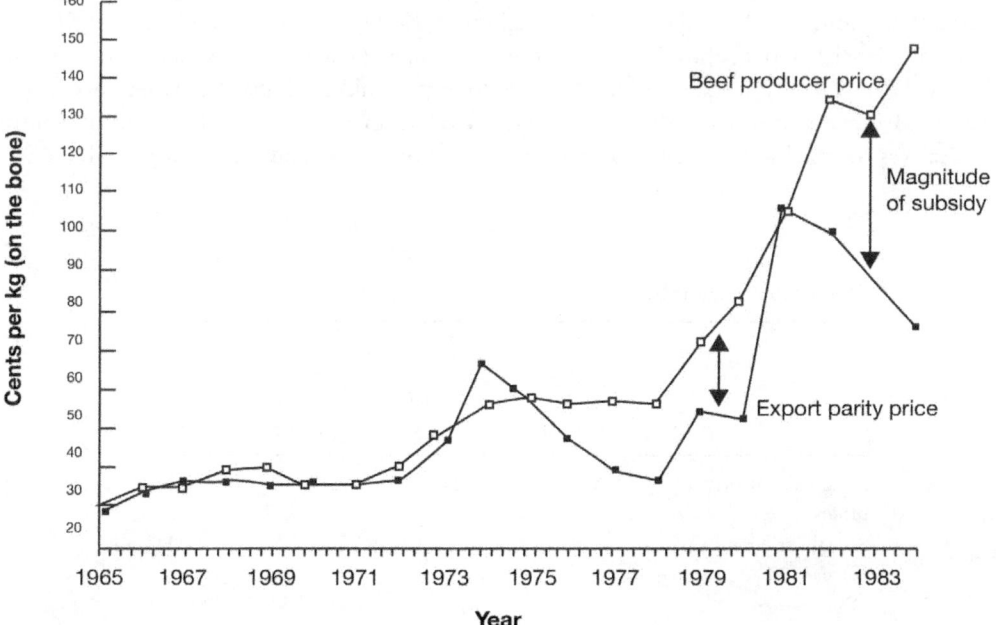

FIGURE 9.6 Using beef parity pricing to assess if commercial beef production was taxed or subsidised

Source: adapted from Child (1988).

robust economic tool that is relatively easy to apply and quickly understood by policy-makers (Monke & Pearson, 1989). As with profitability analysis, PAM uses the basic calculation: profit equals income from sales, less costs of production (Table 9.5). However, to assess if wildlife is a good use of that country's resources it splits costs into two columns: tradable inputs and domestic factors (land, labour, and capital). Comparative advantage is calculated by comparing value-added (i.e. revenues minus tradable inputs, A–B) to the value of local land, labour, and capital used (i.e. domestic factor, C).

By differentiating between private (financial) and social (economic) prices, the third row of PAM measures if markets and policies are distorted in a number of ways, as described in more detail at the foot of Table 9.6. The only technical challenge in PAM is to estimate the real values of cost and sales items, which is simple for goods and services that are traded globally, but requires careful assumptions for domestic factors of production such as land and labour.

In 2010, Jessica Musengezi (2010) applied the PAM methodology to eight properties near Kruger National Park in South Africa. In this area, wildlife enterprises were too diverse (and innovative) to summarise using simple averages, combining hunting, tourism, and the breeding of high-value species for restocking. Property A, for example, supported high-end tourism, charging over US$500 per night. Property D was purchased because the owner liked wildlife, and did not have to worry about profits. Jessica struggled to find any surviving livestock enterprises in this landscape, with the exception of Property G which raised livestock but also managed an abattoir and butchery enterprises to yield profits far in excess of those from pure cattle ranching in the area.

According to gross and net margins, all wildlife properties (except D) were profitable, making an average of R2 000 per hectare but as much as R 8 000. A comparison with beef sector models (it was difficult to get empirical data because livestock in the study had been replaced by wildlife) (Musengezi, 2010) showed that wildlife created two to many times the number of jobs, with jobs being more specialised and better paid, increasing per hectare wage bills some 20–32 times over conventional beef ranching, a finding shared by others (PriceWaterhouse, 1994; Taylor, 2002; Langholz & Kerley, 2006). Wildlife was profitable, with a financial and economic comparative advantage (a technical explanation is given below Table 9.6 to illustrate the PAM methodology). Unlike Zimbabwe, where exchange rates and other controls greatly disadvantaged wildlife in the early 1990s, trade in South Africa was open, although domestic policies (e.g. labour) did disadvantage wildlife by 27%.

TABLE 9.5 Policy analysis matrix

| | Revenues | Costs | | Profits |
		Tradable Inputs	Domestic Factors	
Financial (or private) prices	A	B	C	D
Social (or economic) prices	E	F	G	H
Divergence	I	J	K	L

PAM can also account for environmental externalities by building estimates of the cost of these into social prices.

TABLE 9.6 Summary of PAM results for eight wildlife properties in South Africa in 2010

	Farm							
	A	*B*	*C*	*D*	*E*	*F*	*G*	*H*
Farm size (ha)	5 207	1 700	3 700	2,017	2 800	3 200	14 400	30 000
Big five	Yes	No	Yes	No	No	No	No	Yes
Beds	68	42		12	12	10	n/a	n/a
Ave price	2854	350	260	285	350	250	n/a	n/a
Revenue sources (% of total revenue)								
Accommodation	95	2	79	95	0	5	0	0
Biltong hunting	0	0	0	0	0	17	1	0
Cattle	0	0	0	0	0	0	84	0
Game meat	0	0	0	0	0	1	0	0
Live animal sales	0	92	2	0	100	71	14	100
Retail	6	0	3	2	0	0	0	0
Trophy hunting	0	6	16	0	0	6	2	0
Labour								
Number of workers	194	84	20	19	9	10	75	10
Labour cost (rand/ha)	2 788	600	664	471	238	87	233	14
Labour cost (% of expenditure)	77	21	43	63	51	18	19	18
Gross income (Rand/ha)	8 706	4 936	2 231	105	1 643	1 099	2 906	150
Gross margin (Rand/ha)	8 282	2 886	856	4	1 434	768	1 947	90
Private cost ratio	0.38	0.44	0.13	−1.51	0.34	0.69	0.16	0.16
Domestic resource cost	0.20	0.22	0.59	−1.00	0.26	0.43	0.11	0.08
Effective protection coefficient	0.94	0.88	0.92	0.94	0.96	1.03	0.91	0.90
Profitability coefficient	0.73	0.64	0.14	1.18	0.86	0.55	0.86	0.83
Net policy transfer (% tax on net profits)	27	36	86	−18	14	45	14	17

PAM ratio	Interpretation (for farm A)	Conclusion
Profitability	Turnover of R8 706 and a gross margin (or profit) of R8 282 per hectare	Wildlife is profitable
Private cost ratio = C/(A–B) = 0.380	Every rand of value-added (i.e. A–B) costs 38 cents worth of domestic factors (C)	Wildlife has a comparative advantage (using financial prices)
Domestic resource cost ratio = G/(E–F) = 0.196	Using shadow prices, it costs 20 cents of local inputs for a rand of profit	Wildlife has a comparative advantage (using economic prices)

(Continued)

(Cont).

PAM ratio	Interpretation (for farm A)	Conclusion
Profitability coefficient = D/H = 0.73)	Economic profits are 27% lower than financial profits	Measures net policy transfers. Overall, policies disadvantage wildlife production by 27%
Effective protection coefficient = (A–B)/(E–F) = 0.94	The social value-added (E–F) exceeds the private value-added (A–B) by 6%	Trade policies tilt the economic playing field against wildlife by 6%

Source: adapted from Musengezi (2010).

Land use trends in southern Africa

Land use trends are the most reliable indicator of wildlife economics. A compilation of evidence from the literature (Table 9.7) suggests that wildlife in southern Africa has increased six-fold, while livestock populations on rangelands have halved. The story behind this is insightful.

Livestock numbers in Zimbabwe declined from a peak of 2.7 million heads in 1976 to well under 1 million by 1999, most of which were grain fed, as feedlot technology and grain diets lowered the world price of beef by about a quarter, and as range degradation further reduced livestock viability (as exemplified by Buffalo Range Ranch). Ranches in drylands survived, and often thrived, by switching to wildlife enterprises (sometimes in combination with livestock), using a financing model based on trophy hunting with game meat not yet fulfilling its potential and ecotourism remaining a minor activity. When I began my PhD in 1984, there were a handful of wildlife properties but the value of wildlife rising, and Zimbabwe's hunting sector expanded nine-fold between 1985 and 2000 (Booth, 2002; Bond & Cumming, 2006). By this time, there were more than 1 000 wildlife ranches and at least eight huge conservancies, with a roughly six-fold increase in wildlife on private land. The wildlife economy model was then transferred to communal lands in 1989 through the CAMPFIRE programme (Chapter 11), halting the decline of wildlife in the face of considerable demographic pressures, and allowing the elephant population in communal lands to increase from 4 000 to 12 000 between 1990 and 2002 (Child et al., 2003; Taylor, 2009).

Namibia is similar to Zimbabwe except that, being much drier, it lacks a full spectrum of wildlife enjoyed by savanna countries like Zimbabwe and South Africa (Coe et al., 1976), so hunting and tourism have a lower profit advantage over meat production. Nevertheless, wildlife numbers on private land increased four-fold once ownership was devolved – from 565 000 (1972) to 1.161 million (early 1980s) and to 1.8–2.8 million by 2010. Over the same period, livestock halved from 1.8 million to 0.91 million (Lange et al., 1997; Barnes & Jones, 2009; Schalkwyk et al., 2010). By 2012, three-quarters of ranchers surveyed were using wildlife (92% also had livestock). With similarities to the data presented for Zimbabwe (Figure 9.2), wildlife contributed 42% of the revenues from 29% of the biomass and was increasing on 58% of land surveyed (Lindsey et al., 2011). In Namibia's communal lands, wildlife was decimated during the conflicts that led up to independence in 1990. However, like CAMPFIRE, Namibia confirms that wildlife recovery is not a white man's game, but a result of devolving wildlife ownership to farmers and communities. Namibia has an excellent CBNRM programme, and wildlife increased rapidly from very low numbers to between 150 000 and 200 000 animals by 2010, compared to 121 000 wild animals in parks (Lindsey et al., 2011; NACSO, 2015). Wildlife recovery is less linked to whether land is privately or communally owned, than to whether 100% of the revenues from wildlife are retained by landholders and communities.

South Africa had almost eliminated wildlife outside parks by 1964, with only 200 white rhinos, 19 bontebok, 90 mountain zebra, and 557 000 wild animals surviving. Game farming and ranching took hold in the 1980s, increasing this to between 6 and 18 million animals (du-Toit, 2007; Taylor et al., 2016). Rewilding of land was rapid, with hunting and meat contributing 79% of income compared to 5% from ecotourism (van Hoven, 2015). Private wildlife operations expanded eight-fold from 1979 to 1996, covering 20.5 million hectares or 16.8% of country, and conserving five times as much wildlife as state protected areas (which occupy 6% of country with some 625 000 animals) (Chadwick, 1996; Carruthers, 2008; Dry, 2010). The South African wildlife sector is complicated and diverse. Of South Africa's 10 000 wildlife properties, a third are registered and fenced game ranches, another third combine wildlife with livestock and other enterprises, and another third are conserved for non-commercial purposes (Cousins et al., 2008). The number of wildlife ranchers expanded at 7.45% annually between 1987 and 2005, while the wildlife economy grew steadily at 6% for over a decade despite a stagnant economy (DEA, 2015). South Africa's wildlife economy is now worth at least US$2 billion dollars, with over 65 000 direct jobs. A typical game ranch generates R220/ha compared to R80/ha from extensive livestock farming (Dry, 2010), increasing employment by 4.5 times and wage bills by as much as 32 times (Langholz & Kerley, 2006). As in Namibia and Zimbabwe the expansion of wildlife took place against a backdrop of a contracting livestock sector, especially after the removal of apartheid era farm subsidies: cattle declined from 12.2 million to 8 million, goats from 5.7 million to 2.5 million, and sheep from 39.7 million to 28 million between 1964 and 2005 (du Toit, 2007).

Terms of trade

The price of most agricultural commodities has been declining steadily, with a sharp drop in the price of beef in the late 1970s. Tourism, by contrast, has expanded rapidly for 50 years from well under $50 billion in 1960, to over $900 billion today. Data on wildlife prices are harder to get, but the average price of trophy elephants in Zimbabwe increased rapidly from $250 (1976) to $2 600 (1986) to $10 000 (1990); and elephants now sell for upwards of $20 000 for the trophy alone, and $60 000 plus for the whole hunt (Booth, 2009). Thus, if wildlife is a good bet today, it is highly likely to be a better bet tomorrow.

Wildlife economics and market failure on communal lands

There is no reason that the economics of wildlife differ between private and communal land, because it is traded globally. However, at the same time as wildlife recovered rapidly on freehold land in southern Africa, it began to disappear on communal lands, where the function of economics and markets was severely hampered by open-access property regimes and the non-ownership of wildlife. In a lesson with broad implications, the wildlife authorities in Zimbabwe and Namibia recognised that the problem lay in weak communal institutions, and as soon as they began to return rights and benefits from wildlife to communities, the decline was reversed (Taylor, 2009; NACSO, 2015). Programmes like CAMPFIRE were not developed in vague hope, but were a deliberate attempt to transfer a proven economic model (i.e. profitable wildlife utilisation on private land) to communal lands, which, as we will see in Chapter 11, also depend on appropriate institutions for resource pricing and governance.

TABLE 9.7 Trends in wildlife and livestock populations on private land in southern Africa

Country	Livestock		Wildlife		Source
	Date	Indicator	Date	Indicator	
Namibia	Early 1970s	1 800 000 animals	Early 1970s	565 000 animals	Barnes & Jones, 2009, Lindsey et al., 2011
	2001	910 000 animals 51% decline	2001 2009 2012	1 161 000 animals 1 818 219 animals 2 838 023 animals	
Zimbabwe	1992 1999	139 000 t beef 67 000 t beef 48% decline	1985 1998 1974 1998	$4m hunting income $24m hunting income 600% increase 179 cropping permits 1,000 ranches(?)	Booth, 2002, Mossman & Mossman, 1976, Gambiza & Nyama, 2006, Child, 1988
South Africa	1964 2007	Goats 5 667 000 Sheep 39 717 000 Cattle 12 243 000 Goats 2 500 000 56% decline Sheep 28 000 000 30% decline Cattle 8 000 000 35% decline	1965 2007 1965 2007 1974 2007	4 registered fenced farms 5 061 fenced farms 575 422 animals 18 591 422 animals (+18 016 000) 3 180 000 ha 20 500 000 ha	Dry, 2011, Du Toit, 2007, Carruthers, 2008, Mossman and Mossman, Taylor et al., 2016
Zambia			1997 2012 1997 2012	30 ranches (1 420 km^2) 177 ranches (6 000 km^2) 21 000 animals 91 000 animals	Lindsey et al., 2012
Botswana			1999 2005	17 game ranches 60 game ranches	BWPA, 2005

Total economic value of tourism and parks

The final economic tool that I will present is an analysis of the total economic value of a protected area, using tools and multiplier models developed for North America (Stynes et al., 2000; Stynes, 2005) and modified for Africa and Brazil by several PhD students (Souza & Thapa, 2018; Souza et al., in review). South Luangwa is a typically spectacular African national park. Many such parks are neglected and under-capitalised, and perform nowhere near their potential. However, donors invested significantly in all-weather roads and airports for South Luangwa, the protection of the park's excellent wildlife was also partly funded, and tourism began to grow when a failed state model was replaced by some 650 tourism beds owned by the private sector, and by a policy that the park could retain all its own revenues for management.

To calculate the economic impact of the park, we counted the number of visitors, analysed their expenditure and employment impact, and applied economic multipliers to this expenditure. The park earns $2.91 million from park fees and is said to be self-funded, although this covers only

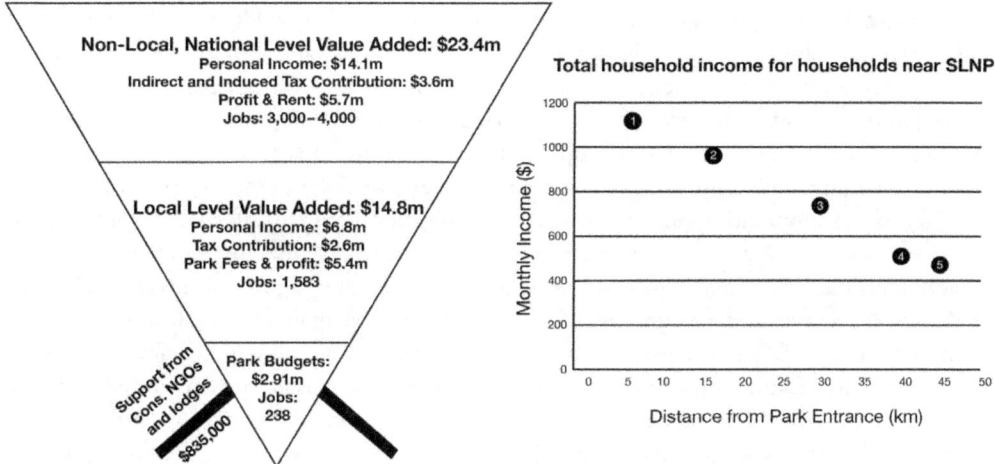

FIGURE 9.7 Total economic value and vulnerability pyramid for South Luangwa National Park
Source: developed with and adapted from Alex Chidakel (personal communications) and Shylock Muyengwa (personal communications).

half its ideal management costs, and the park is probably operating at less than half its economic potential. Nonetheless, South Luangwa generates $14.8 million in added value in the gateway communities, including 1 583 jobs, while the number of local businesses near the park has increased from one in 1983 to over 180 by 2015. However, this is a remote rural area and most inputs to tourism are still imported from the rest of Zambia, where tourism spending from South Luangwa generates an additional $23.4 million in added value and several thousand more jobs.

I have called Figure 9.7 a vulnerability pyramid because the $38 million economy depends on effective funding and management of the park, and governments see park entry fees as a large cash source they can divert elsewhere without considering its economic consequences. Grabbing these park fees threatens the $38 million economic pyramid that depends on good park management, including $2.9 million of park fees and $6.2 million in taxes (before including taxes on profits paid by tourism operators, which are difficult to calculate). Tourism directly employs 1 583 local people, and with dependents indirectly supports as many as 9 000 people in the community. This was validated by a livelihood survey (n = 419), which showed that household income near the tourism hub was over twice that of similar (counter-factual) communities 45 km away, with income declining with distance from the tourism hub (Figure 9.7). The growth of local businesses also tracks the tourism economy, expanding from 1 (1983) to about 250 (2016) (Chidakel and Child, in review). This data supports the claim that parks can be economic engines in remote rural areas, in this case doubling the incomes of households close to the park gate compared to counter-factual households 45 km away.

Conclusions

This chapter has introduced a number of practical economic methods for assessing the economics of wildlife, increasing in complexity from simple income and expenditure accounts, to gross and net margin analyses, return on investment, parity pricing, historical case studies, policy analysis matrices, and economic multipliers. I hope I have illustrated that these tools enable conservationists to make a solid case that wildlife has a comparative advantage, enabling it to provide more income, livelihoods, and employment than simple commodity production, especially when economic multipliers and environmental effects are taken into account. Moreover, terms of trade favour wildlife, so these advantages are increasing, and there is enormous scope to drive the profit curves for wildlife even higher by encouraging trade, product development by private entrepreneurs, and by avoiding differential taxation and regulation of wildlife (associated with its history as a public good).

Economic research of this type informs policy and practice. In Zimbabwe, it gave landholders more confidence to invest in wildlife and validated significant investments in the CAMPFIRE community programme (Chapter 11). At a fine grained level, it showed that the elimination of buffalo in the name of beef export markets, was halving the value of wildlife.[1]

A particular feature of wildlife tourism is that it is a cluster industry associated with economies of scale. Most investments in wildlife tourism occur near other wildlife businesses where roads, airports, supply chains, and a brand are already established (e.g. 'Kruger'), which is why a few recognised destinations often become overcrowded. Developing new tourism destinations is difficult, and requires roads and airports and 'anchor' investments to reach a threshold scale, as well as a spectacular wildlife or scenic resources. This is why so many communities and properties rely so much on hunting to fund conservation, and why the current anti-hunting sentiments that are being used to attack African conservation policies are so counter-productive, and perhaps more harmful to conservation than the illegal wildlife trade.

Because of tourism's limitation, hunting is the most effective economic tool that conservationists have for rewilding drylands (and perhaps some forests). Hunting generates high returns from low offtakes, is successful in environments that are not attractive to tourists, and continues even in war zones. It also requires far lower capital investment than tourism. Thus, by my estimate, we can attribute as much as 80% of the rewilding of southern Africa to hunting, and perhaps as much as 90% outside South Africa and Namibia. This is why we are so concerned about the anti-hunting lobby, and the simplistic and erroneous message it sells. Without hunting, large dryland landscapes will revert to livestock, affecting the incomes of thousands of people. If urban activists were sincere about benefitting wildlife (and even rural people), they would achieve far more by demanding the honest governance of the wildlife sector as measured by the return of benefits to landholders.

The wildlife economy is far more dependent on new and sophisticated institutions than domestic species, including sound policies for wildlife ownership and use. Nobody will invest in commodities that they do not own and which are vulnerable to external threats to markets. Moreover, wildlife is mobile and fugitive, and needs landholders to come together to manage externalities, and to reap ecologies (larger areas are more ecologically sound) and economies (larger areas are more profitable) of scale. Cooperation, however, depends on reliable institutions, policies, and regulations. Thus, when governance breaks down, people revert to low value commodities such as crops and livestock.

The research summarised in this chapter makes a strong economic case for wildlife. Wildlife provides a more sustainable pathway to economic development in complex and unproductive environments like drylands and forests because it dematerialises production. By using wildlife multiple times (tourism, trophy hunting, meat) to provide high-value luxury products (tourism, trophy hunting), it earns far more than primary commodities like beef per unit of environmental input (Figure 9.8). Wildlife produces more (income) from less (environmental inputs), thus breaking the direct link between economic outcomes and environmental use that is the underlying

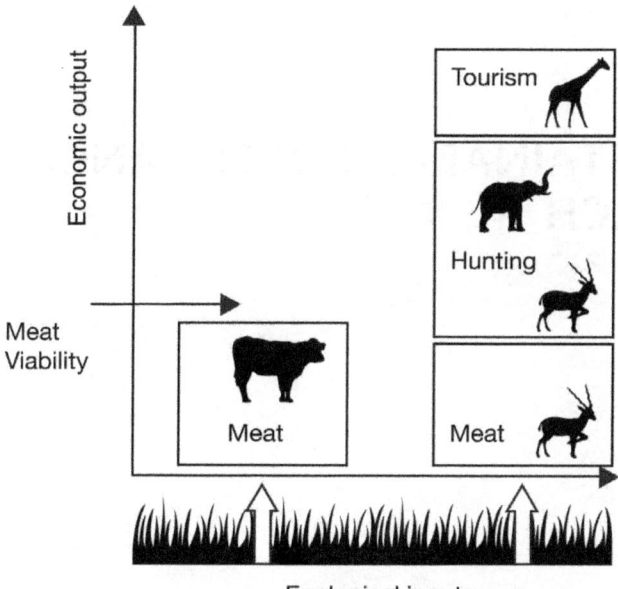

FIGURE 9.8 Dematerialising production: replacing the agro-extractive economy with the bio-experience economy

cause of the degradation associated with commodity production (Figure 9.1). Thus wildlife decommodifies the rural economy:

$$Bio-experience\ economy = \frac{Higher\ livelihood\ and\ economic\ benefits}{Less\ inputs\ from\ the\ environment} = Dematerialisation$$

Dematerialisation is not possible with simple agro-extractive commodity production because production is directly linked to environmental inputs. Livestock, for instance, produce more meat only by eating more grass, but meat production is not viable within the environmental constraints of drylands, regardless of whether meat is produced by wild or domestic species (Figure 9.8). This is hardly surprising: it is not logical to produce low value commodities in complex drylands and forests that are unproductive in terms of secondary production. Interestingly, tourism and, particularly, trophy hunting mimic the hunting and gathering lifestyles that have been sustained in forests and dryland for millennia, in ways that simple extractive agricultural systems do not.

Note

1 Buffalo are the only 'big five' animal that can be produced in large numbers; elephants mature over 70 years, while apex predators like lions and leopards are never numerous. If a rancher sells a 'plains game hunt' with kudu, zebra, wildebeest, impala, warthog, and so on, he can market a seven-day hunt costing \$300/day to earn \$2 100 in daily fees. However, adding a buffalo upgrades this to a 10–14 'big game hunt' for \$600/day, doubling or quadrupling outfitter fees to \$6 000 to \$ 8 400. Elephants and lions do the same, but cannot be produced in the same numbers as buffalo and impose high costs.

10

THE SUSTAINABLE GOVERNANCE APPROACH

LEARNING OBJECTIVES

This chapter describes the sustainable governance approach as a mechanism for reintegrating the economic value of wild resources into the economy to rewild landscapes and address poverty. The 'sustainable governance approach' is a grassroots, democratic, market approach that combines four sets of principles:

1. Proprietorship: the devolution of rights, benefits, and decisions for wild resources to landholders and communities.
2. Price: the set of factors that maximises the value of wildlife to landholders and guides 'prices' towards their (monetary and non-monetary) economic value.
3. Subsidiarity: the need to govern the wild economy from the bottom up, so that people who are most affected by decisions are most involved in making them in multi-scalar systems.
4. Collaborative adaptive management: the process of co-learning and co-developing new institutions, systems, and capacities by combining face-to-face interactions, performance data, and conceptual and other information through an adaptive process.

The sustainable governance approach

This chapter consolidates the principles and historical lessons from the preceding chapters as the 'sustainable governance approach'. To maximise the value of wild resources to the people who live with them, and bear their costs, this governance framework combines the concepts of proprietorship and price:

1. Proprietorship – which deals with the just ownership of wild resources, the internalisation of costs and benefits, and the location of rights, benefits, and decisions with the people who bear the costs of living with wildlife.
2. Price – which ensures that the set of values perceived by the landholders and communities making decisions (i.e. financial 'prices') reflect the true economic value of wild resources in land use decision-making.

However, it also needs to account for governance and scale, and the need to adapt to a rapidly changing world, including:

3. Subsidiarity and inclusive governance – the mechanisms that link deep democracy (i.e. that the people most affected by decisions should be most involved in making them) to scale, including the principle that things should never be done centrally if they can be done locally.
4. Participatory adaptive management – the process of co-learning and co-creating new institutions, processes, and capacities by combining data, information, and multi-stakeholder reflection in the management of change, risk, and uncertainty.

Bringing the full value of wild resources into land use decisions

Economic values include all the use, non-use, and abstract values of wildlife (Krutilla, 1967; Emerton, 1999) that accrue as private benefits, as common pool benefits, and as public benefits (Figure 10.1). While the mathematical science that economics has become equates prices with dollars, in the real world people value material things, but also cooperation, altruism, fairness, beauty, ownership, responsibility, nature, and the place where they live. Nonetheless, land use outcomes are driven mainly by the private cost–benefit calculations of landholders, which to some extent include these values (Figure 10.1). Landholders are deterministic of land use outcomes, so to conserve wildlife we need to ensure that these values are internalised by landholders and communities. Thus, the essence of the sustainable governance approach is captured by the black arrows in Figure 10.2 and measures that convert the theoretical values of wildlife into real value for

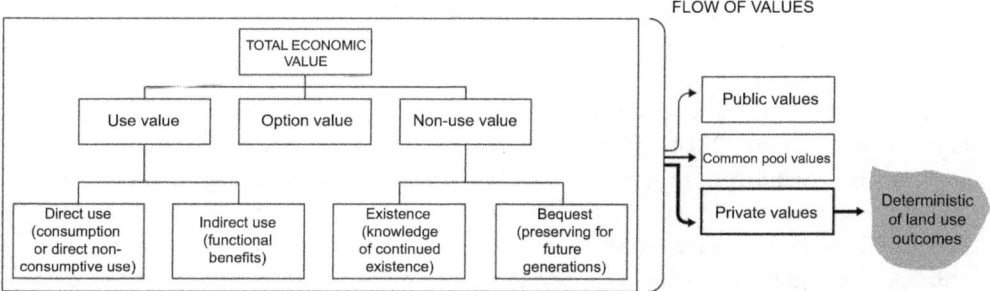

FIGURE 10.1 Private values determine land use outcomes

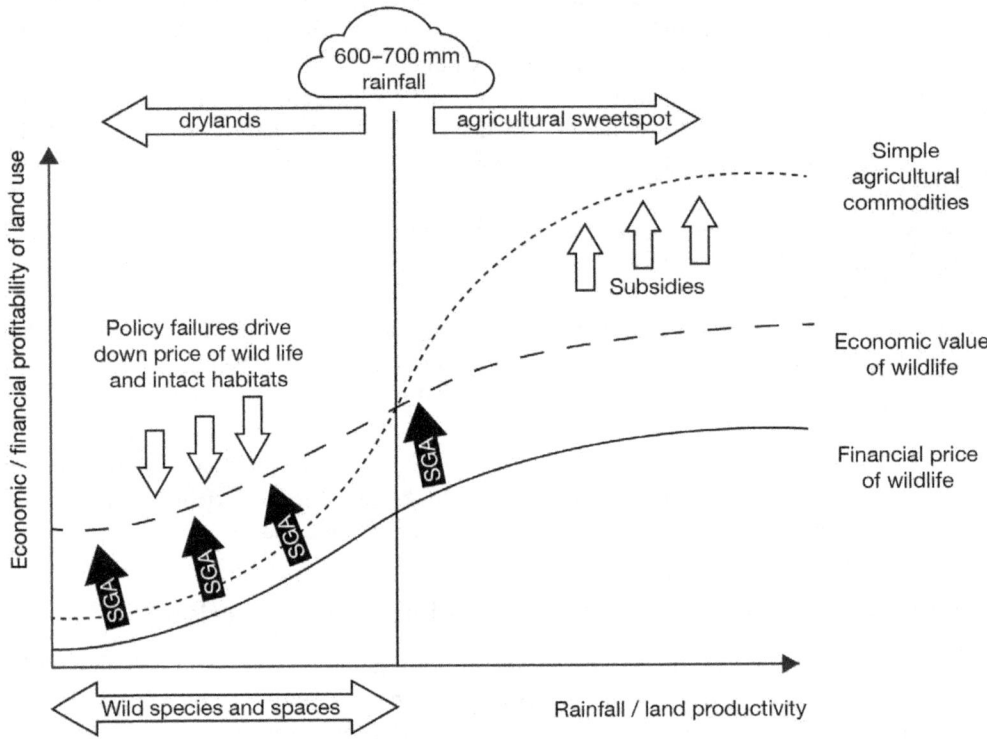

FIGURE 10.2 Goal of sustainable governance approach is to lift financial curve as far as possible to reflect total economic value

Source: modified from a PowerPoint developed by Chris Brown and Greg Stuart-Hill (with permission).

landholders, which, where people are hungry, is often material. Policy reform is then guided by the question: 'Does this action increase or reduce the ability of wildlife to compete for land?' Does it move the private (financial curve) upwards towards the economic (real value) curve?

As noted in Chapter 6, studies are beginning to show that the total economic value of intact ecosystems may exceed that of ecosystems converted to extractive agricultural commodity production (Costanza et al., 1998; MEA, 2005; Ding et al., 2016). The challenge, however, is to convert these theoretical values into the financial signals that guide land use outcomes.

There are two approaches to 'mainstreaming' the value of biodiversity into the economy (Huntley & Redford, 2014). Top-down, regulatory approaches use land use zonation, environmental standards, prohibitions, and so on, and are associated with centralised planning and the public good. The sustainable governance approach, by contrast, is a devolved, market-based approach that puts wild resources back into the economy by returning the rights and benefits to the people who live with them, often marginalised communities who have gained little from their wild resources which the rest of us are accustomed to getting for free.

Conservationists often have a jaundiced view of economic approaches because, as we argued in Chapter 6, resource rights and market rules have been differentially acquired or distorted to favour the few. In other words, these fears are created not by property rights and markets, but by their misuse including, in the example of ungoverned tropical spaces, their near absence. The sustainable governance approach seeks economic efficiency through well-designed markets, combined with strong and just property rights, to ensure that the economic system serves communities and wild resources efficiently and fairly. These outcomes are embedded in the principles of proprietorship and price which seek:

- Ownership that is socially just with strong rights for communities and wild resources, and that also serves to fully internalise the costs and benefits of wild resources and alternative land uses, in the manner described for the efficient allocation of resources through private property rights and free markets.
- Markets that, in the sense developed by Adam Smith, allocate resources according to price and quality, and avoid the distortions associated with the free-wheeling laissez-faire economy and crony capitalism in which the rich get richer.
- To use these markets and property rights to increase the amount and value of environmental goods and services through specialisation, exchange, and trade.
- Collective action to manage the common-pool attributes of wild resources, including subtle externalities and scale.

The proprietorship-price hypothesis

Using the framework of goods, rights, and owners developed in Chapter 5, there is a reasonable alignment between the public nature of national parks and the public ownership and management of them (Arrow (a), Table 10.1), although the last two decades have seen a considerable expansion in the meaning and scope of protected areas (Box 10.1). By contrast, the public governance of off-reserve conservation and wild resources contains serious misalignments, because imposing public outcomes on private landholders contradicts the purpose of this land. The efficacy of the North American Model is somewhat of an anachronism, combining as it does a private/common-pool good with state property and public management (Arrow (b), Table 10.1).

By contrast, the sustainable governance approach aligns the private and/or common-pool character of wild resources (private or common pool) with a combination of private ownership and collective action (Arrow (c), Table 10.1). This is very different to conventional top-down conservation and reverses conventional policies on wildlife ownership and markets (Table 10.2). It entrusts ordinary people with the future of wildlife, on the basis that if they own it (proprietorship), and if it is valuable (price), there is a high probability that they will conserve it, achieving the dual goals of efficient land use and wildlife conservation. Just as it was often middle-class engineers that invented the machinery of the Industrial Revolution, this structure challenges the elitist view that knowledge is produced by scientists and bureaucrats to then be imparted to the less educated. Progress stems, rather, from freeing the inventive and entrepreneurial capabilities of the majority of people.

TABLE 10.1 The alignment of goods, rights, and owners in public and sustainable governance approaches

Goods	Rights	Owners	
Private **WILD LIFE** (c)→	Private *res privatae*	Individuals Corporations	**SUSTAINABLE GOVERNANCE APPROACH**
Common pool	Common property *res communis*	Legally / traditionally defined group	
Public **NATIONAL PARK** (b)↗ (a)→	State property *res publicus*	Public or government	**NORTH AMERICAN MODEL**
	Open-access/non-property *res nullius*	Ownerless	

TABLE 10.2 A comparison of the sustainable governance approach with old-style public conservation

Conservation strategies and policies	Strategy 1: Public ownership and management	Strategy 2: Sustainable governance approach
On-reserve conservation		
Protected areas	Establish state-protected areas to conserve fauna and flora and, later, public access	Expand concept of protected area management to maximise appropriate public goods (i.e. outdoor recreation, wilderness, jobs, economic growth), while ensuring conservation. Protected areas privilege conservation, but: • provide public goods suited to that society; • recognise legitimacy of private and community conserved areas.
Off-reserve conservation		
1. Wildlife ownership	Centralised in the state	PROPRIETORSHIP: Devolve rights to use, sell, and manage to landholders/communities.
2. Commercial use of wildlife	Restrict and/or ban exchange	PRICE: Make wildlife as (commercially) valuable as possible (provided use is humane).
3. Governance of scale and hierarchy	Top-down blueprint planning	SUBSIDIARITY and GOVERNANCE: Manage scale effects of fugitive wild resources by scaling down to inclusively governed landholders, and building 'nested institutions' through upward delegation.
4. Management of change and complexity	Reductionist science, often politicised	CROSS-SCALE ADAPTIVE LEARNING: Use social processes to set goals, and adapt to change and complexity with evidence-based management.

Source: Child, B. this volume

BOX 10.1 PARKS IN TRANSITION

Public management of public protected areas is often effective and avoids the internal contradictions of seeking public goods from the governance of wild resources on private or community land. Nonetheless, the role of parks is changing and we need to reassess who and what parks are for in developing societies. When southern African park agencies met to consider their future, they recognised that:

- Private and community conservation was just as legitimate as state conservation, and sometimes more effective.
- Parks were unlikely to survive if they were managed as economic black holes in developing countries.
- Parks could play an important role as beachheads in the economic rewilding of large landscapes (Child, 2004).

Indeed, Murphree (2002) proposed that parks are common property regimes with a range of owners – public, co-management, private, community – and objectives. This viewpoint was taken up by Grazia Borrini-Feyerabend et al. (2013) who argued that the state does not have a monopoly over the management of parks, and parks managed privately and by communities are equally legitimate and increasingly well managed. Further, most of the growth in global conservation coverage occurs in Category VI protected areas (UNEP-WCMC, 2012), which are often owned by communities, have multiple use values, and therefore fall into the realm of the sustainable governance approach.

Most parks have an economic rationale, and the myth of the Yellowstone Model as untouched by the human economy is not reality (Shelhas, 2001). The purpose of parks (if they are common property regimes) is to maximise societal gain, while ensuring that biodiversity is conserved (Child, 2004). The following efficiency equation is useful for assessing policy options:

Conservation/land use policy = <u>Maximise total economic value to society</u>
while ensuring that ecosystem health
and diversity is maintained

The bottom line is that any use of parks should not jeopardise its ecological health and diversity. However, the uses that we permit and encourage need to be aligned with the values and needs of the society that owns the park, and should maximise these values. Therefore, it is just as acceptable for Africa's parks to proactively create jobs and economic growth as it is for parks in affluent, urban countries to provide wilderness and outdoor recreation. Moreover, parks can become significant engines for economic growth in remote rural areas, especially with appropriate investments in management, tourism infrastructure, and landscape scale (Chapter 9), and a much stronger case can be made for investing in parks based on this economic logic.

Using price and proprietorship to define governance regimes

Figure 10.3 describes four governance regimes for wildlife. These are defined by the interplay between price and proprietorship, and by two threats factors, namely over-exploitation and replacement by domestic species. Different configurations of proprietorship and price result in very different outcomes for wildlife specifically and wild life more generally.

The historical starting point for wildlife conservation is Quadrant 1 – the over-exploitation of wildlife in a frontier economy, including today's illegal wildlife trade. The historical solution – Quadrant 3 – was public management, with state agencies and laws enforcing exclusion. However, Quadrant 3 governance is successful only under specific circumstances, such as those described for the North American Model (Chapter 7). In many places, the capacity to exert public management is overwhelmed or underfunded. Wildlife quickly finds itself in Quadrant 4 – the No Hope Economy – where it is neither owned nor valuable, and is neglected or replaced by domestic species.

Wildlife is not safe where proprietorship (including the capacity for exclusion) is weak: in the No Hope economy it is replaced or neglected. In the Frontier Economy, it is threatened by uncontrolled over-exploitation. These are the 'ungoverned spaces' described in Chapter 4.

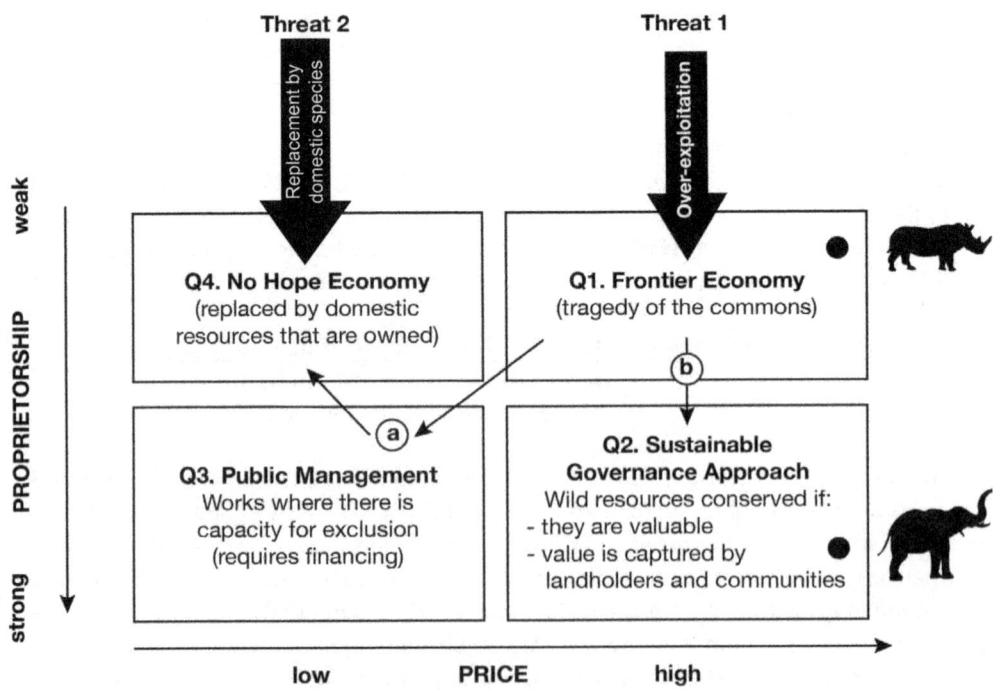

FIGURE 10.3 The interactions between price and proprietorship, and different wildlife economies

Wildlife is only safe where there is clear responsibility and ownership of it, and it is funded (Quadrants 2 and 3), which differ primarily in their funding model: in Quadrant 3 wildlife protection is financed publicly, while in Quadrant 2 it finances itself privately.

This raises the practical question of what to do if resources are in Quadrant 4 or 1.

In Quadrant 4, ecosystem services are under-supplied because little or none of their value is priced, or retained locally, and markets are missing. Payments for ecosystem services, such as water and carbon sequestration, may enable wild resources to compete for land, provided, of course, they are matched with community land rights (Ding et al., 2016) reflecting a move to Quadrant 2.

The Frontier Economy (Quadrant 1) is very much in the news in relation to the illegal wildlife trade, and the poaching of elephant, rhinos, sea creatures, timber, and so on. The typical response to the loss of these high-value species is a call for 'demand reduction', for what are sometimes perceived as a public (or global) asset. Trade bans and nationalisation convert the frontier economy (Q1) to public management (Q3), but only work with reliable financing of public management and exclusion. In poorer countries, the effect of demand reduction has been to move wildlife into the No Hope Quadrant (Arrow a, Figure 10.3), with dire consequences because no-one owns the wildlife and it has no legal value. To make matters worse, demand reduction usually shifts trade into the shadows, so the system remains in Quadrant 1 but spawns criminalisation, which is difficult to reverse.

As I write this, I am sitting in a Mozambican community where civil society is being seriously disrupted by mafia-like criminal syndicates linked to the illegal rhino trade who, like urban drug lords, buy off officials and control society through a mixture of payments, violence, and bravado. In this very same area, elephants, buffalo, lions, and antelope are thriving because they are in Quadrant 2 – benefiting landholders and communities, paying for their own protection, and providing the economic impetus to build a much larger wildlife economy across the landscape. The story with rhinos, however, is catastrophic. CITES has removed proprietorship from landholders, but black-market prices have increased, pushing rhinos into the very top corner of Quadrant 1. Landholders have lost ownership rights and opportunities for self-financing rhino protection, while the high profits they could earn instead empower and embolden criminals. Landholders are further conflicted by the need to risk their lives to protect rhinos that are no longer theirs, through decisions that were imposed upon them, and which cost them a lot of money. Unlike the other species that are thriving on this land, paid for mainly by hunting, it is difficult to see how rhinos will survive. To their tremendous discredit, trade ban proponents have ignored the obvious – that rhino protection and management is expensive. They have promoted a public option (Q3) but, amidst the rhetoric and acrimony of CITES, have completely, and irresponsibly, ignored the need for sustainable financing. At the same time, they have slammed the door on private conservation (Q2), despite its proven effectiveness in recovering rhino populations. Without a working model to finance rhino protection, the fate of rhinos is sealed.

In the absence of piles of public money or philanthropy, banning or restricting commercial use is unlikely to ever be more than a short term and partial solution. Even if trade restrictions do work in the short term (and that is a big if), they must be accompanied by a sustainable financial model to pay for the protection of what has become a nationalised or public resource. However, we know that the costs of policing public resources are high, and rise exponentially if landholders and communities do not participate in making and legitimising these decisions (Wade, 1987). The knee-jerk response towards price reduction (Arrow a, Figure 10.3) is a dangerous pathway with an endpoint that is unclear.

Public management (Quadrant 3) is workable if there are sufficient public funds to achieve exclusion. Well-funded national parks can be very effective, and surprisingly so with low levels of funding, provided the boundaries (i.e. rights of exclusion) are well-recognised and respected, and neighbouring communities and society value them. For example, the hard edges between intensive small-scale agriculture and the forested Kibale National Park in Uganda are relatively intact. Light policing of well-recognised park borders, coupled with the perception that the forest provides rain and other ecosystem services, seems to be enough to maintain its integrity (Goldman, personal communication).

By now it should be obvious that wildlife conservation is only likely to be successful in the bottom quadrants, that is where there is effective exclusion. Given the limits of public or philanthropic funding, relying solely on a public model will constrain wildlife to only that small part of the planet that the public is willing to pay for, which may be small in developing countries where health care and education are prioritised over nature. The obvious solution is to embrace the high price of wild resources as an opportunity (not a threat), and to focus on proprietorship (Arrow b, Figure 10.3), expanding wildlife and wild areas by shifting from a frontier economy into a sustainable one. Devolving rights is highly political and challenging but offers a viable long-term solution; livestock has never gone extinct because it is valuable. Wishy-washy devolution, like allowing landholders to keep only a small portion of wildlife revenues, is unlikely to be transformative. For wildlife to be embraced, landholders and communities should be accorded the same rights (and the same lack of restrictions) over wild resources as they have for livestock or crops, and stronger rights if possible. Once we adopt this approach, driving up price is good for wildlife (Box 10.2), and so therefore is exchange and global trade (Chapter 6).

BOX 10.2 PRICE, PROPRIETORSHIP, AND REGIME SHIFTING

Working in Zimbabwe's CAMPFIRE programme, my friend and colleague, the late Ivan Bond, suggested both price and proprietorship were necessary for a regime shift. Thus, communities might shift to a more sustainable land management regime if they received high benefits from a resource (even if their rights were truncated). They might also shift to a sustainable regime if they had strong rights over a resource, even if the resource had a relatively low value. However, change was most likely where high levels of benefits and strong rights were combined. In practice, moreover, effective proprietorship is comprised of legal rights and the capacity to exert these rights, including exclusion.

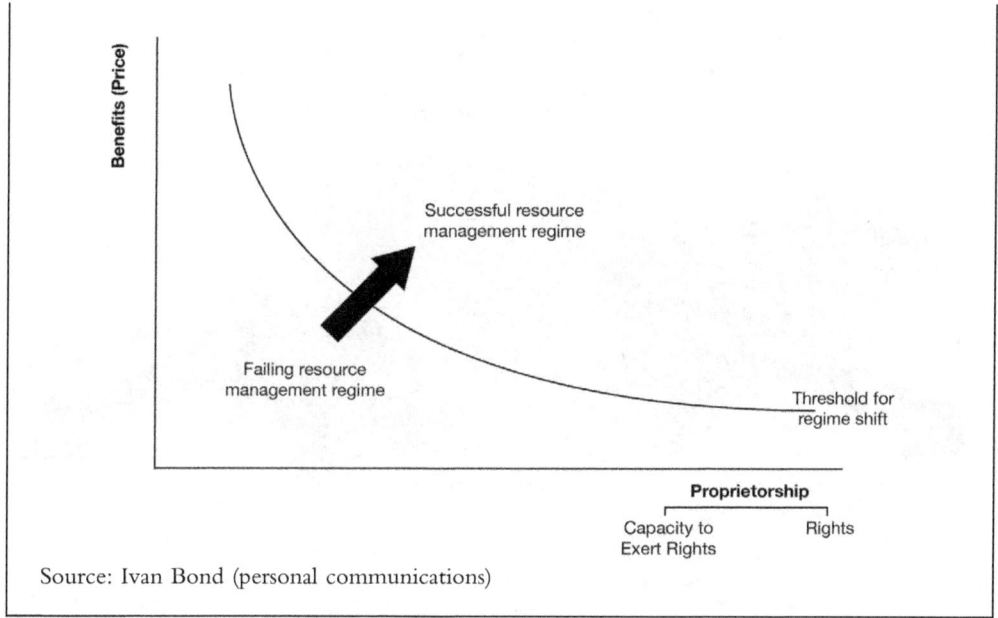

Source: Ivan Bond (personal communications)

The price-proprietorship model is static and economic, and the sustainable governance approach needs to be expanded to deal with governance and scale, as well as building social capital and adaptive resilience (Table 10.2). The principle of subsidiarity, in line with systems thinking (Meadows, 2008) and democratic theory (Tocqueville, 1994), recognises that economic choice and governance should originate with the people, and is best achieved by building nested institutions from the bottom. Similarly, managing risk and change requires transdisciplinary and iterative learning processes that integrate different stakeholders, objectives, disciplines, and datasets (i.e. livelihoods, governance, health, and environment) through collaborative adaptive management. Reductionist and elitist management and science are inadequate in the face of the multiple drivers and interactions affecting society and wild resources today.

Subsidiarity and scale

As shown in Figure 10.4, the costs, benefits, and externalities of wild and domestic species are managed very differently. On the surface, at least, wildlife is 'fugitive' and more complex than domestic species; an impala may live most of its life within the bounds of a single land unit, but a kudu occasionally crosses the boundary, and elephants wander more widely. Historically, the fugitive nature of wildlife has resulted in wildlife governance being centralised, with limited local input. This distorts the wildlife economy, with costs falling to landholders and communities who are seldom the primary beneficiary. By contrast, farmers mostly internalise the benefits from their fields or cows, but are able to shirk the costs of spatial externalities (e.g. excess nitrogen), and of temporal externalities (e.g. the costs of soil degradation, which fall on his grandchildren).

For the same reasons (see also Figure 5.8), the administrative response and control structures for wild and domestic species are also very different (Figure 10.5). Have you ever asked yourself why you meet most natural resource officials in national and regional offices or at conferences, while most farmers are on the farm? As a nationalised resource, the hierarchies

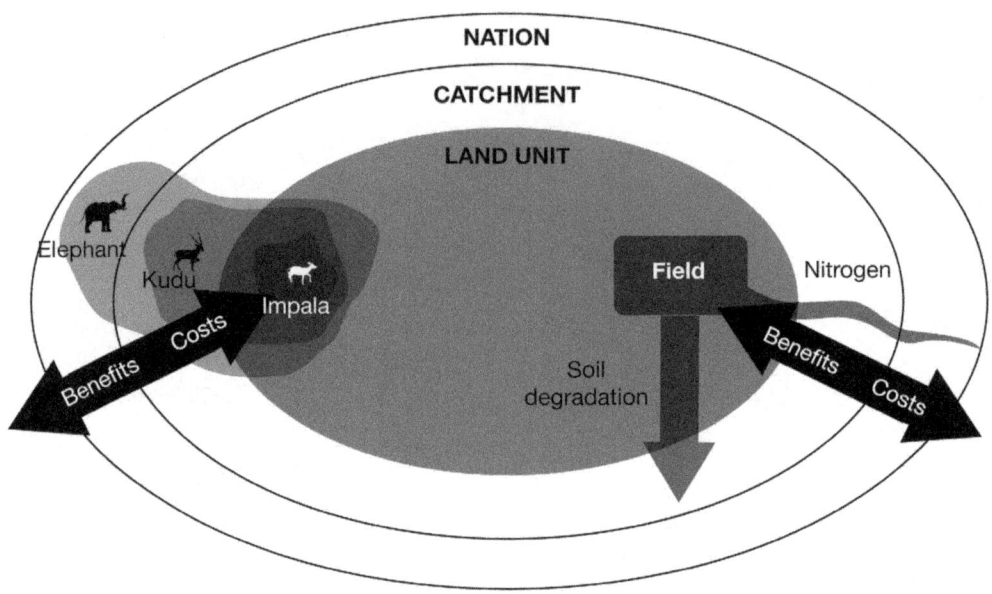

FIGURE 10.4 Distribution of the costs and benefits of wildlife and domestic species across scale

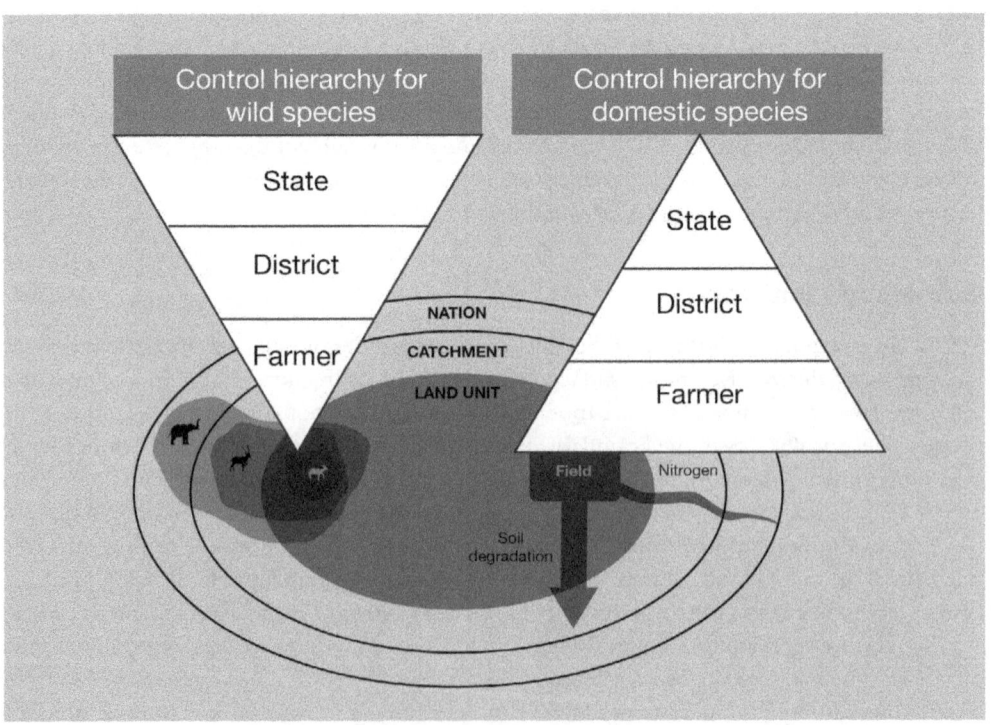

FIGURE 10.5 Comparing control hierarchies for wildlife and farming

of scale governing wildlife are built clumsily by expropriation from above (Murphree, 2000), and most wildlife trade-offs are highly politicised. By contrast, the control of domestic species is highly decentralised and most transactions take place in the economic (not the political) marketplace. Farmers are quite able to deal with complexity, being close to the action. However, natural resource agencies quickly become over-burdened with the complexity that could and should be handled lower down, as central ownership undermines local systems of responsibility and adaptation (Ostrom, 1990).

The centralised governance of wild systems (Figure 10.5) runs contrary to 'systems thinking', which suggests that hierarchical systems should evolve from the bottom up, with the upper layers serving the purposes of the lower layers (Meadows, 2008). It also contradicts the principles of subsidiarity and, indeed, democracy, which implies that discretionary power should be exercised by, and originate from, the smallest or least central unit of authority (Handy, 1994; Tocqueville, 1994). Consequently, the sustainable governance approach emphasises the devolution of rights to wild resources to individual production units (such as villages and farms), and the devolution of the powers of self-regulation to groups of these units, following two rules:

1. The level of regulation should match the scale at which the externalities occur. In the examples given, regulation is not required for impala (because the costs and benefits are largely internalised), but collective regulation is required for the externalities associated with elephants and nitrogen according to the levels at which these occur (see also Figure 5.2).
2. Layers of regulation should be built parsimoniously from the bottom up through a process of delegated aggregation – thus maintaining ultimate authority in the citizenry. These are Murphree's insights, and are elaborated below.

This approach transforms the role of the centre, which steps forward to design systems of property rights, markets, collective self-regulation, and the monitoring of these, and steps back from day-to-day management.

The process of scaling

Ostrom's principles acknowledge the importance of nested enterprises and that local commons are embedded in larger systems. However, we turn to Murphree (2000) for the most astute analysis of how to match the fugitiveness of resources to jurisdictional scale. The essence of Murphree's insight is that the rights to wild resources should be scaled right down to the bottom, and then scaled back up to the correct level, not by static design, but through the process of 'delegated aggregation'. In practice, rights to, say, wildlife and forests are fully devolved to landholders or communities, who can then delegate powers upwards when economies and ecologies of scale suggest that this is prudent. In Zimbabwe, for example, elephant hunting generated significant income for the exemplary CAMPFIRE village of Masoka. However, these elephants often wandered beyond the boundaries of Masoka to raid crops in other villages. The people in Masoka agreed that they should share a significant proportion of the fees they earned from elephants with their neighbouring communities, because if they didn't these communities would take matters into their own hands and kill elephants. Consequently, they created a multi-village institution for managing and sharing elephant revenues. They delegated some of their rights upwards to a higher level of aggregation, but could always reclaim these rights if they were mismanaged. Note that they did not delegate rights to the whole district, or province, or nation, but only to the level required to internalise the costs and benefits of elephants.

Building institutions through delegated aggregation internalises costs and benefits at the right scale and fosters democratic accountability (Chapter 5). Moreover, rights and government 'originate' in the people who can always reclaim these rights, or hold the people to whom they have delegated responsibility accountable, thus creating 'constituent accountability' (Murphree, 2000). Delegated aggregation differs fundamentally from the undemocratic process of building scale through expropriation from above. Further, building scale from the bottom results in an authority pyramid similar to that for domestic resources (Figure 10.5), with many more functions retained at lower levels. The system is 'parsimonious', allowing upper levels to focus on doing far fewer things better. I run a training exercise in which I ask communities to write the functions related to natural resource management (e.g. managing fields, water points, antelope, elephants, money, projects) on cards. Next, I ask them to put these cards at the appropriate jurisdictional scale (i.e. household, village, area, province, nation). I ask them to do this in two ways: as functions are currently allocated, and as they ideally should be allocated to internalise costs and benefits at the right scale. Invariably, the current allocation accumulates functions towards the top of the system (as in the left half of Figure 10.5), while the ideal allocation clusters functions at the bottom (as in the right half).

The example of Zimbabwe's system of natural resource governance (Chapter 8) illustrates how this theory can be operationalised in ways that entirely fulfil Ostrom's principles of collective action and Murphree's rules for scaling (Table 10.2). The Parks and Wild Life Act devolved rights to use wildlife and other natural resources to individual land units (landholders and, later, to communities). However, just as the state is inadequate at internalising costs and benefits, so is pure privatisation and the market. Ostrom recognised this in her Nobel lecture entitled: 'Beyond markets and states: polycentric governance of complex economic systems' (Wall, 2017). Collective action, reciprocity, and sanctions are often needed to manage the natural complexity that overwhelms both the market and the state, not by replacing these systems but by complementing them. In Zimbabwean landscapes, the Natural Resources Act of 1941 was successful because it empowered landholders to democratically self-regulate the spatial and temporal externalities associated with wild resources. It ensured that decisions were well informed through additional investments in extension and education about soil erosion, grazing management, and so on. The apex organisation of the conservation movement was the democratic Natural Resource Board rather than the more standard government department (Child & Child, 2015). A major contribution of Wall's (2017) book, *Elinor Ostrom's Rules for Radicals*, is that he highlights Ostrom's implicit belief in 'deep democracy', an emphasis she was never explicit about. It is fascinating that Zimbabwe's system of local natural resource governance stopped working almost as soon as it became administrative rather than democratic, even though these changes looked very minor on the surface (Chapter 8). This does not mean that the state does not have a key role. In the Zimbabwean system, government retained considerable background powers, but adopted a facilitative rather than an impositional regulatory culture (Nickerson, 1994). It inspected the status of natural resources in each community annually (through the Lands Inspectorate), but always in partnership with landholder committees. This carefully crafted system of state,

TABLE 10.3 The structure of cross-scale governance regimes in Zimbabwe

Scale	State functions and legal authority	Consequences (the numbers refer to Ostrom's (1990) principles for common property management)
Individual property	Parks and Wild Life Act of 1975: devolves rights to manage and benefit from wildlife to landholder	1. Clear rights and boundaries 7. Recognition of local rights
Collective action by landholders (i.e. Intensive Conservation Areas)	The Natural Resources Act, 1941: devolves power of self-regulation to landholders sharing the same catchment	2. Set rules locally and appropriately 3. All people participate in making the rules 4. Effective monitoring by landholders themselves 5. Graduated sanctions 6. Conflict-resolution mechanisms 8. Nested institutions (following Murphree's principles of delegate aggregation, constituent accountability, and jurisdictional parsimony)
Other meso-level functions	Extension (government) Inspection (government)	Cross-scale learning (through Intensive Conservation Area groups) Co-monitoring of impact by state and landholders
National functions	Natural Resource Board (democratic) Natural Resources Court	Democratic Natural Resources Board strongly influences policy Courts provide legal arbitration of last resort

market, and collective functions, described in Table 10.3, greatly improved the quality of environmental management on private land (Whitlow, 1988).

Defining roles across scale

I will use these principles and examples to provide some operational structure to the (often vague) idea of cross-scale and polyvalent governance (Figure 10.6) because, in practice, the definition of roles across scale is essential for effective governance, but is often confused.

Starting with governance, wild resources are privatised at the land unit level (i.e. private rancher, village producer community). At the meso-level, externalities are controlled through collective self-regulation (as described above), and these democratic structures also provide an important mechanism for cooperation and cross-scale learning. Ideally, they should also aggregate upwards into political organisations at the macro-level that, like a farmer's union, advocate for a favourable policy and regulatory environment, and play a technical role in supporting their members.

The role of the state lies in the provision of policy, some services, justice, and security. These are usually administered at national level and operationalised at the meso-level through justice (e.g. police, judiciary), service (e.g. health, education, extension), and oversight (e.g. natural resource) agencies. Learning operates across scales, although research is usually centralised because of economies of scale.

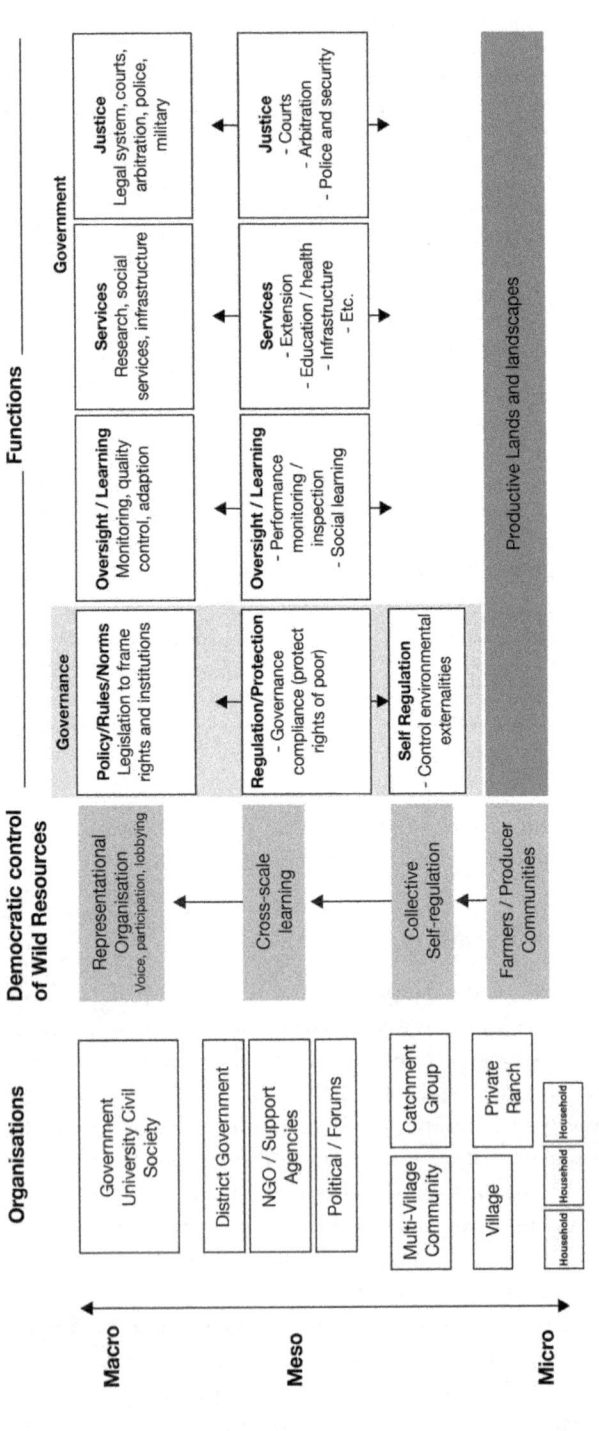

FIGURE 10.6 Polyvalent governance and the allocation of roles across scale

Governance is highlighted in grey in Figure 10.6, and is our focus. Recalling Williamson's four-layer economy in Box 4.5, the design of rules at Level 2 determines the structure of property rights, markets, and organisations at Level 3. The starting point for the sustainable governance approach, and CBNRM, is the devolution of rights for wild resources to landholders and communities. However, as we will elaborate in the following chapters, CBNRM is only effective where communities are also well governed. The Zimbabwean legislation was exceptional at promoting self-regulation amongst land units. However, the need to design rules for inclusive governance within community proprietorial units was not recognised at the time. This is an oversight shared with almost every CBNRM programme on the planet. In some cases, it is assumed that community leaders and elections provided for sufficient levels of participation and accountability. In others, it is assumed that all community members affected by rules and decisions would participate in making them, face-to-face. Both assumptions are wrong. Chapter 4 explained why inclusive governance is not the natural state of affairs in most communities in forests and drylands.

Jumping ahead of the narrative, one of the most important functions of CBNRM policy and support organisations, therefore, is to design the conditions for inclusive governance, and to enforce compliance with these conditions. Applying the words 'enforce' and 'comply' to participatory and democratic processes is a controversial statement – until one recognises that CBNRM needs to be rigorous and disciplined, and that the purpose of democratic compliance is to break the stranglehold of top-down, personalised, and often autocratic governance, and to replace them with inclusive governance based on rules and participation. Indeed, the purpose of these rules is to protect Ostrom's deep democracy, by ensuring participation and transparency in governance, and protecting the rights of marginalised groups, including women. Where devolution is not accompanied by clear and well-defined rules, and systems to ensure that communities comply with these rules (see Chapter 14), systems break down, or are deliberately broken, to provide space for elite capture and even corruption in a manner usually ascribed to nations, but of equal importance in communities (Chabal & Daloz, 1999; Chayes, 2017). Devolution and community development is a conceptually and managerially rigorous process, rather than a touchy-feely one.

Ensuring that governance complies with the 'rules' of CBNRM is not always easy. Many of us are working in these areas precisely because the system is broken and characterised by top-down decision-making, extractive and personalised governance, and low social capital. This raises the question of who designs the rules and manages compliance with them. Theoretically, governance compliance is a state function. However, the situation is seldom ideal, and we need to be innovative about designing the rules for effective governance and then monitoring and 'imposing' compliances with these rules to maintain the integrity of the system. In addition to strengthening the rules in ungoverned spaces, an important goal is to rebuild social capital and trust as communities learn to use, trust, and work within these rules for inclusive governance.

Cross-scale adaptive management

This introduces the topic of social learning and adaptive management, and many similar terms such as action research and transdisciplinary co-learning. In complex non-linear and rapidly changing social-ecological systems, with multiple world views and stakeholders, we need to leave behind us blueprint planning and the primacy of the technocratic elite to consider progress and policy as a more democratically managed experiment (Murphree,

1999). Thus, the fourth aspect of the sustainable governance approach is collaborative, cross–scale, adaptive management and the co-creation of institutions, systems, and capacities.

Adaptive management (Figure 10.7) is a rigorous learning process, not trial and error (Martin, 1999). It starts with local people setting their own objectives, strengthened with additional viewpoints that incorporate broader ecological, economic, and social implications. Transformation usually requires a theory of change that, at the very least, includes governance, economic, and environmental principles, forged by the understanding of primary stakeholders (i.e. landholders) with assistance from other stakeholders. Adaptive management is the empirical testing of this development hypothesis. Inductive co-learning occurs experientially through implementation, supported by the monitoring of key metrics of performance including livelihoods and the economy, the environment, and governance.

As we move from authoritarian to democratic governance of wild resources, we have a lot to learn about iterative feedback processes and the formation of social capital. In my experience, it is important to involve most or all of the community in decision-making to ensure decisions are supported by the best available data and to enhance the process of learning through high quality technical facilitation (see Figure 10.8). Ah-ha moments result in cognitive change at the level of individuals and/or communities, and cognitive change can result in communities taking on challenges themselves. It may also be that taking on these challenges is the source of cognitive change.

Nonetheless, effective adaptive management invariably combines face-to-face participation with the use of visualised data. Face-to-face interactions enhance problem solving in complex social ecological systems (Ostrom, 2007, 2009b). Visualising data (e.g. bar charts, maps) enhances understanding and communication, including the concept of 'distanciation'

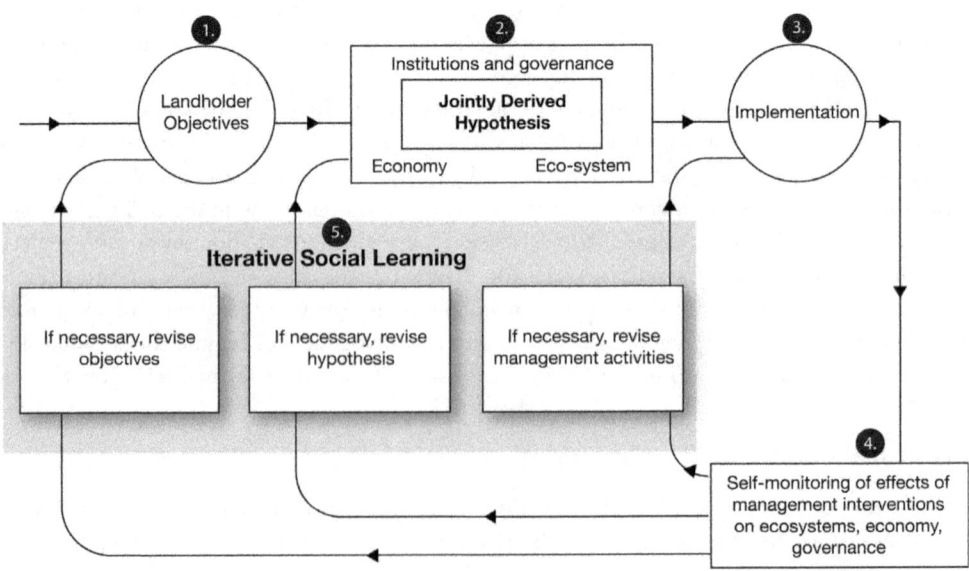

FIGURE 10.7 Collaborative adaptive management
Source: adapted from Martin (1999).

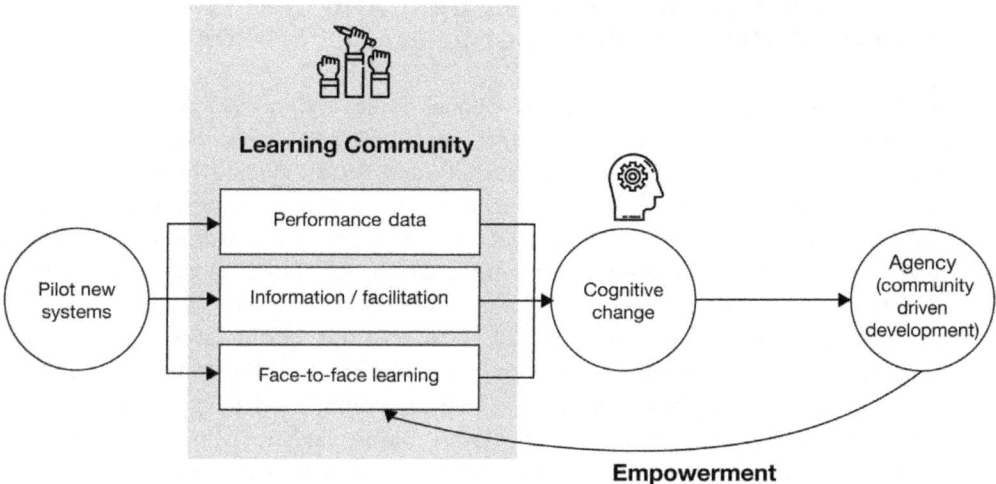

FIGURE 10.8   Developing learning communities: using action to catalyse cognitive change and agency

(van der Reit, 2008; Belay, 2012). In community workshops, I have noticed that the use of graphs and maps shifts people cognitively towards analysis and logical problem solving (using their frontal cortex), with much less complaining and personalisation of issues (their limbic, primitive, and emotional brain).

Many communities have been marginalised in decision-making, yet sustainability may ultimately depend on building communities into learning organisations (Senge, 1990). Given the limited experience and capacity to aspire in many communities, learning processes need to be carefully nurtured around practical tasks, including the under-standing and analysis of data. As illustrated in Figure 10.8, cognitive change and participation then leads to empowerment and, hopefully, agency, as communities absorb and take over the management of the system. Examples of this process are provided in Chapters 11 and 15.

Conclusion

The sustainable use approach represents an alternative conservation and development paradigm adapted to the changing economic nature of wild resources (Figure 5.5) and the emerging threats to them. As developed here, the sustainable governance approach conceptualises the 'use it or lose it' philosophy pioneered in southern Africa over six decades in the language of institutions and economics. Its credibility as a workable paradigm comes from the rapid, even spectacular, recovery of wildlife, growth of the wildlife economy, and rising political acceptance of wildlife as these ideas were applied. In conceptualising these ideas through my own lens of economics and governance, I draw attention to exceptional (but under-recognised) scholarship of a loose regional community of wildlife scholar-practitioners and 60 years of policy experimentation upon which I have drawn so heavily (Box 10.3).

BOX 10.3 A QUICK GUIDE TO THE ORAL AND GREY LITERATURE INFORMING THE SUSTAINABLE GOVERNANCE APPROACH

Like Murphree, I call attention to the exceptional quality and innovation of a vast oral and grey literature concerning sustainable use and CBNRM developed by scholar-practitioners, forged through inductive learning and largely ignored by international scholars (Barrow & Murphree, 2001). This conceptual learning is captured in policy and legislation, many grey reports, and some conceptual writing that is widely shared in the region, but far less so in peer-reviewed journals which are seldom readily accessible to practitioners and policy-makers.

Policy and legislation at the national, regional, and global levels written by Graham Child, Archie Frazer, Bernabie de la But and others in the 1960s and 1970s encapsulates many of these principles, which are summarised by Jones and Murphree (2001). From the 1980s, Rowan Martin, Malan Lindeque, Chris Brown, and Brian Jones, in particular, incorporated these principles into national policy (e.g. Zimbabwe's *Policy for Wildlife*, Minister-of-Environment-and-Tourism, 1992), southern Africa's regional approach to conservation (SADC, 1999), as well as into the resolutions of CITES and IUCN (Martin, 2009a) including IUCN's sustainable use principles (mentioned below).

Graham Child's *Wildlife and People* describes the emergence of a new philosophy for wildlife management well before many of the more academic theories that I use were conceived (Child, 1995). Many of these principles were developed by scholar-practitioners working collaboratively and regionally through IUCN-SASUSG over many years. Two booklets set out the path-breaking theory of change that guided transformational change and policy at the time (SASUSG, 1996, 2003). These were partly infused into IUCN's sustainable use principles (CBD, 2004) which, in the manner of negotiated compromises, lists issues rather than sets out a cohesive theory.

Under continual threat from international special interest, despite their success in recovering wildlife and pioneering CBNRM, these scholar-practitioners captured the lessons of 60 years of experience in five edited volumes:

- *Parks in transition* – which contemplates realigning parks and society in developed countries (Child, 2004).
- *Evolution and innovation in wildlife conservation* – which documents the rich history of state, private, and community conservation in the region (Suich & Child, 2009).
- *Community rights and contested land* – which addresses the challenge of elite capture of wild resources (Nelson, 2010).
- *Responsible tourism* – which addresses the role of the tourism economy in development (Spenceley, 2009).
- *Practical tools for community conservation in southern Africa* – which captures regional experience in CBNRM (Child & Jones, 2006).

These scholar-practitioners also influenced a number of edited books at the time (i.e. Hulme & Murphree, 2001; Oldfield, 2003; Fabricius et al., 2004; Dickson et al., 2009). A particular influence on this scholarship was Professor Marshall Murphree – southern

Africa's intellectual guru – at the Center for Applied Social Studies at the University of Zimbabwe. Murphree infused an understanding of politics and scale into the discussion about wildlife governance and policy, especially once it began to centre on rural communities. Murphree's pithy statements were soon engrained into regional norms as Murphree's Laws, which are well worth reading (Martin, 2009b).

Having taken scientific arguments to CITES, with partial success over many years, the southern Africans learned that, despite the rhetoric about 'scientific management', decisions in such forums are made largely along emotional lines. Thus, a major factor in reversing the ivory ban at the 1997 CITES meeting was to shift the narrative from sustainable use towards poverty eradication and communities, and to change the medium of communication using, for instance, community theatre instead of relying only on scientific data, which was mostly ignored (Guhrs et al., 2006).

In summary, the sustainable governance approach proactively reverses centralised, blue-print conservation and recognises that simple sops and single-factor decisions like banning the ivory trade only make complex problems like elephant conservation worse. Its basic objective is to 'maximize the benefits from wildlife to the people on whose land it lives' (Child, 1995), recognising that if wildlife has a comparative advantage this is good for wildlife and good for people. However, this approach is contentious. It highlights divisions over whether the state or market is best situated to manage the environment (Duffy, 2000). It challenges the centrality of public institutions, including NGOs, in the governance of wildlife. It disputes the supposition that wildlife is a global resource and the primacy of global meetings over grassroots democratic choices over wildlife. It offends those opposed to the use and monetisation of wildlife, who reply not with working alternatives, but by branding the approach with loaded, ugly (and perhaps meaningless) terms like commodification, neo-liberalism, or consumptive use. Thus, southern Africans often found themselves almost alone in global meetings making the case for sustainable governance, protected only by their track record.

Reversing the disappearance of wildlife in southern Africa was neither easy nor accidental. It took decades of effort, courage, and thoughtfulness, and can be reversed far more rapidly than it takes to develop the strong institutions necessary to take it forward. Decades of hard work, for example, were undermined by the crude recentralisation and hunting bans in Botswana (Mbaiwa, 2017), and by the reversal of the democratisation of wildlife in Zimbabwe by a predatory elite (Mapedza & Bond, 2006; Rihoy et al., 2007; Muyengwa & Child, 2017). Nonetheless, when applied properly, the sustainable governance approach is effective. It has four critical elements: proprietorship, price, devolved governance and subsidiarity, and the co-creation of institutions, systems, and capacities through evidence-based, collaborative, adaptive management.

11

KINDLING CBNRM

History and lessons from CAMPFIRE in Zimbabwe

LEARNING OBJECTIVES

This chapter describes the conceptual evolution of Zimbabwe's pioneering CAMPFIRE programme. It describes how:

1. Early initiatives failed because devolution was partial.
2. Critical assessments of these failures led to a much bolder model of devolved community wildlife management.
3. Empowerment and adaptive learning only really began once communities were 'given the keys' to wildlife through the legal acquisition of rights.

It then describes three participatory tools for operationalising the devolution of rights, namely:

4. Open, competitive marketing.
5. Participatory revenue distribution and governance.
6. Participatory adaptive management of quotas and hunting.

This led to rapid progress in community conservation, including participation, benefit sharing, and wildlife conservation.

However, CAMPFIRE was a pragmatic compromise between the intent of the wildlife department (i.e. private-community proprietorship) and political constraints. The chapter shows that:

7. When an approximation of the intended conditions (i.e. CAMPFIRE principles) was implemented, progress was rapid.
8. Without strong rights, and the protection of these rights, communities are ultimately vulnerable to elite capture at all levels. Even communities that experienced the multiple benefits of inclusive, participatory governance were unable to prevent local elite capture without help.

This provides critical lessons for the design of CBNRM institutions.

The emergence of an idea

Zimbabwe's CAMPFIRE programme was a major turning point in global conservation (Borgerhoff Mulder & Coppolillo, 2005), kick-starting the CBNRM movement in southern Africa. It showed that the sustainable governance approach could be applied to wildlife in rural communities, provided communities obtained the same use rights as private landholders and were well-governed. Unlike most CBNRM programmes reported in the literature (Arnold, 2001; Charnley & Poe, 2007), CAMPFIRE was essentially economic – 'getting prices right' through local proprietorship and exchange, a fundamental shift in governance that was also deeply political (Murphree, 1995). In describing how CAMPFIRE evolved, including devolving to communities the rights and capacities for generating benefits, sharing them, and managing wildlife, CAMPFIRE provides a surprising number of lessons about governance. CAMPFIRE did not just emerge in the late 1980s, but was the culmination of several decades of policy experimentation and learning (Table 11.1)

The Sebungwe region in Northern Zimbabwe

In the 1970s, the remote, northern Sebungwe region of Zimbabwe had abundant wildlife in communal areas considered unsuitable for human settlement (Child, 1995b). The area was sparsely populated by Tonga people, a culture of fishermen displaced by the rising waters of Lake Kariba, and dependent on food aid. The area has four protected areas

TABLE 11.1 Historical events leading to CAMPFIRE

1960s–1990	Devolution of wildlife to private landholders demonstrates viability of wildlife economy.
1970s	Parks director proposes community wildlife conservation in 1971. He persuades Parliament to allow revenue from protected areas excised from Tribal Trust Lands to be returned to communities (i.e. Chirisa, Dande, Malipati) in the mid-1970s.
Late 1970s and early 1980s	Park ecologists concerned about rapid human in-migration into wild areas propose the economic use of wildlife to (1) soften hard edges of parks and maintain corridors between them and to (2) promote rural economic growth. Martin and Taylor compile the 'Sebungwe Land Use Plan' which promotes conceptual development, but shows that integrated technical planning seldom works due to bureaucratic challenges.
1982–1984	Project WINDFALL (Wildlife Industries New Development for All) returns revenues from elephant culling and safari hunting to district councils (but not to communities).
1984/1986	Rowan Martin's seminal CAMPFIRE document advances the conceptual foundation for CAMPFIRE.
1985–1989	Pilot programmes (i.e. WINDFALL, Nyaminyami, Masoka, Mahenye) provide lessons and promote non-traditional partnerships (aid workers, wildlife biologists, social scientists, rural politicians, big game hunters) that galvanised CAMPFIRE.
1989	CAMPFIRE proper begins when Nyaminyami and Guruve acquire 'appropriate authority' after several years of preparation, rapidly followed by ten additional districts (with little groundwork.)

(Continued)

TABLE 11.1 (Cont.)

1989–2003	The wildlife agency mandates NGOs (WWF, CASS, ZimTrust) to manage CAMPFIRE, forming the technical CAMPFIRE Collaborative Group. They in turn establish the CAMPFIRE Association as a representative community organisation.
1992	The wildlife agency transfers leadership to CAMPFIRE Association.
1992–2003	Funding from USAID ($22.5 million) and Norway ($2.1 million) supports technical development and capacity building.
1996	Exodus of technical capacity from wildlife department, including key people who promoted CAMPFIRE (Nduku, Martin, Cumming, Taylor, Pangeti, Child).
2003	Final evaluation of USAID project shows CAMPFIRE is still innovative and working effectively, but is hampered by macro-economic decline and currency devaluation.
2003 –	Zimbabwe becomes a personalised and extractive economy with a culture of impunity, resulting in the re-exertion of elite control at community level. Somehow CAMPFIRE limps on, despite economic decline, the loss of technical skills, and the sidelining of NGOs including those supporting CAMPFIRE.

(Chizarira, Chirisa, Matusadona, and Chete), with the history of Chirisa Safari Area as Tribal Trust Land being acknowledged through an agreement that it would be managed to benefit local people. In the 1980s, a flood of environmental refugees from south-eastern Zimbabwe began to threaten this wilderness (rejected for settlement only 20 years previously) when the area was opened up by roads to support tsetse eradication, spraying large amounts of DDT, dieldren, and endosulphan, and funded by the European Union.[1]

Prominent park ecologists (Rowan Martin and Russell Taylor) convened government agencies to develop an integrated land use plan to respond to the influx of tens of thousands of people from degraded communal lands in the south (Martin, 1981). The 'Sebungwe Plan' was of a high technical quality and greeted with enthusiasm by stakeholders, but fell victim to inter-ministerial committees. This convinced the wildlife agency that technical top-down planning didn't work, even in a country as capably administered as Zimbabwe (Child, 1995b, p 158).

WINDFALL

In the late 1970s, aerial photographs showed that expanding elephant herds were transforming large areas of closed canopy Miombo woodland in the Sebungwe parks into grasslands (Martin et al., 1992; Cumming et al., 1997). At the same time, the highly respected Natural Resources Board (NRB) (introduced in Chapter 8) was making strenuous efforts to stop woodland destruction countrywide. Consequently, the NRB warned the wildlife department to address deforestation and biodiversity loss driven by elephants (and fire) pending court action. With a legal mandate to protect ecosystems, not individual species, the wildlife department initiated elephant culling operations (Child, 1995b). Complete herds of female elephants were shot, often in under 45 seconds, and all products were recovered, with hide being far more valuable than ivory. What is important to our story is that 380 elephants were culled in Chirisa in 1980, and 372 in 1981, earned US$463 000 from culling, plus $160 920 from safari hunting (a net benefit of $2.50 per hectare per year).

With his usual foresight, Graham Child had laid the groundwork by persuading parliament to earmark wildlife revenues from former African Areas (like Chirisa) to their associated communities well before independence (Child, 1995b, p 155). It might seem that persuading a white settler state to pay wildlife income to black communities was the biggest challenge, but this was painless compared to persuading a former British colony to depart from decades of centralised financing. This agreement provided the legal mechanism for WINDFALL (Martin, 1989), and the wildlife department worked hard to access this earmarked money to construct schools, clinics, and other projects.

The CAMPFIRE document

Taking advantage of the political space provided by the new Mugabe regime, the Parks director, Graham Child, amended the Parks and Wild Life Act[2] in 1982 to give communal areas the same rights over wildlife as private landholders. While Graham Child drove the new philosophy of devolved wildlife management, he credits Rowan Martin[3] for recognising the weaknesses in WINDFALL and for crystallising the Department's rather ill-defined ideas as CAMPFIRE (Child, 1995).

Martin lamented that the enormous technical efforts behind the Sebungwe Land Use Plan did little 'except, perhaps, some education of the planners' noting that 'local people had already found the best arable lands (without any help from the planners)'. In a statement epitomising CAMPFIRE, Martin argued that 'it became clear that one cannot impose a plan from the top' (and listed eight major planning efforts doing exactly that).

Martin's (1986) self-critical examination was typical of the CAMPFIRE learning environment. Although he had initiated WINDFALL, he quickly recognised that it did not create a direct link between wildlife and benefits. Projects were chosen by district officials, not local people, favoured populated areas rather than wildlife communities, and were indistinguishable from projects implemented by government as part of its normal responsibilities. Martin recognised the catch-22 stalemate where government would manage wildlife in communal areas 'pending the readiness of the local communities to take over'. Like teaching teenagers to drive, there can be no progress until communities are 'handed the keys', with some supervision and coaching. A believer in local control, Martin criticised hand-outs:

> We tend to say 'look what we have done for you' and this is followed by the presentation of a cheque, which bears no immediate relation to any conservation act by any group or individual. One cannot do someone else's conservation for them and pay them not to be involved.

Change would not occur, he said, if wildlife income was paid to the District Council and not to the man who 'lives cheek-by-jowl with wildlife [who] cannot directly relate any benefit he receives to any particular animals which were shot in his area'. The links between wildlife and benefit were further degraded by the long delays in the release of payments by the Treasury.

Perhaps for the first time, Martin's CAMPFIRE document applied economics and institutional economics to the challenge of community wildlife management. Quoting Garrett Hardin's *Tragedy of the Commons* and Stroup and Baden's (1983) emphasis on wildlife proprietorship,

Martin recognised that rural people favoured cattle and cotton because they owned it, but disregarded the economic values of wildlife because they did not. Martin puzzled over how to devolve the ownership of wild resource to communities living in communal areas. His colleague, Norman Reynolds from the ministry of economic planning, had spent the war years as a conscientious objector learning about collective village governance in India (e.g. Wade, 1987b) and introduced Martin to the concept of a 'Village Company'; Martin recognised the foundation of CAMPFIRE needed to be in voluntary community units (Box 11.2).

The big idea behind CAMPFIRE, like private conservation discussed in the preceding chapters, was to strengthen community proprietorship. Internalising the costs and benefits of land use (and directly addressing the open-access nature of communal lands) would allocate rural resources more efficiently. This required small-group tenure and rights of exclusion, plus mechanisms, such as prices or shares, to allocate resources between competing uses on, for example, common grazing land. CAMPFIRE was about improved resource allocation, rather than wildlife conservation. Therefore, new economic institutions should include all resources (land, water, grass, trees, wildlife), although the Department believed wildlife was competitive in drylands. Community participation would be voluntary (and demand driven). The major practical challenge was the process of delineating community groups, and their governance, with a series of field or policy experiments leading to a growing understanding of governance and participation, sorely lacking in WINDFALL.

Nyaminyami: making wildlife tangible for local people

Communities in Nyaminyami relied heavily on food relief after their relocation from Lake Kariba in 1955, yet large amounts of money from hunting and tourism were being paid to central government. Metcalfe, working for Save the Children, recognised this inequity (Metcalfe, 1993a, 1993b) and organised the Nyaminyami Wildlife Management Trust to capture wildlife benefits. He was supported by the local park ecologist (Russell Taylor), who believed that wildlife could pay more than marginal agriculture in these rugged drylands. In a major breakthrough for CAMP-FIRE, Nyaminyami received 'appropriate authority' in 1989 and became the first district to receive 100% of wildlife income directly. Paradoxically, Nyaminyami never adhered to CAMP-FIRE's devolutionary principles. The Trust was strengthened by donors to manage wildlife and its benefits 'on behalf of the people' (Murombedzi, 1991) and never applied the principle of devolution to its own communities, illustrating how hard it can be to unlock initial mistakes.

Guruve: replacing top-down approaches with local empowerment

Two hundred kilometres to the east, Professor Marshall Murphree was working with the Masoka community in the hot, arid Zambezi Valley (Murphree, 1994b; Taylor & Murphree, 2007). Murphree was concerned that top-down land-use plans imposed by the Mid-Zambezi Valley Resettlement Programme was disrupting the original communities (Derman, 1990) and hoped that devolving wildlife income would empower communities against this. Guruve District was also granted appropriate authority in 1989.

Working at community (rather than district) level, Murphree used his strong relationships to persuade the Masoka community to consolidate its settlements inside a fenced zone to reduce wildlife conflict and protect land for wildlife (Taylor & Murphree, 2007). The idea of self-governance captivated two leading local politicians (Ephraim Chafesuka and Taparendava

Maveneke), who would go on to provide energetic political leadership to CAMPFIRE, promoting its ideals to the ZANU-PF government.[4] Adopting what would become known as the principle of 'producer communities', Murphree (2005) credits insightful leadership in the district council for paying wildlife revenues back to the communities in which they were earned (rather than equally to all communities), and for avoiding the 'paternalistic and instrumental' norm of spending wildlife revenue on behalf of the people. With great maturity, the council gave the entire \$47 000 to Masoka for the people to allocate for themselves. Each household took \$200 for themselves, and the balance was used to upgrade their school. This changed attitudes to wildlife: 'We see now (said the chief) that these buffalo are our cattle. We are going to farm them' (Murphree, 1997, 2005), which they did. Murphree explains, in his eloquent way, how wildlife revenue empowered the local community to govern themselves, using income shrewdly for food in drought years and infrastructure when food was plentiful. Poaching and burning became anti-social activities. For the first time, too, communities recognised that natural resources were scarce and needed to be planned, protected, and managed. Instead of encouraging immigration (to improve the likelihood that the government would give them infrastructure), Masoka tried to discourage it, not always successfully.

Stockil and the Mahenye community

At the other end of the country, the Shangaan people of Mahenye were traditional hunters and were virtually at war with Gonarezhou National Park across the Save River. Shadrack, an infamous poacher born in the village, was killing the mighty elephant bulls of Gona-re-Zhou almost at will. In a single two-week period, the department's elite anti-poaching unit recorded 80 poaching cases linked to the village.

Clive Stockil, hunter and game rancher, had grown up in the area as the son of a missionary, and understood the Mahenye people. Some 60 Shangaan families had been relocated from the Park in the early 1970s, and the people believed that if they killed all the animals (which were rightfully theirs), there would be no tourists, and if there were no tourists the park would be abandoned and they could go home.

Stockil sketched out an alternative future with the community. He argued that Shadrack left the meat of poached elephants to rot, which was wasteful and not the tribal way. If Stockil shot two elephants, and gave the community the meat and a big cheque, would they stop poaching? The community agreed to consider this, and the Park director immediately (in 1982) gave Mahenye a quota of several elephant bulls for trophy hunting. After guiding his clients to shoot the elephants, Stockil delivered the meat to the community and paid the fees to the government. Being the WINDFALL period, the money took three years to percolate through the layers of central and local government. Mahenye's people showed great faith and patience, and the money built their first classrooms. Before this, children were taught under a tree modified to make a blackboard. The community lived up to its promise; the crack anti-poaching unit apprehended less than ten poachers, all for minor fishing offences.

As CAMPFIRE progressed, Mahenye rapidly became a model programme (Rihoy et al., 2007). Stockil convinced the community of the importance of key riverine habitats, and seven families voluntarily abandoned these for wildlife. Quotas slowly increased, and with it income. By 1990, Mahenye was being paid directly for hunting and tourism, and sharing the money equally, taking some for household cash and some for building a grinding mill and a clinic. In addition to the hunting, I helped the community negotiate terms for a high-end tourism lodge, which provided some 36 permanent jobs, an access road (rather than a bush track), and electricity. Traditional fish drives were reintroduced, and the community set up its own hunting club using a quota of small buck to retain traditional hunting skills. As the films about Mahenye and Shadrack illustrate (Box 11.1), CAMPFIRE was indeed working.

Mahenye was the only ward in Chipinge District with wildlife. However, the Chipinge councillors, outnumbering Mayenye 31 to 1, argued that wildlife money should be shared equally between all 32 wards in the district. Mahenye's councillor was wily and wise. Yes, he agreed, it was only fair that the district should share its bounty evenly. Of course, Mahenye would willingly share their wildlife dividend with the other 30-odd wards. And they were very much looking forward to their portion of the maize, cotton, and tobacco from agricultural land in the rest of the district. From this point, Mahenye retained 100% of its wildlife revenue, less a 15% levy retained by the district council to cover management costs.

BOX 11.1 USEFUL DOCUMENTARIES FOR TRAINING

Three films provide valuable training aids to the story of the wildlife economy and CBNRM. *Save Valley Conservancy*, produced by Simon Taylor in 2000, describes how private land-holders switched from wildlife to cattle and back to wildlife (Chapters 8 and 9).

In 1990, the South African Broadcasting Corporation's *50–50* featured Mahenye community, including the challenge of living with elephants in fields, and the transition to CAMPFIRE. This documentary shows the community role playing the process of choosing a commercial hunting partner, and then presents a live revenue distribution process. Clive Stockil, the hunter, places piles of money on the table in front of the whole community, pointing to the animals that he was paying for – $10 000 for an elephant over here, and $150 for a warthog over there, even more than for a cow! Having debated their choices over several days, the community then comes up one by one to get their cash dividend, and pays the agreed amount back into buckets for mutually decided projects. We then see these projects, and the high level of account-ability as the council's financial manager is grilled by the community about how the money given to him to make certain purchases was used, or not used.

The third film, *Finding Shadrack*, was made independently of CAMPFIRE, ten years later (also by Simon Taylor). In tracing the story of the notorious poacher, Shadrack, once a Robin Hood-style hero, it shows how CAMPFIRE transformed people's antagonism to wildlife. They now view poaching as anti-social and wildlife as an important economic option.

Implementing CAMPFIRE as a national programme

With a philosophy of empowering the wildlife industry to manage itself, the wildlife department built a coalition of three NGOs to implement CAMPFIRE. This small but invigorated partnership – the CAMPFIRE Collaborative Group – met monthly to share information and decisions. Roles were clearly defined, with ZimTrust organising communities, WWF providing environmental and economic expertise for managing wildlife, and the Center for Applied Social Sciences (CASS), University of Zimbabwe, providing socio-economic research, monitoring, and advice, and many critical social insights (e.g. Bukamuri et al., 2009; Martin, 2009b).

The Group then assisted the communities to form their own CAMPFIRE Association, and in 1992 took the bold move of making the technically inexperienced but politically astute CAMPFIRE Association the lead agency. This prescient move gave CAMPFIRE political credibility and momentum, and kept it alive after the weakening of the wildlife department in the mid-1990s; but sometimes it frustrated the technical agencies. Later, Africa Resources Trust was created to monitor and influence an external environment that was often hostile to sustainable use, while *Action Magazine* integrated CAMPFIRE materials into school curriculums. The Ministry of Local Government was also a member of this partnership. CAMPFIRE was entirely home-grown and rapidly became an internationally respected programme (Borgerhoff Mulder & Coppolillo, 2005; Measham & Lumbasi, 2013).

Lighting the CAMPFIRE

In 1989, a workshop was held to share lessons from the first two districts. This brought together an eclectic and dynamic mixture of white game rangers, rural politicians, and freedom fighters on the opposite side in the bush war, as well as academics, socialist development workers, and community leaders. These were all Zimbabweans who would lead the programme through the next decade. Hearing about CAMPFIRE, several additional communities gate crashed the Makuti workshop. They had not received the preparatory help provided to Nyaminyami and Guruve, but were eager to start. With Rowan Martin, a proponent of adaptive management, making decisions for the wildlife agency, ten new districts were immediately granted appropriate authority (i.e. wildlife rights). This boldness was inspired,[5] as 'authority is a prerequisite for responsible management and should not be held out as a reward for it' (Murphree, 2000; Martin, 2003). Communities immediately got hunting quotas, sold hunting concessions, and began to spend the money, while being held accountable to the CAMPFIRE principles (Box 11.2). As the new CAMPFIRE Coordinator, I was responsible for designing and managing this rapidly growing programme (Table 11.2).

Zimbabwe had a three-tier system of local governance, with districts comprising about 20 wards, and each ward having about ten villages. From the outset, the Ministry of Local Government opposed villages and wards from becoming legal entities. This created the legal problem that appropriate authority applied to district councils, but CAMPFIRE aimed at 'producer communities'. This wildlife department had already learned, through WINDFALL, that centralisation at the local level didn't work. It therefore persuaded district councils that the wildlife economy would not work unless districts devolved benefits and decision-making to communities, just as the wildlife agency had devolved rights to them. Many professional administrators in districts immediately saw the value of devolution and bought into the spirit of the 'CAMPFIRE Guidelines' (Box 11.2) and the Parks and Wildlife Act and Policy.

While the wildlife agencies and district councils remained technically strong, CAMPFIRE was quite successful in ensuring that districts devolved 50% of their income, and often much more. The system wobbled when Zimbabwe slipped into economic decline and desperate councils retained CAMPFIRE income to pay their own salaries. However, political action by several leading communities and the CAMPFIRE Association led to a nationwide agreement that hunting concession holders would immediately and directly pay 50% of their fees to the producer community (CAMPFIRE-Association, 2016). However, this locked in a 50% tax on wildlife, with the silver lining that district councils maintained a financial interest in wildlife.

BOX 11.2 THE CAMPFIRE PRINCIPLES (1991)

The Department of National Parks and Wildlife Management (DNPWLM) laid out the CAMPFIRE principles, which are replicated in abbreviated form as follows (DNPWLM, 1991):

DNPWLM policy recognizes that landholders are better placed to manage wildlife on their land than the Department provided certain conditions are met. The DNPWLM has therefore granted 'appropriate authority status' to certain District Councils provided these Councils have stated their intent to follow the principles embodied in the CAMPFIRE concept. Appropriate authority status effectively gives District Councils the same rights to manage their wildlife as enjoyed by commercial farmers except that quotas must be approved by DNPWLM. To achieve these three objectives [improved managerial capacity, human well-being, and sustainable use of environment], several principles must be followed.

Principle 1: Return benefits to producer communities[6]

Councils are required to return at least 50% of gross revenue from wildlife to the community ... who must be fully involved in the process of choosing how to spend this money. Councils should retain no more than 15% ... [for overheads] and spend no more than 35% for wildlife management ... moreover this expenditure should be in the producer community that earned it. The DNPWLM will look with favour on Councils that (1) return more money to producer communities than these minimum figures and (2) involve communities intimately in making decisions.

Principle 2: Producer communities should be small and homogeneous

DNPWLM encourages Councils to define producer communities [as] 100 to 200 households because this is large enough for ... wildlife, and small enough that all households can be involved ... and accountable.

Principle 3: Full choice of expenditure

Producer communities must be given the full choice of how to spend their money, including both projects and cash payments ... and take the necessary steps to ensure producer communities participate fully in these decisions.

Principle 4: Accountability

Councils should keep producer communities fully informed of, and involved in, CAMPFIRE.

Principle 5: Open, competitive marketing

Hunting and photographic concessions must be marketed competitively using such means as auctions or tenders.

Principle 6: Avoid unfair taxation of wildlife

Wildlife should be taxed in the same manner as other resources. Like cattle or crops, benefits should be given to producers (households) in full.

A brief comment on donor financing

USAID and Norway provided important funding for the CAMPFIRE support agencies (and, indeed, for CBNRM in southern Africa). However, donor funding is often managed in a top-down manner reminiscent of WINDFALL, and is then usually counter-productive. When USAID handed Martin (from the wildlife agency) a typical top-down proposal, he offered them a choice: allow CAMPFIRE to rewrite the proposal (which he did over the weekend), or leave. In retrospect the high-quality technical support funded by these grants was exceptionally important, with high levels of peer accountability for performance. Interestingly, communities that received technical support but relied largely on their own wildlife income often did better than communities that received top-down grants.

The purpose of CAMPFIRE was to empower communities to make their own decisions. Donor financing, however, is not suited to these goals. Donors were unable to give communities discretionary block grants, but funded local projects much in the manner of WINDFALL, discouraging self-reliance and encouraging donor dependency. While this locked some districts into a donor dependence mentality, more confident and self-reliant districts were better at using donor money. We repeatedly argued that the key to CAMPFIRE was devolution, and that we should leverage further devolution by linking grants to metrics of devolutionary performance. Donors were unable to take up this challenge, or perhaps even to understand it, and defaulted towards funding projects through central mechanisms that they controlled.

The enabling environment

The process of implementing CAMPFIRE, greatly simplified, consists of five steps. Step 1 is to create an enabling environment. Martin (1994) contends that the devolution of use rights to communities through the legislative pen was almost enough on its own – job done. He is certainly correct that this is a necessary and foundational condition. However, he forgets the critical role that he and his colleagues played in promoting and protecting these rights at the national, district, and community levels. The literature often refers to stable, long-lasting communities where common property management emerged organically from within the collective (Wade, 1987a; Ostrom, 1990). However, in most places, including CAMPFIRE, the conditions for collective action regimes to emerge are far from perfect (see Chapter 4). Rights coupled with technical support played a critical role in building social capital, systems, and capabilities in communities that had been subject to exploitation and top-down rule for decades and even centuries.

CAMPFIRE then progressed rapidly through the co-development of processes that strengthened communities' rights and capacity in three important areas: step 2 the right to earn money from wildlife, step 3 the right to govern and spend this money with full discretion, but in ways that build inclusive governance and equitable benefit sharing, and step 4 the right to manage their wildlife sustainably to achieve community goals. After describing how these rights were operationalised, I will illustrate the iterative learning process of developing institutions, social capital, and technical capacities through adaptive management and partnerships between professionals and communities.

Tool 1: earning money

With wildlife under such threat, CAMPFIRE maximised wildlife income by replacing the government administrative pricing systems with open, competitive marketing. The engine behind this process was accurate records on wildlife quotas and prices (Figure 11.1). On behalf of the wildlife department, I approved all community quotas and collected accurate records of the fees earned by each community, such as those illustrated for Gokwe (in Z$ and US$) in Figure 11.1. I used the data from all 12 districts, which had between one and three large hunting concessions, to compile an annual performance index by dividing income by the value of the quota for each concession. The height of the bars shows the rising value of all the concessions. This quickly identified the community concession earning the least from their wildlife and, working with WWF (especially Ivan Bond), we assisted them to resolve these problems. The impact of improved marketing on the whole of CAMPFIRE is shown top right (see Child, 1995a for detailed results).

The first task was to work with district councils to optimise their hunting quotas and areas, and to re-advertise the concessions. In the first district in which we experimented with this method,

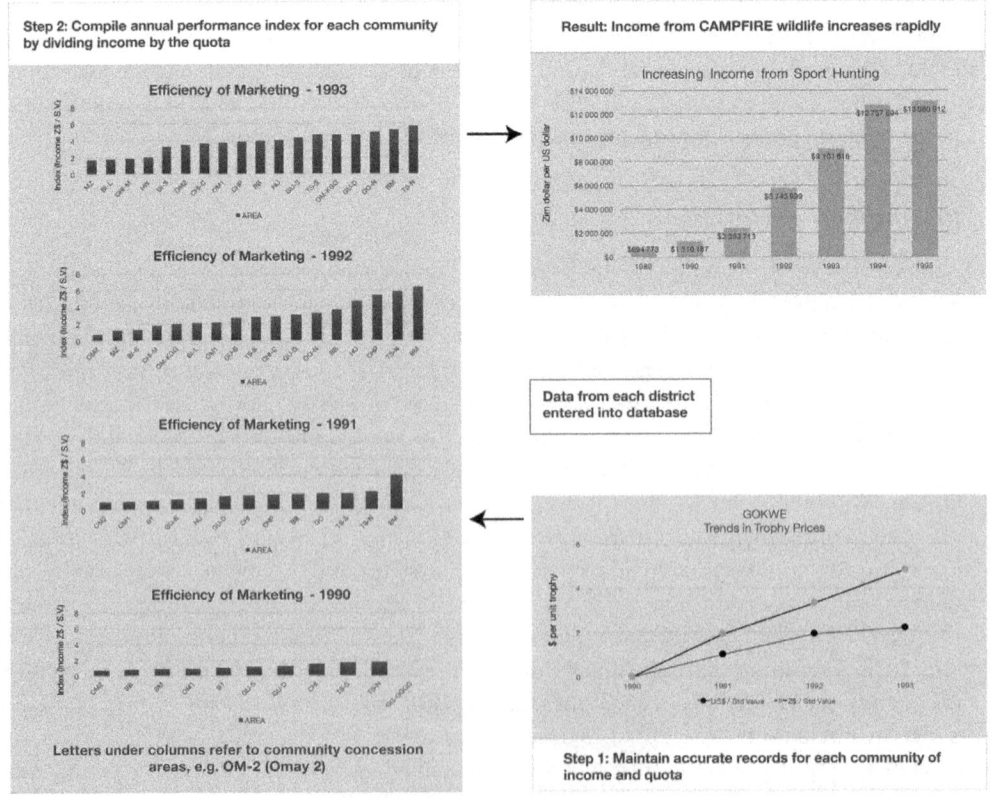

FIGURE 11.1 Using simple databases to optimise the marketing of hunting in CAMPFIRE
Source: Child (1995a).

we received 12 postal bids, assisted the councils to analyse these transparently, and increased the price of the concession from \$108 000 to \$290 000. However, we were not satisfied with the levels of participation, transparency, and empowerment achieved by evaluating postal bids.

The next iteration in developing community marketing was to use postal bids to shortlist the top offers, but then to train 10–20 community representatives to interview the outfitters in person. Working as pairs of facilitators, we took the community through three phases of training. Using a seminar and lecture format, we provided the community with a detailed technical and financial explanation of the safari hunting business, from top to bottom. Next, the community brainstormed criteria for selecting the best outfitter, designed questions for each criteria, and then practised interview techniques. Criteria included money (including the importance of having a minimum guaranteed price in the contract), the professional hunter's track record, knowledge of the area, relationship with the community, character and trustworthiness, language skills, intent to deliver meat to the community, provision of local employment, and attitude towards helping people in need. With one facilitator acting as a hunting outfitter and the other coaching the community, the community leaders practised interviewing questions and techniques for each question. This often raised important issues such as how to manage weakening exchange rates, whether to accept benefits in kind as opposed to cash, and the importance of a minimum guaranteed price.

The next day, each hunting outfitter was interviewed for up to two hours (Figure 11.2). Each community member asked the questions they had practised, probing further, and often bargaining the price upwards – 'Mr Stockil, we notice that you are offering \$XX for elephants when we have seen some elephants are earning \$YY in other districts. Knowing that we are interviewing other hunters, are you sure that is the highest price you can offer?' After the interviews, the offers and attributes for each hunter were summarised on cards on

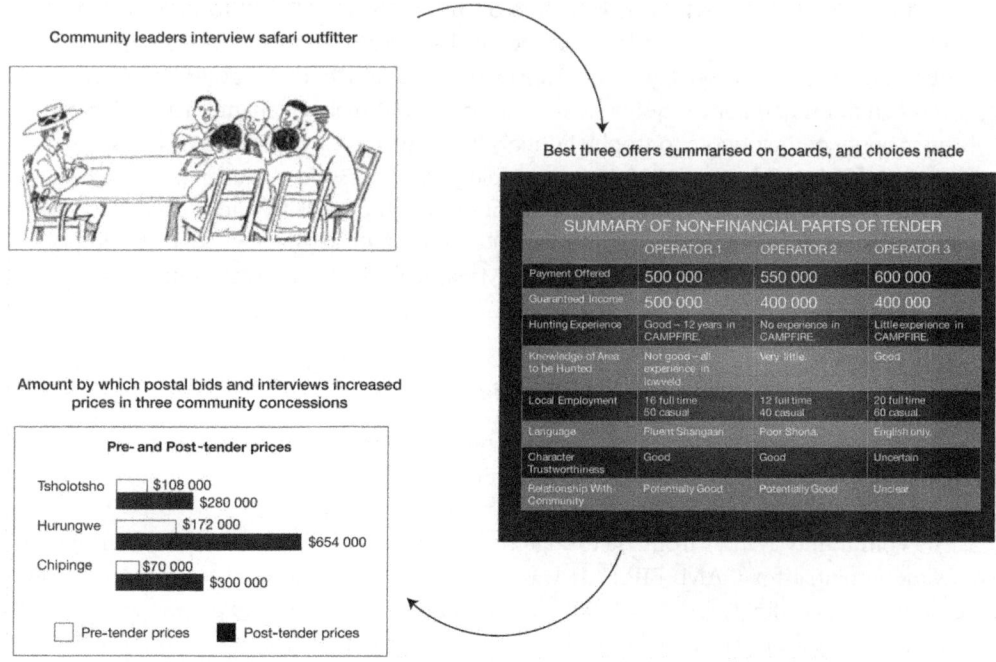

FIGURE 11.2 Participatory negotiation of hunting concessions
Source: WWF-SARPO (2017).

a large board (shown in Figure 11.2). This resulted in a long (and highly educational) discussion about trade-offs, such as total price versus guaranteed price, price versus the likeability of the hunter, and so on, until the best offer was finally chosen.

Not only did this process earn a lot more money (illustrated by the real results from three districts, see Figure 11.2), but it transformed the relationship between communities and 'white hunters', with participants often clapping and laughing spontaneously in excitement following their negotiations. Across-the-table negotiations helped community leaders to enrich their understanding of 'their' hunting business. This process was extremely empowering and economically effective, and is described in detail elsewhere.[7] More drily, data in Figure 11.1 illustrate a rapid increase in wildlife income, from more (and better designed) hunting concessions from 1990 and 1993 (x-axis). Animal lovers will be pleased to know that purposefully utilising wildlife and giving it the highest value possible reduced the number of animals killed, especially elephants (Box 11.3), particularly where communities were fully involved in setting and allocating offtake quotas.

BOX 11.3 REDUCING ELEPHANT OFFTAKES THROUGH HIGH-VALUE SUSTAINABLE USE

CAMPFIRE depended heavily on sport hunting, providing 92.5% of CAMPFIRE revenues, compared to 1.7% from tourism, and only 1% from hides and ivory (because of the ivory ban) with elephant hunting earning 64% of all CAMPFIRE revenues (Bond, 1994).

Surprising as it seems, this was good for elephants. Without sport hunting, it is likely that CAMPFIRE communities would have replaced over 12 000 elephants, buffalo, and many other animals with fields and livestock. Before CAMPFIRE, some 300 elephants were killed in communities each year, including 50 trophy bulls and 200 to 300 to protect crops and life. CAMPFIRE halved the number of elephants killed to 150. Using population estimates and trophy quality trends (incidentally, a much more accurate method), communities set an annual quota of approximately 150 male elephants. They allocated the majority of their quota to trophy hunting (because a trophy elephant bull was worth a hefty US$10 000 even in 1990 terms), keeping less than 30 for problem animal killing. Elephants are a nightmare for local people, but through CAMPFIRE and the values given to them, populations increased from 4 000 to 12 000 in the next decade despite increasing land use pressures (Taylor, 2009; Mazambani & Dembetembe, 2010).

Tool 2: spending money and governance

The second tool, participatory, activity-based budgeting, was first developed in Beitbridge, and the 'revenue distribution process' is a highly effective way of developing inclusive governance and social capital.

The community leaders from Beitbridge invited me to explore the district with them to assess the potential for CAMPFIRE. It was clear that a bold approach was needed because people were exceedingly poor and wildlife has almost disappeared. To generate revenue, I obtained a small quota of three elephants, a lion, a handful of buffalo, and a few other animals, and assisted the district to sell them to a local safari outfitter for $100 000. The animals shot in Chikwarakwara community earned $60 000 in the first year.

On a camping trip around the district, staying at schools, water points, and in the forest, I built strong relationships with the council chair (Nare), a dynamic young councillor (Mulaudzi), and two council executives (Sibanda and Modeme). In the evenings, we discussed the economic puzzle of why low-value livestock was replacing high-value wildlife, turning this into a role play that I often use for training (Box 11.4). We agreed that the traditional practice of using wildlife income for public projects would neither save the wildlife nor empower the people. We needed to radically change the system by treating wildlife as a private good, and trusting ordinary people with cash and their own decisions. Opportunities to change systems are rare, and one of the most nerve-wracking meetings of my life was the full council meeting called to discuss how CAMPFIRE would be implemented. Speaking in animated Venda, my camping companions sold this vision to their colleagues. In two breakthrough decisions, the council resolved that 100% of wildlife revenue would be paid to the 'producer community' where the animal was shot (not shared equally between all 32 wards). Moreover, wildlife would be treated like cattle, and the community was free to use the wildlife income as they wished, provided this was agreed by the whole community.

BOX 11.4 ROLE PLAY ILLUSTRATING THE DIFFERENTIAL OWNER-SHIP OF CATTLE AND WILDLIFE AND ITS FINANCIAL CONSEQUENCES

During the formative stages of CAMPFIRE, the late Ivan Bond and I used role plays to discuss new wildlife policies around the puzzle that $100 cows were replacing $1 000 buffalo. We called up two members of the community to act as cattle and wildlife respectively, using cattle horns or elephant dung as props. Cattle earned $100 so we gave the 'cow' one banknote to hold up. Buffalo earned $1 000–2 000, so we gave the actor ten notes.

The community agreed that the biology of cattle and wildlife was similar. Both were large, four legged ruminants, which converted grass into meat and lived in herds; cattle however were tame, while buffalo were dangerous and wild.

We then asked the community why they favoured $100 cows over $1 000 buffalo. The audience quickly concluded that they favoured cattle because they owned them, whereas wildlife was owned by the state. To illustrate the differential ownership of cattle and buffalo, the actor farmer sold his cow and put the $100 in his pocket to use as he wanted.

Then we asked what happened to the money from wildlife. People had no idea. So we role played this, passing all $1 000 to the actor representing the government. Next, we asked if the new CAMPFIRE programme would work if this $1 000 was paid to the council (as was currently the norm) and gave the actor representing the district council all ten banknotes. If the district council was good, the community said, it would invest this money in a school or a clinic, but probably in another area. However, we are poor, they replied, and this is unlikely to change our land use decisions.

Then we posed the big question. How could we encourage them to use their land for valuable wildlife, like private ranchers were doing? This question flummoxed them, and they were unable to describe an alternative future. So we facilitated new scenarios. What if the money from wildlife came directly back to their village? This would be much better, the community agreed, especially if it provided social services like schools and clinics. But they couldn't eat schools or clinics, nor could it buy important things like sugar, oil, or school uniforms, so they would probably still keep cattle.

What if we treated wildlife exactly like cattle? Pointing to the farmer with the $100, we gave him $1 000 from wildlife to use freely. If we shared money from wildlife exactly as we did for cattle, would this change their minds? Yes, they said, under these conditions they would be more favourable to wildlife because it was now theirs, but they still liked cattle for ploughing and a form of savings.

This role play illustrates how much the political economy of wildlife differed from that of livestock, and the need to transform this if wildlife was ever to become a competitive form of land use. Its economic logic led directly to the participatory revenue distribution processes described in the text, where households get wildlife income as cash and then 'tax' themselves for community projects and management.

The next day we set off for Chikwarakwara, and spent four days meeting with the community (described in detail by Child and Peterson, 1991). In the evenings, camping in the fig tree forest on the banks of Rudyard Kipling's great, grey, green, greasy Limpopo River, we set goals for the next day's meeting and planned how to communicate them effectively. My strategy is to empower and train local leaders to manage the whole process, and to sit unobtrusively at the back of community meetings, intervening only when essential. In the morning, under a large baobab tree, Nare, Mulaudzi, and Modeme led the community through a process designed to link wildlife to its benefits, and to involve the whole community in decision-making (see also Chapter 16).

Step 1 was to provide the community with basic information. The new CAMPFIRE policy was explained using role plays (Box 11.4) and drawings to emphasise local ownership of wildlife and its benefits. Step 2 was to define community membership. This was a long process, but the community eventually agreed on a clear definition of a household and ratified the membership list, after reading it out and modifying it several times. In step 3, a list of the animals hunted was carefully listed on a flip chart, together with their prices. People were volubly surprised that a warthog was worth more than a cow, and struggled with the decimal points for a $10 000 elephant. The fourth step was deciding how to use the money. The community had the full choice of how to use their money, one of which was cash, but it would be wise also to consider projects. These choices were written on a flipchart in what was to become a standard 'activity-based' budget format – household cash, community projects, wildlife management, and administration – with the figures being crossed out and changed as the debate progressed. Figure 11.3 shows the Mahenye community actively involved in making their budget way back in 1990, with a dedicated school teacher, Mr Masangu, adjusting the numbers on the flipchart as the debate progresses, using the standard format illustrated in Box 16.8. We were careful to stretch this process over several days, so people and families had time to discuss the trade-offs between private (cash) and collective benefit at home and in their own spaces. Most of the talking in the meeting was done by men, but we suspected that women would have their say at home. Moreover, the learning process is actually more important than the money and should not be rushed. Once the community decided how to spend their money, they elected an executive committee to implement their decisions. To reverse authoritarian tendencies, the role of the committee was defined (a) to bring people together to make decisions and (b) to implement community decisions, rather than making decisions for the community. As we emphasised, the community

FIGURE 11.3   Participatory activity-based budgeting

was the 'boss'. The committee was instructed to report to the whole community regularly to confirm they were implementing the collective work plan and budget.

We then returned 130 km to Beitbridge town to withdraw the money. This was blocked by the district administrator on the basis that he, not the community, was responsible for development and spending decisions. A tense political standoff followed, including phone calls to high ranking officials on both sides. The local leaders diplomatically suggested that the administrator might be stepping beyond his authority by challenging the rights given to local people by Section 65 of the Parks and Wild Life Act of 1975. If he wanted to cancel the cash payments, they continued, then it was his responsibility to inform the people in Chikwarakwara that this was his decision. The administrator relented, the council withdrew the money from the bank, and three days later the district administrator opened the ceremony in person.[8]

With the whole community dressed in their finest, Mr Nare and Mr Mulaudzi opened the 'revenue distribution ceremony' by grandly carrying in the $60 000 in bank notes, and placing it on the table for all to see. Speeches, presentations, songs, and plays reinforced the community's newfound rights to wildlife. Each member was then called up to receive their full wildlife dividend in cash – $400 ($60 000 income divided by 149 members) – which was a lot of money for people who often had no cash at all. Figure 11.4 shows the district administrator counting out $400 to a pregnant women.

An important message conveyed by this process was that members were entitled to 100% of the wildlife dividends in cash, but also had the responsibility to tax themselves to provide collective benefits. To illustrate this, buckets were placed on the table and labelled according to the projects selected – $170 for a grinding mill and $30 to support the school. Members then counted off bank notes into the plastic buckets for projects as agreed (their 'tax') (Figure 11.5), before signing the required paperwork, or rather thumb-printing it as most were illiterate. Finally, to again symbolise lines of accountability, the council chair introduced the newly elected committee to the people, pointing to the cash in the buckets and outlining what it was for.

This process was designed to encourage learning and to symbolise inclusive governance including participatory decision-making and accountability. The cash payments personified equal ownership of wildlife by 149 members of the village and their discretionary right to use this money as they chose, including to 'tax' themselves to provide projects that they understood

FIGURE 11.4 Each person gets their whole share in cash

FIGURE 11.5 Money is paid back into buckets for projects (tax)

and owned. This was a marked advance that contrasted with WINDFALL-like arrangements, where councils deduct money and build projects on behalf of communities, without the money ever getting to the people or being understood and allocated by them. A few weeks after revenue distribution, another ceremony celebrated the opening of the grinding mill, which the community had proven quite capable of acquiring and installing themselves. A year later I arrived a day early for the next annual general meeting (there were no cell phones in those days). With time on my hands, I talked to a number of people including a number of grandmothers, who, to a person, understood how much money they had received, how it was being used, and who was responsible for it. The grinding mill was running well.

Tool 3: the right to manage

The third set of tools are much broader in scope and strengthen the rights and capacities of communities to manage wildlife themselves. These adaptive learning systems were developed through a highly effective process of participatory technology development (Goredema et al., 2006) using games, exercises, and role plays. I will illustrate the process by describing the management of high-value hunting by communities, noting that the same processes can be applied to human wildlife conflict, fire management, campsites, fisheries, project management, and so on.

Wildlife is difficult to count, and a system of triangulation and adaptive management was designed to set quotas. Communities were trained to undertake walking transects and to record and analyse their data. Aerial surveys were done by outsiders, but the results were demystified using a game that involves flying toy aeroplanes over maps, counting the 'animals' (represented by different coloured beans), and learning about sampling and statistics. Communities learned to measure all manner of trophies, to record and analyse offtake records that hunters were required to keep, and to plot long-term trends on graphs. They set up systems for monitoring community wildlife conflict and protocols for reacting to incidents. This culminated in the annual quota-setting meeting. Data were summarised on a set of boards, with columns for previous quota, hunting success rates, walking counts, aerial surveys, trends in trophy quality, hunters' advice, community opinions, offtake percentages, and so on. The community debated the data for each species at length, adapting the overall quota slightly up or down. They then allocated their quota between different uses (trophy, meat, problem animal, own hunting), making trade-offs between the income and benefits from different uses. The resulting quota document was submitted to the wildlife agency for approval. This process reinforced the value of wildlife, linked it to landscape management, and ensured that the community knew what to expect from the hunter and the council (Taylor et al., 1997). These tools are captured in a remarkable set of manuals.[9]

Through a process that combined data and experiential learning, communities rapidly took on direct responsibility for managing their wildlife and its habitats, setting and allocating quotas, monitoring and recording hunting activities, employing game guards, counting animals, managing problem animals and electric fences, managing fire, ensuring that settlements or dogs did not disturb key hunting areas (identified on maps where animals were counted or killed), and so on. CAMPFIRE did not manage wildlife on behalf of the community. It threw them in the deep end with sufficient assistance to co-develop local management systems.

The performance and 'success' of CAMPFIRE

Once the enabling legal conditions were in place, CAMPFIRE's rapid progress can be measured by wildlife populations, participation, community income, political acceptance, and innovation (Murphree, 1997). Within five years CAMPFIRE made major progress, with over 80 000 households benefiting directly from wildlife (Table 11.2). CAMPFIRE began to generate revenues in Zimbabwe's most marginalised communities, and much of

TABLE 11.2 The rate of growth of the early CAMPFIRE programme

Year	Districts with appropriate authority	Number of wards getting benefits	Number of households	Benefiting population
1989	2	16	7 861	55 000
1990	12	30	22 040	155 000
1991	12	66	51 938	365 000
1992	12	70	70 610	495 000
1993	12	70	68 798	480 000

Source: Child et al. (1997).

this money was devolved to producer communities, who took on the challenges of implementing projects, managing hunting, protecting wildlife, human wildlife conflict, and so on. This stopped and even reversed the loss of wildlife in the face of powerful demographic pressures in remote communal lands (Taylor 2009). Politically, CAMPFIRE was endorsed at the highest levels, and the national CAMPFIRE movement grew stronger as over 20 additional districts, which saw the potential for local control over forests, water, and even minerals, joined the 12 original wildlife districts. CAMPFIRE promoted the concepts of local ownership and benefit regionally and globally, and led to a significant and useful literature (though I am sympathetic to Nuulimba and Taylor's (2015) frustration that progress made by communities is almost matched by academic criticism, which can be self-serving rather than problem solving). Success at village level reflected a shift from personal to impersonal, participatory governance and the 'rule of law'. The transformative power gained through negotiating contracts, sharing revenues, implementing projects, and managing wildlife can only be partly captured, even by the best writing (Murphree, 1994a, 1999; Rihoy et al., 2007; Muyengwa & Child, 2017) and film (Box 11.1).

I was reminded of this remarkable progress when, seven years after I left the programme, I led the final evaluation of 14 years of USAID funding (Child et al., 2003). Despite the unsettled Zimbabwean economy (we carried suitcases of money into the field to buy groceries), CAMPFIRE still had an energy about it at all levels, was driving devolution, and was innovative commercially. Data on income, wildlife trends and sustainability, ecotourism projects, meat distribution, employment, expenditure on wildlife managements, human wildlife conflict, and so on was being collected reliably to show CAMPFIRE moving in the right direction despite Zimbabwe's travails. This conclusion was shared by others (Mutandwa & Gadzirayi, 2007; Taylor & Murphree, 2007; Taylor, 2009) and, to this day, data continues to be collected (Mazambani & Dembetembe, 2010; CAMPFIRE-Association, 2016).

Progress in reforming the political economy of wildlife

Table 11.3 shows CAMPFIRE's progress in transforming the political economy of wildlife in terms of the devolution of authority and the rights to sell, benefit, and manage wildlife, as defined by Murphree(1994a). WINDFALL gave communities projects, but made only a tiny paternalistic step beyond conventional top-down public management. CAMPFIRE represented a giant step, but still fell well short of the vision of fully independent and tenured village companies with the same rights over wildlife as enjoyed by private

TABLE 11.3 Phases in the devolution of proprietorship

	Colonial period (1890s–1970s)	WINDFALL 1978–1989 First Generation CBNRM	CAMPFIRE 1989–1997 Second Generation CBNRM	CAMPFIRE (vision) Third Generation CBNRM	Domestic plants and animals
State	• Regulation • Authority • Right to sell • Right to benefit • Right to manage	• Regulation • Authority • Right to sell • Right to benefit • Right to manage	• Regulation	• Regulation	• Regulation
District Council		• Benefits (approved projects)	• Authority/exclusion • Right to sell • Right to benefit (50%) • Right to manage	• Regulation	• Regulation
Producer community	• De facto management (poaching)	• Projects from above	• Right to benefit (50%) • Right to manage	• Authority/exclusion • Right to sell • Right to benefit (100%) • Right to manage	• Authority/exclusion • Right to sell • Right to benefit • Right to manage

landholders, or over domestic resources, mainly because of the Ministry of Local Government's refusal to allow communities to become legally incorporated as independent entities.

We can also trace the process of devolution by 'following the money'. As CAMPFIRE Coordinator in the wildlife department, I visited all the districts regularly and 'audited' community finances.[10] I was most interested in what the expenditure told me about devolution, and shared these data publicly, especially at regular countrywide meetings of the CAMPFIRE Association (Table 11.4). Most pleasingly, the amount of money getting to communities increased (from 47% in 1989 to 74% in 1993) at a time when government support to councils was faltering. Moreover, the amount controlled and managed by the communities themselves increased from 17% ($189 278) in 1990 to 66% ($4.4 million) by 1993 (Table 11.4). Tracking the proportion of money devolved to communities became a standard component of district and national reporting at the CAMPFIRE Association (Figure 11.6). Unfortunately, tracking participatory use within the communities did not (because it takes time to collect and validate this data) and this lack of transparency allowed the local elite capture discussed below to continue 'undetected'.

Community income and differential taxation

Using the data still being collected by the CAMPFIRE Association, we can track the ups and downs of the programme in a turbulent time in Zimbabwe's history (Figure 11.6). CAMPFIRE income grew rapidly between 1989 and 1994, then steadily until 2001, with the major perturbations after that reflecting inflation and exchange rate fluctuations that are impossible to get to grips with. The income reaching producer communities increased from under 40% to 59% (and 74% if two non-conforming districts are excluded), which suggests that the measures taken to avoid 'aborted devolution', in the absence of statutory devolution (i.e. persuasion and full involvement of quality district officials), were moderately

TABLE 11.4 Measuring progress towards fiscal devolution by 'following the money'

CAMPFIRE Income and Expenditure 1990–1993

	1990	1991	1992	1993
Income (Z$)	1 658 025	2 030 243	5 511 594	6 620 210
Wildlife	1 089 855	1 975 243	5 511 450	6 597 353
Other	568 170	55 000	144	22 857
Expenditure (Z$)	935 906	2 095 290	5 505 815	6 996 073
1. Council retention	392 769	604 741	1 577 918	910 617
	36%	31%	32%	14%
2. Wildlife management by council	32 957	274 084	526 855	1 204 562
	3%	14%	10%	18%
3. Community benefits	510 180	1 216 465	3 220 042	4 880 894
	47%	62%	58%	74%
a. Projects managed by councils	320 902	122 500	167 247	521 56
	29%	6%	3%	8%
b. Devolved community income	189 278	1 093 965	3 052 796	4 359 332
	17%	55%	55%	66%

Source: B. Child, original data.[11]

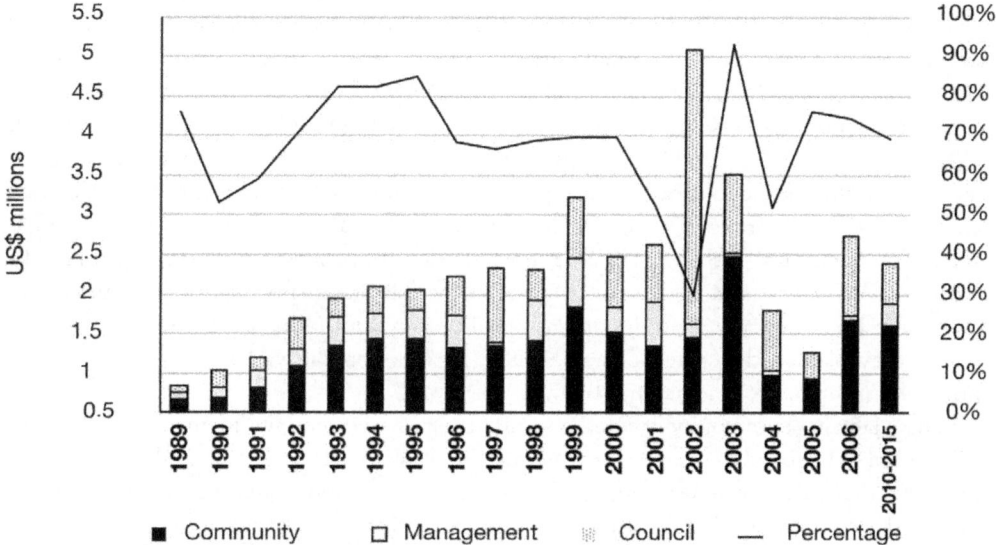

FIGURE 11.6 Allocation of CAMPFIRE revenues over time

Sources: based on data compiled from Taylor (2009); Mazambani & Dembetembe (2010); CAMP-FIRE-Association (2016).

effective. The decline in the communities' share in the late 1990s reflects the loss of professional oversight in the wildlife agency (Duffy, 2000). Remarkably, CAMPFIRE self-corrected after 2005, despite the loss of the officials and NGOs that started it.

Nonetheless, Taylor (2009) is correct in lamenting that the 50% retention of revenues by councils is a massive tax on the wildlife, with producer communities retaining just under 50% of revenues in the 24 years for which records are available between 1989 to 2015. Hill (1996) commented that CAMPFIRE is 'not only a wildlife program; it is very much a rural taxation program' with councils keeping half the wildlife revenues, but also expecting communities to fill in for government responsibilities by providing schools, clinics, and water-points. Although the CAMPFIRE champions warned from the very beginning that differential taxation would under-cut the future of wildlife (the same taxes are not applied to domestic crops and animals) they were never able to correct these percentages, and passed on these concerns to the Namibians who wrote the legislation to ensure that 100% of wildlife revenue was returned directly to community conservancies.

Aborted devolution

However, Figure 11.6 does not tell us what was happening inside the community. By 2000, Mugabe's power was threatened, the rule of law gave way to neo-patrimonialism, and people in power began to plunder the state, including CAMPFIRE, with a culture of impunity (Mapedza & Bond, 2006; Rihoy et al., 2007). Community leaders connected to the ruling hegemony exploited this culture of impunity to personalise many of the CAMPFIRE benefits. This undermined carefully built systems of impersonal accountability (Muyengwa & Child, 2017) reflecting the 'politics of disorder' and state capture (Chabal &

Daloz, 1999), albeit at a very local level. Without the protection provided by wildlife professionals and NGOs, local elites were quick to latch onto the fruits of CAMPFIRE (Rihoy et al., 2007). Ordinary people greatly appreciated the benefits of inclusive governance, but were unable to prevent its collapse as three brief case studies show.

In Masoka community, impersonal and participatory CAMPFIRE governance was well-liked by the community and led to remarkable gains in many spheres of devolved management, including household cash, many projects, employment, participatory administration, and financial transparency (Taylor & Murphree, 2007). By 2016, however, Muyengwa found a community distressed by the breakdown of democratic governance and elite capture (Muyengwa & Child, 2017). Silenced by the macro-political environment, they were only able to express their discontent in subtle ways, including through the spirit world where they blamed the spate of people killed by lions on the violation of CAMPFIRE's principles (Matema & Andersson, 2015).

The Mahenye community suffered a similar fate. As a community member said, 'CAMP-FIRE used to be for all the people, now it's a family business' (Rihoy et al., 2007). Rihoy et al. provide a much deeper understanding of the processes of impunity and recentralisation than is possible here. Local people fully understood that their rights were being abused, but with no-one to protect their rights, they were powerless to prevent the collapse of participatory budgeting and equitable benefit sharing. Land was allocated by the chief and a disgruntled community member explained that '[challenging the leadership] would result in losing land and even being chased from the area' (Rihoy et al., 2007). This echoes the argument in Chapter 2 that property rights underpin people's rights to engage in inclusive governance.

Zimbabwe's growing neo-patrimonial personality cascaded down to the local level, allowing politically loyal local gate-keepers to personalise CAMPFIRE benefits. Indeed, mirroring colonial policies of indirect rule or even feudalism, Mugabe restored the legal powers of traditional leaders through the Traditional Leaders Act of 2001, undermining CAMPFIRE through recentralisation, personalisation, and political impunity (Mapedza & Bond, 2006). With a decline in the transparency of the participatory governance so carefully cultivated by the wildlife department, CAMPFIRE became vulnerable to the 'politics of resource appropriation' with the political elite and their private sector allies taking a disproportionate share of wildlife value through patronage and other shrewd measures (Murphree, 2005). Nonetheless, Rihoy et al. (2007) were optimistic about the capabilities and resilience of rural people who, having tasted inclusive governance, would return to it when circumstances allowed.

While the pernicious effects of Mugabe's rule on communities are stark, global conservation forums can have similar effects. CAMPFIRE and its sister CBNRM programmes in southern Africa allocated enormous energy and resources to countermand threats to global markets (e.g. trophy hunting, elephant hide), drawing efforts way from wildlife conservation and poverty reduction (Hutton et al., 2005). Nonetheless, the rights of communities continue to be appropriated as we see with the recent hunting ban in Botswana, which may have satisfied an anti-hunting lobby, but reversed decades of investment in CBNRM. I measured a 40% decline in livelihoods in one community that were 'rewarded' for taking wildlife management seriously by being unceremoniously dumped into destitution overnight (Onishi, 2015). It was clear that blocking legal use would force communities into poaching and the shadows of illegal markets (Mbaiwa, 2017) and, indeed, a flurry of recent newspaper articles are reporting higher levels of poaching in Botswana. I make this point because undermining the rights of local people to own and benefit from their wildlife in undemocratic global forums may be as bad for people and wildlife as corruption and even the illegal wildlife trade.

Conclusion

CAMPFIRE demonstrated that community conservation can become effective remarkably quickly once simple principles are followed and well-designed, and inclusive institutions are in place. CAMPFIRE worked best when rights were devolved to a local level, and demonstrated the multiple advantages of face-to-face governance. It made significant progress in valorising wildlife and modelling local systems for wildlife management. Like private conservation, it showed that privatising wildlife (at the community level) out-performed conventional public management and the conditions in which this occurred.

CAMPFIRE achieved 'second generation CBNRM' (Table 11.3), allowing significant devolution and substantial progress but never reaching its goal of 'third generation CBNRM', where communities have the same rights for land and wildlife as for domestic plants and animals and the private sector, with the sole exception that they are managed as a collective. As a pragmatic compromise, made necessary by the government's refusal to recognise communities as independent entities, devolution was based on conditionality rather than title. This ultimately compromised how CAMPFIRE was implemented and led to a number of challenges (Murphree, 1995).

CAMPFIRE was implemented through Zimbabwe's unusually effective system of district governance. As we discuss in detail in the next chapter, participatory CBNRM greatly outperforms representational forms of governance, although in CAMPFIRE the relatively good performance of representational systems (Figure 12.5) can be traced back to strong oversight by district councils and their commitment to devolution. Nonetheless, the transformation of wildlife from being a public asset to a private-community asset was only partially successful, with excessive taxation (about 50% of wildlife revenues) by district councils. This situation improved while programme oversight was strong, but resource appropriation increased as oversight weakened. Nonetheless, CAMPFIRE did demonstrate examples of healthy contestation over money and authority between community and district, even during Zimbabwe's hard times. Tilly (2009) argues that democracy emerged from a similar dynamic between the rulers and the ruled in England as feudalism was transformed towards liberal democracy; and an optimist would see CAMPFIRE evolving in this direction had Zimbabwe remained a state based on the rule of law.

The CBNRM literature is somewhat unclear about what it means by democratic decentralisation (Ribot et al., 2010) and often conflates deconcentration to local adminis-tration with devolution (Chapter 5). Despite my admiration for the efforts, commitment, and bravery of many district council officials, who did so much to make CAMPFIRE work, and have not yet given up, we should not confuse community wildlife production with the rural administration of wildlife as a public asset.

Perhaps the most striking lesson of CAMPFIRE is that personalised and extractive governance rapidly re-exerted itself at the community level when the state became a sophisticated network bent on self-enrichment (Chayes, 2017). Communities were not strong enough to resist internal forces of elite capture without the protection offered by the CAMPFIRE Collaborative Group. We will never know if taking the full step to entitling individual communities with resource ownership could have avoided this.

How would CAMPFIRE have evolved if Zimbabwe's later political circumstances been less hostile to the concept of self-governing communities? Zimbabwe and Namibia always worked closely together and, heeding advice from CAMPFIRE, Namibia ensured legislatively that 100% of wildlife income was devolved to community conservancies. Like CAMPFIRE, Namibia understood the value of quality performance-related data, and after 25 years it is clear that Namibia's 82 registered community conservancies are making substantial progress in terms of wildlife recovery, community wildlife management, and income generation (NACSO, 2015). A powerful graphic (Box 13.5) shows that the substantial investment in building Namibia's national CBNRM programme had high economic returns. The programme 'broke even' in year 12 (2002) but by 2015 total economic returns ($60 million) exceeded support costs ($10 million) by a factor of six (NACSO, 2016, p 63). This year-12 break-even point coincides with the time that CAMPFIRE (and Zimbabwe) began to be seriously affected by economic and governance challenges. Having personally seen the early signs of exponential growth when I reviewed the programme in year 13, one wonders what might have been. As it happens, CAMPFIRE has survived nearly two decades of hyperinflation and political mis-governance, albeit in a much diminished state, and we will have to wait and see what emerges, if and when Zimbabwe recovers.

CAMPFIRE and sister programmes in the region, especially Namibia, demonstrate that we have the knowledge and tools to implement CBNRM with a very high likelihood of success, especially if we are tenacious and accept that rebuilding social capital and recapitalising degraded ecosystems takes at least 20 years. However, implementing CBNRM, like most worthwhile challenges, is not a matter of wishful thinking and good intention. It needs to be done properly, with a sound understanding of its underlying concepts. Progress requires disciplined adherence to a set of governance and economic principles, including locating natural resource governance and benefit at the lowest level – the rural village level. When CBNRM does not work, the diagnosis will invariably point towards weak devolution of rights to communities, elite capture within the community, failure to ensure that the wild resource in question is viable, and the absence of data-rich processes that involve communities in learning how to manage for themselves. If we are prepared to do CBNRM properly, with a sufficient commitment of resources including professional skills, there is a high probability of success.

Notes

1 The parks director, Graham Child (1995), rails at this environmental destruction: 'misdirected bureaucratic accountability ... is a recurrent theme. Bitter experience has shown that technical agencies and their policies become locked into inappropriate programmes, which persist long after their economic rationale has disappeared. [Tsetse control, buffalo eradication, veterinary legislation, etc.] has greatly prejudiced wildlife in an effort to prop up the viability of livestock, although wildlife is environmentally and economically superior. To avoid such artificial distortions, institutions guiding land use must become responsive to economic forces rather than the traditional power base to which ... bureaucratic agencies owe their existence. This implies more appropriate institutional arrangements to integrate local people into the decision-making process, with the word "appropriate" implying improved resource accountability' (Child, 1995).

2 When drafting the Parks and Wild Life Act in the early 1970s, Fraser and Child had provided for communal area communities to become appropriate authorities with rights equivalent to the mangers of private land (Child, 1995). This was (again) blocked by the politically powerful Ministry of Local Government. The next two near-breakthroughs were prevented by changes in government during the political transition from white-minority Rhodesia to independent Zimbabwe. However, immediately post-independence under the new Mugabe government this

clause (the famous Section 65) was applied to communal areas with a full amendment of the Parks and Wild Life Act in 1982. Policy documents prepared in anticipation several years previously emphasised the delegation of appropriate authority status to local communities. However, communities were governed by various local government acts and the ministry of local government refused to allow local communities to become legal entities. The wildlife department had to be content (very reluctantly) with devolution to district councils – a strategic compromise (Murphree, 2005). The return of CAMPFIRE revenues directly to district councils short-circuited the two-year Treasury process, but nonetheless fell short of the 'CAMPFIRE principles'.

3   Child had strong-armed the civil service into employing Martin, who was an electrical engineer and not a biologist. Martin was soon making his own telemetry devices to track and study elephants in Chirisa but, having actively promoted the Sebungwe Planning process and WINDFALL, recognised their shortfalls and wrote the path breaking 'CAMPFIRE document' (Martin 1986).

4   Ephraim Chafesuka, the Member of Parliament for Guruve, became the first chair of the CAMPFIRE Association and it was not long before Taperandavha Maveneke, the district administrator, became the dynamic executive officer of the same association. Using their political skills and connections in the ruling party, these swashbuckling individuals played a major role in the political acceptance of CAMPFIRE, right up to the President.

5   The wildlife agency ensured that CAMPFIRE began immediately in these communities, even though formal authority took several years to pass through the legal system.

6   The department issued a guidance letter that 'recommended' devolution of no less than 50% of income to producer communities, allowing councils to retain 35% for management and 15% as a council tax. This was a mistake, allowing too much money to be retained. Later, the department tried to insist that producer communities retained 80% of revenues, which was a successful strategy in high performing districts, but not everywhere. This provided an important lesson to the Namibians, who devolved 100% of wildlife revenues to community conservancies, an important reason for the continued success of their programme.

7   The step-by-step process used for developing hunting and tourism concessions and participatory negotiation are described in the excellent WWF manual series – 'Marketing Wildlife Leases'. The detailed results of this process are recorded in Child (1995). See also Child and Weaver (2006).

8   This description is greatly foreshortened. The reader can get a much better sense of the excitement elicited by this breakthrough in the detailed description of this event compiled by the late John Peterson and myself (Child and Petersen, 1991).

9   Ivan Bond, Lilian Goredema, Charlie Mackie, Russell Taylor, Norman Rigava (all working for the WWF Southern Africa Regional Programme Office), and Steve Thomas (ACTION Magazine) developed an exceptional set of training manuals. The 'Wildlife Management Series' includes manuals on CBNRM generally, managing safari hunting, marketing wildlife leases, quota setting, counting animals, problem animal reporting, fire management, electric fences, financial management, and project planning. Several of these are accompanied by manuals aimed at 'training of trainers'. These manuals provided comprehensive information in a highly digestible format, although Bond later found that it was even more valuable to use these as a foundation for assisting communities to write their own local manuals and posters.

10   Detailed reports describing each district compiled by myself are available from the Graham and Brian Child collection at the University of Florida.

11   This is data that I collected in my (friendly) 'audits' of councils as CAMPFIRE Coordinator. It led the CAMPFIRE Association to collect national datasets (with the help of the WWF) using district accounts that continues to this day. This early data is more accurate than the national datasets, but also excludes two districts that were not following the CAMPFIRE principles: Nyaminyami never recovered from the initial centralisation in the experimental stages of CAMPFIRE, while Chiredzi's CAMPFIRE programme was plagued by corruption at council level, which was later resolved.

12

DOES IT TAKE A VILLAGE?

Is there a difference between participatory and representational governance?

LEARNING OBJECTIVES

This chapter builds the case that participatory, face-to-face governance is essential for successful CBNRM. The chapter:

1. Describes the implementation of participatory governance in the Luangwa Valley, Zambia, and compares its performance against representational governance.
2. Discusses the strategy of 'imposing' participation in non-inclusive societies, and monitoring compliance with rules that protect marginalised groups including women.
3. Uses data from regional surveys to illustrate community attitudes to participatory versus representational governance.
4. Uses community finances to illustrate a threshold (Dunbar's number) above which informed participation and equitable benefit sharing decline precipitously.
5. Uses economic games to suggest that performance differences are structural and not cultural.
6. Relates participatory and representative governance to the challenge of scale.
7. Suggests that sequencing governance by scaling down before scaling up retains the benefits of participatory governance, but also allows for economies and ecologies of scale associated with representational governance.

The chapter concludes by asking if widespread participatory governance is logistically possible, and proposes that genuine progress is difficult without it.

Participatory democracy and scale

The foundational conditions for CBNRM (Chapter 13) are reflected in Ostrom's first and third principles: community rights – to personal safety and property – and face-to-face participatory decision-making. Although she never said so outright, Wall (2017) suggests that deep democracy and self-governance lies at the very core of Ostrom's work, a case which this chapter builds independently. The literature pays a lot of attention to participation (Pimbert & Pretty, 1997; Reed, 2008). Yet the mechanisms of deep participation, including small community size (Wade, 1987a), face-to-face decision-making (Murphree, 1994a) and the power of participatory or deliberative democracy (Aragonès & Sánchez-Pagés, 2009; Boulding & Wampler, 2010) are seldom given the credence they deserve, nor are they followed in development practice. My experience has convinced me that Ostrom's third principle is foundational: that all people affected by rules (or decisions) must participate in making them (Ostrom, 1990). CBNRM initiatives that rely on representational forms of governance, which most do, are ultimately flawed. Using a detailed case study (Luangwa) and empirical data from CBNRM programmes in several countries in southern Africa, I will make the case that genuine participation results in major improvements in performance and is a necessary condition of citizenship and multi-dimensional poverty reduction (Sen, 1999), but that it depends on the scale and form of governance.

Most of the literature describing the theoretically unexpectedly weak performance of CBNRM discusses the failure of decentralisation from state to community ('aborted devolution') (Nelson, 2010). This issue is obvious and needs no further elaboration. However, even when benefits do get to communities, most initiatives fail to secure the essential public goods of informed participation and benefit sharing. These outcomes are predictable, and therefore manageable; small communities that conform to strong rules of transparency, accountability, and participatory governance are highly likely to succeed. Unfortunately, CBNRM is usually implemented through multi-village representational committees, with a high probability of failure, through top-down structures that perpetuate the mechanisms of indirect colonial rule. We need to persuade NGOs, donors, and even government agencies that it may seem easier to work with the leaders of large (representatively governed) communities, but that CBNRM programmes governed this way have clay feet.

In my own experience, the differences between working with single-village participatory democracies and multi-village representational governance are stark. Therefore, in researching this chapter, I have been surprised that these differences are hardly, if at all, identified in project documents or the literature on decentralisation and CBNRM. The purpose of this chapter is to highlight these differences and provide a conceptual understanding of within-community governance, or 'micro-governance'.

What does the literature on participatory democracy and micro-governance tell us?

The conservation literature is beginning to discuss issues of democratic decentralisation and local democracy (Ribot, 2003; Larson & Ribot, 2004), with an emphasis on downward accountability through electoral processes, representational forms of governance that establish mini-natural resource agencies within district councils, and plural forms of governance that include co-management and NGOs (Ribot et al., 2010). Decentralisation, however, has often not provided the 'democratic dividend' anticipated (Crook & Manor, 1998), and in some places less inclusive forms of governance have re-exerted themselves (Ribot, 2007). In other examples, decentralisation appears to have been effective, including for forestry (Arnold, 2001; Agrawal, 2005), wildlife (Jones & Weaver, 2009), and other common-pool wild resources (Ostrom, 1990).

Weak performance results from a poor conceptualisation of democratic decentralisation. Thus donor projects, and even the literature, emphasise electoral accountability (a weak concept) rather than true democratisation and self-governance. Too much faith is placed in elections as a mechanism of accountability, yet elections are not even mentioned in Dahl's seminal definition of the criteria for democracy (Dahl, 1989) (Box 14.4), and those of us who work in rural areas know that 'elections' are not necessarily democratic.

Further, many of us working with wild resources represent public agencies, and think through a regulatory lens rather than, for example, perceiving CBNRM as a process in devolution and management of wild resources through privatisation, markets, and collective regulation. We tend to bureaucratise the governance of natural resources, shifting away from Ostrom's conceptualisation of common property management as consisting of groups of local (private) users who regulate themselves. Almost all of Ostrom's examples were of individual fishers, irrigation farmers, and water and forest users coming together to regulate themselves through collective action. This is very different from seeing community forest or wildlife management as a mini-government agency (even an elected one) regulating the use of resources by its constituents. In thinking 'beyond markets and states' Ostrom accepts the validity of private resource allocation, but also accepts that markets are crude and that collective action is often necessary to incorporate complexity, social values, sense of place, externalities, and other things that people value into resource trade-offs (Wall, 2017).

Participatory democracy

Key theoreticians of common property governance and CBNRM imply that participatory governance is important, emphasising that 'most individuals affected by the operational rules can participate in modifying the operational rules' (Ostrom, 1990) and the importance of scale and face-to-face deliberation (Murphree, 1994a). Having cited these two gurus, and their conviction that participatory democracy has powerful advantages, I note that it is not widely mentioned in the literature, and is even less widely practised, even in CBNRM communities where it is essential for rebuilding social capital and managing complex resource trade-offs.

In mature democracies we tend to forget that democracy means rule by the people, and that our systems originate in these principles. Ancient Greece is the quintessential example of participatory democracy, even if it struggled with the issue of scale and the tyranny of the masses. In Switzerland, direct democracy improves communications between citizens and representatives, encourages citizens to be better informed, and gives politicians less scope to pursue their personal interests. Moreover, participation favours common interest over self-interest, including a greater willingness to pay taxes because public services are provided more efficiently and are better understood (Feld & Kirchgässner, 2000). The cornerstone of American democracy was town hall democracy, giving all citizens a say in their own affairs (Tocqueville, 1994). In modern times, the effectiveness of participatory budgeting was highlighted by the experiments in Porto Alegre in Brazil. These suggested that participatory democracy has four major advantages and anticipates our findings in CBNRM (Marquetti et al., 2011): it strengthens democracy, as participants learn about their rights and responsibilities, it improves the use of public resources and reduces corruption, and its distributional effects improve the lives of the poor. Participatory budgeting promoted more efficient, transparent, and accountable administration of public resources, 'an outstanding achievement in itself' (Marquetti et al., 2011), with positive distributional effects targeting resources towards the neediest groups (Besley et al., 2005). Participatory democracy is also said to be transformative – making people more tolerant, reciprocal, better engaged in moral discourse and judgement, and more engaged in evaluating options (Warren, 1993). Indeed the literature on discursive or deliberative democracy suggests that participation encourages people to learn about issues together, learn about their rights and abilities to participate in governance, and, by becoming engaged in policy-making, people change the way they see the world and change themselves (Marquetti et al., 2011). Participatory democracy is built around informed face-to-face dialogue, rather than elections.

Lessons from Luangwa

The CBNRM programme in the Luangwa Valley in Zambia provides a quasi-experimental study of the effects of scale and forms of participation on the performance of CBNRM.[1] This project went through three phases. It began as a typical top-down integrated rural development project, working through the six chiefs and associated representative committees to do things for the people (1989–2005). It was then transformed into a bottom-up project through 43 democratic Village Action Groups (VAGs) (1996–2002), before it was again recentralised and managed through six Community Resource Boards. Assessing the performance of CBNRM as it flipped from representational to participatory governance, and back again, highlights the differences between these systems.

The Luangwa Valley is remote and neglected, with two superb national parks and extremely poor communities. In the late 1970s and early 1980s, poachers killed nearly 100 000 elephants and every single rhino. President Kaunda, a lover of wildlife but suspicious of his own wildlife agency, asked Norway to fund the semi-independent Luangwa Integrated Resource Development Project to manage the 9 050 km^2 South Luangwa National Park and uplift 50 000–70 000 people in six closely related Kunda chieftainships in the 4 500km^2 Lupande Game Management Area (GMA) (Gibson, 1999). With a large budget and strong leadership, the Luangwa Integrated Resource Development Project (LIRDP) brought poaching under control, provided infrastructure for tourism, and embarked on an ambitious integrated rural development project in the community. Forty per cent of the revenue from tourism in the Park and hunting in the GMA was used top down, largely as directed by the six chiefs (Gibson & Marks, 1995; Dalal-Clayton & Child, 2003). External donor reviews criticised this top-down process. Participatory rural appraisals showed that ordinary people did not understand the project or its benefits (Dalal-Clayton & Child, 2003), and perceptions about wildlife remained strongly negative (Balakrishna & Ndlovu, 1992; Wainright, 1996).

Fiscal devolution

I joined this project in late 1995 and introduced participatory governance through a one-page policy that laid out the roles of the different levels in the community and their fiscal entitlements (Table 12.1) which proved to be an excellent way of designing CBNRM interventions. Somewhat surprisingly, the five ministries overseeing LIRDP accepted this radical proposal (despite strong resistance from the six chiefs). This transformed community governance. All (100%) income from wildlife in the community area was paid directly into the community bank account (the park now retained its own revenues), and was divided as follows:

- The 43 villages ('doing' level) would get 80%, with complete discretion over use of this money provided the decision was collective.
- The six Area Development Committees received 4% of revenues, accepting that their function was coordination, and money spent on meetings and allowances adds little value.
- The chiefs got 1% of income as a personal and non-accountable payment.[2]
- The remaining 10% remained in the bank account as a buffer for contingencies.

Malama, the smallest and most remote community, was chosen for the first revenue distribution because of the splash that sharing a lot of wildlife income with a few households would make. Following the model described for CAMPFIRE, the community reached agreement over how to share their money after four energetic days of meetings. Senior officials from the wildlife agency and Norwegian embassy made the long journey to attend the distribution ceremony, which was the first cash distribution in Zambia. The community gathered with considerable excitement in the deep shade of huge mango trees. However, Chief Malama then pointedly drove away past the meeting in his old Land Rover, boycotting the ceremony despite the excitement of all the villagers and the presence of high-level officials. This nearly halted the policy of fiscal devolution in its tracks.

Fortunately, the community facilitators were running similar meetings in the Msoro communities where Chief Msoro, an ex-civil servant sometimes at odds with his fellow chiefs, encouraged revenue distribution to proceed. Burned by the setback in Malama, the first cash payout in Chivyololo Village was low key, but our luck had turned, and it was recorded quite by chance by a journalist from the local radio station. Word spread rapidly through the Valley. Popular pressure ensured that it was not long before all 43 VAGs received their benefits, and a month later Chief Malama sheepishly requested LIRDP to facilitate cash distribution in his community.

The process of participatory revenue allocation followed the methods developed in Beitbridge and Mahenye (Chapter 11), but was more rigorous and intentional, using cash distribution to institutionalise new norms of governance, including impersonal and written rules.

Each community was given a standardised constitution to which they made some changes. The constitution emphasised inclusive governance and that wildlife (and wildlife revenues) belonged to the people collectively. Therefore, all decisions (especially budgets) would be made face-to-face by the community, with a quorum of 60% of members and never less than 60 people. The role of leaders was to bring people together to make decisions, and to implement these decisions, reversing decades of authoritarian top-down approaches. The new constitution stated key principles of inclusive governance and wildlife

TABLE 12.1 Policy guidelines for community-based wildlife management agreed at LIRDP's Inter-Ministerial Review and Policy Committee in 1996[3]

Definition of roles	*Income*
Chiefs (6) • Patron (non-executive, non-administrative role) • Overall advisors, and maintain traditional values • Neutral arbitration • Guide decisions on broad land use issues	**Purpose**: personal use (non-auditable) 6% of wildlife income from Lupande GMA (1% to each chief). An additional ZK1.5 million was agreed after a long wrangle. About $2 500.
Area Development Committees (ADCs) (6) COORDINATING LEVEL • Coordinate development plans for area and implement large multi-VAG projects • Oversee implementation of plans in VAGs • Oversee wildlife management and safari hunting in area	**Purpose:** Administration and coordination. • 4 % of wildlife income from Lupande GMA (divided between the six ADCs) Where the ADC is given responsibilities for larger projects (e.g. secondary school) or activities (e.g. employ game guards), money must be voted to it from VAGs.
Village Action Groups (VAGs) (43) ACTION LEVEL • All people prepare, prioritise, and agree all plans and budgets in general meetings • Choose how to allocate wildlife revenues to projects, activities, or household cash (*tyolela*) at general meetings • Hold quarterly general meetings to report on performance • Plan, implement, and monitor VAG level projects and activities • Maintain bank account and financial records • Manage wildlife at the local level (e.g. employ game guards or punish poachers according to by-laws)	**Purpose:** equivalent to income from crops or livestock except that the community must decide its use. • 80% of wildlife income from Lupande GMA May be used for any purposes decided by the community, including household needs (cash), projects, and activities
LIRDP • Monitor performance of CBNRM (finances, wildlife, institutional development) • Develop managerial capacity of community institutions through extension and training • Ensure compliance governance conditions	**Purpose:** institutional development, governance compliance, capacity-building • No income from Lupande GMA

ownership (two pages), though it was mostly devoted to procedural clarity, including mechanisms for calling committees or leaders to account, and even replacing them (15–20 pages). Because constitutions were 20-page documents, underlying principles and rights were reinforced and clarified with flip-charts (see Figure 12.1) and role plays (Box 11.4) at every general meeting in 43 villages for the next six years, highlighting the shift to a bottom-up approach and associated procedures.

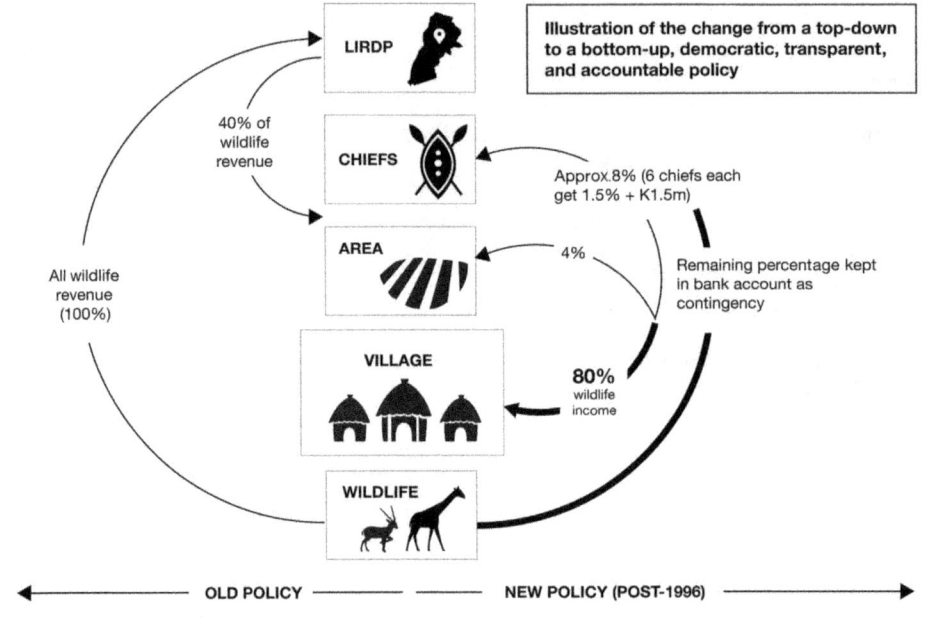

LIRDP evolved from First Generation CBNRM (left) to devolved Second Generaton CBNRM (right) in 1996.
In First Generation CBNRM:
- Devolution is partial
- People and communities are 'subjects'. Decisions are made by middle-level government officials and especially chiefs
- Participation is low, with few projects and no cash benefits

Second Generation CBNRM Projects:
- Generates real grass-roots participation and empowerment by devolving revenues
- Encompass principles that ensure full participation in democratic, transparent, and accountable system

FIGURE 12.1 Comparison of old and new policies used in community training in Luangwa

The annual general meeting was carefully structured into three parts: general information, performance reporting, and decision-making. It initially took four days. General information was provided about wildlife policy, events, and important issues like HIV/AIDS. The constitution was always reinforced using role plays and flip-charts. Second, the community received, discussed, and approved reports on finances, projects, wildlife management, offtake, and income, with the community facilitation team helping to prepare information and verifying (or refuting) the accuracy, honesty, and clarity of community reports. Reports always included finances and a list of wildlife offtake and income, supported by posters of animals and their prices. Interestingly, 60–80% of people knew the exact price of buffalo and lions, except in Malama (40% understanding) where we had forgotten to distribute the posters. The membership list was also updated and ratified.

Third, decision-making was centred around the process of 'participatory activity-based budgeting' (Chapter 16). This used a standardised flip-chart with four categories: (a) cash; (b) community projects or activities; (c) wildlife management; and (d) administration. With as much crossing-out and discussion as possible, the community eventually agreed on their budget. As in Beitbridge, the process of making financial decisions was managed over several days to give people time to absorb and discuss information.

Elections were the very last step in the annual general meeting. They were held every year to 'practise' and institutionalise democratic accountability. This was the last activity, on the basis that the community could choose better leaders once it knew how well the money and projects from the previous year had been managed, and how it wanted to use its new money. Elections get far more prominence than they deserve in many community programmes in the mistaken understanding that they provide democracy and accountability, yet they are held infrequently following the logic that it is expensive to train leaders only to see them voted out. Projects that focus on a few well-trained leaders risk accountability problems, asymmetric power relationships, and leaders 'over-staying'. Community-based management (as opposed to committee-based management) requires building the capacity of the whole community (the 'followership'), and giving considerable attention to avoiding entrenching leaders and power asymmetries. My strategy is to hold all meetings and trainings where anyone in the community can see what is happening, and can wander in and out if they wish too. Financial training, for example, targets at least 20 trainees in each village, with one becoming the treasurer and 19 additional sets of eyes watching the money. It is important for communities to have an office, but a serious mistake is to undertake training or decision-making in such a small and inaccessible space. It also seems that better leaders rise to the fore in transparent and meaningful systems.

Annual meetings are insufficient for building social and organisational capital, or for institutionalising the norms of rule-based governance and accountability. Standardised quarterly general meetings should be held on a monthly or quarterly basis. Key activities are to 'control' expenses and projects. These need to be standardised and institutionalised. Thus, financial control includes four steps. First, the treasurer uses a flip-chart to present a comparison of expenditure and budget using a standard format. The accuracy of this information is then confirmed (or refuted) by an independent audit report provided by the community facilitator. Third, the community is required to discuss and take corrective action where expenditure does not match the budget. Finally, the financial report is approved (or not approved) by the community, and decisions are recorded in writing. I was somewhat hesitant to introduce double-entry book keeping, but training community members in financial management, and auditing 43 VAGs quarterly, was less onerous than it seems. I place limited faith in financial auditing to deal with problems like false receipts and exaggerated prices or the misallocation of funds. I have much more confidence in the social accountability provided by general meetings, when this is combined with simple, visual reporting of income and expenditure, and a comparison of expenditure with the agreed budget.

Building devolved institutions with tough love

Societal progress, and the transition from 'Big Man' rule to inclusive rule of law, depends on effective rules and third-party enforcement of these (North, 2003, 2005). Moreover, effective devolution and participation is a rigorous process that follows 'loose-tight' processes (Peters & Waterman, 1982) or tough love (Handy, 1994). In Luangwa, we designed and imposed rules that entrenched Ostrom's principles that all people affected by rules and decisions should participate in making them. These rules did not emerge from within the community and met resistance from leaders, especially traditional leaders. However, they ensured that all decisions, especially about money, were made by the whole community, clarified roles and accountability, and protected ordinary community members, especially women and marginalised groups, against the tyranny of their leaders. They also promoted decision-making that incorporated information and transparency, with clear procedures for dealing with situations where this was not the case. One of the most important functions of the project was to monitor if communities were complying with these principles. The annual payment was only released to the villages once their compliance with the rules was certified; that is that decisions were made by all the people, using appropriate information, with clean finances that followed their budget, general meetings were held regularly to monitor finances and progress with projects, membership lists were up to date, and elections were held (Box 12.1).

Imposing rules is controversial, and the chiefs challenged this system all the way up to ministerial level. The minister instructed the project to hold a referendum to resolve the issue. The first village ballot delivered such overwhelming support for the new system (110 votes versus 7) that further polling was called off by the chiefs, who were embarrassed by the excited and highly energised community that was determined to exert its rights. People strongly supported the shift from personalised to impersonal governance (Acemoglu & Robinson, 2012) although it had been 'imposed' on them. Community capacity improved rapidly once communities were entrusted with financial management, wildlife patrols, building schools, and so on, supported by transparent performance monitoring systems. Capacity building was much more effective when linked to the uptake of responsibility than as an expensive (and ineffective) training exercise.

Ten locally recruited community facilitators were quite adequate to train and monitor governance, finance, wildlife, and projects in 43 villages, with each facilitator supporting about five VAGs. I inherited a demotivated team with little discretionary responsibility,[4] who were doing many things on behalf of the community rather than encouraging the communities to do things for themselves. The teams' confidence and capacity was developed using loose–tight principles. Each year, the community support team spent a week together setting goals and principles, using cause and effect problem trees to develop a theory of change, and developing a work plan and budget with clear deliverables and indicators. These were peer-reviewed quarterly. Consequently, all staff participated in, understood, and were committed to the goals and philosophy of the CBNRM project. We strived not to do anything for the community, but to get them to do everything for themselves. Staff were not judged on how busy they appeared to be, but on whether they delivered measurable results. The key outputs were self-governed, transparent, effective, and equitable VAGs, and improved income through protecting and managing wildlife. The latter included patrolling and monitoring by 78 village scouts, wildlife dams, quota-setting, problem animal reporting and management, environmental education and theatre, and electric fencing. Community facilitators lived in the community, saving considerable costs compared to a centralised team and gaining a much better understanding of what was going on. They were expected to carry out two activities per day, which they planned and tracked with a calendar. Project management meetings were centred around the systematic monitoring of performance using standard forms for community meetings, village finances, community projects, and village scouts. As the ultimate clients, we interviewed three members of each village quarterly (i.e. $n = 500/\text{year}$) to assess their understanding of, and satisfaction with, wildlife, finances, projects, governance, and so on. At quarterly team meetings, this data was used to peer-assess performance against targets, to plan future work together, and to develop a learning organisation. The data also informed the annual village compliance report (Box 12.1), which was the trigger for the release of the annual payment.

BOX 12.1 CERTIFICATION OF VAG COMPLIANCE AND APPROVAL OF RELEASE OF FUNDS

I hereby confirm the following:

Criteria for evaluation of governance compliance	Sign
1. This VAG held at least four general meetings during the year at which matters were openly and transparently discussed and which were well attended.	
2. The financial accounts of this VAG are accurate, follow the budget, and no money has been misused (or if misuse has occurred acceptable corrective action has been taken). (Before approving this, you should (a) be convinced that adequate and responsible corrective action has been taken and (b) summarise the problem and actions below).	
3. The finances and other matters of this VAG were properly presented and approved by the community at the AGM.	
4. A membership list was updated and approved by the general community.	
5. Elections were freely and fairly held and a newly approved committee is now in place to receive the wildlife income.	
6. Projects and activities were properly presented, allowing the community to choose sensibly.	
7. The choice of projects and approval of the budget was done by the community in general meetings, without coercion.	

....
Approved (signature)   Name   Title

Attachments:

1.   AGM Minutes (by community)
2.   AGM and QGM Reports (standard report compiled by facilitators)
3.   Financial statement (income, budget, actual expenditure, cash carry-over)

By copy of this assessment, approval is given for the utilisation of 2010 funds ($ ......)
by .... Village
Notes ....

How did the community use its money?

Over six years, the 43 VAGs paid out 47% of their income as cash dividends, 34% in projects, 10% in wildlife management, and 8% for administration. Figure 12.2 shows how the money was spent, with an increasing allocation to wildlife management (which is good) but also to administration, which needed to be carefully controlled. Similar graphs, but at

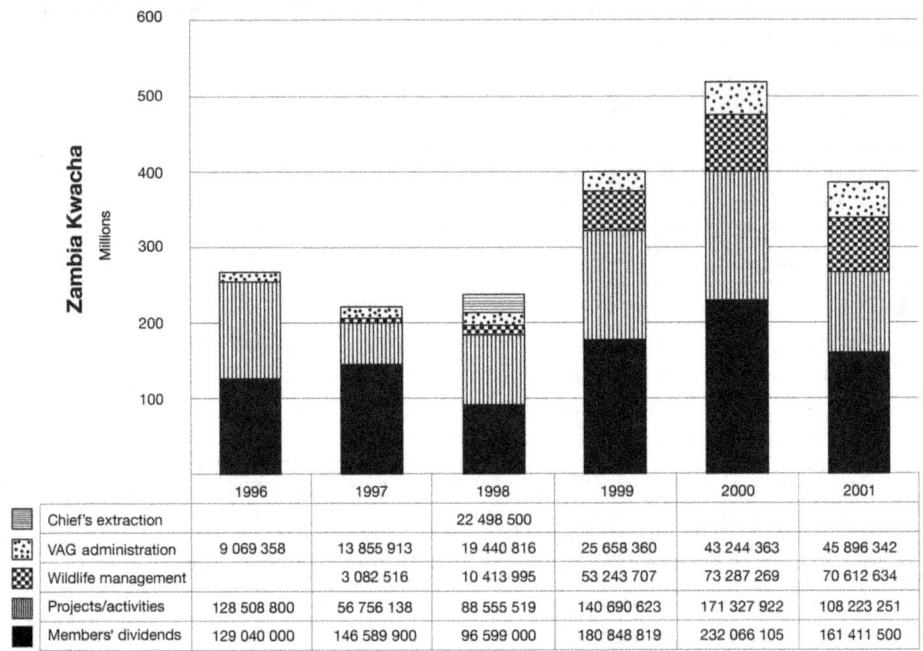

		1996	1997	1998	1999	2000	2001
	Chief's extraction			22 498 500			
	VAG administration	9 069 358	13 855 913	19 440 816	25 658 360	43 244 363	45 896 342
	Wildlife management		3 082 516	10 413 995	53 243 707	73 287 269	70 612 634
	Projects/activities	128 508 800	56 756 138	88 555 519	140 690 623	171 327 922	108 223 251
	Members' dividends	129 040 000	146 589 900	96 599 000	180 848 819	232 066 105	161 411 500

FIGURE 12.2 Allocation of expenditure in 43 Village Action Groups in Lupande GMA (1996–2001)

the village level, were even more valuable for tracking governance. Most, but not all, communities were sensible, favouring cash in drought years and sometimes allocating all their money for an important project (Dalal-Clayton & Child, 2003, pp 118–119). Cash allocations seemed to mirror the growth of social capital and accountability. In communities vulnerable to elite extraction, for example, cash was favoured, and five villages that shared a particularly 'selfish' chief allocated 63% of their benefits as cash compared to 40–46% in the other communities. This reflects the constant battle of recouping the money 'borrowed' by chiefs, which was mostly successful, except in 1998 when one chief died having taken community money to cure himself. Navigating the tensions between chiefs and 'subjects' over money was tricky. We resisted the pressures to intervene and get caught in the middle, and adopted a strategy of extreme financial transparency with regular public financial reports that listed how much money chiefs and other leaders had 'borrowed'. While we were not privy to the internal machinations of the community, the practice of extorting money from communities suddenly ceased in 1999.

During this period the Kwacha was unstable, varying from ZK1 207 (1996) to ZK1 862 (1998) to ZK3 848 (2001) per USD.

The comparative performance of the top-down and bottom-up phases

Bottom-up second-generation community-based management outperformed top-down first-generation committee-based management by several orders of magnitude. Figure 12.3

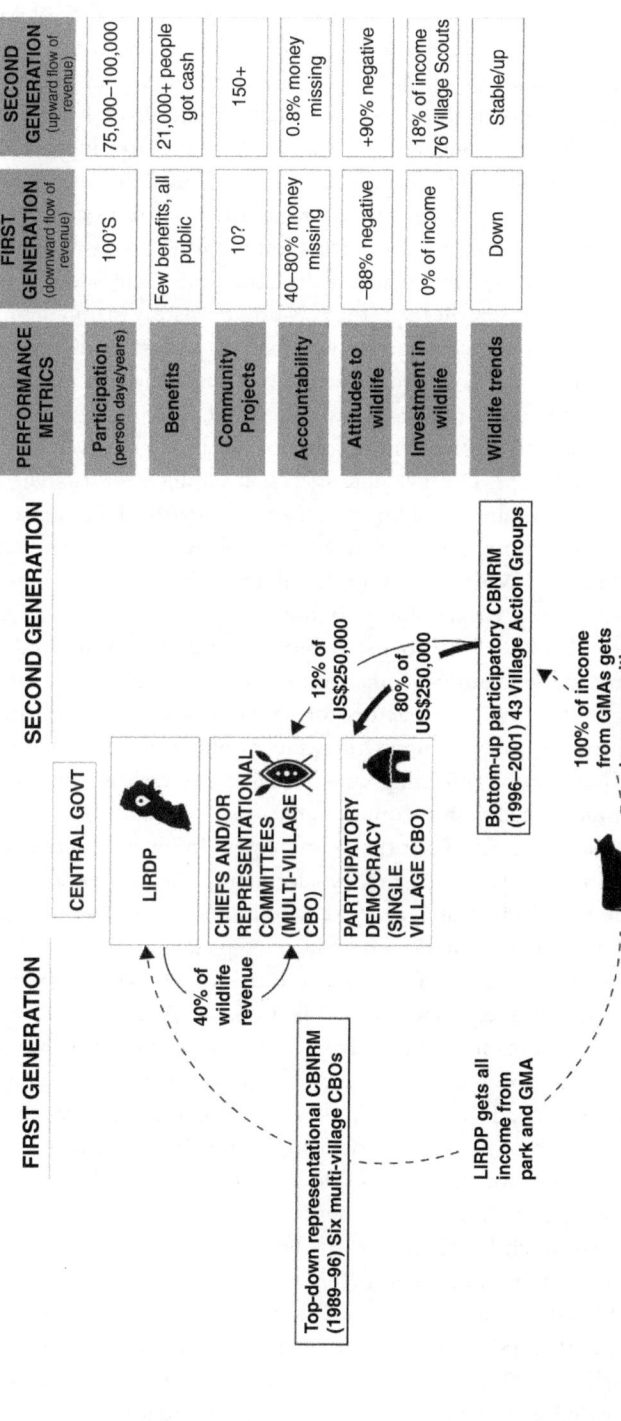

FIGURE 12.3 Comparing performance of top-down (1989–1995) and bottom-up (1996–2001) phases in CBNRM in Luangwa

illustrates the top-down (1989–1995) and the bottom-up phase (1996–2002), and provides comparative performance metrics.

In the bottom-up phase, communities engaged in tens of thousands of participation days, having been largely excluded in the top-down phases when decisions were made for them by the chief and his indunas. Over 21 000 people received a cash dividend from wildlife each year, compared to zero in the top-down phase. Officials expressed concern that this was wasteful, because local people are ignorant and lazy and would turn their money into alcohol. The data tells a different story. The communities constructed over 150 projects including 24 teachers' houses, the renovation or construction of 38 school classroom blocks, 27 clinics or health posts, 11 water well projects (often with several wells), an electric fence, and 45 other projects such as food relief in drought years, and soccer and sewing clubs. There were no more than ten projects in the top-down phase,[5] mainly grain storage shelters (which were not valued by the community who had no excess grain to store), schools, and meat harvesting and bus transport projects that were not viable and sold by the leaders without financial accountability (Dalal-Clayton & Child, 2003, p 105).

What explains the 15-fold increase in the number of projects, given that the community used 47% of income as cash? The example of Nsefu clinic is illustrative. During the top-down phase, the clinic remained stubbornly at the foundation level for a number of years despite many time-consuming planning meetings facilitated by the project and associated government departments. Within six months of their first payout, however, the Nsefu community completed the clinic plus two houses for medical staff. Quite clearly, the community made and knew how to use a budget, and the local need for a clinic was strong. The community used money much more parsimoniously than when LIRDP built projects for them. It still hired contractors for key jobs and purchased tin roofs and steel door and window frames, but procurements were carefully vetted. Community volunteerism, such as making bricks or collecting sand, saved scarce cash for important needs, as less than half the community had cash income of any sort.

Perhaps the most stunning result was the improvement in financial accountability. Quarterly audits showed that less than 1% of the money allocated to villages was not accounted for (i.e. ZK3.35 million out of ZK400 million in 1999), and in most cases measures were underway to recover missing money. This compares to audit discrepancies of 40% in the ADC accounts, which were vulnerable to the chiefs and also lacked the intense scrutiny of money allocated to the villages[6] (Dalal-Clayton & Child, 2003, p 166), and between 40–80% where other Community Resource Boards in Zambia have been analysed (Malenga, 2004).

The community also took wildlife management seriously, incrementing the allocation of money to this purpose: 0%, 1%, 4%, 13%, 14%, 18%. Within three years, 78 village scouts were employed by the community and patrolling regularly, while Msoro community built two dams to supply water for wildlife and designated a new wildlife area without help or prompting from the project. Anecdotal evidence suggested that poaching was becoming less acceptable; 'meat' was much harder to get in villages, or could only be found 'at night', and animals were reported in areas not seen before.

Attitudes to wildlife were also transformed. During the top-down phase, Balakrishna and Ndlovu (1992) found that 88.2% of respondents did not support safari hunting (a euphemism for the project). By 1997, 97.8% of respondents felt it was necessary to conserve wildlife and 93.4% said it provided a benefit to them, through household income (49.6%), projects (12.1%), and employment opportunities (17.2%) (Dalal-Clayton & Child, 2003, p 116).

Recentralisation

After 2002, the CBNRM programme was recentralised. The new Zambia Wildlife Authority (ZAWA) was required to become financially self-sufficient, but was funded largely by hunting in community areas. It forced through a policy to retain 50% of trophy revenues from community areas but, in the absence of transparent records, actually kept far more. This essentially destroyed the Luangwa programme. Not only did ZAWA return very little money, but the money it did return was paid to the six Community Resource Boards, devastating the village system so carefully established (Lubilo & Child, 2010).

Five years later, the governance of these communities was reassessed to measure the effects of recentralisation (Lubilo, 2007). Table 12.2 compares community understanding and satisfaction for the three phases of the programme (top-down, bottom-up, top-down again). Financial transparency plummeted from 72% in the bottom-up period (1996–2001) to 20% after being recentralised (2007), even using a much looser definition of 'understanding of finances'.[7] Only one Community Resource Board was able to provide financial records (itself an important indicator), and these showed that 63% of income was absorbed by administration and 28% for wildlife management by village scouts, leaving only 9% for community benefit compared to 82% in the bottom-up phase. Community support for the programme declined from over 90 % to 39% and poaching increased: 'It's the government's wildlife now,' said one community member. 'They get the money from it and they should look after it.'

It is revealing that the small and remote Malama community bucks the general trends. It was the weakest community during the bottom-up phase, with members scoring 50% for understanding of finances on the assessment tool compared to an average of 72% (Dalal-Clayton & Child, 2003). However, by 2007, 56% of community members in Malama still understood their finances (compared to 20% overall) and support for CBNRM was 57% (compared to 39% overall), although Malama suffered the most from problem animals. I speculate that this is because Malama was small (200 households) and much less affected by recentralisation than large communities.

Does it take a village?

Although these communities made remarkable progress, the data hardly captures the full richness of the unleashed human spirit as people made their own decisions, received cash, built projects, and gained confidence. Having seen this transformation with my own eyes, I have no doubt that CBNRM works. However, my experience was limited to communities within the bounds of Dunbar's number (100–300 households), being naturally small, or deliberately designed to be small and accountable. However, in my new life as an academic, visiting other programmes or assessing them through the literature, it slowly dawned on me that my sample was far from the norm. Most CBNRM communities were far larger, with representational rather than participatory forms of governance. Performance was 'uneven' or disappointing as they struggled with financial accountability, elite capture, meaningful participation, and equitable benefit sharing (Dressler et al., 2010; Gruber, 2010).

Interestingly, the difference between participatory and representational governance is not generally mentioned in the literature, except by Ostrom and Murphree who emphasise face-to-face accountability. Yet, the top performing CAMPFIRE communities (i.e. Mahenye, Masoka, and Chikwarakwara) all practised participatory governance (Chapter 11). The

TABLE 12.2 A comparison of attitudes and understanding of community finances in different phases of the community programme in Luangwa

Community	Number of VAGs	Number of households getting cash in 2000	Understanding of community finances		Support of programme		
			1997–1999	2007	Top-down (pre 1996)	Bottom-up (1996-20-02)	Top-down (2007)
Jumbe	9	4 436	80%	16%			42%
Kakumbi	5	3 811	93%	12%			17%
Malama	3	510	50%	56%			57%
Mnkhanya	11	5 023	53%	14%			37%
Msoro	10	4 087	66%	0%			
Nsefu	5	3 221	78%	11%			38%
	43	21 088	72% $n = 851$	20% $n = 451$	-88%	90%+	39% $n = 428$

Source: data from personal records and Lubilo (2007).

cross-sectional data from Luangwa also showed that face-to-face governance outperformed representational governance by a factor of 10 or more. Was this a general finding? Is it important? Is this why many CBNRM initiatives flounder? The rest of this chapter addresses these questions.

The governance dashboard

To understand how ordinary people were participating in and benefiting from CBNRM, I developed an action research process called the 'governance dashboard' with a number of my PhD students (Child et al., 2014). We applied these surveys in Zimbabwe, Zambia, Namibia, Botswana, South Africa, Mozambique, and Tanzania ($n = 1\ 500$). Our first finding was that participatory CBNRM was uncommon or even rare, with only three single-village CBNRM organisations in the 25 sites sampled. Nonetheless, there were fundamental differences between participatory and representational governance; 88% of people in a typical small village with participatory governance trusted their leaders with money, compared to 22% in a large village with representational governance (Figure 12.4). These differences held for many variables that measured CBNRM performance, including satisfaction with AGMs, clarity of decision-making, knowledge of the price of wildlife, knowledge about wildlife income and expenditure, and even how many people knew the name of their elected leaders. For instance, 85% of respondents from single villages attended AGMs compared to 34% in multi-village communities.

Collective action associated with CBNRM provides four public goods. Scaling up provides ecologies and ecologies of scale, whereas scaling down is associated with informed participation and equitable benefit sharing (Figure 12.7). CBNRM in southern Africa was performing very well in terms of increasing wildlife populations and community income

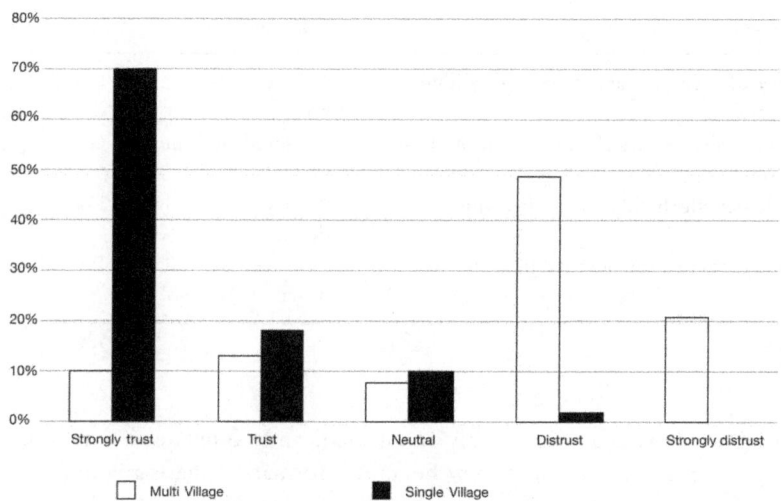

FIGURE 12.4   Comparison of trust in small and large communities

(Child et al., 2003; Taylor, 2009; NACSO, 2015). However, the dashboard surveys showed that the satisfaction of ordinary members, informed participation, and equitable benefit sharing were quite low in larger communities with representational forms of governance, especially compared to small villages (unpublished data summarised in Table 12.3). Even ignoring the space opened up for corruption, committee-based governance allocated more money to central functions that often benefited committee members (e.g. sitting allowances) and less to direct community benefits.

Following the money

To study this further, I compared community benefits (defined as the money allocated to community cash and projects) to money used centrally for wildlife management and administration. This distinction can be fuzzy, and game-guard salaries, for example, are classified as centralised expenditure because they only benefit a few people in a community directly. The data are revealing (Figure 12.5). Small communities allocated 60%, and often as much as 80%, of wildlife income as community benefits. There is a clear tipping point at about 250 to 500 members, with larger communities having representational forms of governance and centralising 80–90% of wildlife income. In well-managed programmes (e.g. Namibia) much of this expenditure was reinvested in game guards and wildlife management, but the lack of community benefit was a sore point amongst members in several communities. More often, however, the money was wasted. Audits of eight communities in Zambia showed negligible community benefit (6%), some commitment to village scouts (25%), but three-quarters of income being wasted in administration costs and 'sitting allowances' (35%) or not being accounted for (40%) (Malenga, 2004). Similarly, surveys showed that ordinary people in multi-village

TABLE 12.3 Performance of 20 CBNRM communities in southern Africa

Provision of public goods through collective action	Scale dimension	General rating of performance
1. Ecologies of scale (i.e. landscapes and wildlife populations)	Achieved by scaling up	Good
2. Economies of scale (larger areas earn more money per hectare)	Achieved by scaling up	Good
3. Equitable benefit sharing ('spending money')	Achieved by scaling down	Poor
4. Informed participation/democratisation	Achieved by scaling down	Poor

Source: Child et al. (2014).

communities in Botswana were highly unsatisfied, and complained that 'only the Trust was "eating"', with limited community benefit, information sharing, and participation. In the same macro-environment, the small Sankuyo community was highly satisfied with its governance, allocating 60% of wildlife income to benefits, including cash pensions for elderly people, employing people from each household in campsites or as game guards and cleaners, providing houses for destitutes, water for households, transport, and so on. These arrangements emerged from within the community, with little governance oversight or training (personal data and observation).

CAMPFIRE, interestingly, is an outlier. I have already discussed CAMPFIRE's high performing small communities (e.g. Mahenye, Masoka, Chikwarakwara, three of the single villages illustrated in Figure 12.5). But how well did larger communities perform? There is no complete dataset because the importance of within-community expenditure was not understood at the time. However, I was able to extract financial data from my records for three of the 12 districts.[8] These three districts devolved 73% of wildlife income to producer communities, of which the communities spent 75% as cash or community projects. Tsholotsho and Binga managed CAMPFIRE at the ward level (i.e. representational governance). Tsholotsho communities got 64% of wildlife income which they used for cash (26%) and projects (39%). The use of the 61% allocated to communities in Binga is not specified. By contrast, Hurungwe district paid revenues to the nine wildlife producer villages rather than to the five wards in which they were situated. These villages allocated 95% of wildlife income to community benefits (60% cash, 35% projects), matching the pattern in Figure 12.5. The higher level of community benefit in CAMPFIRE villages and wards, compared to representational governance generally, can be attributed to Zimbabwe's relatively strong district administration, with dedicated officers in many district councils being committed to devolution (Goredema et al., 2006).

Economic games

As part of a training workshop in Maun, Botswana, I simulated participatory and representational budgeting with 20 community leaders, district officials, and NGO personnel. To role play a representational budget process, the group elected three people to represent them, who went to their 'office' (corner of the room), and developed a budget on a flip-chart

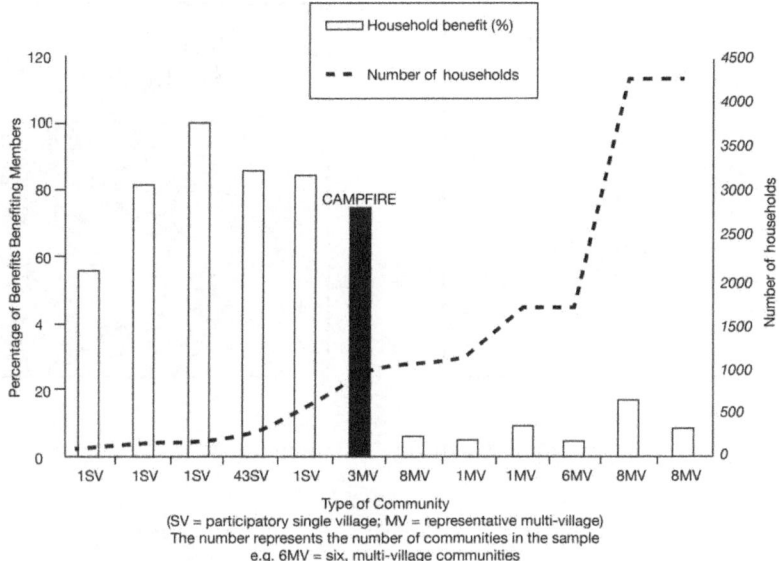

FIGURE 12.5 Relationship between size of community and allocation of wildlife income to community benefits

Source: personal data collected over many years in Zimbabwe, Namibia, and Zambia.

with no input from the community. The elected leaders then presented the budget back to the community in the 'annual general meeting'. This is where ordinary community members are supposed to make changes to the budget, but few changes were made because the 'committee' either defended its decisions or ignored suggestions. The participatory budgeting simulation involved exactly the same people, who sat in a circle and made a budget on a flip-chart. The budget was messier, the conversation more animated and interactive, and the budget written up on the flip-chart was full of changes and crossings-out in response to community discussions.

The results were strikingly different, although they were made by the same people, in the same room, on the same day. In three workshops, representational budgeting allocated an average of 36% to community benefits, compared to 72% for participatory budgeting (Figure 12.6),[9] matching the field data (Figure 12.5). Satisfaction with participatory budgeting (77%) was much higher than representational budgeting (20%), which was felt to be frustrating, non-inclusive, and disempowering.

Participatory and representational governance

The distinction between participatory and representational governance is not new. According to de Tocqueville, a democracy is where everyone meets together to represent themselves (as in township government in New England), whereas a republic is where people's interests are represented by elected persons (Tocqueville, 2000). The dichotomy between organic, traditional communities (*Gemeinschaft*) and modern, mechanical, impersonal society (*Gesellschaft*) has a long history in the literature, developed by such luminaries as Ferdinand Tönnies, Max Weber, and Émile Durkheim. The early literature

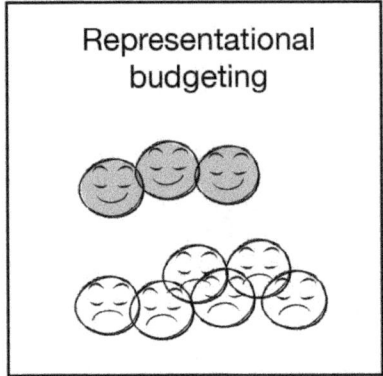

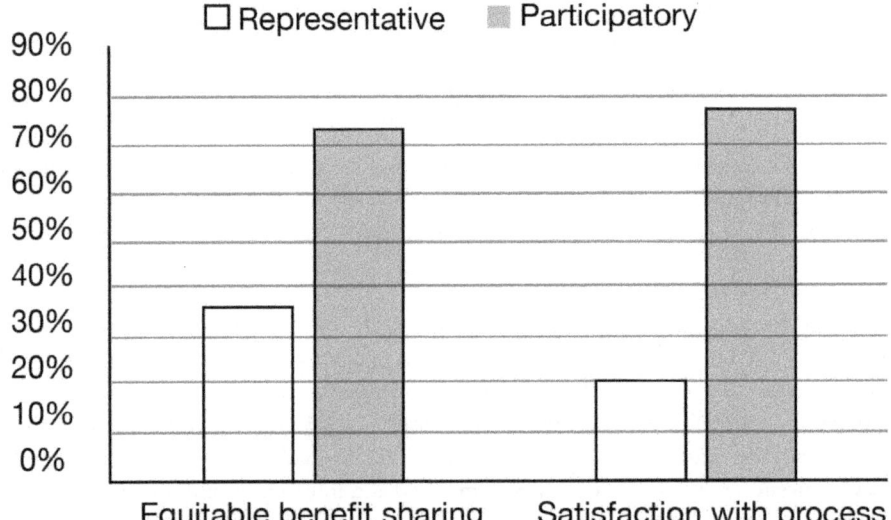

FIGURE 12.6 Comparison of financial allocations and satisfaction with participatory and representative budgeting processes

on CBNRM and the commons also emphasises participation and small communities (Wade, 1987b; Ostrom, 1990; Murphree, 1994a; Martin, 2009a) and a disproportionate number of successful CBNRM sites are, indeed, participatory (Bond, 2001; Castello et al., 2009; Measham & Lumbasi, 2013).

Yet most CBNRM projects practise representational governance, with weaknesses in terms of equity and inclusiveness. This makes it surprising that the distinction between participatory and representational governance is seldom recognised in theory or in practice, even in articles specifically discussing scale (Batterbury & Fernando, 2006). CBNRM projects funded by donors invariably adopt representational governance. Is this why CBNRM is said, so often, to be in crisis, with ubiquitous challenges of elite capture, financial mismanagement, and low levels of participation?

Representational CBNRM

Representational CBNRM is where large communities, often consisting of multiple villages, elect a central body to manage themselves. Well-managed representational CBNRM can result in rapid recovery of wildlife and improving wildlife income. However, representational governance is vulnerable to elite capture. In South Africa, Musavengane and Simatele (2016) find that 'the current decay in traditional systems of governance is rapidly resulting into high levels of corruption and unfair distribution of benefits of their natural resources'. The centralisation of benefits, elite capture, and low participation are ubiquitous problems in community conservation (Arnold, 2001; Ribot, 2003; Poffenberger, 2006; Charnley & Poe, 2007; Frizen, 2007; Tacconi, 2007; Larson & Soto, 2008), and are also widely reported for district decentralisation processes (Crook, 2003; Frizen, 2007; Grindle, 2007).

Representational CBNRM supposedly favours managerial efficiency over participation (Overdevest, 2000). Many projects adopt representational forms of governance because they think that face-to-face participation is logistically challenging for communities of several thousand adults, or that large communities can afford to employ better managers. They may also prefer to work with a few elected leaders, rather than community members as a whole. The primary mechanisms of accountability are a single annual general meeting (AGM), and elections held every two or three years. This is problematic because AGMs and elections often reflect the trappings of democratic accountability rather than its substance (Dahl, 1998; Tilly, 2009). In large communities, it is not uncommon for AGMs to be stage-managed to avoid core issues, such as financial accounts being presented poorly or fudged, or procedures being manipulated to protect leaders whom the community clearly want to remove from office (Child et al., 2014). The size of these communities prevents most people from attending or participating in meetings effectively. Leaders are able to manipulate information and the agenda to circumvent accountability. AGMs and elections can become ceremonial, providing the veneer of downward accountability, participation, transparency, and accountability, but not its true substance (Mupeta, 2008). Thus, multi-village CBNRM reflects 'committee-based' rather than community-based management. As Jon Anderson et al. (2013) wryly states: 'the rich get richer, and the poor get committees'. Committee-based management is vulnerable to low levels of participation and benefit sharing, and high levels of elite capture.

Participatory CBNRM

Participatory governance is where the whole community (or at least most of it) meets together to discuss issues and make decisions, and instruct (rather than be instructed by) the committee on matters of implementation. In the field, participatory governance may not look too different from representational governance, with anything from 100 to 300 rural people meeting under a tree. However, the data in this chapter suggests that participatory governance is profoundly different, with quantum improvements in the public goods we have called equitable benefit-sharing and informed participation. Participation builds social capital[10] (trust, financial accountability, collective action, understanding, and following the rules), as well as physical capital (more projects) and natural capital (less poaching, more control over use of wood, fish, and so on).

There is a fundamental difference between participatory community-based management and representative committee-based management (Figure 12.7). We need to be deeply cognizant of these differences and their effects when designing CBNRM interventions.

Is it a question of sequencing?

CBNRM can only be sustainable if it develops social capital, trust, shared norms, and the understanding and commitment necessary for reliable collective action. One argument is that representational governance accelerates recapitalisation of the resource base, and that participation can emerge later. But when do representational forms of CBNRM begin the process of forming deep social capital, rather than undermining them? This mirrors the debate about the linkages between authoritative government, economic growth, and democratisation. In England, democracy emerged over several hundred years as the rulers grudgingly gave the ruled more rights in exchange for support, especially in wars (Tilly, 2009). We can envisage a similar process in CBNRM, as the leaders co-opt the followers to conserve wildlife and the benefits on which the elites depend, while the followers leverage this process to expand their rights. However, this is far from guaranteed, as England is an outlier and early adopter (Chapter 2) and most extractive regimes persist for centuries and even millennia. Hence Murphree's Law, much quoted by CBNRM practitioners in southern Africa – there is an in-built tendency at any level in bureaucratic hierarchies to seek increased authority from levels above and resist its devolution to levels below (Murphree, 1989).

Why, then, is participatory democracy so rare in CBNRM and in development more widely? The technical justification is that small communities face significant diseconomies when it comes to obtaining technical and administrative skills (Overdevest, 2000). A second reason is logistical: support agencies prefer to work with fewer, larger communities. Third, the importance of participation and pro-poor benefit may be recognised more in rhetoric than in practice, with limited understanding of the careful preparation needed (Wall, 2017), while implementing agencies favour clear-cut outcomes such as conservation, income, and concrete projects compared to much fuzzier social processes and social capital formation. Perhaps the real reason is that governments, and even facilitating agencies, simply don't trust ordinary people with something so important as managing themselves (Wall, 2017). A more cynical argument is that empowering communities threatens the people benefiting from their resources (Chapter 3). Finally, many commentators wonder if it is realistic to promote face-to-face processes of decision-making and accountability at the necessary scale. My counter-argument is that poverty, mismanagement, and elite capture are an outcome of processes (and states) that are democratically hollow. There is no shortcut. We need to build social capital and institutions from the bottom up and, moreover, genuine progress is the result of unlocking the potential of each and every person, not just a few leaders.

Having our cake and eating it

As noted briefly above, CBNRM collective action provides four sets of public goods:

- Economies of scale, with larger areas being more valuable in terms of the wildlife economy, and perhaps also for biodiversity and ecosystem services.
- Ecologies of scale, with bigger environments generally being more functional than small ones.
- Equitable benefit sharing, through more local and more equitable distribution of costs and benefits (including distributive, economic, and social justice).
- Informed participation in decision-making, which is a critical component of accountable democracy (Dahl, 1998) and results in greater freedoms and choices for people (Sen, 1999).

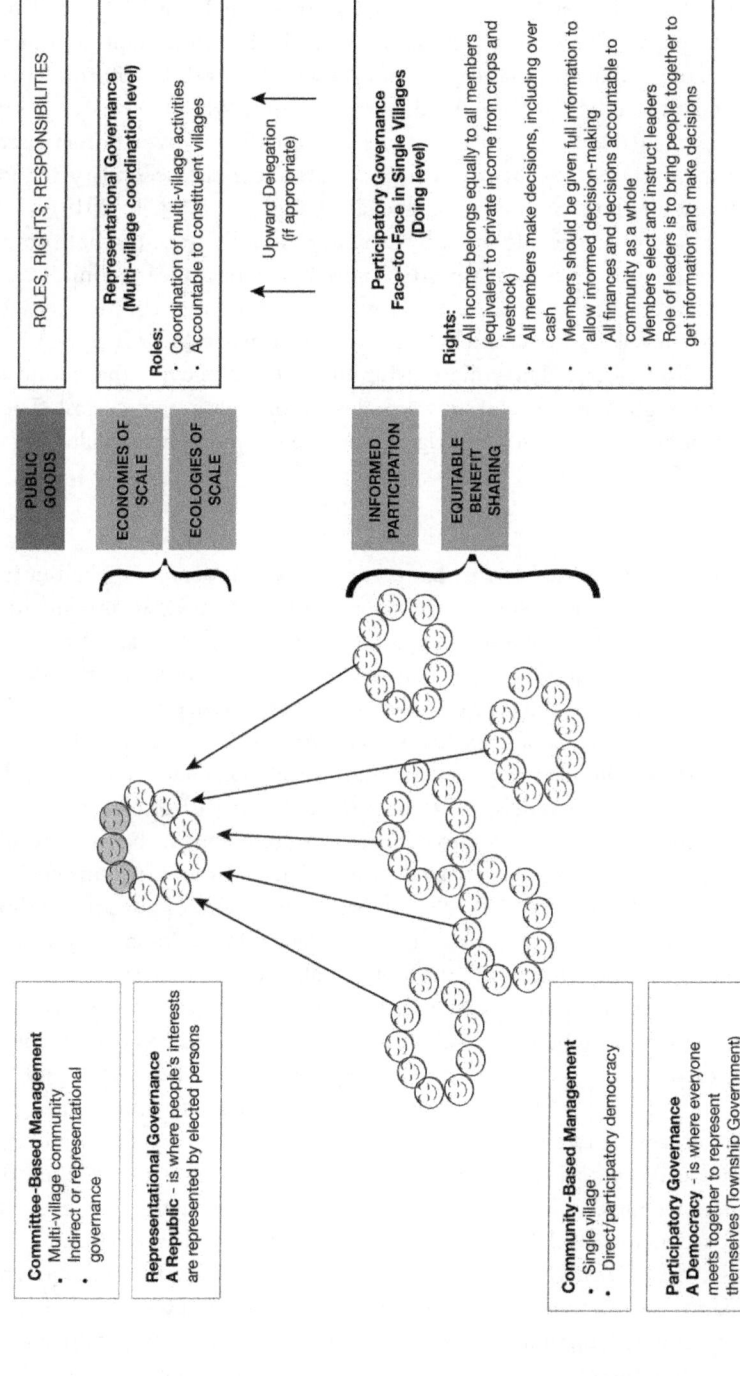

FIGURE 12.7 Participatory and representational governance, scale, and public goods

We face the paradox that economies and environments are better managed at larger scales, but human interactions are better managed at smaller scales. This paradox can be solved by devolving rights to the lowest level, and then delegating some tasks upward where this is sensible (Chapters 10, 14). This is illustrated in Figure 12.7, and the policies adopted in the Luangwa CBNRM programme (Table 12.1) where rights and benefits were devolved to villages, (building social capital and the public goods of informed participation and equitable benefit sharing), with upward delegation only where necessary. Thus, villages in Luangwa employed their own village scouts and built their own primary schools, but they delegated the authority to manage village scouts and build secondary schools to area committees because of economies of scale (see also Child and Wojcik, 2014).

Thus, we can have our cake and eat it by building carefully from the bottom up. Higher scales, like area or district wildlife or forestry committees, may have technical capacity, but they are often democratically hollow, with weak roots in the community. Too many functions at higher levels also overwhelm management capacity, and lengthen and weaken accountability feedback loops. This follows Murphree's (2000) advice that the scale paradox can be solved through a process of upward delegation, with the critical Tocquevillian implication that rights and governance originate in the people (Tocqueville, 1994).

Conclusions

The common response to these ideas is that participatory democracy is nice but impractical, and governance has to be representational to manage landscapes at any significant scale. I disagree. For sustainability, it is just as important to rebuild social capital as to rehabilitate natural capital, and this requires assembling hierarchical systems from the bottom up to ensure that choice and governance originate in the people. Once the basic building blocks are in place (i.e. inclusively managed villages), scale can be built quite easily.

This requires a mind shift that recognises communities as independent jurisdictional units with the same rights as private landholders, rather than as the lower part of bureaucratic hierarchies. Most CBNRM initiatives have drifted into structures that resemble Weberian bureaucracies and the indirect rule and even feudal governance of old. There is merit in returning to Norman Reynolds's proposal at the birth of CAMPFIRE that the basic unit of governance should be the 'Village Company', in which each member is a legal shareholder. Constituting communities as village companies (with the same rights as private landholders, but attached to collective title), emphasises the entrepreneurial and productive nature of CBNRM, and not just its regulatory functions. Recognising members as legal shareholders reflects CBNRM principles. I have suggested that these need to be enforced through guidelines, legislation, and compliance monitoring. However, we could get to a similar point by also embedding these principles in modern and enforceable corporate legal documents. This provides additional protection to the rights of 'shareholders', especially minorities and marginalised groups, who can legally defend their rights in commercial courts, and not just through policy or administrative arrangements.

The main challenges to CBNRM are failed devolution and internal governance. This chapter suggests that CBNRM will remain disappointing in terms of equitable benefit sharing and participation unless it is rooted in well-designed face-to-face processes of decision-making and accountability. Rules for inclusive community governance may need to be imposed, and definitely need to be protected by external compliance management and monitoring. Implementing participatory democracy is probably easier than many people imagine (Child and Dalal-Clayton, 2003), but even if it is difficult, do we really have an alternative? Can we succeed in either conservation or development if we do not create robust institutions that empower ordinary people to participate in and manage their own lives, with equitable ownership and benefits from wild resources?

Notes

1   I was recruited to manage this project from 1996 to 2002, and this chapter uses this experience and data, much of which is available in a less digested format in Dalal-Clayton and Child (2003).
2   Having been the primary beneficiaries of the old top-down system, the chiefs strongly resisted these changes for several years (a battle worthy of its own book). According to traditional protocol, chiefs were neither auditable nor accountable (remnants of the divine right of kings). The new policy sidestepped the lavish dance that the project had done previously, which prevented it from auditing chiefs, so money could not be fully accounted for.
3   I have edited this table from the original to shorten and clarify it.
4   In the previous phase of the project, these staff had followed instructions and appeared, on the surface, to lack competency. However, once engaged in work planning and performance reporting, and once they understood their jobs, they performed at high levels. Twenty years later, most of the wildlife agency's community facilitators originate from this team, with the tragedy that several of the best passed away at a young age.
5   There are no records of what projects were built by communities in the initial top-down period, and there was also conflation between projects funded with Norwegian money and self-funded community projects. I tried to identify these projects using participatory mapping, and also through many visits to these communities. There were no more than ten community-funded projects in the top-down phase.
6   Two chieftainships accounted for the majority of unaccounted money – Kakumbi and Nsefu: (a) ZK24 million was collected by Hon. Chief Kakumbi from VAGs and not paid to Kakumbi ADC (the ADC could not produce records nor explain the whereabouts of this money); (b) there was an unauthorised but recorded loan of ZK10.5 million to the late Senior Chief Nsefu from Nsefu ADC; (c) ZK10 million was unaccounted for in Nsefu ADC (the treasurer was implicated in making false records).
7   In 1997–1999 community members were interviewed to assess if they understood roughly 0%, 20%, 40%, 60%, 80%, 100% of community wildlife finances. Members were expected to understand community income, allocation, and wildlife prices. In 2007, community members were asked only if they had been provided 'all', 'some', 'none', 'didn't know' information about community finances.
8   As CAMPFIRE Coordinator, I 'audited' all 12 districts twice a year, compiling data on finances, wildlife, quotas, observations, and so on. This analysis uses these departmental reports (which are available on request).
9   If anything, the results underestimate differences between participatory and representational processes. Much of the training emphasized community benefit, and may have biased results in this direction. Further, these communities were required by government to employ qualified managers, which imposed higher fixed costs and thus reduced equitable benefit sharing.
10   In an email exchange with Regis Musavengane, he expressed considerable concern about the loss of social capital in post-colonial communities. Thus, said Regis: 'Based on the empirical evidence I got from 2 rural communities in Northern KwaZulu Natal here in South Africa one can easily see that it's a long walk to redress the fruits of oppressive eras. I found out that Community Leaders do not respect the villagers and in-turn the villagers have lost their trust in leadership structures. This is so worrying ... here we now ask ourselves if this is not self-inflicted oppression. A local oppressing a local – this challenge isn't easy to deal with'.

13

THE GLOBAL EMERGENCE OF CBNRM PRACTICE AND THEORY

LEARNING OBJECTIVES

This chapter introduces some of the key threads in the broad concept of CBNRM. It:

1. Describes the increasing recognition of the interplay between people, environment, and sustainable development, as demographic growth and markets overwhelm the capacities of public management.
2. Provides a brief introduction to the literature about communities, conservation, development, and decentralisation.
3. Presents vignettes of some of the better-known examples of CBNRM, including community forest management in Asia and Latin America, community-based fisheries in Asia, and community wildlife management in Africa.
4. Defines CBNRM as applying to wild resources, as they make the transition from public to collective and private governance.
5. Defines CBNRM as a model with four attributes:

   a. The devolution of proprietorship of wild resources to communities.
   b. The valorisation of wild resources to maximise community benefit.
   c. Governance within the community, but within a larger framework of meso- and macro-level institutions.
   d. The management of natural resources.

The rise of community conservation

Community conservation has been on the rise for several decades (Dressler et al., 2010), following an increasing recognition of the interplay between people, conservation, and sustainable development (IUCN, 1980; Bruntland, 1987). There have been major investments in decentralised natural resources management, community based forest management, community conservation, Integrated Resource Development Projects, biodiversity mainstreaming, and CBNRM by the World Bank, Global Environmental Facility, USAID, Norway, the Department for International Development, UK, and many others including most environmental NGOs. This has spawned a large, but often disconnected, literature, with very little agreement on exactly what constitutes community conservation. For instance, some authors see market-based or neo-liberal approaches as problematic (Barrett & Arcese, 1995; Dressler et al., 2010), whereas others view them as the solution (Barrow & Murphree, 2001). Some see conservation and development as a zero-sum game (Mcshane & Wells, 2004), while others argue that conservation and development are synergistic (Whande et al., 2003) because institutions governing wild resources are so inefficient that it is easy to get more of both with reform (Child et al., 2012).

The enthusiasm for community conservation waxes and wanes, with a general consensus that it is a good idea but is not delivering the goods. Initial enthusiasm was dampened by a rash of poorly conceived projects that were seldom successful (Brandon & Wells, 1992; Mcshane & Wells, 2004), and benefits failed to materialise (Tole, 2010). This has been blamed largely on a reluctance to genuinely devolve proprietorship (Barrow & Murphree, 2001; Ribot, 2002; Nelson & Agrawal, 2008; Alden Wily, 2009; Tole, 2010) where agencies adopt the language of devolution but not the practice (Ribot, 2003). Although the literature advocates decentralisation, its analysis of the processes, practicalities, and performance of the mechanisms of decentralised resource governance remains perfunctory. Decentralisation, it is agreed, is a good thing, but how exactly to do it is less clear. There is occasional discomfort about how decentralised units of accountability (i.e. the communities themselves) function, including concerns about marginalisation of poor people (Cleaver, 2005) and women (Tole, 2010), the problem of elite capture (Frizen, 2007), and complex power issues within a community (Agrawal & Gibson, 1999), but the literature seems not to have completely put its finger on this issue.

There is also a concern that CBNRM is expected to deliver within the time frames of poorly conceived and externally imposed projects (Barrow & Murphree, 2001), without a longer and tenacious approach that seeks to develop inclusive and participatory governance and social capital. Finally, there is not much debate about how CBNRM institutions, rules, and norms emerge. Should they be imposed, or will they emerge organically? Most examples of successful and long-enduring commons are associated with stable communities with high levels of social capital (Wade, 1987; Ostrom, 1990). Community approaches can be successful where governments are genuinely committed to devolution and pro-poor approaches, and are prepared to challenge local elites to encourage equity, transparency, and accountability (Crook and Sverrisson, 1999, quoted in Tole, 2010). However, these are not the conditions facing CBNRM in much of the tropics. How does that participation, democracy, and accountability arise, given that social capital, trust, and reciprocity, which are norms essential for collective action in the first place (Tole, 2010), are often missing? CBNRM, as defined later, also requires shifting from personalised and extractive governance towards inclusive, impersonal governance through mutually agreed rules, and this transformation is not always welcomed by entrenched elites (Wilshusen, 2008).

Community-based conservation encompasses many ideas. The eloquent and valuable literature on many aspects of CBNRM has not yet come together as a comprehensive theory. It tends to be disciplinary and scholarly, rather than geared to the resolution of real-world problems. I have recently reviewed dozens of GEF biodiversity projects (worth hundreds of millions of dollars). These show a rapid increase in community approaches for protected area buffer zones, sustainable land and forest management, agro-biodiversity, and so on. The understanding of governance frameworks and economics for taking on these challenges, however, is nowhere near the level of biological knowledge.

What is CBNRM?

The term CBNRM emerged in the 1980s as a label for donor interventions, and is not very useful because it can be interpreted in so many ways. Thus, Tole (2010) suggests that community-based forest management 'implies a bottom-up approach, one that involves a designation of power over forest resources to local people, granting independence and leverage to communities to decide how their forests will be managed and for what purposes'. Dressler et al. (2010) consider CBNRM to be an 'incremental social process of assisting impoverished communities to set priorities and make decisions for developing natural assets and social equality to reduce livelihood vulnerability and improve conservation'. Ribot is concerned that the warm and fuzzy connotations of CBNRM often enable central agencies (including NGOs) to maintain their extractive powers behind a smokescreen labelled 'participation' (Ribot, 2003). Martin (2009a) argues for a broad vision of CBNRM that incorporates: (1) sustainable use and sustainable development; (2) complex systems thinking; (3) the right of community to full entitlement over resources; (4) the full range of resources and not just wildlife; (5) non-monetary and monetary values within the concept of economic benefit; (6) legitimacy through international treaties; and (7) issues of climate change, health, and poverty alleviation, rather than conservation per se. CBNRM is a wide palette indeed.

The emergence of community conservation in the 1980s and 1990s

Ideas about sustainable development gained traction with the publication of the World Conservation Strategy (IUCN, 1980) and the Bruntland Commission (Bruntland, 1987). At about the same time, conservationists began to take people more seriously because they were worried about the loss of wildlife and the future of parks as ecological islands in a sea of people. The World Parks Congress in Bali in 1982 was an important turning point, with Zimbabwe having a significant influence on the IUCN's uptake of the guiding principles. In Holdgate's[1] preface to Graham Child's (1995) *Wildlife and People*, he mirrors the emerging sentiment in southern Africa:

> if wildlife and protected areas are to survive, they must be socio-politically acceptable, economically viable and ecologically sustainable. Where land is scarce, people have to see that some of it is best used as wildlife habitats − best in the sense of being most beneficial to them. Otherwise they will turn habitats into pastures and replace antelopes and elephants with cattle. And who can blame them?

The goal of CBNRM and the sustainable governance approach, indeed, is summed up by Child (1995, p 235):

The Zimbabwean experience suggests that if wildlife is permitted to contribute meaningfully to their welfare, people will not be able to lose it in their battle for survival. If wildlife does not contribute significantly to their well-being, people will not be able to afford to preserve it, except as a tourist curiosity in a few protected areas.

Community approaches emerged spontaneously across the world, with surprising little reference and learning from each other. A short summary of some of the more important programmes is provided in Boxes 13.1 to 13.5. While I will argue later that governance within communities follows a set of ubiquitous principles, the pathway to get there is highly contextual. Thus, community forest management in Latin America was initiated by social movements as a response to the loss of land by indigenous and local communities (Box 13.1). In Asia, communities were more sedentary with stronger land rights and village governance (Agrawal, 2005), and participatory forest management materialised as an anti-dote to weaknesses in top-down forest management (Box 13.2). Community fisheries management in South East Asia was mooted in the 1990s, mostly in response to failed centralisation, and in recognition of the historical capabilities of traditional systems (Box 13.3). Community forest management in Africa may have been moderately successful in Tanzania (Nelson & Blomley, 2009), and widespread in West Africa through donor aid (Ribot et al., 2010), but without genuine rights and tenure it has never provided the anticipated benefits. Nonetheless, the application of CBNRM principles to a REDD+ (reducing emissions from deforestation and degradation) forest project in Tanzania suggests the community forest management is workable provided communities acquire rights and benefits (Box 13.4). Community wildlife management in southern Africa has already been discussed at length, but was initiated by wildlife administrators who changed the rules in anticipation of a different future. An important lesson from the current front-runner, Namibia (Box 13.5), is that rights coupled with high-quality support have a high return on investment, but not within the lifespan of most donor projects – it took the Namibia programme 12 years to break even after rights were devolved, but after 25 years benefits exceeded costs by a factor of six.

BOX 13.1 PROPERTY RIGHTS AND SOCIAL MOVEMENTS IN COMMUNITY FOREST MANAGEMENT IN THE AMERICAS

Mexico provides an important example of 'privately held communal property' (Porter-Bolland et al., 2012) with some of the longest experience in community forest management (Bray et al., 2006; Porter-Bolland et al., 2012). After the 1910 revolution, some 17 million people living in communities gained formal (but non-transferable) title rights to their land (*ejidos*) (Arnold, 2001). Members own their houses and individual plots, but communal resources are owned and managed through communal governance structures in which all members legally participate face-to-face (Charnley & Poe, 2007). For several decades (the 1940s to the 1970s), communities had severely truncated commercial rights. Timber production was controlled by the government through logging concessions to industry companies, resulting in restricted benefit streams and some forests being converted to agriculture and livestock. In the 1970s, peasant organisations and policy reformers in government campaigned for stronger local rights to manage and benefit from forests. This culminated in the 1986 Forestry Law, which transferred decision-making powers over forest harvesting to the *ejidos*, and a new Forestry Law in 1992 that reduced unwieldy bureaucratic involvement. Governance happens through village parliaments, some of which are successful and some of which are not. Many communities now manage their forests quite successfully through community forest companies and joint ventures. There is a significant literature about *ejidos*. Opponents suggest that communities are corrupt, and the *ejidos* are associated with poor productivity and land degradation. Proponents note that poor initial performance was often due to outside influence, especially top-down planning and restrictions of the rights of *ejidos* to use their resources (especially forestry) to best advantage, plus low levels of extension and support. They suggest that while the system is not perfect, and governance arrangements require further analysis, local community ownership generally works quite well (Bray et al., 2005; Wilshusen, 2008).

Key concerns where governance is not well managed are elite capture and few benefits (Arnold, 2001), corruption and conflict, and constraints imposed by a heavy and unstable regulatory frameworks (Charnley & Poe, 2007).

In the 1970s, in the Brazilian State of Acre, a social movement of traditional forest dwellers led by Chico Mendes fought for community land rights (Schmink & Wood, 1992; Brown & Rosendo, 2000; Cronkleton et al., 2008). 'Indigenous reserves' and 'extractive reserves' are now a very significant component of Brazil's protected area estate covering millions of hectares. They give local communities *de jure* tenure rights to their land and aim at securing traditional forest-based livelihoods. Indigenous lands now occupy one-fifth of the Brazilian Amazon and (even with people living in them) have been more successful at conserving forests than other approaches, including some protected areas (Nepstad et al., 2006). Indeed, a literature is emerging globally to show that forests managed through forms of governance that include community management and co-management are as or more effective than the hegemonic state control of the past, especially if rules are designed and applied locally (Ostrom & Nagendra, 2006). However, these systems are limited by the fear of giving communities full commercial rights, especially for wildlife. Low-value commercial uses of wildlife take place in the shadows, and are damaging wildlife populations, rather than enhancing them as happens in southern Africa and with, for example, exclusive and commercial village management of pirarucu, a huge Amazonian fish (Koziell & Inoue, 2006; Pinheiro, 2018).

BOX 13.2 PARTICIPATORY FOREST MANAGEMENT IN ASIA

In Asia, community forestry was initiated in the 1970s to adapt forest management to the needs and interests of local people through greater devolution and local participation (Arnold, 2001). In an excellent analysis, much informed by field experience, Arnold traces the initial replacement of local forests by the state, the later inability of the state to manage large areas of forest effectively, and the reintroduction of decentralised community forest management. As with public wildlife in Africa, Asia's forests degraded under centralised industrial management (de facto open access), which also excluded communities from a critical part of their livelihood strategies.

A number of countries acknowledged that centralised management had failed to protect forests or provide livelihoods. Helped by the donor community, they initiated community forest management in countries ranging from South Korea (community woodlots) to India's social forestry programmes (from 1976), and Nepal's community forestry which addressed deforestation in upland watersheds. The idea of participation advanced from passive to much more inclusive interpretations, although often more towards co-management (e.g. joint forest management) than true devolution (Arnold, 2001). As discussed in Chapter 12, Arnold points out that 'decentralisation' through local leaders or local government usually means that benefits and control pass to local elites rather than the majority.

An example of stronger devolution was the development of forest user groups in Nepal, which was well backed by legislation. Arnold is much more positive about the results of this programme, including stronger democratic participation and improved land and forest management, but still expressed concerns about possible elite capture and marginalisation of women. Nonetheless, the outcomes of this programme have been favourable for most stakeholders (Birch et al., 2014).

Arnold concludes very much as did the proponents of CAMPFIRE (Chapter 11). Without a legal basis for the transfer of power to communities, participation is more apparent than real, and results are decidedly mixed, especially where the transfer of rights excludes the most valuable resources. Arnold also identified the widespread problem that local institutions are often not accountable to their constituents:

> Devolving control or decision-making powers to bodies that do not have accountable leaders is likely to give power over the resource to particular individuals or groups of individuals within the community, effectively privatising use rights in their favour. It thus risks defeating the social objectives of community forestry.
>
> *(2001, p 71)*

BOX 13.3 COMMUNITY FISHERIES MANAGEMENT IN SOUTH EAST ASIA

In the mid-1990s community-based fisheries also began to emerge, especially in Asia. As with both wildlife and forestry, the trend in Asian fisheries after World War II was for the state to take over their management. This diminished the role of traditional and informal local management, which had run fisheries quite sustainably under a certain set of conditions. Centralisation resulted in a rapid deterioration of local fisheries under distant, under-staffed, and under-funded national government fisheries agencies (Pomeroy, 1995). Nationalisation and privatisation of fisheries to corporate interests did not solve the problem of over-exploitation, yet often deprived local people of livelihoods. This led to the re-emergence of community-based resource management, and the need to recreate local management because many indigenous systems had been weakened or disappeared, except in Indonesia where they endured and performed well.

The Philippines had a long history of traditional fisheries rights and allocation, though this was centralised to national and municipal governments until 1991, when new devolutionary policies led to an exponential increase in community-based marine protected areas and locally managed marine areas. Success was attributed to a community-driven 'bottom-up' approach, with important advances in empowering local fishing communities but with more mixed results in the health of local fisheries (Maliaoa et al., 2009). Similarly, in Thailand, centralisation did not prevent biological over-exploitation or conflicts between small-scale and commercial fisheries (Pomeroy, 1995). Thus, in 1993, the Department of Fisheries began experimenting with community-based fisheries management by granting rights to local communities. In Vietnam, fishermen historically had local fishing associations called *van*, but these were banned in the 1970s, resulting in the over-exploitation of coastal fisheries.

The most enduring community-based fisheries were in Indonesia, but these were not legally recognised in the 1990s (Pomeroy, 1995). However, later some traditional systems such as Panglima Laot in northern Sumatra (whereby communities develop rules and regulations for their fishing grounds) became formally recognised. Indonesian fishermen were able to define their boundaries, to set and change rules, and to manage local fish effectively. They did struggle to manage open sea fish because of the challenges of dealing with powerful external stakeholders such as commercial fishing companies and higher level state agencies (Cinnera et al., 2012).

In reflecting on 20 years of experience in community fisheries, Allison et al. (2012) argue that both security of person and security of property are important for the governance of sustainable fisheries. While the more standard argument about security of tenure and economic efficiency holds, these are not the only insecurities faced by fishermen – basic human rights and personal security were often as or more important to insecure and vulnerable people. They conclude:

> First, and most fundamentally, our counterargument is that securing human rights is integral to improving fisheries governance and management outcomes in many of the world's fisheries. Strengthened rights are essential to reduce vulnerability and increase adaptive capacity, which in turn underpin social–ecological resilience.

BOX 13.4 CBNRM AND REDD IN TANZANIA

In Tanzania, CBNRM has recently been applied to the challenge of deforestation and REDD + programmes, adopting principles directly from the experience of community wildlife management in southern Africa (Morgan-Brown, 2014). Villages in the Lindi area in southern Tanzania were assisted to develop zoning or land use plans. The purpose was to set aside community forests, and REDD payments would be made according to measured performance in reducing deforestation. Initially, communities were reluctant to participate because, historically, community-based forest management only benefitted village government activities, village leaders, and local elites (and were often associated with corruption), while poor households incurred a disproportionate share of the costs of setting aside forests.

Villages were only willing to expand their forest reserves after they were provided with estimates of potential REDD cash earnings, and agreements that communities could pay cash dividends to individual community members if they so wished. After receiving their cash dividends, several villages decided to expand their reserves in the hopes of obtaining even larger payments in the future, despite their initial scepticism and mistrust of the process.

Trial REDD payments ranged from $2 000 to $30 000 per village, and were calculated using satellite remote sensing and statistical models to estimate how much each village had reduced deforestation compared to the average rate of deforestation in the area (i.e. the expected emissions reductions in village forest reserves). All registered members in a village became eligible for annual cash dividend payments from REDD. In some villages, this included up to three children per family, with this money being collected by the mother. Dividends proved to be a powerful incentive for forest conservation, reducing deforestation by 30% in the first year (Morgan-Brown, 2014). Research suggested that even very small cash dividends contributed to improving livelihoods, especially for women and the very poor. Dividends also improved childhood health and education because children were eligible for dividends through their mothers. Interestingly, 44% of people surveyed stated that someone in their household used their cash dividends for entrepreneurial activities, including agriculture, beekeeping, and small business.

BOX 13.5 CBNRM AND WILDLIFE IN NAMIBIA

Community wildlife management in Namibia and Zimbabwe co-evolved with considerable cross-scale learning. The park directors from Zimbabwe and Namibia were good friends (Chapter 8) when they co-initiated private wildlife conservation in the 1960s and 1970s, with white commercial farmers gaining rights 'huntable game' (buffalo, kudu, oryx, springbok, warthog, etc.) through the Nature Conservation Ordinance of 1975.

In white settler states, however, progress in devolving rights to rural communities took longer. Garth Owen-Smith and others spent a number of years preparing the ground for CBNRM, largely by funding community game guards. Pointedly, CBNRM in Namibia only blossomed once communities gained legal rights through the Nature Conservation Amendment Act of 1996 and began to generate substantial financial and economic benefits (Skyer & Saruchera, 2004). The Namibian government was keenly aware that CAMPFIRE was flawed because it devolved authority to district councils, not communities (Jones & Murphree, 2001). Consequently, rights were devolved to communities that volunteered to form conservancies, who received 100% of the income from hunting and tourism once they satisfied a narrow set of conditions – defining their boundaries and members, and developing a constitution with an equitable benefit distribution plan.

Important features of the Namibian programme are 100% revenue retention by community conservancies, the well-coordinated and high quality support provided by a number of NGOs including an exceptional system of community monitoring called MOMS (management orientated monitoring systems) or the event-book system (Stuart-Hill et al., 2005, 2007).

Wildlife was nearly wiped out during the independence wars, but recovered rapidly after the introduction of CBNRM. I highly recommend examining Namibia's annual state of the conservancies report to absorb the magnitude of the achievements of this programme along many axes (NACSO, 2015) – wildlife numbers, recovery of endangered species, land area conserved, participation, benefits, knowledge, improving management, and others.

Namibia's picturesque scenes and narrower range of wildlife make it much more amenable to tourism than most other community wildlife projects in Africa. Nonetheless, viability requires a judicious combination of hunting and tourism, and economic simulations showed that a hunting ban would be deleterious to most of the communities and wildlife involved (Naidoo et al., 2016). Local people are angry and cynical about external attempts to impose a non-use ideology that has very large costs to communities and conservation. They are deeply offended by the arrogant, simplistic, and anti-democratic mistrust of local landholders and communities, regardless of their enormous efforts and achievements, including the imposition of ideas that are culturally imperialistic – such as imposing non-hunting cultural norms on Bushmen!

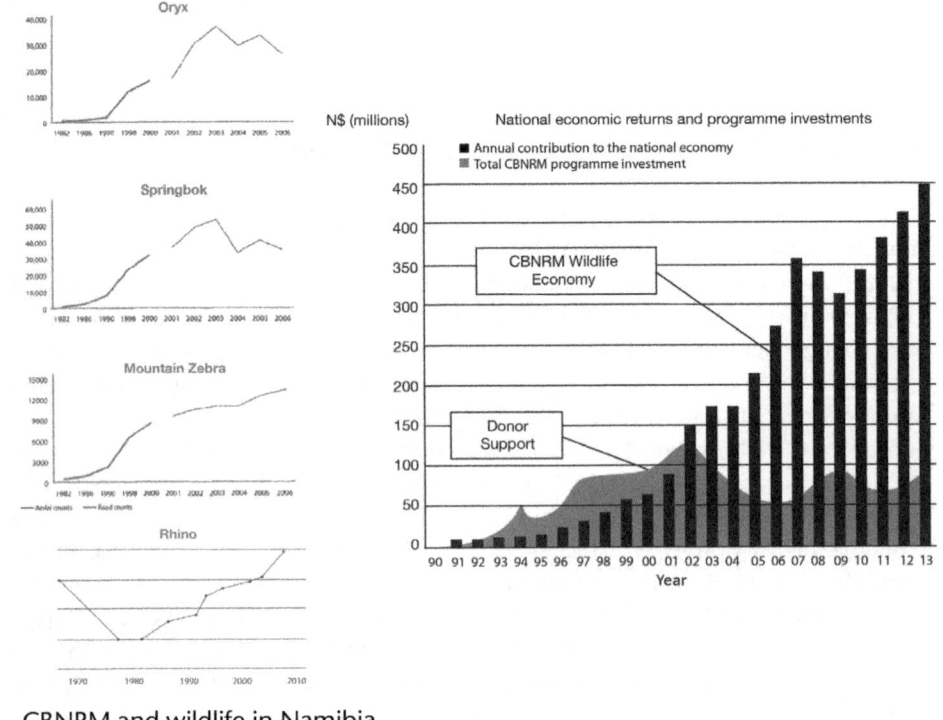

CBNRM and wildlife in Namibia
Source: NACSO (2018).

Since Namibia initiated CBNRM in the late 1990s, the number of community owned businesses and joint ventures has increased and in 2017 stood as follows:

- Hunting increased from 5 to 56 concessions;
- Tourism lodges increased from a few to 54 (plus 20 under construction);
- Tourism and hunting earned N$132.8 million for communities;
- There were 171 natural resource enterprises;
- Community conservation facilitated 5 350 jobs in 2017;
- CBNRM provided 500 000 kg of meat for local communities in 2013 – 2 million high protein meals;
- The contribution of CBNRM to Namibia's net national income was $7.11 billion.

The recovery of wildlife, and the rapidly growing national economic impact, is illustrated in the figure.

CBNRM as a melding of ideas and experiences

With WWF initiating its 'Wildlife and Human Needs' programme in 1985, conservation NGOs recognised the opportunity to tap into development financing. This led to a decade of integrated conservation and development projects, mostly with disappointing results and with little sign that lessons from earlier projects were incorporated into later ones (Mcshane & Wells, 2004).

At the same time, ideas about democratic local governance (Blair, 2000), learning organisations (Korten, 1980; Senge, 1990), participatory development (Chambers, 1983, 1994), and decentralisation (Rondinelli et al., 1989) emerged, and were also taken up by foresters, especially in South Asia (Arnold, 2001). Arnstein (1969) categorised participation as ranging from non-participation, to tokenism, to citizen power, ideas later absorbed by the conservation literature (Pimbert & Pretty, 1997; Barrow & Murphree, 2001).

By the 1960s, businesses were also rethinking their approaches, and the shift from top-down to bottom-up management preceded CBNRM by several decades (Micklethwait & Wooldridge, 1996). Yet, to its detriment, the CBNRM literature seems largely unaware of theories of organisational development despite the parallels. Like wildlife governance, early organisational management was top down following Henry Ford and Frederick Winslow Taylor's rigid ideas about 'scientific management'. Later, these ideas were softened by Mary Parker Follett and others from the human-relations school, who defined management as 'the art of getting things done by people' (Micklethwait & Wooldridge, 1996). Peter Drucker recognised that companies were social systems as well as business organisations, and introduced the idea of management by objectives (Drucker, 1973). By the 1970s, Peters and Waterman (1982) began to 'democratise' management, encouraging flatter and decentralised organisations, including the concept of 'loose–tight' management (Chapter 12), where day-to-day management is delegated (and relinquished) to teams that are given a lot of freedom and space for local decisions provided they achieved 'tight' goals (Handy, 1994). Again like CBNRM, these changes built on Douglas McGregor's (1960) conceptualisation of humans as being self-motived (Theory-Y) as opposed to requiring constant supervision

(Theory-X). Implementing CBNRM is about change management, from the simple 'freeze-unfreeze-refreeze' ideas of Carl Lewin, to John Kotter's (1996) useful eight-step model of change (Figure 8.1). Lewin also developed important ideas about democratic values and society, and about learning through Action Research (Burnes, 2004). Finally, it is necessary to be cognisant of Abraham Maslow's (1943) hierarchy of needs when working in rural communities.

An important influence on CBNRM was common property thinking. Garrett Hardin outlined the dangers of open-access property regimes in his influential 'Tragedy of the Commons' paper (Hardin, 1968) and started an important debate about collective action. This discussion was transformed by Ostrom's masterpiece, *Governing the Commons*, who demonstrated that the human capacity for local collective action and ingenuity could go a long way to finding decent economic solutions for common-pool resources provided certain principles were followed. She must have got it largely right because her principles have stood the test of time (Ostrom, 2009), encouraging a deluge of case studies and supporting literature, even if many donor and conservation projects are in too much of a rush to understand and incorporate them.

The importance of market-based conservation is probably the area that has most been neglected. Its application is largely confined to southern Africa, and there is downright opposition to the idea of bringing environmental products to the market in many places. Thus, the strong push for a rights-based approach in Amazonian indigenous and extractive reserves (Nepstad et al., 2006; Ding et al., 2016) is coupled with a very mixed message when it comes to the rights of indigenous and local people to use wild resources profitably. As we noted in Chapter 5, the aversion to valorising wildlife though the marketplace, and restricting wildlife to low-value, local subsistence uses, is ideological, but intellectually curious. Conservation approaches are inconsistent. They allow fish to be killed in many protected areas, but not mammals. They oppose the dollarisation of wildlife, yet invest heavily in certification, markets, and payments for non-timber forest products, agro-biodiversity, value chains, and ecosystem services including a huge global investment in REDD carbon payments (Frost & Bond, 2008).

Defining CBNRM

CBNRM has received considerable attention from donors and academics since the 1980s. While there has been a lot of thoughtful writing about CBNRM, and significant commitments of development assistance, as a whole CBNRM is neither well conceptualised nor theorised. Given so many threads, CBNRM is difficult to define. Attempts to shake out the key principles of it by analysing the literature tend to provide lists of factors important to success (Cox et al., 2010; Dressler et al., 2010; Gruber, 2010; Tole, 2010; Brooks et al., 2012), a detailed analysis of an interesting but narrow aspect of it (Porter-Bolland et al., 2012), or important case studies (Koziell & Inoue, 2006; Measham & Lumbasi, 2013). Academic critiques tend to ignore the considerable body of practical experience (Barrow & Murphree, 2001) while many authors analyse CBNRM from their disciplinary standpoint. Consequently, CBNRM is not yet framed by what we might call a development hypothesis or a theory of change, which is the objective of this chapter.

Barrow and Murphree (2001) propose a typology for community conservation ranging from park outreach, to co-management, to CBNRM and citizen control. They agree that CBNRM is about participation, but the need for self-mobilisation and empowerment should not be confused with consultation or the thin end of co-management (Arnstein, 1969; Pimbert & Pretty, 1994). Indeed, CBNRM's endpoint is the discretionary economic and political choice enjoyed by most people in prosperous societies. CBNRM seeks genuine entitlement over land and natural resources, rather than weaker forms of participation. In other words, it is essentially about 'private-community ownership'.

CBNRM can become overwhelmingly complex, to the point of paralysis. This complexity is greatly reduced once our focus is clear. My interpretation of CBNRM is that it is about community empowerment (economic and political) over land and wild resources in ungoverned spaces. Forest and dryland communities need security of person and property as necessary conditions for multi-dimensional development, poverty reduction, and economic efficiency (Sen, 1999). CBNRM, in essence, represents a Glorious Revolution (Chapter 2) for the disempowered people in the world who live disproportionately with wild resources. Likewise, CBNRM brings wild resources into the economic realm on a level playing field by governing them with the same advantages that have long favoured domestic plants and animals.

CBNRM is about inclusive governance and the careful crafting of institutions. Inclusive governance has political and economic dimensions (Figure 2.1). CBNRM extends to rural communities the right to benefit from and engage in political and economic choices. CBNRM also optimises the allocation and use of wild resources by internalising the full costs and benefits of land use by wild and domestic resources. The goal is to transform social ecological systems into new basins of attraction (Gunderson & Holling, 2002; Walker et al., 2004) where (a) greater protections, freedoms, and choices unlock the inherent potential of individuals and communities (Sen, 1999); (b) collective action is effective and just; (c) resources are allocated to their highest value uses (Smith, 1776); and (d) raw materials are turned into more highly valuable goods and services (Beinhocker, 2006).

In these ways, CBNRM is an antidote to the deinstitutionalisation of rural areas described in Chapter 4. It is a subset of the sustainable use approach, but has the additional challenge of cultivating collective action in communities that have been disenfranchised and traumatised (Chapter 4). CBRNM, therefore, is the process of reinstitutionalisation, beginning with the devolution of genuine proprietorship to land and wild resources. It directly addresses the challenge that ungoverned spaces and communities are the losers in a 'dual economy', allowing a modern sector and elites to enrich themselves by extracting 'surpluses' without paying the full costs of their actions (Figure 3.2).

Which resources does CBNRM apply to?

There is considerable debate about which resources CBNRM applies to, including community wildlife management, community forestry, and community-based fisheries, to give a few examples. Many of the ideas behind CBNRM emerged out of common property theory, including large scale irrigation practices in Asia and even groundwater management in the USA (Ostrom, 1990). There is an increasing association of CBNRM with non-timber forest products that are valuable to rural households (Shackleton et al., 2007). Agriculturalists apply it to ecosystem services in programmes such as LandCare (Prager & Vanclay, 2010). And it has been suggested as a mechanism for implementing payments for ecosystem services in relation to carbon and water (Bond et al., 2009). Intuitively, CBNRM deals with wild resources like wildlife, fish, forests, carbon, wetlands, and rangelands that are scattered, fugitive, mobile, overlapping, ecologically complex, and associated with considerable uncertainty and variability. It is community-based because we cannot simply draw a line around these resources and allocate the management to a single individual as economic theory would have us do; their geographic and temporal variability requires that they are managed at a collective scale.

Like the sustainable governance approach, CBNRM should not be associated with any one resource. It is about improved resource governance and allocation, and indeed many of its principles could apply equally to the collective action needs of impoverished urban communities. However, CBNRM also recognises that wild resources are no longer public goods (Chapter 5) and provides an institutional framework for privatising wild resources at the level of community land units, and for the collective self-regulation of externalities. That CBNRM is about environmental justice is well recognised. That it is deeply economic is not recognised. Yes, CBNRM gives local people control over the resources that are rightfully theirs. However, it also brings wild resources long excluded from wealth-creating economic systems back into the economy in ways that create value and are sustainable. These economic aspects (Chapter 6) are largely ignored by the CBNRM literature. Thus, I concur strongly with Hulme and Murphree (2001) that CBNRM is a radical approach to conservation that is both neo-liberal and focused on environmental justice, encompassing three major conceptual strands (Hulme & Murphree, 2001):

- The state should devolve proprietorship, including the responsibility for and benefits from managing wild resources, to the communities that live with them (Murphree, 1991).
- Natural resources should be exploited as profitably and sustainably as possible to achieve both conservation and development goals (SASUSG, 1996).
- The neo-liberal concepts of markets, property, and exchange should play a greater role in giving communities discretionary choices, in shaping incentives for conservation, and in allocating resources to their highest valued uses.

Thus, CBNRM is about putting wildlife back into the marketplace on equal terms, not locking it out of the ebb and flow of normal human transactions. Moreover, Murphree (personal communication) has argued that 'conservation' also has the meaning of generating the most from the least; it is about efficiency. On a planet that cannot afford to use resources wastefully, CBNRM promotes efficient resource tradeoffs, rather than favouring any one species or outcome through ideological or non-economic preferences. Farming should be promoted in suitable arable areas; CBNRM provides a mechanism for the sustainable use of wildlife and wild resources in marginal areas where they have an economic comparative advantage.

Thinking of CBNRM as a complex policy experiment

The present political economy of conservation localises costs and nationalises or globalises benefits and authority (Wells, 1992). CBNRM seeks to reverse this inequity by devolving the management and benefits of wild resources from a centralised bureau-technical elite (public management) to the hundreds of thousands of communities who live with these resources.

From the preceding synopsis of the literature, it is clear that CBNRM is complex. There is biological complexity involving the use of wild resources and ecosystems, their sustainability, and the non-linear relationships between them. There is economic complexity, ranging from foundational theories like property rights, pricing, and resource allocation, to the practicalities of maximising the values of wild resources (Chapters 5 and 6). There is social complexity in the definition of communities and the heterogeneity and forces within them (Agrawal & Gibson, 1999). And there is the ever complex challenge of governance – the formation and stewardship of rules at multiple levels through political processes.

The way we handle complexity is critical. We need to act, without complexity becoming an excuse for non-action or the cause of operational paralysis. It is therefore useful to think of CBNRM as an adaptive policy experiment, which includes a model or hypothesis of the system we are managing, transformational knowledge of how we might change it, and objective knowledge of the end results that we are seeking (Lang et al., 2012).

CBNRM addresses multiple barriers. The high (economic) values of common-pool wild resources are not reflected in land use incentives and outcomes, and we need to learn to ensure that the full worth of wildlife, forest, fish, carbon, and water is reflected in the day-to-day decisions of the people who live with wild resources. Second, we need new configurations of private and collective action to manage complex and fugitive wild resources. Third, the poorest people live with the richest biodiversity, so we cannot solve conservation problems without also addressing multi-dimensional poverty. And finally, institutional theory suggests that progress will be slow or non-existent until we overcome man's long history of top-down extractive institutions. Freedom, innovation, and prosperity will flow from a future in which inclusive, democratic, and bottom-up institutions cement people's rights to decide the direction of their own lives, establish effective rules and institutions, and retain the rewards of their action and innovation.

A simple model of CBNRM

As a subset of the sustainable governance approach, CBNRM has the additional challenge of developing communities as collectively managed land units (i.e. micro-governance). CBNRM's theory of change includes four operational elements (Figure 13.1):

1. Proprietorship – devolving proprietorship to communities in as full and secure a form as possible to transform wild resources from a public good into private and private-collective goods.
2. Price – maximise the value of wild resources to landholders and communities. This requires strengthening markets and property rights, and removing barriers that under-price wild species. At a finer scale, social mechanisms are also needed to internalise the costs and benefits of non-market costs and benefits.
3. Inclusive micro-governance – we should not forget that communities themselves also need to be properly governed. The importance of inclusive within-community governance will be elaborated on in the next chapter because this subject is badly neglected. In addition, micro-governance is embedded within systems of meso-governance and macro-governance which will also be discussed at the end of the following chapter.

4.   Community management of wild resources – natural resource management is endlessly contextual, including multiple aspects of forest, wildlife, fisheries, land, agriculture, and ecosystem services management. In this book, we will therefore limit ourselves to two issues; providing communities with the incentives to manage resources reliably, and the process of developing natural resource management capacities through experiential and adaptive learning in the application of genuine rights and benefits. While these processes are ubiquitous, resource management skills are highly specific. This is described in Chapters 11 and 15.

This defines CBNRM as a form of privatisation operationalised through private-community tenure. CBNRM emphasises economic production and control, but also inclusive governance of the collective. As noted earlier, CBNRM is about sustainable production, and should not be conflated with regulation and the re-creation of mini-public agencies at the local level (Chapter 5).

The Figure 13.2 expands this four-step concept. As shown at the bottom of the figure, inputs to the system (natural resources) are regulated by the system's governance and management characteristics to provide outputs in the form of livelihoods and economy, sustainable resource use, and system resilience and adaptability.

The 'inputs' to CBNRM are collective (fugitive) wild resources, such as wildlife and tourism, forests, grazing commons, ecosystem services, fish, and carbon, rather than individually owned domestic resources (Figure 13.2). As we argued in Chapter 3, ungoverned spaces are used so inefficiently that new institutions can sustainably increase outputs by a factor of ten or more (Figure 3.4). This reinstitutionalisation starts with (1) the

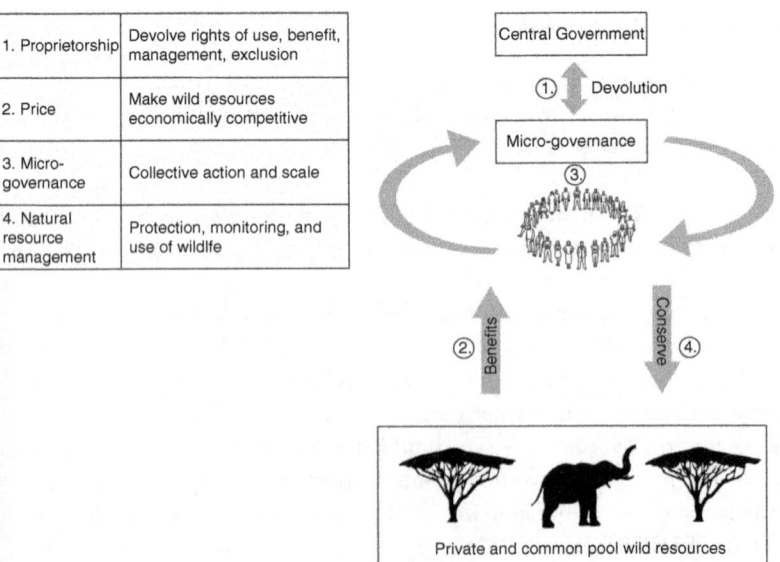

FIGURE 13.1   The four steps in CBNRM: proprietorship, price, micro-governance, and resource management

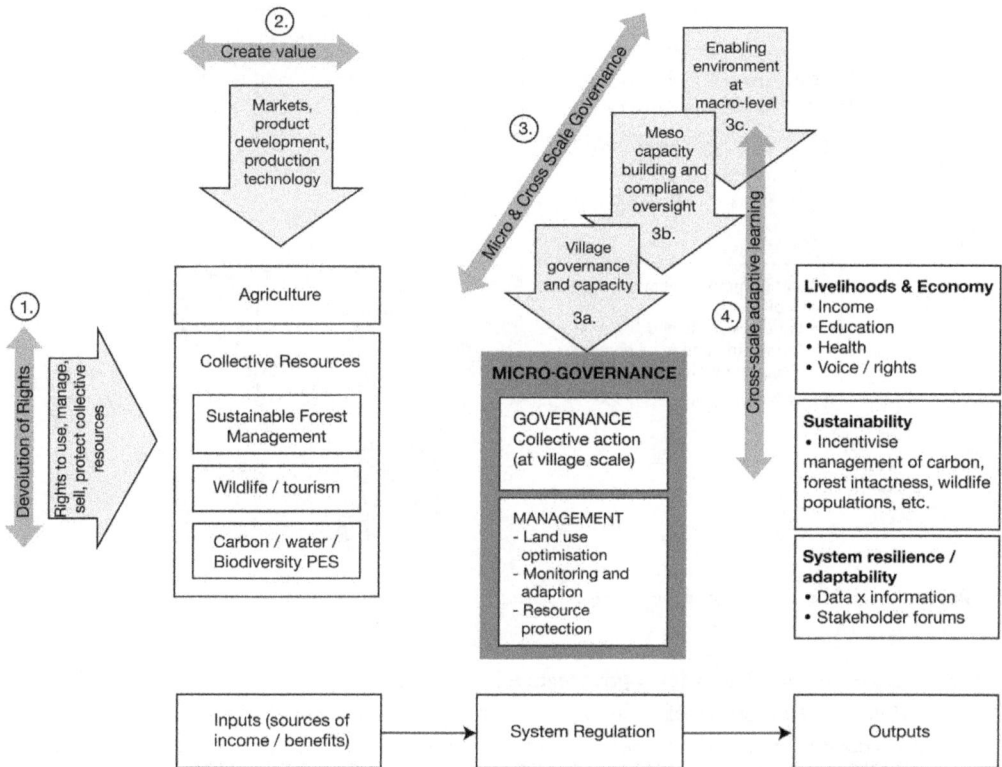

FIGURE 13.2   An expanded model of CBNRM and the sustainable governance approach

devolution of the rights to manage and benefit from wild resources to communities and (2) the process of valorising wild resources and getting prices right.

However, the economic and political transformation of rural villages requires much greater scrutiny of what we mean by micro-governance (3a), and how to support this at the meso-level (3b) and through an enabling macro-environment (3c).

At the local level (3a), building community institutions requires an inclusive community process to make and steward the rules (i.e. governance), but also the management of land and resources. I have illustrated the relationship between governance and management in Figure 13.3, noting that the latter includes economic resource allocation through land use planning and zoning, active resource management, monitoring and adaptive management, and the protection of the resource from non-entitled users (i.e. exclusion). Participatory governance is the essence of CBNRM, and is expanded theoretically in the next chapter with practical guidelines and tools in Chapter 16.

It is at the meso-level that most interaction with communities occurs (3b in Figure 13.2), but these functions are often too important to be delegated to middle-level staff. The functions of CBNRM support agencies include building the capacity of communities to govern and manage themselves, including the capacity of communities to market and manage their resources, as well as the introduction of performance monitoring systems and adaptive management. Procedural monitoring is also critical to enforce compliance with

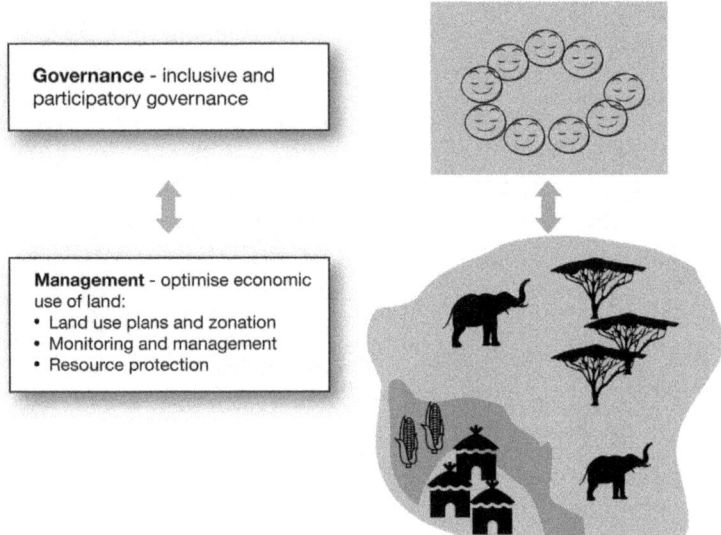

FIGURE 13.3 The relationship between governance and management in CBNRM

CBNRM principles. Meso-level governance includes cooperation between individual villages where there are economies of scale, including for landscape planning, managing law enforcement, and the configuration of tourism and hunting concessions that are large enough. This may be done by multi-village organisations, or in polyvalent forums that include local government, NGOs, and other stakeholders. Cross-scale networking and peer-to-peer learning also occurs at this scale.

The highest level of governance is often referred to as the enabling environment (3c), and includes legislation, research, sector-coordination, the development of knowledge, financing, and legal protection (Figure 10.6). It is difficult to bring CBNRM to scale if the enabling environment is not conducive.

The final component of this model is cross-scale adaptive management (4 in Figure 13.2) which, like governance, is a big box. This includes adaptive learning within each level, and of the system as a whole, with government and support agencies as much in need of new capabilities as communities. Learning and adaptation occurs through various combinations of effective performance monitoring, introspection, and through communities of practice, as described regionally for southern Africa (Chapter 8, Suich & Child 2009), and also in the context of CAMPFIRE (Chapter 11).

Conclusion

CBNRM began to emerge at a critical juncture in the late 1970s, roughly at the point where the ability of public management to cope with the increasing demands placed on wild resources by demography and globalisation on wild resources was exceeded. CBNRM has some characteristics of convergent evolution, with the management of forests, fisheries, wildlife, and other resources moving somewhat in the same direction simultaneously across the globe. The underlying change processes, however, were often quite different, reflecting different operational strategies in the complex four-level economic system described by Williamson (2000) and illustrated in Box 4.5. Thus, the devolutionary process in Latin America was driven by social movements, including the Catholic Church. In Asia, the drivers of change tended to be a partnership between British academics and development assistance, while the transformation of the wildlife sector in southern Africa was the result of deliberate policy reform by African wildlife administrators. This spawned a large, erudite, but disconnected literature, with limited cross-fertilisation between foresters and wildlife managers, scholars and practitioners, and scholars in different disciplines. Despite the impact of the writings of Elinor Ostrom, for instance, remarkably few of the projects funded by the plethora of development agencies and international NGOs pay heed to these principles when approaching community conservation. This literature has largely missed the salience of scale and participatory governance. It has missed, or even opposed, the importance of land use economics and the valorisation of wild species, scolding this approach with contemptuous epitaphs like 'neo liberal' or 'commodification'. It also seldom refers to the literature of organisational management, which is also navigating the transition from top-down scientific management to devolved systems of governance. Consequently, CBNRM remains a term that covers many things. In this chapter I have responded to these gaps by defining CBNRM as a subset of the sustainable governance approach, in which communities are the primary units of land management. This concept is illustrated as a simple four-component model that combines proprietorship, price, subsidiarity and community governance, and cross-scale adaptive management. However, the internal governance of communities is an important and neglected topic, which requires a much longer discussion in the next chapter.

Note

1   Martin Holdgate was a highly influential Director General of IUCN, and Graham Child served as an IUCN Councillor for Africa. Interestingly, IUCN's motto 'in harmony with nature', appears to derive from Zimbabwe's wildlife agency, where it was embossed on the coat of arms proudly worn by game rangers since the early 1970s, underneath a waterbuck's head. Similarly, the three circles of sustainability also appear to be derived from Zimbabwe because the term 'socio-politically acceptable, economically viable and ecologically sustainable' used by Holdgate in this quote is quintessentially Graham Child's, and one of his favourite phrases.

14

THE APPLICATION OF THEORIES OF MICRO-GOVERNANCE TO CBNRM

LEARNING OBJECTIVES

This chapter describes micro-governance within the community. It:

1. Identifies elite capture as a major impediment to the transition from extractive to inclusive governance.
2. Emphasises the importance of inclusive governance for social and economic progress.
3. Frames the design conditions for inclusive governance, drawing heavily on Ostrom's principles for effective common property regimes and Murphree's CBNRM laws.
4. Recognises that CBNRM takes place in less than ideal conditions, and reorganises and expands Ostrom's principles (by reference to the literature on democracy, complex systems, scale, new institutional economics, and market economics) in this context.
5. Concludes with a conceptual model to guide the implementation of CBNRM, which reorganises Ostrom's principles as foundational, emergent, and framing conditions within an experiential learning cycle.

CBNRM, elite capture, and the transition from extractive to inclusive micro-governance

Devolving rights for wild resources to communities is a necessary condition for CBNRM, but only becomes sufficient when coupled with effective governance within the community. Rights that are clearly defined and recognised by the authorities (Ostrom's seventh principle) provide a protective eggshell around the community. However, what happens inside this eggshell is critical but often neglected as CBNRM seeks the profound transition from the extractive, personalised governance of history toward inclusive governance and rule-of-law associated with prosperity. The importance of inclusive governance is discussed in Chapter 2, with the vignettes in Chapter 13 showing that within community governance tends to be overlooked in CBNRM, resulting in elite capture. This chapter develops the concept of 'micro-governance' to provide guidance for how to proactively create functional communities. CBNRM protects wild species and generates benefits. However, the essence of CBNRM lies in inclusive governance, and we seek to promote the public goods of informed participation and equitable benefits, while dodging the scourge of elite capture.

Elite capture, which so often undermines CBNRM (Frizen, 2007; Ribot et al., 2010), is a 'natural' feature of extractive, centralised regimes across history (North et al., 2009). Elite capture features in Egyptian and Chinese civilisations (Fukuyama, 2011). It is a key feature of European feudalism and modern 'traditional' leadership in Africa, where elite landholders in the form of lords, dukes, and chiefs are gatekeepers for a personalised state. This 'natural' extractive hierarchy even reasserts itself in communities that have experienced the considerable benefits of inclusive governances, as we saw in the latter stages of CAMPFIRE in Zimbabwe (Chapter 11, Muyengwa and Child, 2017). However, global norms are changing. Many modern societies no longer accept the divine right of kings or feudal governance. This challenges us to apply the same standards of inclusive governance to CBNRM, not least because many communities are trapped in the backwaters of extractive and authoritarian governance and neo-feudalism. CBNRM seeks the virtuous conditions of impersonal 'rule of law' in which people are equally entitled to keep the fruits of their labour and imagination, and to participate in public affairs (Chapter 2).

This chapter defines the conditions necessary for inclusive, efficient, sustainable micro-governance, or governance within the community, as well as the practicalities of designing and protecting these conditions. As we saw with CAMPFIRE (Chapter 11) and Luangwa (Lubilo & Child, 2010), it is not good enough to set up conditions for inclusive local governance. Strong and persistent measures are required to protect these conditions until inclusive local economic and political institutions become deeply entrenched and normalised. Simply insisting on elections every three years or so is not the answer because elections are easily manipulated and seldom hold 'elected' leaders accountable (Dahl, 1989). Indeed, meaningful elections that provide genuine accountability are an outcome of inclusive governance, rather than a pathway towards it.

The two gurus of micro-governance

We can trace a significant proportion of the theory of CBNRM back to two gurus – Elinor Ostrom from the USA, and Marshall Murphree from Zimbabwe. Ostrom and colleagues like Wade, McKean, Agrawal, and Gibson have done a remarkable job elucidating the principles of common property management (Wade, 1987b; Agrawal & Gibson, 1999; Mckean, 2000), which are encapsulated in Ostrom's eight design principles (Box 14.1). Ostrom (2000, p 35) summed these up by saying:

> When the users of a resource design their own rules (Design Principle 3) that are enforced by local users or accountable to them (Design Principle 4) using graduated sanctions (Design Principle 5) that define who has rights to withdrawal from the resource (Design Principle 1) and that effectively assign costs proportionate to benefits (Design Principle 2), collective action and monitoring problems are solved in a reinforcing manner.

Three further design principles relate to (6) 'access to rapid, low-cost, local arenas to resolve conflict among users or between users and officials', (7) 'recognition of the right to organise by a national or local government', and (8) the necessity of multiple (or nested) layers of governance for common-pool resources. More recently, and shifting the emphasis from static design principles to adaptive capacities, Ostrom emphasised the importance of face-to-face interactions for resolving complex social ecological problems (Ostrom, 2007, 2009b), and that user monitoring may be as or more important than the type of ownership (Ostrom, 2009c).

BOX 14.1 OSTROM'S DESIGN PRINCIPLES FOR LONG-ENDURING COMMON PROPERTY RESOURCE INSTITUTIONS

1. *Clearly Defined Boundaries*
   The boundaries of the resource system (e.g., irrigation system or fishery) and the individuals or households with rights to harvest resource units are clearly defined.
2. *Proportional Equivalence between Benefits and Costs*
   Rules specifying the amount of resource products that a user is allocated are related to local conditions and to rules requiring labor, materials, and/or money inputs.
3. *Collective-Choice Arrangements*
   Most individuals affected by harvesting and protection rules are included in the group who can modify these rules.
4. *Monitoring*
   Monitors, who actively audit biophysical conditions and user behavior, are at least partially accountable to the users and/or are the users themselves.
5. *Graduated Sanctions*
   Users who violate rules-in-use are likely to receive graduated sanctions (depending on the seriousness and context of the offense) from other users, from officials accountable to these users, or from both.
6. *Conflict-Resolution Mechanisms*
   Users and their officials have rapid access to low-cost, local arenas to resolve conflict among users or between users and officials.
7. *Minimal Recognition of Rights to Organize*
   The rights of users to devise their own institutions are not challenged by external governmental authorities, and users have long-term tenure rights to the resource.
   For resources that are parts of larger systems:
8. *Nested Enterprises*
   Appropriation, provision, monitoring, enforcement, conflict resolution, and governance activities are organized in multiple layers of nested enterprises.

Source: adapted from Ostrom (1990 and 2009a)

The second guru, Marshall Murphree, a professor at the Centre for Applied Social Research at the University of Zimbabwe, is more focused on the practical challenges of CBNRM governance. Learning inductively by working with communities, he played a major role in shaping the CBNRM narrative in southern Africa where his pithy statements are part of CBNRM lore (Martin, 2009b). His work is highly complementary to Ostrom, but he introduces a much greater appreciation of economics and the mechanisms of cross-scale governance. Murphree never considered academia to be his audience, so his ideas are less entrenched in the peer-reviewed literature. His impact is better measured by the practical uptake and conceptualisation of his ideas within CBNRM in southern Africa, but also in global institutions such as IUCN and the Convention for Biological Diversity where his phraseology echoes clearly in, for example, the Addis Ababa Principles of Sustainable Use (CBD, 2004). I have summarised Murphree's principles in Box 14.2, categorising them as his 'economic' principles (which relate to incentives, costs, and benefits), and principles for scale and the uptake of responsibility by small communities in whom he had great faith.

Both Ostrom and Murphree emphasised that face-to-face governance lies at the heart of collective action and CBRNM, and one of the puzzles of common property scholarship and practice is how widely their foundational insights are ignored. The confusion

---

BOX 14.2 MURPHREE'S PRINCIPLES FOR VIABLE COMMUNAL PROPERTY REGIMES

Rules relating to internalising economic costs and benefits:
1. Effective management of natural resources is best achieved by giving it focused value for those who live with them.
2. Differential inputs must result in differential benefits (producer community).
3. There must be a positive correlation between the quality of management and the magnitude of benefit.

Rules relating to devolution and scale:
4. Tenure over natural resources should be delegated to the lowest level of social scale possible.
5. The level at which benefits accrue should be the level at which management occurs.
6. The unit of proprietorship should be the unit of production, management, and benefit.
7. The unit of proprietorship should be as small as practicable within ecological and socio-political constraints.
8. An internal legitimacy endogenously derived but also sanctioned by the state is likely to produce a more robust base for organisation.

The relationship between group size and the resource base:
- Large groups with weak resource bases are unlikely to succeed.
- Small dispersed groups with large valuable resource bases will have difficulty acting in cohesion.

The design of small local jurisdictions:
- The fewer members the better.
- The closer they live together the better.
- The more they interact together on a daily basis the better.

Sources: Murphree (1994b, 2009); Martin (2009b)

may well lie in the challenge of scale. Ostrom had immense confidence in small communities (Wall, 2017) and understood they were part of nested institutions and polyvalent governance but never really described how small communities and larger systems were linked. The principles for building scale and nested institutions are elucidated elegantly by Murphree (2000) in his profound paper, 'Constituting the Commons' (see Box 14.3). According to Murphree, healthy configurations of scale are not designed top down, but emerge from sequencing, by first devolving rights to the grassroots and then allowing some of these rights to be aggregated upwards. Murphree calls this scaling down then scaling up. Scaling down ensures that resource rights are firmly entrenched at the level of individuals or communities. Communities then hold the power to delegate rights upwards, if and when they feel this is prudent, through the process of 'delegated aggregation'. Building scale through upward delegation ensures that rights and government 'originate' in the people, as happened with America's town hall governance (Tocqueville, 1994). Because communities and their members can always reclaim these rights, they can therefore hold the people to whom they have delegated responsibility accountable, leading to 'constituent accountability'. This differs fundamentally from the process of building scale through expropriation from above. Moreover, building scale from the bottom is far better at matching resource fugitiveness to jurisdictional scale. As noted in Chapter 10 and Figure 10.4, many more functions are retained at lower levels, and upward delegation is parsimonious, allowing upper levels to focus on doing fewer things better (i.e. jurisdictional parsimony).

BOX 14.3 MURPHREE'S LAWS FOR SCALING

Devolution needs to be complemented by a process for creating 'nested sets' of institutions to match the scale of resource fugitiveness to the scale at which most costs and benefits are internalised. Murphree's insight is that for 'scaling up' to be sustainable, this involves first 'scaling down'. He develops three principles for linking functional, ecological, and jurisdictional scales:

- **Delegated aggregation.** To meet the ecological imperatives of larger scales, management institutions (i.e. nested institutions) must be expanded through a process of aggregation rather than expropriation.
- **Jurisdictional parsimony.** Management institutions should match the scale of the specific resources to be managed (i.e. internalising all or most of the related costs and benefits) and should be no larger than necessary.
- **Constituent accountability.** To reach the desirable situation, where local groups influence the allocations of entitlements through the political process, local jurisdictions must become a significant political constituency of the state, and one to which the state is accountable. In other words, governance of wild resources should originate in local communities and be accountable to them.

Sources: Murphree (2000); Martin (2009b)

Foundational conditions for CBNRM

Ostrom's principles have stood the test of time (Ostrom, 2009a). However, after 25 years we can rearrange and expand them as a dynamic theory of change, rather than as a list of very sensible ideas. Moreover, our goal is not simply to understand what an effective common property regime looks like. It is to proactively create new common property regimes in the less than ideal circumstances of ungoverned spaces, where communities are not homogeneous, where levels of trust and social capital are low, and where wildlife resources are public rather than community assets.

With this in mind, I have rearranged Ostrom's principles in Figure 14.1, and expanded them through reference to the literature on market economics (Chapter 6), new institutional economics and scale (Chapters 4 and 10), deep democracy (Dahl, 1998; Dunbar, 1998; Wall, 2017), and complex systems, adaptive management, and participatory learning (Chapter 15) (Holling, 1973; Martin, 1999; Gunderson & Holling, 2002; Walker et al., 2004).

In my analysis, Ostrom includes four principles that are foundational, and four principles that are emergent properties of these. Foundational condition 1 is recognition of the rights of communities to organise, to which I have added the need to ensure that community governance is inclusive. Foundational condition 2 is secure community rights with defined membership and boundaries, as discussed in the context of proprietorship (Chapter 5), economic allocation (Chapter 6), and inclusive governance (Chapter 2). Foundation condition 3 is that resources have value. The need for the benefits of collective action to

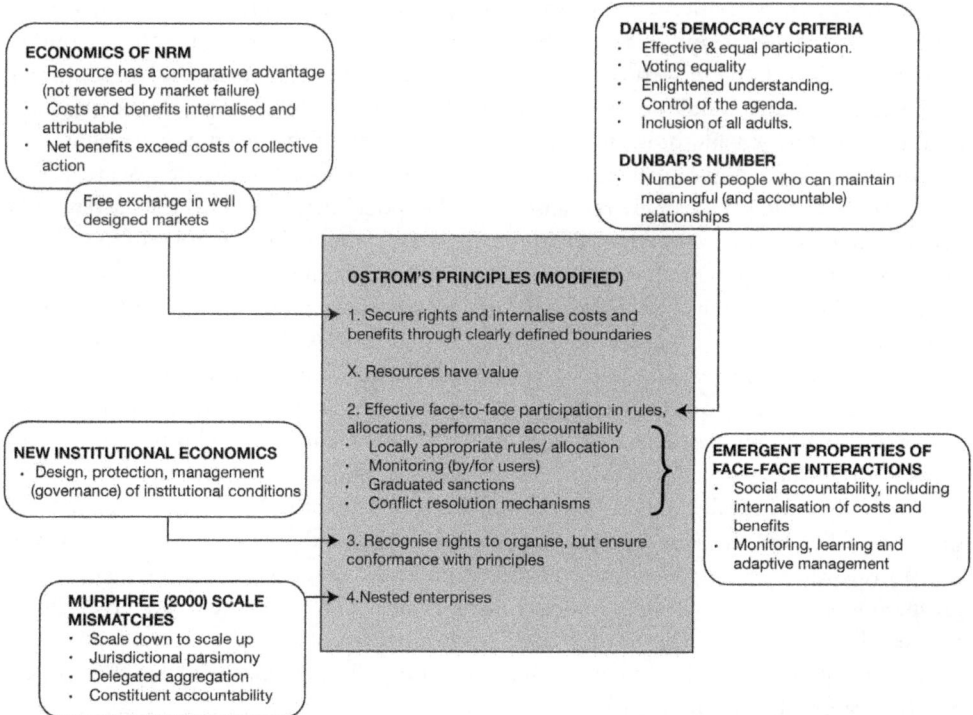

FIGURE 14.1 An expanded explanation of Ostrom's principles

exceed its costs is certainly noted by common property scholars (Wade, 1987a), but these economic criteria are not explicitly included in Ostrom's original conceptualisation, yet are critical to successful CBNRM. They mirror the emphasis on 'price' in the sustainable governance approach (Chapters 6 and 10) and Murphree's 'economic' principles (Box 14.2). Foundational condition 4 is effective face-to-face participation in the formation of rules, making of decisions, and transparency and accountability of the system. Trusting people to make the right decisions for themselves lies at the core of Ostrom's writings (Wall, 2017), and the importance of face-to-face decision-making (and therefore of scale) is emphasised repeatedly by Murphree (Box 14.1), yet, as mentioned above, receives little attention in the practice and scholarship of CBNRM.

Ostrom also highlights principles which we can paraphrase as locally appropriate rules, monitoring, graduated sanctions, conflict resolution mechanisms, and nested enterprises. I have labelled these as emergent principles, because they are usually outcomes of learning processes within the community. These principles act to improve the accountability and the adaptive capacity of the system, including the internalisation of costs and benefits.

Foundational condition 1: long-term tenure rights and the legitimacy of local regimes

Ostrom recognises that rights, boundaries, and local organisation require external legitimation (principle 7) but she never focused on this problem (Wall, 2017). Effective CBNRM requires that the community has long-term tenure rights to the resource, and the rights to devise their own institutions, which, according to Ostrom, should not be challenged by external governmental authorities. While I concur completely with Ostrom about the importance of inclusive governance, many historical factors prevent inclusive governance from being a natural condition of communities left to their own devices. These conditions need to be developed and protected, and should be linked to the devolution of rights, with requirements that communities comply with the conditions for inclusive governance. In these ways, CBNRM is strongly reminiscent of the model that underpins the current era of personal freedom and prosperity (Figure 2.4), which emphasises security of person and property and the importance of inclusive democratic governance.

Informed participation is far from automatic even in small communities (see Chapter 11). In the real world, democratic decision-making within CBNRM communities may need to be imposed and held in place through systems of compliance management. Indeed, the design of inclusive micro-governance, and its maintenance, usually requires external support and enforcement. This raises the question of who sets the rules, and who protects them in the transition from personalised to impersonal forms of resource governance (North, 2003). Theoretically, and ideally, this is the role of the state (Chapter 2). However, ungoverned spaces exist because the state has not designed such rules, and may even be reluctant to do so. Writing these rules is not difficult (although most examples are from inclusive and prosperous states) and the rules of micro-governance are ubiquitous and predictable – I based the first community constitutions I developed on those from a local sports club. However, managing the process of establishing an enabling environment for these rules is highly contextual, and often extremely challenging.

Foundational condition 2: clearly defined boundaries

The starting point for CBNRM and collective action is to define boundaries and rights of wild resources and communities. It is virtually impossible to conceptualise or implement CBNRM in the absence of clearly defined resource rights, and the importance of rights for community empowerment is well recognised in the literature. However, in practice, too many short cuts are taken in defining who owns these rights, with a tendency to allocate them to a community as represented by a committee. Community rights need to be much more clearly defined, in the manner of individual shareholders of a village company. Also under-recognised is the contribution of exclusion and the internalisation of costs and benefits towards economic function.

Foundational condition 3: resources have value (added principle)

CBNRM is predicated on unlocking the true value of wild resources, so I have added this principle to Ostrom's list. CBNRM can only succeed where resources have a positive value, after covering the costs of collective action. The need to maximise the value of wild resources, and to ensure that this value accrues to the people who live with wild resources, has already been discussed at length (Chapter 8). Moreover, members who get 100% of wildlife income, and have the right to spend it as they choose, will value wildlife more highly than if wildlife benefits are treated as non-discretionary public goods (i.e. you can only buy classroom blocks with your wildlife money).

Foundational condition 4: most individuals affected by rules or decisions can participate in modifying them face to face

Ostrom recognises the importance of inclusive governance and democracy when she suggests that 'most individuals affected by operational rules can participate in modifying the operational rules'. Yet, despite being a political scientist whose husband also worked on local democracy, she never states this explicitly – perhaps it was so obvious to her that this was not necessary. But it is not obvious to all, and most CBNRM programmes make the serious mistake of assuming that decision-making by an elected leadership is a substitute for informed participation by the full membership. This is wrong thinking. I have explicitly referred to Dahl's (1989) criteria for democracy (Box 14.4) in Figure 14.1 to show that inclusive governance requires effective participation by enlightened participants. Dahl, a pre-eminent scholar of democracy, does not include elections in his criteria, re-emphasising the point that elections are a defective mechanism for accountability. I introduced Dunbar's number, and the deeply human processes of reciprocity in Chapter 2, to explain the importance of face-to-face accountability in the case studies (Chapters 11 and 12), especially in the early stage of the transition to inclusive governance. What is so surprising, given the scholarship of Ostrom and Murphree and the obviousness of the case that I have built (especially Chapter 12), is that both scholars and project designers pay so little attention to the effects of face-to-face decision-making on social process, decision-making, and the emergence of social capital and accountability.

We see this in the conflation between community-based versus committee-based governance. Most CBNRM programmes are designed as representational meso-level institutions in which the needs of local people are 'represented' through people they elect (Figure 12.7). Where this is well managed, it can result in significant gains in environmental management, wildlife populations, community income, and others, as with the national Namibia CBNRM initiative (NACSO, 2015). Unfortunately, representational forms of CBNRM governance seldom result in equitable benefit sharing or community participation (Chapter 12), while periodic elections also provide only a weakened form of accountability. Consequently, representational CBNRM programmes are plagued by problems of elite capture and, increasingly, financial mismanagement. These are the classic symptoms of extractive and centralised governance, which can create economic growth for a time, but lack innovation and usually succumb to struggles between elites – this is described in detail in Acemoglu and Robinson's book, *Why Nations Fail: The Origins of Power, Prosperity, and Poverty*, and also by North et al. (2009).

Representational CBNRM, or centralisation at the local level, lacks the inclusive political institutions essential for long-term sustainability, and is highly unlikely to 'evolve' towards a more participatory state. By contrast, face-to-face CBNRM delivers critical public goods in the form of informed participation, equitable benefit sharing, and protection of marginalised groups, including women. It follows that well-governed villages are the bricks out of which larger governance edifices are constructed, managing economies and ecologies of scale through delegated aggregation as described in Figure 12.7.

BOX 14.4 DAHL'S FIVE CRITERIA FOR DEMOCRACY

1. Effective participation

Citizens must have adequate and equal opportunities to form their preference and place questions on the public agenda and express reasons for one outcome over the other.

2. Voting equality at the decisive stage

Each citizen must be assured his or her judgments will be counted as equal in weight to the judgments of others.

3. Enlightened understanding

Citizens must enjoy ample and equal opportunities for discovering and affirming what choice would best serve their interests.

4. Control of the agenda

The demos or people must have the opportunity to decide what political matters actually are and what should be brought up for deliberation.

5. Inclusiveness

Equality must extend to all citizens within the state. Everyone has a legitimate stake within the political process.

Note that Dahl does not include elections in these criteria, as meaningful elections are an outcome of democratic accountability, and not a means to obtaining it.

Source: Dahl (1998)

Emergent properties of CBNRM

In reanalysing Ostrom's principles, I have suggested that four are foundational and occur at the boundary between community and authority. Once communities have the rights and incentives to manage wildlife, and follow participatory processes in doing so (i.e. the foundational conditions), the next four principles listed by Ostrom (locally appropriate rules, monitoring, graduated sanctions, and conflict resolution mechanisms) emerge adaptively through experiential learning. These principles are more internal to communities. I only separate them because they emerge through a process of on-going citizen-to-citizen learning and negotiation, and can be improved continually. This requires (1) sound processes of micro-governance, (2) high quality facilitation, training, and capacity building, and (3) monitoring.

The second point of elaboration is that CBNRM is much broader than the management of common-pool resources, and requires two sets of rules – first for the governance of the community itself, and second for the management of resources. The rules for self-governance and financial management are foundational, and fall under 'all people affected by rules must participate in making the rules'. Small communities the world over have common evolutionary roots in clans of hunter-gatherers, so rules for inclusive governance usually are ubiquitous. By contrast, the rules and systems for managing natural resources (and for managing community development, including social projects like schools, health, and water) are context specific. Therefore, local rules for managing water, wildlife, forests, fisheries, non-timber forest products, aquifers, grazing, and so on are tailored through a process of informed, face-to-face decision-making and adaptive learning, and honed through a process of community experience combined with high quality facilitation and data.

Locally appropriate rules for managing wild resources

The least descriptive of Ostrom's principles is number 2 which she variously defines as 'congruence between appropriation and provision rules and local conditions' (Ostrom, 1990) and 'proportional equivalence between benefits and costs' (Ostrom, 2009a). My interpretation is that she is searching for rules that create accountability and internalise costs and benefits in proportion to effort. These are Murphree's principles that 'differential inputs must result in differential benefits' and 'positive correlation between the quality of management and the magnitude of derived benefits' (Box 14.2), as well as the principle of producer communities described in Chapters 11 and 12.

Monitoring

Monitoring (principle 4) is essential for controlling performance (e.g. finances, governance, community satisfaction), and for managing projects and natural resources. Monitoring should be conducted by the community or by the people accountable to the community (Ostrom, 1990). When used appropriately, this data contributes to community learning, social accountability, and the internalisation of the costs and benefits related to resource use.

Graduated sanctions

The principle of graduated sanctions (principle 5) implies that punishment for breaking rules is proportional to the size of the crime. For instance, communities often deal with poaching of small animals through traditional means, while elephant poachers are sent to national courts and prisons. In later writings, Ostrom finds that even simply making known the names of people who get caught can be enough to control systems. Especially in small groups, humans have a remarkable capacity for controlling wrongdoers and for reciprocity. We see this in the Luangwa example, where the insistence on providing communities with accurate financial records (including loan's to chiefs), eventually prevented chiefs from expropriating community money. Transparency was enough to correct the system through some form of social process to which outsiders were not privy.

Conflict resolution mechanisms

Ostrom discusses the need to resolve conflicts (principle 6) in ways that are rapid and low-cost, including ensuring that rules are clearly understood. Conflict resolution takes a number of forms, including establishing clear rules in the first place that define how decisions can be made or challenged, as well as brokering conflicts and even legal solutions.

For community governance, for example, a combination of clear and oft-repeated rules, backed up by reliable monitoring of compliance criteria, is quite effective for preventing conflicts in the first place. I have illustrated this in the process of institutionalising rules, rights, and constitutions (Chapters 15 and 16). Thus, community constitutions describe procedures for making decisions and resolving common conflicts, including an independent pathway for ordinary members to hold their leaders to account. For example, communities should be able to call extraordinary general meetings at which they require office bearers to present information, especially financial informa-tion, and where they can sanction wrongdoing, including removing corrupt or non-performing leaders. The combination of effective monitoring (e.g. of finances, patrolling, projects) and transparent discussions in open forums goes a long way in pre-empting conflict.

Through the introduction of rule-based governance, rights, and boundaries, CBNRM is in and of itself a process of bringing past problems, inequities, and conflicts to the fore, and resolving them, including the transition from open-access resource management to private-community tenure. This usually requires external assistance including conflict resolution and quality facilitation. Legal redress is also valuable, for example for clarifying and strengthening communities' rights to wild resources.

Cascaded institutions

Village-level communities, at the scale of Dunbar's number and practising participatory democracy, are the bricks from which CBNRM is constructed. Ostrom (1990) suggests that governance activities are organised in multiple layers of nested enterprises (principle 8). Returning to Murphree,

> Generally, the smaller a regime is the more effective and efficient it will be. Increases in scale complicate communication and decision-making, and beyond certain levels regimes must bureaucratise with attendant costs. Compliance inducement shifts from low-cost modes of moral and peer pressure to the high cost methods of policing and formal coercion. Increase in scale erodes the sense of individual responsibility.
>
> *(Murphree, 1997; Martin, 2009b, p 19)*

Once effective village-level institutions are in place, nested institutions can be built upwards from there, requiring additional bureaucratisation. Building upwards from thousands of villages seems daunting. However, what are the alternatives?

Many administrative hierarchies in post-colonial states (such as district councils and district administration) are democratically hollow because these higher-level organisations are constructed from above, and are not effectively rooted in the citizenry. This differs, for example, from New England in the north-east USA, where the principles of democratic participation were normalised through town hall meetings, and people's democratic rights run deep (Tocqueville, 1994). Our goal is to develop civic accountability in ungoverned spaces. This may seem naively optimistic and overwhelming, but is precisely why it is so important: there are no shortcuts to progress and re-creating citizenship, while accountability is the antidote to the over-centralisation of wild resources and the tragedy of open access (Figure 5.5).

Devolving the rights for wild resources to communities (i.e. denationalisation) matches the increasingly private nature of wild resources and the regimes for governing them (Chapter 5). Moreover, building jurisdictional scale from the bottom creates much better alignment with the fugitive characteristics of resources (Figures 5.2 and 10.4), so local civic action becomes the foundation for the polyvalent governance that has allowed the slight recovery of biodiversity in inclusive states (Chapter 3). In CBNRM, therefore, rights and functions should be devolved to individuals or the village level, while a few functions with major scale advantages are then 'delegated upwards'. A simple example of this is the village scouts in Luangwa. On average, two village scouts were employed and paid by each of the 43 village action groups, but the management of groups of about 20 scouts was delegated upwards to the six multi-village committee. If the latter did not perform, the VAG could reclaim its rights (i.e. constituent accountability). This mechanism also kept the number of functions delegated to higher levels to a small number (jurisdictional parsimony) including, for example, secondary schools but not primary schools.

Enlightened participation and capacity building

In the examples that Ostrom uses to support her case, she describes these rules in place, but pays far less attention to processes by which stakeholders initially negotiated them (Ostrom, 1990). Since our purpose is less to study working situations, but more to create new ones, we need to pay attention to the process and conditions out of which the principles of collective action emerge. Implementing CBNRM effectively requires good design but, like self-managed teams in business, equally important is a culture of entrusting communities with rights, and insisting and assisting them to make their own decisions within a framework that holds them accountable for performance. Participation in rule-making lowers the transactions costs of managing compliance with rules, remembering that without social legitimacy enforcement is impossibly expensive (Wade, 1987a) as we see in many rules governing wildlife. Participation also encourages locally designed rules to be custom-made to local circumstances.

My emphasis on face-to-face deliberation, scale (Dunbar, 1993), and democratic process (Dahl, 1998) is closely aligned with both Ostrom and Murphree (Murphree, 1994b; Ostrom, 2007), but perhaps more emphatic because their advice is largely overlooked. The wider literature on CBNRM certainly pays lip service to citizen control, democratising wild resource management, self-mobilisation, and the delegation of rights and power to communities (Arnstein, 1969; Pimbert & Pretty, 1994; Fabricius et al., 2004, p 31), yet the mechanics and deep preparation for inclusive governance are seldom mentioned, including in project documents. Lack of attention to these details explains, perhaps, why CBNRM is more promise than performance.

There are practical ways for designing community structures and processes to fulfil these requirements. Foremost, the community must be small enough for face-to-face deliberation, and should meet regularly to make and review decisions in general meetings. The management of community money is an efficient way to institutionalise informed participation and equitable decision-making, and also the participatory adaptive management process where everyone is entitled to quality information, and fully involved in making decisions and reviewing the progress of these decisions. I have emphasised the process of activity-based budgeting and participatory quarterly financial reviews (Chapters 11, 12, 16) to illustrate this process. In practice, these principles should be incorporated into procedural rules, using a bill of rights, constitution, and compliance monitoring. At the core of these rules is the principle that all people affected by decisions should participate in making them. The decision-making body is all people sitting together. The role of an elected representative is not to make a decision for the people, but to bring the whole community together to make their own decisions collectively, and to bring well-organised information to this process. Community representatives are instructed by, and accountable to, their constituency.

It is tempting for agencies implementing CBNRM to work through representational governance because of the false appearance that it is easier, logistically, to work with a few 'qualified' or elected people to get things done. Ultimately, however, the opportunity costs in terms of fostering inclusive self-determination and social capital are high. Thus CBNRM communities need to be configured as a 'learning organization' (Senge, 1990) with 'enlightened participation' and the incorporation of reliable information into face-to-face forums for resolving problems, including complex ones (Ostrom, 2007).

CBNRM communities gain agency when they do things for themselves, and purposeful experiential learning is more effective when it includes the key elements shown in Figure 10.8. The acquisition of information is important, as is external facilitation in obtaining, understanding, and using this information. Communities learn a great deal from exchange visits, extension, training, the provision of basic information (Chapter 15), and so on. However, real learning (cognitive change) occurs when they work together to integrate this information into solving empirical challenges, sometimes with external facilitation. I have illustrated community resourcefulness (or agency) as emerging from cognitive change (Figure 10.8), but cognitive change, equally, can be an outcome of doing things for themselves as communities develop much more positive self-paradigms. As we saw in the Chikwarakwara in CAMPFIRE (Chapter 11), communities were initially sceptical of the opportunities offered by CBNRM, but the action of benefit sharing and building a community grinding mill kicked off a cognitive transformation, rather than the other way around. This may even be generally true: good policy is more likely to be a response to the lessons of pilot projects than from cognitive deliberation in the capital city.

In my experience, communities learn faster when they apply data, especially visualised data, to practical management problems (i.e. experiential, adaptive management). This requires systematic monitoring of key variables by the community or by people accountable to the community (Ostrom's fourth principle), but also the analysis of this data.

The Namibian MOMS system (Management Orientated Monitoring System) is an exceptionally well designed system that promotes learning and empowerment through a locally owned data collection and management information system (Stuart-Hill et al., 2005). Basic data requirements are identified by the community, rather than the dictates of external agencies. The system is built up in bite-sized pieces, starting with simple monitoring systems that track practical issues, especially community finances and governance, the performance of micro-projects, human wildlife conflict, and animal numbers.

Data collection is not enough. This data needs to be presented in formats that are accessible to the community, regularly, with an emphasis on visualisation and simple graphs, models, and maps (Van Der Reit, 2008). The collection, management, visualisation, and use of data in social learning is a field of knowledge that is still developing. Visualisation of data using maps, graphs, and charts appears to 'distanciate' people from the emotional content of the data, leading to cognitive problem solving (i.e. solving problems in the analytical frontal cortex) rather than processing the problem emotionally through the primitive – limbic – part of the brain. This explains why CBNRM meetings and focus groups that use data and graphs are often much more productive than discussions that seek community opinion in the absence of data, which can become personal rather than problem-solving. Visualisation and 'distanciation' are described by Van Der Reit (2008).

The glue that brings this all together is collective problem-solving and implementation that combines monitoring, data, face-to-face learning, planning, action, reflection, and adaptation. In the latter part of her career, Ostrom argued that face-to-face interactions might be a critical ingredient in solving complex or 'wicked' problems (Ostrom, 2007), while in her Nobel acceptance presentation she also suggested that monitoring can be more powerful than property rights. In practical terms, adaptive learning is ideally incorporated into CBNRM top to bottom, from setting quotas to distributing revenue and project management. This can be as simple as using work plans with the following headings: agreed goal, status, problems faced, corrective action.

At a theoretical level, CBNRM incorporates new ideas of complex systems thinking, resilience, and adaptive management (Holling, 1973; Gunderson & Holling, 2002; Meadows, 2008). Thus Murphree states (quoted by Martin, 2009b, p 11):

> We need to see process as an end as well as a means, and to accept that the core objective of CBNRM is increased communal capacity for adaptive and dynamic governance in the arena of natural resource use. It is about local capacities to handle change and to negotiate the human impact on nature from past to future. It is as much about resourcefulness as it is about resources … The core objective of CBNRM is increased communal capacity for adaptive and dynamic governance in the arena of natural resource use.

So far, we have focused on the community itself. However, CBNRM is as much about building the capabilities and management culture of the meso- and macro-levels that set communities on the pathway to self-reliance as it is about the communities themselves. This is not a simple task, and invariably requires a team of professionals capable of developing cross-scale processes that have the ultimate goal of empowering communities. It is no coincidence that both CAMPFIRE and Namibia had such teams. Long-term professional and personal relationships between communities and experienced scholar practitioners are an important component of successful CBNRM programmes, both for structuring adaptive learning systems and for incorporating scientific and technical experience from elsewhere. The development of collaborative adaptive management, with communities at the centre, was briefly described for CAMPFIRE in the context of managing hunting and wildlife management (Goredema et al., 2006; Rigava et al., 2006). Additionally, much relevant technical data (such as the findings from experiments in other countries) is beyond the reach of local people, communities are often isolated from new ideas and knowledge, and they cannot aspire to something they don't know about. They also lack the experience and exposure to integrate or judge the data. Incorporating this requires sensitive 'capacity-building' where professionals co-develop appropriate local systems for monitoring and adaptive management, and communities generate knowledge through experimentation (see Chapter 15).

The CBNRM theory of change

For motivations ranging from poverty reduction to community empowerment and environmental conservation, there has been an intensifying search for empirical strategies and theoretical principles for achieving community conservation in many forests and drylands since the 1970s (Chapter 13). This final section provides an overall framework, or theory of change, for CBNRM (Figure 14.1). It builds on the conceptualisation of CBNRM in Figures 13.1 to 13.3, together with inclusive micro-governance, which I have illustrated in Figure 14.2 as governance 'within the eggshell', where the community and its rights are clearly delineated, and all people affected by decisions participate in making them (Ostrom, 1990; Dahl, 1998).

Following the proprietorship-price hypothesis, CBNRM requires the devolution of proprietorship for wild resources to communities, combined with measures to ensure that the value of these resources is maximised and internalised by the community (top left of Figure 14.1). However, inclusive village governance is rare in rural areas (Chapter 4), just as liberal democracy and broad prosperity were rare anywhere before 1800. Therefore, inclusive governance, including the protection of women and marginalised groups, needs to be developed. This seldom arises organically within communities, so that the design and enforcement of participatory governance are often external inputs to the system. Consequently, I have suggested that 'governance compliance' is a critical input to the CBNRM process. This was illustrated by the Luangwa example (Chapter 12), where the 'imposition' of inclusive governance was willingly adopted by ordinary community members if not their leaders. The challenge of shifting from extractive to inclusive governance regime in the face of the natural propensity for elite capture (at multiple levels, including within the community) should not be underestimated, including measures to prevent the re-exertion of extractive forms of governance.

Most CBNRM communities have been treated badly by history, shattering traditional systems of resource governance and often leaving communities with low social capital. Therefore, social and associational capital invariably needs to be re-created in communities by applying external inputs to a well-structured governance system, while simultaneously strengthening the technical and managerial skills of the communities. We have the knowledge to develop inclusive micro-governance with a high probability of success, and there is no technical excuse not to do the job properly. I provide guidelines for institutionalising new norms of inclusive micro-governance through practice, repetition, and compliance management in Chapter 16.

In many communities, the resource base has also been depleted, and the recovery from an open-access regime to a system that can sustain benefit flows often requires rehabilitation and recapitalisation of the environment. In the case of wildlife, a highly effective approach is the external financing of village scouts to monitor and protect high-value resources, because this combines recapitalisation of the resource base, capacity building, empowerment, and job creation.

Returning to governance within the eggshell, the adoption of inclusive governance should not be negotiable, because it is the well-spring out of which CBNRM capacities emerge. However, as CBNRM communities begin to manage their own affairs and natural

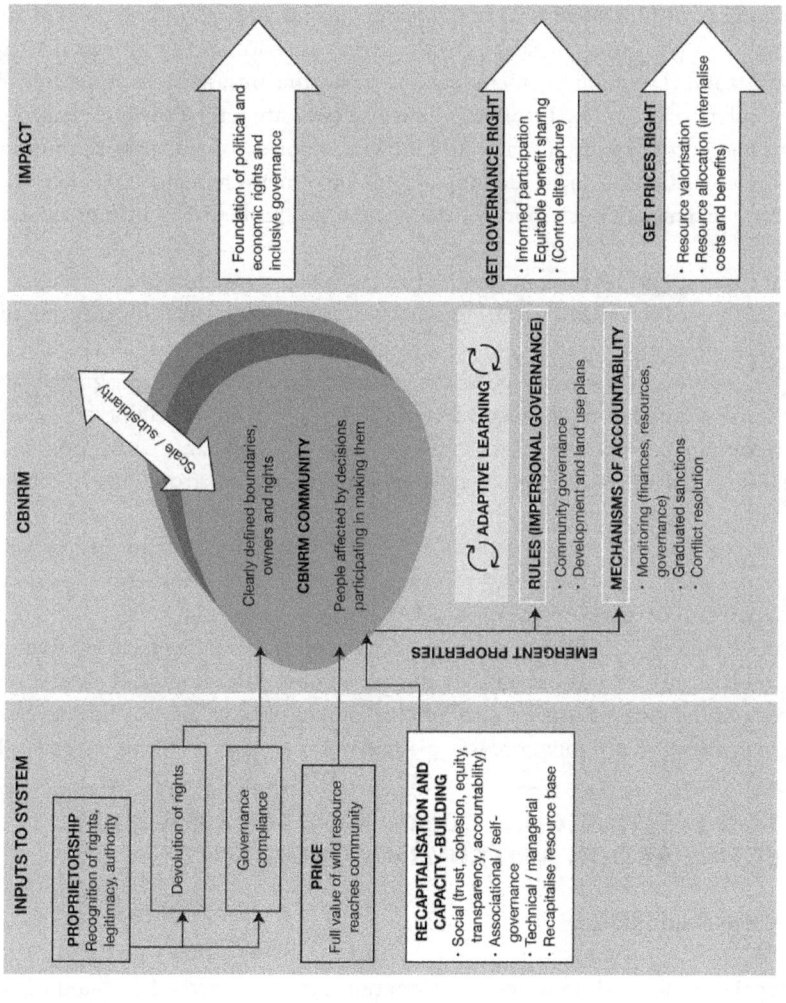

FIGURE 14.2 The process of building micro-governance and common property management through CBNRM

resources we need to differentiate between two set of rules (bottom centre of Figure 14.2). The first set of rules entrenches the principle that 'all people affected by decisions participating in making them'. These rules take the form of an individual bill of rights and a village constitution, and are institutionalised and normalised through repetition and compliance. However, effective natural resource management requires that communities begin to plan and control their own development through 'locally appropriate rules', including rules of use for natural resources, land use planning, zoning, and control.

In Figure 14.2 the 'rules' are illustrated separately from the 'mechanisms of account-ability', although both improve through experience, introspection, and external facilitation. Monitoring lies at the heart of adaptive management. Monitoring is linked to graduated sanctions and conflict resolution, including cases like the misuse of community finances or breaking natural resource rules. Together, these processes strengthen the internalisation of costs and benefits and accountability for action. Likewise, informed participation, monitoring, and accountability contribute to social learning and adaptive management, and improve the social and economic efficiency of the system. This organisational resilience is institutionalised through:

1. Standardization and repetition, codified in rules and procedures;
2. Monitoring – of compliance (for governance) and performance (for projects and natural resource management);
3. The presentation of monitoring data to the community with maps, graphs, and figures to create 'enlightened understanding' transparency and accountability;
4. The resolution of complex problems through the regular, transparent, face-to-face mechanisms that Ostrom emphasises later in her career (Ostrom, 2007, 2009c).

The 'scale/subsidiarity' arrow acknowledges the importance of the process of building capacity across scale, and of matching resource fugitiveness to jurisdictional scale by scaling down and then scaling back up through delegated aggregation.

Figure 13.1 defines CBNRM as having four major components, including generating benefits, devolving rights and benefits to the community, developing inclusive community governance, and the local management of wild resources. The least well understood of these is inclusive governance, which therefore is explained in this chapter and summarised in Box 14.5.

BOX 14.5 INSTITUTIONAL CONDITIONS FOR EFFECTIVE COLLECTIVE ACTION AND CBNRM GOVERNANCE

Foundational conditions

1. **Proprietorship and boundaries** – Communities have clearly delineated boundaries and strong proprietary rights (Schlager & Ostrom, 1992) over wild resources.
2. **Price and value** – Wild resources approach their economic value, making wild life a competitive land use option.
3. **Subsidiarity and micro-governance** – All people affected by resources participate in decisions about resources (Ostrom, 1990), preferably face-to-face and supported by information (Dahl, 1998; Ostrom, 2007); these conditions are protected and enforced (Child & Wojcik, 2014).

4. **Inputs to recapitalise communities**

   a. Self-governance, social and associational capital.
   b. Technical and managerial capacities.
   c. Rehabilitation of ecologically and economically sustainable resource base.

Emergent properties

5. Impersonal, participatory governance (rule of law) is institutionalised.
6. Effective rules of use emerge (to manage land and resources effectively and sustainably).
7. Mechanisms evolve to internalise the costs and benefits of resources including monitoring (by the community themselves), graduated sanction, and low cost mechanisms for conflict resolution (Ostrom, 1990).

Framing conditions and subsidiarity

8. The rights of communities to own resources and make decisions are recognised by authorities (Ostrom, 1990; Murphree, 2000).
9. **Scaling** – Proprietorship should be anchored as low as possible to match the scale at which the externalities occur, and nested institutions should be built parsimoniously from the bottom up through a process of delegated aggregation, maintaining ultimate authority in the citizenry (Murphree, 2000).

Impact

10. Get prices right (valorisation of wild resources and efficient resource allocation through internalisation of costs and benefits).
11. Get governance right (informed participation and equitable benefit sharing displaces personalised and extractive governance and elite capture).
12. Contribute to an inclusive and economically efficient society.

Defining clear conceptual goals and processes is critical for effective CBNRM, but implementation has the additional requirement of aligning the management culture around these conceptual mechanics. CBNRM requires effective and disciplined technical and managerial capacity building, and the application of these high quality inputs for at least 10 to 20 years. Projects that expect this process to happen in a top-down manner within a five-year project cycle, and without professional implementation, are symptomatic of the underlying problem of disempowerment and disappointment, not a solution to it. The early CAMPFIRE programme and the ongoing national CBNRM programme in Namibia are iconic examples of CBNRM for a reason. They combine sound conceptualisation and long-term partnerships between dedicated and experienced technical experts and communities.[1]

(I have made a big deal of community proprietorship throughout the book, and have mentioned, but not really dealt with, the problem of personal security, especially in places where communities are subjected to criminality and violence. Given that the Glorious Revolution required 'security of person and of property', skipping over security of person is a substantial oversight. Indeed, one of the core functions of the state is the provision of this security, a fact recognised very early on by Thomas Hobbes. In the absence of a functional state, and safety, any progress that relies on moderately effective institutions, including CBNRM, becomes exceedingly difficult. Therefore, the first step towards CBNRM in some places needs to be the provision of security. In a programme in which I am currently working, for example, we are attempting to provide a more secure environment for CBNRM using village scouts which, somewhat like neighbourhood policing around the world, is an important step, but it is impossible to achieve security with reliable support both legally and militarily from state security forces.)

If we dare to dream a little bigger, CBNRM has a higher purpose. These 'impacts' are illustrated on the right side of Figure 14.2. CBNRM is an antidote to the ungoverned spaces created by deinstitutionalisation. The underlying cause of loss of wildlife and wild places in areas co-inhabited by poor people is also weak or missing institutions – for land, for people, and for wild resources. Thus, getting governance right strengthens community participation and rights, and can contribute in a bottom-up way to building a more accountable state. Moreover, wild resources are private or private-community assets, not public resources as is commonly assumed, often with considerable value if brought back into the marketplace. It follows that reintegrating wild life (writ large) into the economy through effective proprietorship and pricing provides an alternative pathway to both wildlife conservation and economic development. The combination of rights, valorisation, and the internalisation of the costs and benefits of wild resources (through rules of use, monitoring, and graduated sanctions), gets prices right and contributes to a larger and more sustainable economy.

This will change the political economy of the wildlife economy. As we saw in southern Africa, wildlife administrators reshaped markets and regulations, enabling wildlife to pay for itself so that communities made the 'right' decision about resource allocation and became part of the growing wildlife economy. This shifted the burden for paying for wild resources away from the state. However, it also changed the role of the wildlife agencies from policing compliance with top-down regulations, towards creating positive incentives for conservation (Murphree, 2004). Globally, CBNRM communities will adopt the bio-experiential economy when they acquire rights to wildlife resources, and when policy-makers develop markets that pay communities for bio-diversity and other ecosystem services. This is likely to accelerate the development of missing or partial markets for ecosystem services. It may also focus minds on more pragmatic solutions. For instance, the current REDD+ payment system seems way over-complicated, with high transaction costs including consultancy expertise and fees. Simply paying per pixel for carbon might well be far more effective.

At the local scale, dysfunctional systems of governance and economy lock many social-ecological systems in many dryland and forest areas in a reinforcing cycle of social and environmental decline (Chapter 4). CBNRM deliberately reverses these conditions, mimicking the conditions underlying the Glorious Revolution by getting prices and governance right. If scaled sufficiently, proper implementation of CBNRM (with the emphasis on *proper*) might enable communities to share in the age of prosperity, contribute to the broader adoption of inclusive governance, and conserve forests and drylands and the wildlife they live with.

Note

1 In both cases, interestingly, self-confident leadership was necessary to acquire external funding while 'protecting' these programmes from some of the preconceptions and unworkable ideas that are often imposed on communities when projects are designed by well-meaning but inexperienced people who are too far from the action.

15

IMPLEMENTING CBNRM

LEARNING OBJECTIVES

This chapter provides a framework and checklists for implementing CBNRM including:

1. A four-stage process for implementing CBNRM.
2. A CBNRM viability analysis including economics, governance, support and policy reform, and checklists for framing these analyses.
3. A situation analysis, including participatory rural appraisal, livelihood surveys, and the adaptive governance dashboard process.
4. The main components of a CBNRM programme: governance, economics and benefits, natural resource management, security and services, compliance and performance tracking and adaptive management, and the enabling environment.
5. The main requirements for programme support and adaptive management.
6. A brief introduction to capacity building in the context of CBNRM and adaptive learning.

Four stages for implementing CBNRM at a new site

With considerable theory and several case studies behind us, this chapter suggests a four-stage process for implementing CBNRM systematically at a new site (Figure 15.1).

The first step is to assess if the proposed CBNRM programme is viable at site level, and to judge if negative answers can be reversed by modifying the enabling environment.

The second step is to understand and build relationships with the community and related agencies. This includes motivating a change coalition around a viable vision of an alternative

Implementation Process

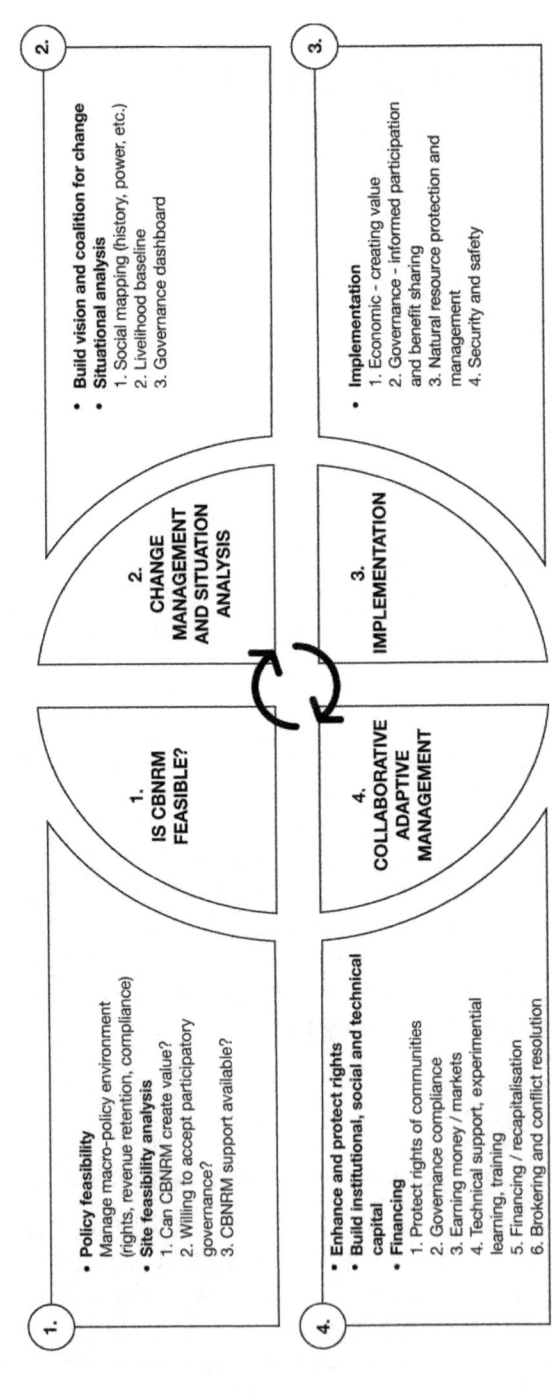

Policy feasibility
Manage macro-policy environment (rights, revenue retention, compliance)
Site feasibility analysis
1. Can CBNRM create value?
2. Willing to accept participatory governance?
3. CBNRM support available?

Build vision and coalition for change
Situational analysis
1. Social mapping (history, power, etc.)
2. Livelihood baseline
3. Governance dashboard

Enhance and protect rights
Build institutional, social and technical capital
Financing
1. Protect rights of communities
2. Governance compliance
3. Earning money / markets
4. Technical support, experimential learning, training
5. Financing / recapitalisation
6. Brokering and conflict resolution

Implementation
1. Economic - creating value
2. Governance - informed participation and benefit sharing
3. Natural resource protection and management
4. Security and safety

1.
IS CBNRM FEASIBLE?

2.
CHANGE MANAGEMENT AND SITUATION ANALYSIS

3.
IMPLEMENTATION

4.
COLLABORATIVE ADAPTIVE MANAGEMENT

FIGURE 15.1 Four stages for implementing CBNRM at a new site

future, combined with a situation analysis to gain a better understanding of the community history, livelihoods, geography, and other issues.

The third step is implementation proper. This usually requires at least ten years, and includes establishing sound systems of rights and governance, creating value and an economy to support livelihoods and management costs, and improving the management, monitoring, and protection of natural resources. Having recently worked in a hotbed of rhino poaching and criminality, I am now convinced that a fourth activity is crucial – to ensure that the community is safe and secure.

Finally, CBNRM is unlikely to succeed without the capacity to enhance and protect community rights, and to build institutional, social, and technical capacity at all levels. In addition, the natural resource base will often need to be rehabilitated and recapitalised. All of this costs money.

Assessing CBNRM feasibility

There is no point embarking on CBNRM unless it creates value, is well governed, and is well supported. This raises three critical questions at site level plus a fourth, broader question about the enabling policy and support environment. Thus:

1. Can a CBNRM project genuinely benefit the community, environment, or both?
2. Is the community willing to adopt inclusive, participatory governance at the right scale?
3. Is there a long-term partner(s) willing to oversee governance and support implementation?
4. If the answer to any of the above is no, can these conditions realistically be changed?

If the answer to any of these questions is 'no', CBNRM is unlikely to succeed. Have the courage to walk away, rather than wasting money or disappointing community expectations. Good intentions and great needs are not enough. Moreover, the temptation to accept money to assist a community where the intervention is not sustainable defrauds both yourself and the community.

However, the answer to these questions, especially the questions about viability and governance, is often 'no' because of policy failure. This can be reversed through policy reform and market development, but 'moving the curves' (Figure 6.2) is never a trivial undertaking.

Question 1: is CBNRM economically viable, or can it be made to be viable?

Assessing whether CBNRM can be viable can be quite technical, and involves good economic judgement. Because this is so seldom done, I have provided a 'checklist' that describes how to approach the question (Table 15.1). The top half of the table is a financial analysis that asks if, as things currently stand, the resource is viable. The bottom half is an assessment of whether the resource potentially has an economic comparative advantage, plus the more practical challenge of translating this into a financial advantage through policy reform or market development. These terms are explained in Chapter 6.

In line with Maslow's hierarchy of needs (Maslow, 1943), CBNRM is most effective where wild resources provide tangible benefits at the individual or small community level,

TABLE 15.1 Example of checklist approach for a financial and economic assessment of CBNRM resources

Name of resource	Wildlife	Livestock	Trees	Bushmeat
Financial Analysis. With current prices and policies, is the resource in question viable?				
Precedent on private land: Do you know of a comparable example where this resource is economically viable and ecologically sustainable?	Yes, many examples of viable private and state wildlife enterprises	Jury is out on viability of livestock in absence of subsidies, including environmental subsidies	Only commercial plantations of exotic species (not appropriate)	Viable as by-product of hunting/tourism. Stand-alone meat commodity production seldom/never viable in drylands
Precedent on communal land: Do you know of a comparable example where this resource is economically viable and ecologically sustainable on communal lands?	Yes, wildlife-based CBNRM successful under some conditions	Livestock are abundant, but usually as privately owned animals on open access grazing	Charcoal production is common, relies on open access resources so sustainability questionable. Nascent examples of REDD payments, but with donor support (Morgan-Brown, 2014)	Bushmeat is common, but sustainability highly questionable in open access conditions
Current status of the resource:	Wildlife (5% of carrying capacity) is very depleted and will need to be rehabilitated	Cattle stocking is reasonable (60%)	Community agrees that trees are significantly over-harvested	Wildlife very depleted through 'poaching'
Economic analysis and market failures. Can we make the resource viable by fixing institutions of proprietorship, markets, and scale?				
1. Creating value by correcting property and policy failures				
Non-ownership (and open access): With proprietorship, can the community add value by (1) planning and managing the resource, (2) rationing use, (3) sharing benefits, and (4) excluding outsiders?	Devolution of proprietorship can convert wildlife from a non-resource into a high value resource	Livestock are privately owned, and the problem is with open-access use of range	With proprietorship and assistance, the community can plan to replace unsustainable exploitation with a sustainable strategy (e.g. for charcoal)	With proprietorship, community is likely to shift wildlife from low value (bushmeat) to high value (hunting, tourism) uses

Differential taxation/regulation: If the community does not get 100% of the market value for the resource, can this situation be improved?	Yes, the current policy that the community only gets 5–20% of value of wildlife trophy price is not viable	The community already gets 100% of the price of livestock, but uses them for cultural not commercial reasons	The community are price takers for charcoal, and the middlemen get most of the added value (so this can be addressed)	Because use is illegal, prices are low
Market restrictions: Can removing bureaucratic restrictions on markets create sustainable value?	Badly set quotas and nonsensical markets greatly lower value	There is little public intervention in livestock markets	There is little public intervention in (illicit) charcoal markets	Veterinary restrictions limit meat sales
2. Creating value through product development and missing markets				
Product and market development/investments: Can the community create value by developing products and markets?	Very large potential to increase value through hunting or tourism investments	Difficult to judge, and gains probably small	The possibility of raising prices through organising markets (e.g. for charcoal) is worth assessing	Markets for wild meat are improving, and there is a need for meat in the community
Payments for ecosystem services: Can the community access markets for carbon, water, biodiversity, and other ecosystem services?	Yes. Seek out philanthropy to support wildlife, especially rhino protection	Livestock, if overstocked, reduce ecosystem services	Yes, this is theoretically possible, but the barriers for a small project to access nascent global carbon markets are high	No
Conservation easements and biodiversity offsets: Can the community access payments for biodiversity through easements and offsets?	There is a willingness to pay for wildlife, but this comes with strings attached	No	Some scope for REDD+ activities	No

(Continued)

TABLE 15.1 (Cont.)

Name of resource	Wildlife	Livestock	Trees	Bushmeat
3. Creating value through scale				
Is a larger area likely to generate significantly more benefits per unit area than a smaller (fragmented) area?	Yes (large wildlife areas are very valuable)	No	Possibly, if REDD is viable	As a by-product of hunting and tourism
Conclusions				
Status	The wildlife sector is booming on private land in southern Africa. Some national parks are generating a significant tourism economy. There are several successful community-based wildlife management programmes.	Commercial dryland beef production has declined for many years, especially since government subsidies were removed. Cattle are abundant in many communal areas, but we cannot trust these economic signals because cattle are a quintessential example of a tragedy of the commons – individually owned livestock on open-access grazing are associated with widespread environmental deterioration.	The private sector occasionally sells wood and charcoal, but normally as a by-product of land rehabilitation. There are nascent examples of payments for carbon sequestration (REDD+), but there are high transaction costs, with some risks in obtaining long-term market access (Morgan-Brown, 2014).	The private sector sells a significant amount of bushmeat, often as processed meats (biltong, sausages). This is always a by-product of hunting and tourism businesses. There are no examples of viable stand-alone bushmeat production on private or communal lands. The extensive bushmeat trade in Africa reflects a tragedy of the commons (there is sometimes a premium on bushmeat) and is a major threat to wildlife sustainability.

	Wildlife	Cattle	Trees, firewood, charcoal	Bushmeat
Conclusion on viability	Wildlife is a viable use of drylands if policy constraints are removed and communities retain 100% of revenues	Cattle are an option in drylands, but the jury is out on financial viability and environmental sustainability. CBNRM should focus on good pasture management (but this may require subsidies)	Trees, firewood, and charcoal production can definitely be managed better. It is difficult to know if markets for these products are sufficient to cover the costs of management systems. The sale of carbon should be considered, but the price and reliability of markets and the costs of validation schemes need to be analysed	Bushmeat production is a low-value use of wildlife with high sustainability risks. However, it is an important by-product of hunting and tourism, as meat is highly valued by communities
Potential benefits: Will an investment in CBNRM create significant value?	Yes. High potential to unlock wildlife economy	Low potential to increase livestock production. A focus on range improvement may be worthwhile	Worth considering, but high barriers. Could use charcoal more sustainably	Yes, by shifting wildlife to high value uses and adopting sustainable rates of harvesting
Costs, barriers, and risks	Need to address macro-issues like ownership and global markets	Risks include elite ownership and encouraging environmental over-use	Solutions require local collective action and/or development of global markets	Need to combine local management with macro-issues of ownership and markets

especially in communities vulnerable to food deficits. Rather than using financial models, a simple and robust approach is the 'private sector test'. If the private sector is doing it, it is likely to work; if the private sector is not doing it, the enterprise is highly likely to be risky or not viable, and there should be very strong justification to proceed. Start by compiling knowledge of economic options from your own experience, consultants, or the literature as I have done for land use options in drylands in southern Africa in Table 15.1. Using this information, I am very confident that communities can improve their livelihoods through well-managed wildlife enterprises, but would hesitate to embark on a CBNRM programme that relies on livestock or game meat production, because I am suspicious that gains must come at the expense of the environment. Carbon and sustainable forest management have potential, but need a careful viability analysis, including secure markets and prices. Obviously, the goal is to combine multiple uses.

Too often, donor-funded projects combine the double challenge of establishing CBNRM and needing to develop new products. This is naïve and even irresponsible. Product development is often a 20-year process, even with private-sector capital and energy. Even worse, money is often wasted through 'commercial' projects that cost the donor $2 and yield the community $1 often under the guise of 'alternative livelihoods',[1] which proposers would never attempt with their own money.

Creating value by addressing market and policy failure

The financial benefits of wild resources are often well below their inherent economic value. Lifting the financial curve towards the economic curve requires addressing market and policy failures, especially non-ownership (and open access), differential taxation and regulation, market restrictions, and missing markets, as shown in the bottom part of Table 15.1.

The Namibia CBNRM programme (Box 13.5) is a good example of policy reform. It made wildlife valuable to local people by:

- Devolving ownership of wildlife to community conservancies.
- Maximising sustainable hunting quotas, marketing hunting and tourism opportunities, and encouraging investment in wildlife production (e.g. infrastructure, restocking, wildlife protection).
- Removing bureaucratic requirements and fees and, most importantly, ensuring that 100% of wildlife benefits get to communities.
- Engaging internationally and scientifically to prevent bans on wildlife products and trophy hunting from reducing the value of wildlife to local people (Naidoo et al., 2016).

None of these steps is trivial.

Creating value by developing products and addressing missing markets

In the short term, communities can work with NGOs and the private sector to develop products, markets, and value chains for wild resources, including new tourism products and craft markets. The difficulty of achieving this is usually under-estimated.

In many systems, ecosystem services have considerable theoretical value, but markets for these services are often immature or missing. Developing these often requires investments far above the level possible for a CBNRM site or even a country. Payments for ecosystem services, for example, are economically sensible and theoretically sound, but are challenging to develop – even today few communities can access carbon payments despite the investment of billions of dollars in REDD+. In some cases, short-term philanthropy has filled in for 'pending' markets, with donors financing pseudo-carbon payments, or paying some members in communities to protect wildlife. With wildlife in particular, scale and branding can add considerable value, such as private concessions or community conservancies around well-known national parks.

The concluding notes in Table 15.1 provide a summation of the financial and economic potential of wild resources, and a judgement of if and how CBNRM can become viable, and which barriers to address. This need not be an arduous or complex undertaking. Indeed, simple analyses are invariably better, using professional judgement and back-of-the-envelope calculations. However, identifying corrective policy actions requires some skills in institutional economics and the economic management of the respective resources. For southern Africa, Table 15.1 suggests that the return on investment in wildlife is high (which is why community wildlife conservation has been successful), whereas the return on investment in livestock and forestry is questionable (there are relatively few examples in this region).

If the financial and economic analysis suggests that there is a realistic probability that CBNRM can deliver real benefits to a community, the next step is to assess if the institutions for governing wild resources and benefits are in place, or can be developed.

Question 2: is there a commitment to inclusive, participatory governance?

Chapters 12–14 make the case for participatory face-to-face governance. We assess if this is likely by asking three questions (Table 15.2):

- Geographically and demographically, can the community be divided up into units or sub-units of some 150–250 households or members (i.e. Dunbar's number) that can meet together easily and regularly?
- Politically, will local leaders and officials allow this to happen?
- Will higher levels of authority, including donors, allow or encourage devolved and inclusive community governance (and 100% revenue retention)?

When initiating CBNRM, it is useful to analyse the forces opposing or supporting participatory governance to know what is coming. Governments and donors are familiar with top-down and representative models of governance, and may often need to be persuaded of the value of inclusive governance and scale. Even experienced professionals are ignorant of the differences between participatory and representational governance, and readily accept the former once they understand its advantages. In other circumstances, resistance is moderate, based on comfort with the status quo and a general reluctance to change the way things are done. However, resistance can be intense where there are entrenched interests. At the local level, resistance usually comes from the

TABLE 15.2 Checklist to assess the forces for and against participatory governance

Evaluation question	Likely/possible/ unlikely	Will there be resistance? How will you overcome it?
Geographic/demographic criteria. ✓ Does the number and distribution of people in the community easily allow all members to meet face to face on a regular basis with (150–250) households? ✓ If the community is a lot larger than 150–250 households, can it easily be split into sub-communities of this size?		
Likelihood of internal support/resistance to participatory governance. Will leaders or officials support (1) face-to-face decision-making and benefit-sharing including (2) full discretionary choice in the use of wildlife income? ✓ Traditional leadership ✓ Influential modern leaders ✓ Multi-village community representational organisations (e.g. wards, conservancies)		
Likelihood of legal, policy, or administrative resistance to participatory governance. Are any of the following organisations likely to oppose (1) participatory decision-making and (2) structuring of the community at the correct scale? ✓ Government agencies at district or provincial level ✓ Government agencies at central level ✓ NGO agencies involved with this resource or other community activities ✓ Other		
Conclusions As a CBNRM support agency, do you believe you have the skills to navigate the above environment to put in place an inclusive system of governance? Alternatively, are there champions that have the drive, motivation, and skills to do so with a high probability of success that you can support?		

traditional leadership (see Chapter 12) or the elites benefiting from representational organisations through perks, sitting allowances, and even corrupt activities. It is not unusual for institutions to be locked in dysfunctional and self-serving configurations for decades and even centuries (Chapter 2, North, 1990). This is why getting the scale and rules of CBNRM correct at the beginning is important, through policies and practices that ensure that rights and benefits are devolved to small 'producer' communities at sub-chieftainship levels.

Another constraint to inclusive forms of governance are mid-level government officials, but they can also be its greatest champions once they understand the power of genuine participation and benefit sharing (as in many CAMPFIRE areas, Chapter 11). The same is true of NGO staff, who can be remarkably resistant to devolution, perhaps reflecting the top-down culture of service delivery, or fear of working with the toiling masses rather than a few well-spoken leaders who are cooperative and speak English. NGOs often want to be seen to be doing things for communities, and have no faith in the abilities of uneducated communities to do things for themselves.

Assessing the operational space for inclusive governance is contextual, and as much art as science. Table 15.2 is self-explanatory, and shows how this can be done systematically. Also, don't take no for an answer. Developing inclusive governance usually requires small communities-of-practice with the vision, skills, and tenacity to navigate through multiple layers of governance. They may need to take advantage of gaps between government policy documents and reality, and to find committed administrators to work with. In countries with decentralisation policies, officials often oppose empowering communities. In countries without decentralisation policies, key individuals in government may be willing and able to champion devolution.

Participatory governance will not emerge organically from good intentions, nor will better institutions emerge through an evolutionary process. Participatory governance is a product of careful and tenacious strategy, and some luck. There are probably as many potential strategies for catalysing participatory governance as there are CBNRM projects. I list several examples below.

In CAMPFIRE there was initial resistance to entrusting villages with money by key government agencies and even some partners. To overcome this, I worked in several progressive sites, including Beitbridge and Mahenye, to demonstrate the effectiveness of participatory governance, and then used the experience and data to encourage adoption by other sites. Given CAMPFIRE's underlying philosophy of participation, this approach was adopted more broadly, but certainly not by all districts.

In the Luangwa Valley Zambia, we were able to replace top-down systems with participatory governance because of the poor performance of the latter, and because the donors wanted demonstrated progress and had financial leverage. Consequently, our rather bold proposal for participatory governance was included in administrative guidelines and project documents almost without comment. However, when the individuals that supported true devolution left their posts a few years later, the situation was recentralised (Lubilo & Child, 2010).

I have also experimented with a highly decentralised approach with several of my African PhD students, especially Shylock Muyengwa, Rodgers Lubilo and Patricia Mupeta. Governance dashboard surveys in several countries in the region revealed that individuals in representational situations often expressed high levels of dissatisfaction with leaders, benefit sharing, participation, information flows, and the like. Part of the dashboard process is a commitment to sharing the results with the community within two weeks of the survey. Consolidated results highlighted this 'dissatisfaction with the status quo', and data feedback sessions had to be handled very sensitively to channel dissatisfaction into positive outcomes rather than conflict and chaos. This change process was successful in

Wuparo Conservancy in Namibia. Wuparo agreed to operate as a federation of several villages rather than a single multi-village organisation, with the level of revenue sharing, community benefits, and member satisfaction increasing immediately (Lubilo, 2011). Although this approach of revealing membership dissatisfaction to unlock the status quo showed promise, change needs significant support from the parent CBNRM programmes, which are sometimes too set in their ways to drive the necessary change.

In my experience, data and anecdotes from pilot demonstration sites are far more effective at invoking change than an intellectual discussion about the relative merits of participatory and representational governance. A viable strategy in a greenfield situation where government policy is weak is to build a coalition with key officials and partners to pilot participatory governance, using the revenue distribution process if possible. With time and persistence it is then possible to weave positive field results into policy reform.

In CAMPFIRE, it was relatively easy to introduce new forms of governance on the back of the wildlife economy, because wildlife represented a new resource with few entrenched interests at the local level (Chapter 11). However, we also attempted to introduce a system of tradeable shares for collective grazing, such that each member of the community was an equal beneficiary. This faced strong resistance from cattle owners who were benefiting from a commons tragedy and who tended to be the more powerful members of a community.

Question 3: can the CBNRM project be effectively supported?

The third question concerns the long-term capacity, tenacity, and skills necessary to build community systems and capacities, and to influence and protect the enabling environment. This is a complex question, half of which is really about motivation and commitment, personnel quality, and the fallacy of the short-term project cycle. The excellent Namibian CBNRM programme (Box 13.5) highlights the necessity of long-term quality support in bright neon. Effective CBNRM requires a significant investment ($10 million annually in this example) even when managed as effectively as the Namibians have done. Nonetheless, CBNRM was neither consolidated nor did it break even in as many as three typical project funding cycles. However, by years 20–25 the returns on investment on a well-managed CBNRM project are demonstrably high. In addition, support agencies need to be able to provide skills that match the implementation model, including governance, economics, and the development of systems for managing natural resources.

Question 4: can the policy environment be changed to attain conditions for success?

Having identified the economic and governance constraints to CBNRM (Tables 15.1 and 15.2), the next step is a realistic assessment of if and how these constraints can be addressed, what resources this will require, and the likelihood that this can succeed.

Build vision and coalition for change and situational analysis

If a CBNRM project is potentially feasible, the next step is to use a well-articulated and realistic vision of an alternative future to bring together a coalition of people able and

TABLE 15.3 Structure of quantitative livelihood survey

	Module	Simple survey	Time (min)	Detailed survey	Time (min)
1	Demographics	• Number of adults and children in a household • Ethnicity • Time in location	5	• Demographic structure • Education by age class • Occupation of each person • Time living at home • Ethnicity • Time in location	10–20
2	Assets	• Type of housing	1	• Full list of assets	5
3	Amenities and services			• Source of fuel • Access to water • Type of toilets • Access to health care	3–5
4	Nutrition and food security		1	• Food security and shortages (day, month, year) • Nutrition (food groups consumed)	5–10
5	Social capital			• Membership and participation in collective organisations • Trust of leadership, each other • Sources of information	5
6	Household production			*Nature-based production* • Natural resources (food, medicines, materials) • Collective NRM (wildlife, forestry)	5–20

(Continued)

TABLE 15.3 (Cont.)

Module	Simple survey	Time (min)	Detailed survey	Time (min)
			Production from domestic plants and animals	
			• Livestock	
			• Crops	
			Transfer payments	
			• Outside employment and remittances	
			• Grants, charity, and gifts	
7 Household consumption	• Weekly expenditure	2	• Expenses (weekly, monthly, annual as appropriate)	5–10
8 Trends in services and livelihood activities			• Changes in livelihood activities	5
			• Changes in services and health	
			• Changes in natural resources	
9 Risks and vulnerabilities	• List major shocks	5	• Score list of shocks	5–10
10 Attitudes to parks and wildlife			• Score positive and negative attitudes to park, park managers, wildlife, etc. (often based on Social Assessment of Protected Areas – see Franks and Small, 2016)	5
11 Community governance	• Score positive and negative on attitudes to key aspects of governance	5	• Detailed survey of governance ('governance dashboard')	5–10

willing to initiate change. This can range from the efforts of a small NGO with a few local leaders and officials, to a national level programme.

On balance, I would recommend undertaking a situation analysis to describe the target community. This costs money, but quantitative and qualitative data improves the design of interventions and proposals, and often corrects preconceptions or builds empathetic approaches towards people who are hungry, lack education, and have low social capital. Three tools are especially useful: simple participatory rural appraisal, livelihood surveys, and the governance dashboard process. The processes of vision and coalition building, and the situation analysis, are reinforcing and can be combined.

A simple participatory rural appraisal is useful for meeting and talking to groups of local people in a productive way ('focus groups'). At its most basic, this should include developing an understanding of the community's history in their own words, using a time line to illustrate traumatic events (war, drought), infrastructure, changes in livelihoods, and so on (an 'event matrix'). A spidergram is useful for understanding how they classify habitats, and how they use each habitat. Mapping the ownership and use of the area is especially valuable, preferably on real maps, although sketches on flip-chart paper will do. Finally, assessing community fears and priorities is revealing (Merz, 2014).

I have gradually become convinced that a quantitative livelihood survey, with a sample of 100 to 400 households, or more, is usually worth the $20 000–40 000 price tag. The cheapest way of doing this is to combine survey expertise for design and analysis with a local team for data collection (consultants costs twice as much). It is not essential to start a CBNRM programme with a livelihood survey, but it invariably improves design, and provides a baseline to measure impact.

The most time consuming part of the survey is building the relationships to open the door to working with the community by, for example, explaining the value of a livelihood survey to the chiefs, chairpersons, and district administrators. I have outlined modules that can be included in a livelihood survey in Table 15.3.[2] I am still in two minds over whether to skip collecting detailed demographic information on each family member (because this is time consuming), but quantifying household production (which is also time consuming) is invaluable, as is tracking social and associational capital, though these methods are still insufficiently tested.

Conducting the survey requires a list of all households in the community (sampling frame) and rigorously training eight or so local people to collect the data using tablet computers. Once a team is trained, collecting a large amount of data happens remarkably quickly, with a team of eight collecting 40 surveys per day. Each survey should take an hour or less. It is more efficient to use tablet computers (not paper forms) to collect the data, which can then be stored and analysed using open source software (Eguren & Sprague, 2014; Muyengwa et al., 2014). The biggest challenge is the analysis of copious data. This requires a clear idea of the output needed in the form of graphs and tables, and it is irresponsible to do the survey unless someone is committed to analysing and reporting the data, a process that takes about two months. It is disrespectful to collect data without immediately returning it to the communities, although livelihood data is not very exciting to discuss.

Despite working in communities for nearly 30 years, every livelihood survey reshapes my intuitive understandings of the community, is valuable for designing interventions, and is especially valuable for educating donor and private sector partners about challenges and opportunities. In Mozambique, for instance, data on low levels of education in post-conflict communities suggests implementation will be slow. In most communities, documenting food insecurity (Figure 15.2), and how many mothers cannot always feed their children, lends perspective (and humanity) to implementing agencies.

My particular interest is understanding how the livelihoods in each household and community are constructed from wild harvesting, community wildlife or forest enterprises, farming (crops and livestock), employment outside the community, and grants and charity. Collecting quantitative data on production takes time and care. However, the more commonly used frequency data (e.g. do you grow maize?) is often highly misleading in drylands. Although most people farm, crops invariably contribute less than 10% of livelihoods. I much prefer to quantify production (how many bags of maize did you harvest?) for each household following the categories listed above. A surprisingly low income from farming, and high dependence on off-farm income (i.e. remittances, grants, and charity), is typical of many dryland communities in the more developed economies in southern Africa.[3] A very useful fine-scale data product is a bar-chart showing the composition of livelihoods for each household. This invariably results in a Gini-type frequency diagram that shows inequality in villages (especially in livestock ownership). This fine scale household data can be analysed in numerous ways. For example, in Sankuyo village in Botswana, the spreadsheet model showed that banning hunting would drop the 'middle class' that was emerging through the wildlife economy back into chronic food insecurity and near-destitution, which is exactly what happened (Onishi, 2015).

I had always assumed that rural African communities worked together well. However, survey data (Figure 15.3) showed that the associational experience in most communities is linked to the chief and the church, and tends to be autocratic. There is some cooperation in social services, but almost none in economic activities. Social capital also tends to be low, with low levels of trust, including in the leadership. Interestingly, many communities, like most of us on the planet, like wildlife and protected areas (even if they dislike park managers, conventional wildlife policies, and problem animals), and are natural partners in conservation, once they are allowed to participate.

The third tool is the governance dashboard process (Child et al., 2014) which uses a simple survey to ask individuals about governance issues within the community, feeds the data back visually and immediately, and is designed to encourage corrective actions. People are asked if they like and understand their governance structures, if they trust their leaders (especially with money, Figure 12.4), how much information they get, their involvement in decisions, and so on. When used as a baseline, the governance dashboard is largely blank because it is measuring systems that are not yet in place, but it quickly becomes a valuable tool for managing governance adaptively. To stimulate a process of introspection and adaptation, feedback workshops are designed to follow four steps:

1. The data are presented as visually as possible, and carefully explained.
2. The community is asked if they agree or disagree with the data.
3. The community is asked to explain the causes underlying the data, leading to a qualitative discussion that is highly enriching.

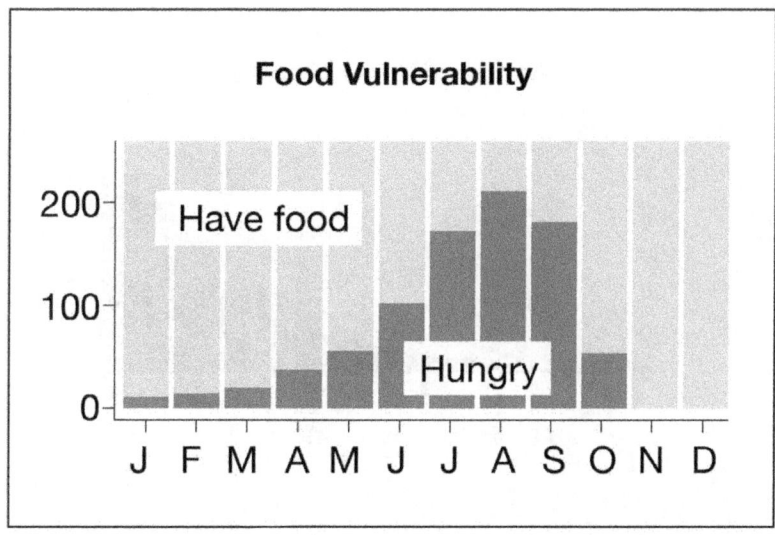

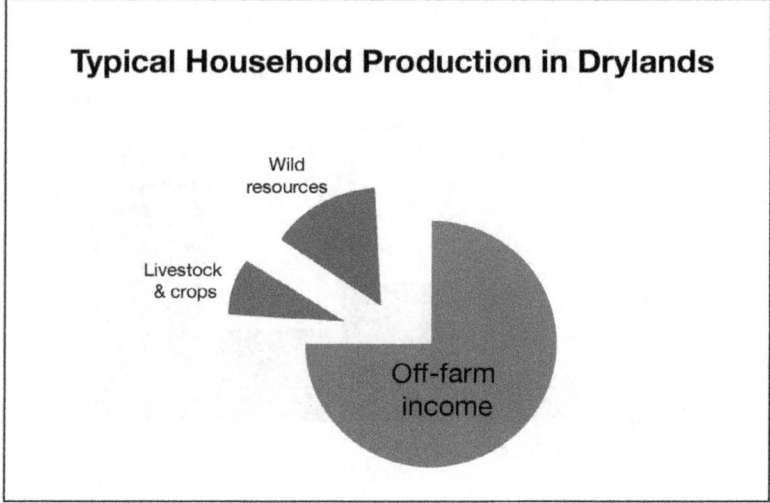

FIGURE 15.2   Examples of data from livelihood surveys: food vulnerability and sources of household production

Sources: Eguren & Sprague (2014); Merz (2014); Muyengwa et al., 2014).

4. The visualised data often bring to the fore issues that people are concerned about, but do not address individually. For example, they find that, say, 86% of them are dissatisfied with the financial management, or 80% of young people like wildlife compared to 35% of old people. People often respond emotionally to governance data, so it needs to be handled carefully, and with preparation, if it is to be used to catalyse corrective social action. For example, the way that the facilitator handles the results '86% of the community does not trust the leadership with money' can lead to a mature debate about solutions, or to intense conflict.

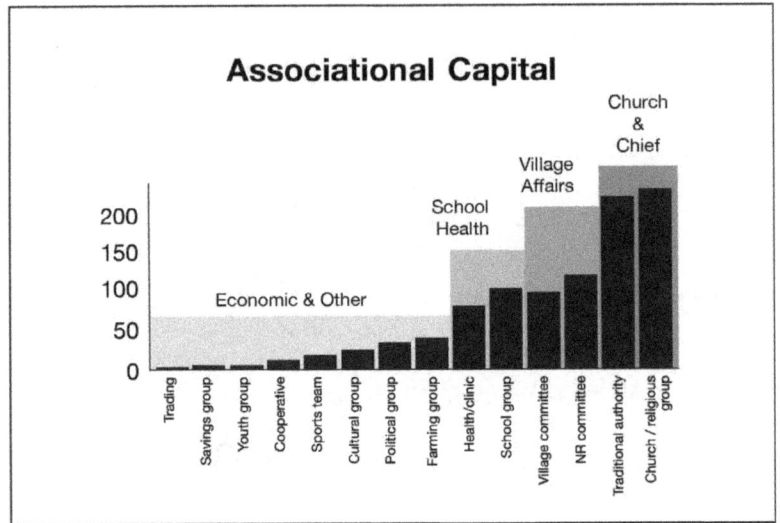

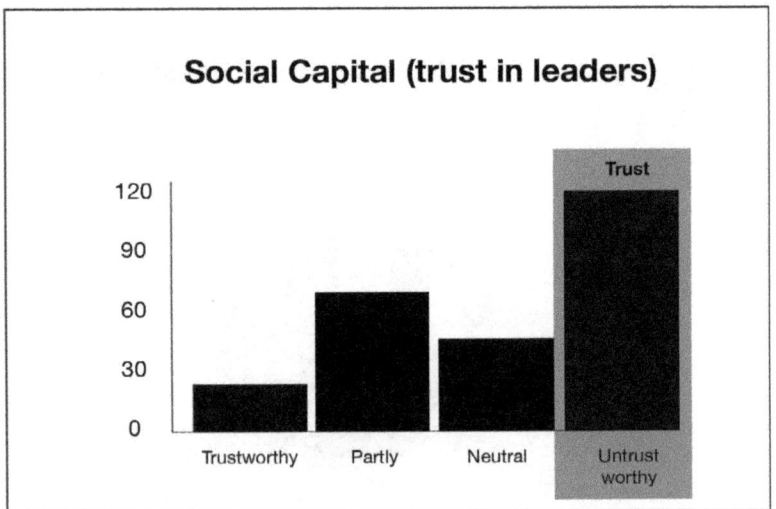

FIGURE 15.3 Examples of data from livelihood surveys: associational and social capital

Source: Eguren & Sprague (2014).

The governance dashboard is a valuable adaptive management tool because governance data (unlike livelihoods) responds rapidly to interventions, especially in communities with a low baseline of social and associational capital. For instance, a selection of four questions from the dashboard done in the Mangalana community in Mozambique (Figure 15.4) shows that CBNRM is moving in the right direction, improving relationships with the game park, increasing participation in decisions, demonstrating to people that wildlife is important to their future, and encouraging them to poach less than before. A further graph

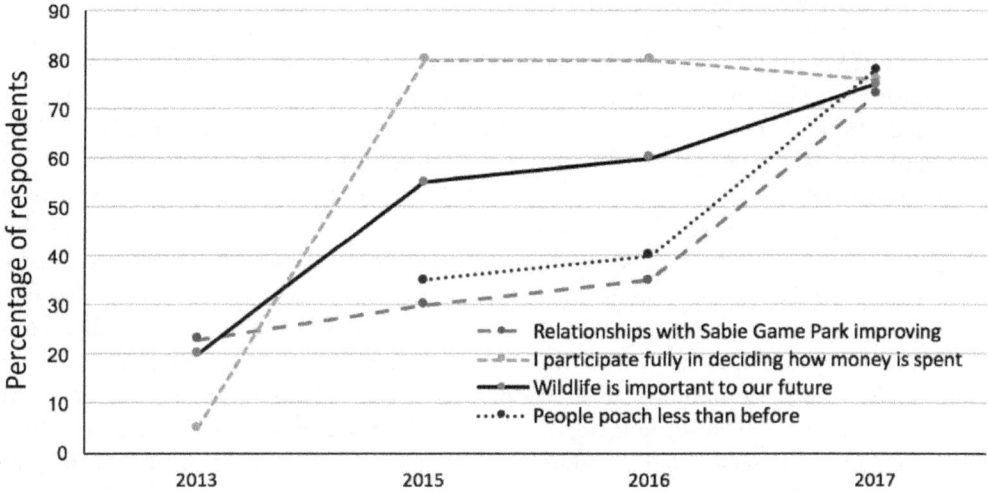

FIGURE 15.4 Example of dashboard data from Mangalane community in Mozambique
Source: modified from data collected by Merz (2014) and Themba (2017).

(not illustrated) shows a rapid increase in trust in the leadership, and then a reversal after two of the five villages mishandled money.

Implementation

Implementation requires its own book. Therefore I have used a simplified results chain and indicators (i.e. part of the logical framework project methodology) to illustrate the scope of a balanced community wildlife programme (Table 15.4). The long-term impact (project goal) is measured in terms of improvements in livelihoods, the rural economy, and the status of wild resources. The means of getting there (project purpose) is through effective CBNRM governance and organisation at the community level, and the capacity to support this. A number of components then contribute to an effective CBNRM programme:

- *Governance.* Having made the case for participatory governance, getting governance right is at the heart of most CBNRM programmes. I have included practical guidelines in Chapter 16 and refer the reader to an earlier manual for CBNRM practitioners (Child & Wojcik, 2014).
- *Benefits and economy.* It is much easier to implement CBNRM if there are tangible benefits, especially household or cash benefits, but benefit generation varies widely depending on the resources in question. For wildlife, I have illustrated the process of 'earning money' in Chapter 11 (Figures 11.4 and 11.5), and I refer the reader to several manuals that describe the process of marketing wildlife effectively (WWF-SARPO, 1997; Child & Weaver, 2006). Indeed, effective CBNRM is like an aeroplane, and seldom flies without two wings – the building of a wild or biodiversity economy to fuel the system (rather than reliance on donor support) and the ownership and governance of this economy by communities. The other components flow from this.

- *Natural resource monitoring and protection.* Once people have rights to their wild resources, and see them as valuable, they are usually receptive to improvements in natural resource management. CBNRM can be applied to a large number of resources, and each of these involves many aspects of management, including monitoring, protection, and planning. This is a huge topic, and I will limit discussion to the capacity-building process (see below).
- *Security, settlements, and services.* While CBNRM should emphasise individual benefits first, communities should also work collectively to invest in services where they consider this to be a good use of their personal income. Many communities in southern Africa live in remote and scattered settlements, which make the provision of modern services including education and health care difficult.[4] Personal and community safety and security is also important. It is nearly impossible to work constructively with a community if they can't feed their children, or are scared. Some communities are vulnerable to criminal syndicates and violence, including illegal wildlife trade, which can be addressed through some combination of local policing and external support.
- *Compliance and performance tracking.* The essence of adaptive management is the tracking of compliance indicators (for governance) and performance indicators (for many other aspects), and the social processes of data visualisation, learning and introspection, and adapting to what the data show.
- *Enabling environment.* External to the community is the need to strengthen community rights and capacities, develop markets for wild resources, remove barriers that disadvantage rural people, research and learn from the process, and adapt policy and practice experientially.

Programme support and adaptive management

Unlocking the conditions for CBNRM to emerge at local level requires that the enabling conditions are created, protected, and enhanced. Site-level CBNRM governance is ubiquitous and follows generalisable principles (Chapter 16). By contrast, enhancing the enabling conditions for CBNRM is context specific, and geography, politics, and history can play out in many different ways in a four-layered social economy (Box 4.5).

Effective CBNRM hangs on the devolution of a full suite of rights for land and wild resources to communities, while ensuring that these rights are governed inclusively. The process of devolving rights is highly political and context specific. Thus, communities in Latin America gained rights through social movements (Level 1, see Box 4.5), whereas CBNRM in southern Africa was designed by administrators who proactively rewrote policy and legislation (Level 2) (Chapter 13). In the absence of well-organised national support, enhancing the enabling conditions is challenging, but can be approached bottom-up using pilot projects (Level 4). We should not forget that rights include the removal of market restrictions and differential regulation and taxation.

Effective CBNRM also requires combatting elite capture and misgovernance at the local level by monitoring compliance with the principles of inclusive governance and sanctioning non-compliance. While the principles that need to be tracked are ubiquitous (Chapter 16),

TABLE 15.4 Results chain and indicators for a community wildlife programme

RESULTS CHAIN	INDICATORS
DEVELOPMENT OBJECTIVE Livelihoods improved through sustainable natural resource management	• Livelihoods improving, economy growing • Wildlife increasing and habitats improving
IMMEDIATE OBJECTIVE Community well organised, with all people benefiting from and participating in wildlife management	• Governance conformance standards defined (bill of rights and/or constitution) and met in all communities (Chapter 16) • Social capital improving (trust, knowledge, participation) • Organisational capacity improving (participation, information, financial accountability, satisfaction with governance) • Management capacity improving in all facets through co-learning
Output 1: Governance Effective, accountable local organisation in place (following the principles of impersonal, participatory governance)	• All people sharing in decisions and benefits (cash, projects, meat, etc.) • 90–100% of wildlife benefits shared at household level (as cash or projects)
Output 2: Economics and livelihoods Wildlife marketed and generating maximum income	• Wildlife marketing is open and competitive • Wildlife sold for 90%+ of world prices • Community retains 100% of wildlife benefits • Monitor jobs, household revenues, community projects, etc.

Output 3: Natural resource protection, monitoring, and management
Natural resources protected and monitored
	• Communities count animals, record hunting offtakes and quality, and set their own quotas
	• Community protect wildlife and natural resources by:
	  • conducting 250 patrol days/month
	  • maintaining poaching catch/effort incidents at < 1/100 patrol days
	• Annual transects of vegetation, wildlife show …
	• Human wildlife conflict monitoring and management shows …

Output 4: Security, settlements, and services	• Criminality in communities is controlled, and people are not living in fear • Services improving and location matches land use plans
Output 5: Compliance management and adaptive monitoring and learning Multi-dimensional performance monitoring and adaptive management system in place at all levels	• System in place for tracking governance compliance, livelihoods and economics, wildlife and natural resources, income and expenditure, etc. • Data feeds into policy and practice (adaptation) at local and national level

(Continued)

TABLE 15.4 (Cont.)

Output 6: Enabling policy environment Communities have rights to use, benefit from, and protect their resources	• Devolved proprietorship with 100% of benefits reaching communities • Income from resources is maximised (i.e. achieves free market prices) and not compromised by fees, regulations, or restrictions • Communities get sensible offtake quotas on time (or set their own quotas) • Research and learning improves policy and practice experientially

the challenge is who does it. Ideally, compliance management is a key state function, but CBNRM is often implemented where state capacities are weak, and CBNRM support agencies may need to be innovative when it comes to leveraging points for compliance. Thus, compliance can be a condition of project support, or can be linked to renewal of quotas, payments of wildlife income, and so on.

CBNRM requires the long-term management of the enabling political and institutional environment. In the short term, NGOs can initiate this process. Namibia's programme is an excellent example of support organisations collaborating through the Namibian Association of CBNRM Support Providers (NACSO) to coordinate policy advocacy and technical inputs for more than two decades. In the longer term, CBNRM associations need the technical and political capacity to operate effectively at the centres of power. CAMPFIRE's rapid progress, and ability to persist in difficult circumstances, was enhanced by developing the legitimacy and capacity of the CAMPFIRE Association to represent local communities and advocate for them.

CBNRM communities require experiential capacity-building to develop institutional, social, and technical capital (see below), as well as the financing and technical know-how for environmental rehabilitation and production. CBNRM is essentially about changing systems, bringing conflicts and trade-offs to the fore, and resolving them, which requires conflict resolution skills, and also the brokering of deals between communities, governments, donors, and the private sector. CBNRM support, financing, and recapitalisation is often provided through support to government agencies and NGOs, and one of the motivations for writing this book is to advocate that these agencies are held much more accountable for improving the success rate of CBNRM initiatives.

Communities are not playthings. Individuals and organisations should carefully evaluate their ability to do good, not harm, before engaging and raising expectations, while funders should expect much higher levels of rigour and performance in the design and support of CBNRM programmes.

Capacity building

This section draws on the work of Marshall Murphree, in a contribution for the World Bank that we wrote jointly (Child & Murphree, 2004). 'Capacity building' figures prominently in the agenda of contemporary development discourse, often in terms of teaching, or the top-down impartation of the knowledge or skills required to carry out the objectives of specific programmes or policies. This is an incorrect interpretation. Capacity building rests fundamentally on experiential learning, to which teaching can contribute, and involves all levels of the system learning from each other. Experiential learning provides people with better insight into the opportunities and constraints of their condition. They also gain 'response-ability' (Covey, 2004), the capacity to make informed judgements regarding alternatives and their consequences. Thus:

> *Community capacity building* is about promoting the '*capacity*' of local communities to develop, implement and sustain their own solutions to problems in a way that helps them shape and exercise control over their physical, social, economic and cultural environments.
>
> *(Department for Community Development, 2006)*

Capacity-building programmes need clear definitions of their targets. Often in CBNRM the main target is the local leadership within the community. Local elites are prime movers and their capacities are clearly an important focus. However, training of leaders reinforces the underlying tendencies towards authoritarian governance and elite capture. It may drive a wedge between leaders and their constituencies, aligning leaders with external agents and creating intra-communal communication barriers, often compounded when the 'business' of CBNRM is conducted in English rather than the vernacular. Experiential learning demands a broader palette, in which all members of the collective are involved. Capacity building should be directed not at leadership, but at the followership and, indeed, the whole system. Building system capacity is a cross-scale phenomenon that should not ignore other important targets for capacity building – the community of external agents who intrude on CBNRM processes at local (and higher) levels, including local government bureaucrats, NGO staff, researchers, and scientific advisors. Their technical capacities may be adequate, but their capacities for facilitation, empathy, and nurturing the capacities of others (as opposed to issuing instructions) are usually, if not uniformly, very low. Thus, the target for capacity building is the whole system, ranging from community leaders and followers, to the external agents that support them.

CBNRM is usually implemented through donor projects, but top-down 'designer development' and 'technology transfer' associated with such projects is antithetical to CBNRM which treats all management as experiential learning. Thus Murphree (2001) suggests that CBNRM redefines both development practices and the role of science and technology as 'invited rather than imposed, directed rather than directive, and facilitative rather than manipulative', and suggests that external agents are themselves appropriate targets for capacity building in the facilitative stances and skills required. This also applies to CBNRM research, which has a moral obligation to contribute as much to the 'subject' communities as the knowledge it extracts from them.

Capacity building occurs largely through experiential co-learning, which can be complex and costly but is cheaper than the alternative. Ticking off project deliverables increases the temptation to return to a teaching mode, and to focus on training the leadership. However, teaching is ineffectual unless linked to clear ends, while the differential empowerment of leaders exacerbates some of the very problems that CBNRM seeks to overcome. This is why, in the Luangwa example (Chapter 12), training always involved both the leadership and followership in locations where ordinary community members could wander in and out. Capacity building cannot be all embracing, and Child and Murphree (2004) propose the following components for a capacity-building approach in CBNRM.

1 Acquisition of basic information

Basic information is the foundation of capacity building, but many communities do not know their rights, the economic value of their resource base, how much they are paid for their resources, and what happens to their money. Capacity building starts simply by making sure this information is available to communities in the vernacular, and preferably also in written format. Appropriate field manuals can be useful. Even more valuable is the process of facilitating communities to make up their own manuals or simple posters by translating generic manuals for their own circumstances and language (Goredema et al., 2006).

2 Development of organisational skills and governance

CBNRM seeks to institutionalise inclusive, impersonal institutions, while simultaneously integrating communities into the larger economy in advantageous ways, especially where value is generated through community-private partnerships. This requires mechanisms for democratic decision-making, but also some degree of bureaucratisation, as functions are defined and responsibility is delegated. Thus mechanisms must be in place for making collective decisions, for assessing and controlling finances and performance, and for appointing and replacing leaders and employees. These mechanisms usually take the form of a set of rules and rights, a constitution, plus ways of ensuring that these rules are followed (i.e. compliance management). This is illustrated in Chapter 16.

3 Acquisition of technical data and skills

The acquisition of technical data and techniques emerges through professional external facilitation and co-learning processes where professional science and local learning and monitoring are combined. Technical information, such as regional lessons or scientific findings, is usually beyond the reach of local people. However, it can be absorbed quite rapidly as information relevant to choices by local people, especially where scientists and community work together to solve problems through a process of experiential learning and adaptive management; if the data are presented as imposed wisdom it is far more likely to be rejected. The processes of negotiating hunting concessions and quota-setting in CAMPFIRE are good examples of co-learning (Figures 10.4 and 10.5). The event-book or management orientated monitoring systems (MOMS) developed by the Namibians (Stuart-Hill et al., 2005) is an excellent example of a bottom-up management information system.

The efficacy of monitoring systems lies in a partnership where scientists work with local people to set up systems to monitor performance and provide ongoing support to the process of collecting, storing, analysing, visualising, and triangulating the resulting data for making decisions. Establishing such adaptive management systems is not a once-off exercise. Over a period of about two years, scientists need to work with the community to identify the variables that the community deem important, and to establish modules for monitoring these data, starting with bite-sized pieces and only adding modules slowly over time. In a direct contrast to the extractive research so often practised by academics, it is important that the community plays a central role in defining what data to collect, and also thereafter owning and analysing the data, and keeping it in its own archive. After co-designing the management information systems, it is necessary to revisit the community on a quarterly or half-annual basis to maintain the quality of the data and processes. This 'auditing' plays a second function of summarising community-owned data for national databases and reporting. Data is remarkably empowering as the community begins to understand and manage its own wildlife, campsites, or businesses, and it is not unusual for community areas to have higher quality data than state or privately managed land. The gold standard for community-based monitoring is the Namibian programme, and I refer the reader to WWF-Namibia for a deep understanding of the nuances of community-based monitoring, and to the NACSO website (www.nacso.org.na/) for a demonstration of the effectiveness of a 20-year investment in building community-based monitoring systems.[5]

4 Appraisal and adaptive skills

Most projects that support CBNRM require external monitoring and evaluation (M&E). However, building monitoring and appraisal into community processes (e.g. through quarterly general meetings) is a critical step in developing adaptive management and social learning and innovation. This is why Ostrom (1990) states that the monitoring must be done by the community, or by people accountable to the community, as we see so clearly in the Namibian MOMS system described above.

This differs from conventional M&E, which is designed to measure compliance with predetermined project activities and outputs, as a mechanism for upward accountability. I am a strong advocate of intensive performance management of community programmes using indicators in a clearly defined results chain (such as the log-frame illustrated in Table 15.4), and of using data regularly and visually in public forums. However, it is vital that the log-frame becomes a tool for adaptive management, rather than a straightjacket, with (formal) processes for changing indicators as implementers learn more about how the system works.

In addition to iterative adaptive processes, CBNRM involves transformational visioning and the conceptualisation of an alternative future. This usually starts with a realisation that the status quo is entrapping the community and that new options for governance and resource management are available. However, communities by their very nature have limited exposure to alternatives, so their capacity to aspire to these alternative futures is severely limited. Simply explaining new options to a community is unlikely to result in radical change. Enabling community representatives to visit peer communities to learn about these alternatives is far more powerful. However, change sometimes only occurs once the community actually tries something. In the Beitbridge community, for example (Chapter 11), no amount of flip-charts, role plays, or explanations was sufficient to break down community scepticism – it was only when the cash arrived and was put on the table in front of them that they began to believe that change was possible. The cognitive shift in the communities occurred not through explanation, but through action.

5 Negotiating skills

Negotiating skills are increasingly important as communities become part of larger systems of control, and seek to negotiate political processes or make commercial agreements with private entrepreneurs. The development of community-private commercial partnerships is a specialised arena of knowledge where communities are likely to need external facilitation to understand, for instance, the nuances and prices of hunting of forestry agreements, as well as the legal context, drafting of contracts, and procedures for tendering and interviewing concession agreements (Figure 11.2, WWF-SARPO, 1997). Nonetheless, local communities grasp the essentials of negotiation remarkably rapidly, and provided they have reasonable data on prices are just as able to negotiate a good price for an elephant as for a goat. However, the tendency to over-complicate financial explanations, and to over-complicate legal agreements, reduces transparency, is unnecessary, and invariably tilts the advantage away from the community. For example, rather than complex profit-sharing agreements for tourism lodges, it is best to base fees on tangible and measurable outcomes, such as dollars per bednight. Similarly, 20-page legal agreements written in non-transparent legalese can usually be cut to two pages of clearly understandable text.

6 Learning transfer

The transfer of knowledge between communities and, indeed, between community programmes is an important part of learning. This can and should be done conventionally through reports, regional meetings, comparison of activities and performance statistics, web sites, and so on. However, by far the most effective means of communication is face-to-face peer learning, such as when members of one community visit another, though this is expensive.

One could easily add further items to these six. However, doing a few things properly is more important than comprehensiveness, because communities that become competent in one aspect will readily expand their skills. Capacity-building programmes should prioritise immediate needs linked to a theory of change (e.g. Table 15.4), while being careful not to overwhelm the community. As designers of CBNRM programmes we ourselves need to gain the capacity to know when to stop!

Building the capacity of the system

While the purpose of capacity building is to improve community governance and management, this usually requires building the capacity of the whole system, including defining the roles and developing the skills of actors at the meso- and macro-levels. Very often, middle-level professionals provide most of the training. This has value but tends to lead to the top-down impartation of knowledge rather than genuine capacity building. The development of cross-scale processes usually requires the support of professionals with personal and long-term relationships with communities, to work up and down the system to clarify roles, provide mentoring and training in these roles, and to work at a higher intellectual level in solving problems. There is no substitute for consistent, persistent, light-touch facilitation (Jones, 1999), often by scholar-practitioners and champions who initiated CBNRM. Confident professionals are more likely to entrust and expect communities to do things for themselves, which may be slower in the short term, but succumbing to the temptation to do things for the community often undermines the development of community self-reliance.

Having sketched the outlines of a strategy for implementing a CBNRM programme, the next chapter elaborates on the process of building institutional capacity and inclusive governance.

Notes

1   Although included in many project documents, the concept of 'alternative livelihoods' is often superficial, without technical details, or judgement of the challenge of making a living in forests and drylands.
2   I have developed this survey over a number of years working with students (e.g. Shylock Muyengwa, Patricia Mupeta, Rogers Lubilo, Leandra Metz, Alexandra Sprague, Antonieta Eguren) in Mozambique, South Africa, Botswana, Namibia, and Zambia. It is currently being finalised as a guidance document for the Scientific and Technical Panel of the Global Environmental Facility.
3   I am personally sceptical that dryland agriculture offers rural people a pathway out of poverty. In surveys in drylands in southern Mozambique, South Africa, Botswana, and Namibia, farming seldom provides more than 10% of household livelihood, whereas off-farm income often provides 70% or more. There is no doubt that conservation farming increases yields several-fold in drylands in southern Africa, but the question is if the value of the extra yields covers the costs of inputs and support. Even in Rwanda, farming provided less than 50% of household livelihood (Mulindahabi, 2017). In remote areas of southern Zambia, where access to a modern economy is more difficult, agriculture also provided less than 15% of household production. By contrast, livestock and natural resources each provided a further 40%, but raised the question of environmental sustainability.

I therefore ask my class to debate the question: is agricultural development an oxymoron? It is true that many poor people depend on agriculture, implying that to help them, investments in agriculture are sensible. However, people may also be poor because they rely on agriculture. While I fully support improved micro-agriculture as a short-term measure, the reality is that these environments are too unproductive and variable to support livelihoods based on simple commodity production, and are damaged by attempts to simplify them (Chapter 3). Wildlife may provide one of the best chances to improve economic yields. However, even the best combinations of farming and wildlife cannot support the level of demographic growth we are now witnessing. The only real alternative is education (combined with health care and the empowerment of women), to enable people to take up livelihoods that are not dependent on direct harvesting of natural resources, as artisans (bricklayers and carpenters) or professionals (lawyers and doctors). These are the 'alternative livelihoods' we should be aiming for, but require long-term investment beyond the scope of most interventions.
4   These settlement patterns bear further investigation. They may optimise resource access. However, they appear to be unsuited to modern times, and may well be a response to previous marginalisation and trauma (Chapter 4), and an attempt to hide from authority.
5   An especially good example of participatory resource management is the participatory technology development process developed in CAMPFIRE and Namibia, and the related set of manuals and materials (Taylor et al., 1997; Rigava et al., 2006). In the 1990s, the WWF-Southern African Regional Programme Office in Zimbabwe produced 12 excellent wildlife management manuals

through this process. Likewise, the Namibian Association of CBNRM Support Agencies provides a number of manuals on its website that have been developed and tested over a period usually exceeding 20 years (www.nacso.org.na/resources/training-manual). I highly recommend the Management Orientated Monitoring Systems developed in Namibia with its thorough principles of engagement with communities, elegant but simple management information system comprising yellow (recording of events), blue (monthly), and red (annual) data management systems, and a large number of tried and test modules ranging from counting animals, to managing crafts and campsites.

16

PARTICIPATORY GOVERNANCE AND REVENUE DISTRIBUTION IN PRACTICE

LEARNING OBJECTIVES

This practical chapter shows how the theory presented in previous chapters (i.e. the importance of inclusive governance) is implemented through the revenue distribution process. The chapter:

1. Emphasises that the foundational principles of CBNRM are that all members of a community (1) are equal shareholders in the natural resource 'business'; (2) must participate in making the decisions that affect them; and (3) must check that these decisions are implemented.
2. Defines the tools necessary to support impersonal institutions and the 'rule of law' including a member's bills of rights, clear rules and procedures, and constitutions.
3. Describes a practical, eight-step process for developing participatory CBNRM institutions through a standard administrative cycle constructed around managing the money transparently and accountably. This includes:

   a. Providing information.
   b. Defining the community.
   c. Establishing rules and procedures for participatory governance.
   d. Defining roles and responsibilities.
   e. Informed and participatory review of performance, especially community finances.
   f. Participatory planning (especially participatory, activity-based budgeting).
   g. Elections.
   h. The revenue distribution ceremony.

4. Introduces basic requirements for community financial management.
5. Emphasises the importance of monitoring compliance with CBNRM principles, and illustrates the basic tools for doing so.
6. Describes how compliance and performance monitoring is the primary mechanism for experiential capacity building.

Developing community citizenship

This chapter describes practical tools for developing inclusive community governance and social capital so that all members of the community:

- Participate equally in collective action.
- Participate in a well informed way.
- Become economic citizens, with equal rights to the benefits from collectively owned resources.

In many rural communities in marginal agricultural areas, people's sense of economic and political citizenship is weak (Chapter 4). Their primary experience with collective action is as passive and subservient subjects of a traditional authority or by attending church. Social capital and collective action is limited, especially in commercial enterprises or natural resource management (Figure 15.3). By contrast, effective societies and economies have inclusive institutions, where citizens have a stake in the community and are able to protect their rights to participation and benefits (Acemoglu & Robinson, 2012). This polarity reflects the difference between extractive and inclusive institutions (Table 16.1). Indeed, the purpose of CBNRM is to develop inclusive governance (and avoid extractive regimes) for reasons of social justice and economic progress. Representational institutions may be effective for natural resource management, and often operate like mini-top-down agencies, but participatory institutions are essential to develop social capital, including:

- Enlightened and effective participation that combines face-to-face decision-making with quality information (e.g. on individual rights, collective processes, finances, projects, and wildlife management).
- Equitable benefit sharing and the optimisation of economic tradeoffs between private and collective expenditure.
- Resource proprietorship, by ensuring that each and every member of the community obtains tangible benefits from the resource, and participates in decisions relating to the management of the resource and the benefits flowing from it.
- Social learning, accountability, and effectiveness by setting goals together and then measuring performance relative to these goals.

The objective of CBNRM is to rebuild the social and institutional capital that has been lost to political and economic marginalisation. The starting point is that all members

TABLE 16.1 Inclusive and extractive institutions

Extractive institutions Rule of man	Inclusive institutions Rule of law
A small group of individuals (elites) do their best to exploit the rest of the population. Characteristics of the institutions: • Rule by man: many decisions are personalised and depend on 'who you know'. • Strong motivation to protect vested interests, keep things as they are, and block socio-economic progress.	Many/most people are included in the process of governing so the exploitation process is weakened or absent. Rule of law: people can protect their rights and have access to opportunity through: • Impersonal rules; • Secure property rights and contracts; • Equal access to information; • Make choices through markets; • Access to education, health, and opportunity for most citizens.

Source: modified from Acemoglu and Robinson (2012) and (Beinhocker, 2006).

of a CBNRM community are joint and equal owners and beneficiaries of collective resources like wildlife, grazing, forests, and water. They are equal shareholders in the 'village company'. Operationally, this requires all members of a community to meet together on a monthly or a quarterly basis to review information and make decisions, somewhat like town hall meetings and direct democracy in New England (Tocqueville, 1994). This is a big step up from the two mechanisms of accountability on which many CBNRM projects hang – tri-annual elections and participation in approving the budget at the annual general meeting – which are unlikely to fulfil Dahl's criteria for democratic participation (Tables 14–4) nor, therefore, deliver accountable, transparent, and equitable participation.

Inclusive CBNRM institutions combine (1) individual rights with (2) collective action for sharing benefits and managing the resource. Four tools are invaluable for inclusive CBNRM governance (Box 16.1). Carefully crafted rules need to be integrated into two formal documents, an individual bill of rights, and a community-level constitution. These principles are institutionalised through participatory financial management, which requires a simple but well-designed financial system. Finally, simple work plans enable communities to plan and track projects and activities.

Introducing participatory governance through fiscal devolution

Most communities with wild resources have a history of centralisation and extractive governance. Introducing participatory governance requires unfreezing the status quo, changing the systems, and refreezing the system in ways that incorporate the principles of devolved proprietorship and participation (Lewin, 1958). Change often starts by making people 'uncomfortable with the status quo' by highlighting impending risks or new opportunities and building a coalition around a powerful vision for change (Kotter, 1996). With CBNRM, it is relatively easy to argue that centralised resource proprietorship is exacerbating poverty and causing

BOX 16.1 TOOLS FOR INCLUSIVE CBNRM GOVERNANCE

Member's bill of rights:

1. Defines members as the primary beneficiaries and decision-makers (i.e. the equivalent of shareholders).
2. States that the role of elected leaders is to bring people together to make decisions (not to make decisions on behalf of members), and to follow community instructions.
3. Defines roles within the community and of supporting organisations (see Table 12.1).

A set of rules or a constitution that:

1. Repeats and reinforces the principles in the bill of rights.
2. Defines procedures for making decisions, checking that decisions are implemented, and taking corrective action, including sanctioning and/or replacement of officials. Key procedures are:
    a. Annual general meetings including:
        i.   Performance reporting.
        ii. Participatory, activity-based budgeting.
        iii. The election of committees or executive employees.
    b. Monthly/quarterly general meetings to compare expenditure to budget (using a variance analysis) and track implementation.
    c. Mechanisms for communities to call extraordinary general meetings to change decisions, check finances, replace leaders and employees, etc.

3. Participatory financial management – an effective and transparent system for managing money including:

    a. Procedures and templates for participatory, activity based budgeting.
    b. Monthly/quarterly public discussions of income, expenditure, and variances.
    c. Financial books and receipts.
    d. Informal monthly/quarterly auditing and an annual formal audit.

4. Action minutes/work plans – to record decisions in the form of a simple work plan (e.g. What does the end point look like? Who will do what, when, with what resources?).

environmental degradation and the loss of wildlife through neglect, poaching, and the illegal wildlife trade. Change requires building relationships with key decision-makers at all levels around a vision in which the wildlife economy and CBNRM address both poverty and the environment. Change is the process of translating this theoretical vision in which into action, demonstrating it works, and expanding and protecting the new approach.

By far the easiest mechanism for developing effective collective governance is by managing money. This money comes from wildlife, carbon, forests, and so on, but can equally be provided by discretionary grants. Participatory governance can then be developed and entrenched by going through the process of allocating and accounting for money frequently and repetitively.

We should not begin to introduce participatory organisations unless and until four things are in place (Chapter 15). First, we need to be sure that we are not selling the community a flawed economic vision, and we need hard evidence that we can unlock new benefits, especially cash, which is invaluable for initiating new systems of governance. Second, we need to be confident that the introduction of participatory governance will not be blocked. Third, we need assurance that someone in authority will support (or drive) the process, with an ideal situation being a coalition of policy-makers and support organisations to develop, capacitate, monitor, and maintain systems of micro-governance. Finally, we actually need to know how to do the job, especially technical knowledge about the design and maintenance of inclusive institutions. This chapter provides this technical guidance and describes how to work with communities to design, implement, and maintain participatory organisations.

Case study of Mangalane community in Mozambique

When I was approached by a private wild reserve in Mozambique (Sabie Game Park) to assist with implementing a CBNRM programme in a post-conflict area, my first task was to assess if the programme was viable. In my head, I went through the questions in Tables 15.1 and 15.2. With a full spectrum of big game on the borders of Kruger National Park, wildlife was a sound land-use option. Moreover, we could immediately unlock direct benefits for the community in the form of a 20% share of hunting revenues. This would enable us to start, but we would need to increase these benefits in the future, for example by changing the policy of sharing 20% to a 100% policy. Second, the community consisted of five village clusters and had no central structure, so we could immediately work at the face-to-face scale. Third, we sought government blessing for the process and, finally, we put together a loose consortium with WWF and the Southern African Wildlife College to provide long-term support to the programme.

The next phase was to discuss this vision with community leaders, who had a long history of conflict with conservation authorities, as this community is a major transit point for rhino poachers entering Kruger in South Africa. We arranged for leaders to visit private game ranches to learn about the viability of wildlife. Importantly, we worked with key government officials to inform the community that it was entitled to get income from wildlife, and to clarify who owned this money and how it was to be used. To understand the community better, one of my Masters students conducted a situation analysis using qualitative and quantitative methods. We learned that almost nobody over the age of 40 had any education due to the civil war, and it was obvious that people were exceedingly poor.

As I described for CAMPFIRE, our goal was to train community members to lead the change process as far as possible. With the 20% revenue in the bank, we were ready to start. We therefore asked the community to select 20 members for training, and spent a week camping with this group of young people, with the objective of preparing the community for its first annual general meeting and revenue process. Wildlife is an unfamiliar business to them, so the professional hunter and game manager described hunting and wildlife management in significant detail, using props such as buffalo skulls to explain that hunting clients wanted trophies (not meat), and that they paid more for bigger horns. Trainees learned about governance, including through role plays. We developed a standardised agenda for community general meetings (Box 16.2). This described (1) wildlife income and prices, policies, participatory governance, and so on; (2) defined procedures for allocating the money and electing committees; and ended with (3) a revenue distribution ceremony. The community participants composed flip-charts, presentations, and role plays for this standard CBNRM agenda, and practised presenting these with each other. We then organised an annual general meeting in each village, knowing that it can take as much as four days, but once the community is used to the system it can be shortened to two days.

BOX 16.2 STANDARD AGENDA FOR GENERAL COMMUNITY MEETINGS

Provision of information and data

1. Information: the provision of information about community rights, new policies, status of the wildlife business, climate change, and so on.

Defining membership, and reinforcing principles and rules

2. Membership: define or update.
3. Define procedures for participatory governance: including (1) rules of the money; (2) members' bill of rights; (3) roles and responsibility; (4) a constitution that incorporates these principles; and (5) compliance criteria.

Standard business meeting

4. Participatory review of performance, including income and past expenditure.
5. Participatory planning, and participatory, activity-based budgeting.
6. Elections: of a committee or individuals to implement their plans.

Symbolic revenue distribution ceremony

7. Symbolic and visual revenue distribution ceremony.

1. Provision of information and data

In Mangalana community, CBNRM was new, and people had little idea of the value of wildlife in this post-conflict community. Mozambique did not have any policies or procedures for CBRNM, except that licence fees from wildlife hunting on Sabie Game Park were paid to the government in Maputo, and 20% of this money was paid back to the community on the border of the park. In village general meetings we explained that the community would soon be getting this money using a flip-chart to illustrate current and future money flows (similar to Figure 12.1). However, we were honest that this would not be much money until the policy was changed to give the community 100% of their wildlife income, as happens in other countries in the region.

The introduction of CBNRM to remote rural communities is often quite revolutionary and takes time to absorb. Well-prepared flip-charts and role plays are important for visualising and clarifying these explanations. It is also useful to reinforce these explanations by providing paper handouts (in the vernacular) so that community members can read and absorb these new ideas in their own language. In Mangalane, for example, community leaders handed out summaries of wildlife prices and income.

2. Defining the community, boundaries, and scale

In Mangalana, preliminary surveys counted 698 adults living in 244 households in five village areas. Because of the importance of scale in CBNRM (Chapter 12), an important decision was to establish the five villages as the primary recipients of wildlife benefits and to avoid serious long-term problems by starting with a single multi-village organisation. In this case, defining the community was easy. This is not always the case, as boundaries are sometimes contested. Moreover, some large communities are represented by a single traditional leader, and it may take skilled negotiation to subdivide these communities into manageable units.

Once boundaries of a community are agreed, the next step is to define the membership of each community. There are generally two ways of doing this. A household is a useful economic unit, but can be difficult to define where members of extended families live in urban areas, if villages consist largely of grandparents and children with a missing adult generation due to urban work or HIV/AIDS, and where there is polygamy. An innovative REDD project in Tanzania (Box 13.4) defined membership as the two adults in the household plus up to three children, allowing each household up to five REDD dividends, with four being controlled by the mother and being used preferentially for food, education, and health care (Morgan-Brown, 2014). A second method is to define membership as all adults in a community over 18 years and resident for more than five years. This is simpler to administer and update, but results in a lot of people getting a small dividend and does not strengthen families as economic units. When defining membership, there will often be complexities that can only be decided by the community. For example, should the schoolteachers who often live in the area for a number of years, but are not original members of the community, be considered as members for the purpose of decision-making and benefit sharing?

There is no right way of defining membership. What is essential is to discuss options with the community, and then to write down a definition of a member, which is then used to compile a membership register. The list of members must be read out to the whole community, allowing people not on the list to have their cases for membership carefully considered. Often they may not be eligible. Sometimes they will have simply been missed. Occasionally, their case will require a slight modification of the written membership definition. Indeed, it is by dealing with these cases that the definition can be honed.

3. Establishing principles of participatory governance

When we started working in Mangalana community in Mozambique, national guidelines for CBNRM governance had not been developed. We therefore worked closely with an appropriate official to develop local CBNRM rules that reflected governance principles. These took the form of simple statements easily understood by the communities (Box 16.3). With very low levels of literacy in this community, we put off the introduction of a bill of rights and a constitution to a later time.

BOX 16.3 CBNRM PRINCIPLES FOR BEGINNER COMMUNITIES

1. Wildlife, and the income from wildlife, like the income from cattle and crops, belongs equally to each and every person.
2. Therefore, just like the income from livestock, people can use it how they choose, including as household cash. The single caveat is that its use must be decided collectively.
3. The choice of how to spend their money must be made by the whole community together, meeting face-to-face (even if they decide to share it as cash).
4. Any money not paid as dividends and used for projects must be managed transparently and according to the wishes of the people. This requires a process of budgeting, book keeping, auditing, and reporting back to the community.
5. The role of the committee is to bring people together to make decisions, not to make decisions for them.
6. The role of the government is not to manage the wildlife and the money, but to check that the community is doing this properly by following the correct procedures.

Money creates temptations, and few rural communities have experience with managing money collectively. Therefore, it is important to establish clear procedures that mirror CBNRM principles to ensure that money is used properly and accounted for. We therefore presented the community with a simple set of rules, called the 'rules of the money' (Box 16.4). Very often these changes are too radical for communities to absorb immediately. The explanation can be strengthened using role play, but the community will only really believe it once they go through the process of making decisions, and getting money and benefits in their pockets.

BOX 16.4 THE 'RULES OF THE MONEY'

1.   Every adult is entitled to an equal share of the money.
2.   Community members always choose how to use the money together, and are completely free in this choice, including the retention of 100% of the income as household cash.
3.   People discuss and agree how to use their money at the annual general meeting, summarising their decision in a budget with the following headings:

     1.   Cash dividends.
     2.   Projects.
     3.   Wildlife management.
     4.   Membership fee/administration (<10%).

4.   Communities must set up a bank account and simple financial books to track expenditure compared to budget (personal loans are not permissible).
5.   The community must receive a monthly (or quarterly) financial report comparing expenditure to budget from the committee. If it is accurate and follows the budget, the community should approve it; if not, they should take action to correct any problems.
6.   The community will get the next tranche of money (i.e. 'the 20%' in this example) only if the rules are followed, if the money is accounted for, and if expenditure follows the agreed budget. Sanctions (such as delaying the payment of money or hunting quotas) will be applied if these procedures are not followed.
7.   The community elects a committee annually. The purpose of this committee is to bring people together to make decisions and to implement the community's financial decisions. The members are the 'boss' and tell the committee what to do, not the other way around.

Roles and responsibilities

In addition to emphasising CBNRM principles, it is invaluable to clarify roles and responsibilities, and reinforce this at general meetings. Box 16.5 illustrates how to do this on a single page, and can be used as a policy document (Chapter 12). In addition, these roles should be provided to the community in the form of posters or booklets,[1] and included in the individual bill of rights and village constitution. Box 16.5 is an idealised example that emphasises the primacy of community members, and follows the principles of delegated aggregation (Chapter 14). It assumes that a national CBNRM association has the mandate and capacity for compliance and performance management. However, this function can be done by the state or shifted to CBNRM support agencies depending on the realities on the ground.

BOX 16.5 DESCRIPTION OF ROLES AND RESPONSIBILITIES IN CBNRM COMMUNITIES AND CROSS-SCALE AGENCIES

Level of governance	Roles and responsibilities	Income
Community members	**Joint owners of the resource** • Entitled to full dividend share of income from resources • Primary decision-makers	90% of income from wildlife
Community organisation (e.g. Village Action Group – VAG)	**Doing level for collective action** Involve all people in decision-making through general meetings to prepare, prioritise, and agree all plans and budgets • Allocate wildlife revenues to projects, activities, household cash, natural resource management (NRM), and administration at general meetings • Hold quarterly general meetings to report on finances and performance • Plan, implement, and monitor VAG-level projects and activities • Maintain bank account and financial records • Manage wildlife at the local level (e.g. employ game guards or punish poachers according to by-laws)	The income that individuals tax themselves for collective projects
Elected VAG leaders	• Bring people together to make decisions • Provide information on all aspects of the programme • Implement instructions given to them by members through general meetings and report on progress • Manage community finances and book keeping • Implement community projects	3% for administrative costs (or through delegated aggregation)
Multi-VAG Coordination Committee	**Coordinating level for larger area** • Maintain bank account and financial records • Coordinate development plans for area • Plan, implement, and monitor large multi-VAG projects	4% for administrative costs (or through delegated aggregation)

(Continued)

(Cont.)

Level of governance	Roles and responsibilities	Income
	• Monitor and oversee wildlife and NRM (though village scouts are employed and paid by each village)	
Traditional authority	**Non-executive, non-administrative role (patron)** • Provide advice and maintain traditional values • Neutral arbitration • Guide decisions on broad land-use issues • Protect the constitution	3% honorarium
Government oversight agency	• Provide guidelines and policy • Monitor compliance with governance principles (or delegate this function to CBNRM Association)	

National CBNRM Association	**Conformance and performance management**
	• Monitor compliance with governance principles including:
	    ○ 90% of income to communities
	    ○ Full community participation in decision-making
	    ○ Money accounted for according to community budget
	• Monitor performance (finances, wildlife, institutional development) of CBNRM
		2% of community income

CBNRM support agencies	**Capacity-building**
	• Develop managerial capacity of community institutions (i.e. design systems and provide training)
	    ○ Build institutional capacity
	    ○ Build NRM skills and systems	

4. Participatory review of performance, especially finances

Communities should hold monthly or quarterly general meetings. The more they practise and institutionalise standard reports and decisions, the more they grow to expect this information to be provided. Annual meetings are insufficient, and far too infrequent to catch problems early.

After reinforcing principles and providing general information, the meeting turns to business. This starts by reviewing community income and expenditure, progress with projects, the performance of game guards, and so on, using standard report templates. In the

AGM, an income statement is written up on a flip-chart (and on handouts) as fully and clearly as possible, as illustrated by Box 16.6 for a wildlife tourism and hunting operation. The community should be trained to expect a detailed income explanation in this format at every meeting, as well as similar reports on expenditure (see below), projects, and so on.

In some situations, the next step is to allocate the money to the communities using an agreed formula. The principle of 'producer communities' is that money should be returned to the village in which it was earned. In the Mangalane community, each person in each of the five villages outside Sabie Game Park was entitled to the same share, in this case $21 (MT650), according to the membership register. This was explained carefully to the community with a flip-chart that resembles Box 16.7.

BOX 16.6 EXPLAINING THE SOURCE OF THE MONEY

- This September report shows that tourism is operating at 55% occupancy (which is reasonable) and earned the community $32 230.
- The hunting season is halfway through, about half the quotas have been utilised, and hunting income currently stands at $95 000

Source of revenue	Quantity	Price ($)	Income ($)
Carry over (from last reporting period)			5 000
Tourism	323 bed nights (55% occupancy)	10/ bednight	32 230
Hunting			
Concession fee	One-off payment	15 000	15 000
Elephant	2 (quota 4)	20 000	40 000
Buffalo	5/12	8 000	40 000
Etc.			
Total income			132 230

BOX 16.7 CALCULATING INDIVIDUAL DIVIDENDS AND SHARES FOR EACH VILLAGE

- Each person should get an equal share of wildlife income.
- There are 698 people in 244 households in five villages.
- Share is MT650 per person ($21).
- Share of each village is calculated below.
- All villages wanted to check the register, so printed list of households and adult members were handed out for the community to check and update.

Village	Number of households	Adults	Share (Mozambican metical)
1. Mukakaza	70	166	107 730
2. Mavanguana	78	264	171 730
3. Ndindiza	24	69	44 884
4. Kostine	23	46	29 923
5. Babtine	49	153	99 525
Total	**244**	**698**	**454 044**

Source: Child, B.

5. Participatory, activity-based budgeting

As its name implies, there are two aspects to participatory, activity-based budgeting.

Participation implies that a quorum of 60% of the community or a minimum of 60 members attends the meeting at which the community allocates its annual revenues, and that the members clearly understand their finances and the choices they are making. Figure 11.3 shows the schoolteacher in Mahenye community in Zimbabwe, Mr Masangu, carefully taking the community through these choices. He uses a flip-chart with a standard activity-based budget with four rows for expenditure – cash, projects, natural resource management, and administration (Box 16.8). The total column indicates the total amount for the whole community. However, people in remote rural communities, especially elderly women, often do not understand these big numbers. Therefore, an extra column is added that pro-rates the budget as it affects each member. In a participatory process, it is normal for figures to be crossed out and adjusted as the community assesses different choices.

BOX 16.8 PARTICIPATORY ACTIVITY-BASED BUDGETING

	Total for community	Per household (n = 165)
INCOME (see Box 16.6)	$132 230	$800
EXPENDITURE		
1. Cash benefits	~~$132 230~~ $66 115	~~$800~~ $400
2. Local projects	$66 115	$400
3. Wildlife management	0	0
4. Administration and overheads (must be < 10%)	0	0

The first step in participatory activity-based budgeting is to calculate the dividend owing to each member. In the example in Box 16.8, each of the 165 members of the community gets an $800 share of the total income of $132 230. Recalling the rules of the money, the first principle is that this is their money and they can all agree to take it all home as cash. However, the second rule is that this decision must be made collectively. Therefore, the facilitator asks if the community would like to tax itself to provide important collective projects like schools, clinics, water-points, and food. In this example, the community decides that each person should be levied $400 for two projects they have been talking about for a long time. The facilitator puts $400 in the projects column, and crosses $400 off the $800 cash dividend, doing the same for the 'total for the community' column. New communities seldom invest in wildlife management or administration in year one.

One of the dangers to watch for is that administrative costs can quickly eat up the whole budget in non-productive expenditures, because leaders are tempted to increase the allowances they get for meetings. To combat this, the rules should cap administration at, say, 10% of total income. Allowances and fees should be agreed and set by the whole community so that they are transparent and locally appropriate. Several African countries have an unfortunate system of 'sitting allowances', and I have seen community accounts where these eat up thousands of dollars with nothing to show. It is bad practice to pay people to 'sit'. A workaround is performance payments, with the treasurer, for instance, getting a set allowance when his or her quarterly financial report is approved. Another danger occurs when community income dribbles in monthly, because there are too many opportunities for the committee to spend the money in unplanned ways. Accumulating the money in a separate account pending a single annual payment and budgeting session is a much safer system.

The budget process contains many opportunities for learning and should not be rushed. At a minimum it should take place over two days, allowing members to list and discuss potential projects and activities, but to then go home and discuss these ideas in non-formal spaces. Women talk far less than men in meetings, so having time to discuss options with their husbands at home before a decision is taken is important. In more mature CBNRM programmes, the community will discuss expenditure choices for several months before the revenue arrives.

The process of planning, budgeting, and choosing projects or activities to invest in should become more sophisticated over time. As communities gain capacity, they should become more demanding of the technical quality of the projects they choose. After getting revenue for several years, the community should no longer simply choose projects by voting them up or down. One approach is for the people proposing the projects to present a work plan and budget to the community, which then selects projects based on the quality of the implementation plan, as well as their needs.

The budget process should be interesting and engaging, using communication and visualisation techniques to ensure that everyone in the community, including the grand-mothers, can understand and follow the process. Especially in the first years of a programme, communication can be improved by sketching pictures of projects, or using monopoly money to role play decisions and the budgeting process. The participatory, activity based budgeting process is exciting and transformational. I have witnessed this myself in several different countries. Theron Morgan Brown is similarly excited by how it energised and educated REDD+ communities in Tanzania (Morgan-Brown, 2014). David Elliot (personal communication) found the process to be transformational, generating community understanding and inclusion in a previously centralised community ecotourism project in the Ecuadorian Amazon.

6. Elections

The last step in the business meeting is elections. I strongly recommend holding elections annually. Practising elections annually institutionalises electoral accountability. It reduces a dangerous dependence on a few leaders, and does not prevent exceptional people who perform well from being re-elected. Moreover, tri-annual elections have a nasty habit of slipping to four or even five-year intervals. Accountability trumps convenience and the counter-argument that it is wasteful for leaders to serve for only one year because of the investment in training them. Indeed, any programme that is training only the leadership opens space for elite capture and power asymmetries. In any case, an inclusive, participatory process de-emphasises the role of individuals and the dubious reliance on electoral accountability. I would also recommend that people standing for elections are asked to explain how they will promote community goals (e.g. job creation, projects, financial honesty) to encourage members to elect people suited to these goals.

7. The revenue distribution ceremony

Once the budget is agreed, the money is 'put on the table' in front of the entire community. In Figure 16.1 we see (1) Mr Nare, the Chairman of Beitbridge community, ceremoniously carrying $60 000 into the community meeting in a basket on his shoulder and placing it on the table in front of everyone. After speeches and explanations, (2) each member gets their entire dividend

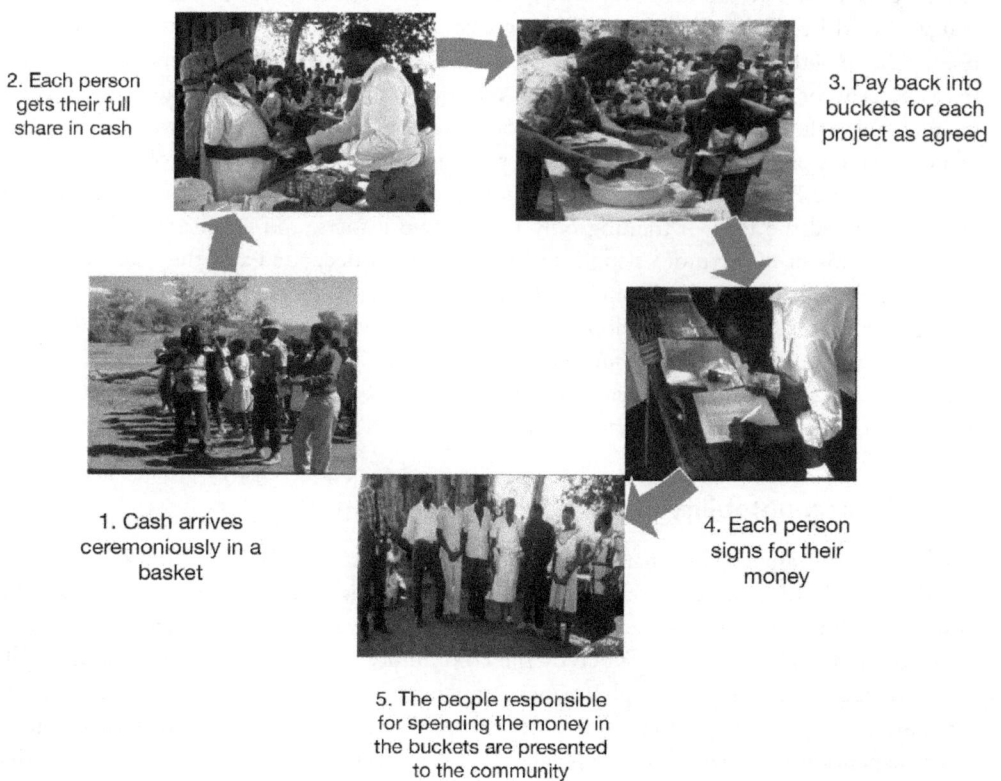

2. Each person gets their full share in cash

3. Pay back into buckets for each project as agreed

1. Cash arrives ceremoniously in a basket

4. Each person signs for their money

5. The people responsible for spending the money in the buckets are presented to the community

FIGURE 16.1 The revenue distribution ceremony

as cash in hand. Next (3) they pay an amount into the two 'project' buckets according to the 'tax' agreed by the community. I have illustrated this with a Shangaan man with disabilities, symbolising that all people in the community are entitled to an equal share. In front of each bucket is a flip-chart paper that names the project, and the amount that each person will contribute to it. For administrative reasons, (4) each person then signs a pay sheet for their money, and this process is transparent even to members who sign with a thumb print. Finally (5), the committee tasked with using the money is introduced to the community and reminded of their responsibilities.

Banking and bookkeeping

Bookkeeping tracks all income and all expenditure in a format that reflects the budget. This is not the place to describe a financial system in detail. However, the community treasurer requires several tools and skills. A receipt recording every transaction must be kept systematically in a file. This can be recorded in a very simple financial log, especially if the amount of money and expenditure activities are small. However, in the Luangwa project we successfully used a standard double entry system in a cash analysis book to keep track of multiple projects in each community.

The important thing is to reconcile income and expenditure on a monthly/quarterly basis, and to present this summary to the community. I have replicated a report from the system used in Luangwa in Figure 16.2 to show what is possible even in remote communities. The report has three sections. The top table tracks income. The second table compares expenditure to the budget agreed by the community, and if variance exceeds 5% the community is required to resolve the underlying issue (see below). The 'balance' table makes sure no money is missing from the bank or cash box. This system was introduced and implemented, including ongoing training and auditing, in 43 villages and six area committees by a team of two people. Clear financial reports provide a solid foundation for transparent reporting and control.

Communities benefit from training in many aspects of financial management. I have already outlined the risks of training only one or two leaders, and emphasised the value of training groups of 20 or more people so that capacity is deepened and they can hold each other accountable. To make financial training fun, I highly recommend the 'CAMPFIRE management money game' developed by the late Ivan Bond from WWF-SARPO. The game is based on Monopoly, but with tourism and hunting concessions, and teaches community members to handle cash, make out cheques and deposit slips, fill out receipts, and record, analyse, balance, and report on finances, and so on.

Financial accountability

There are two aspects to financial accountability. Most people focus on keeping and auditing the books, which is a basic requirement. It is more important, however, to ensure that finances are reported back to the community regularly, in ways that are easy to understand, and in a manner whereby the community can hold the treasurer accountable for following the budget. A simple financial variance analysis on a flip-chart is adequate for this purpose (Box 16.9). It follows the same format as the budget, but adds four columns: actual expenditure, variance (i.e. expenditure less budget), problems faced, and corrective action. If the variance exceeds 5% of the budget, it must be flagged and discussed by the community, with agreement about what corrective action will be taken.

FINANCIAL SUMMARY

Name of Village: ---------------------------
Date report worked out: ------/-------/------
Covering the period: 1 January – 31 March

Money received during this period (Specify)

Source	Amount
1. Annual revenue from hunting	7,000,000
2.	
Total money received	7,000,000

Money spent during this period (Specify)

Expenditure	This quarter	This year	Budget	Variance
1. Household cash dividends a. HH cash 210HH @ 20,000 each	4,200,000	4,200,000	4,200,000	0
2. School rehabilitation a. Cement/paint 850,000	850,000	850,000	1,000,000	150,000
3. Wildlife management	0	0	1,200,000	1,200,000
4. Administration a. Allownace 25,000 b. Transport 50,000 c. Stationery 80,000	155,000	155,000	500,000	345,000
TOTAL	5,205,000	5,205,000	6,900,000	1,695,000

Balance:

Cash at start of quarter:	7,000,000
Less expenditure	5,205,000
Less Money in Bank	1,750,000
Less money is cash	45,000
Balance:	0

------------------------------   ---------------------------
Chairman   Treasurer
Approved by community: -- (details of vote)
Date: --------/-----------/---------------

FIGURE 16.2   Community financial summary

BOX 16.9 FORMAT FOR A MONTHLY/QUARTERLY FINANCIAL VARIANCE ANALYSIS

Activity	Budget agreed by the community	Actual expenditure	Variance	Problems faced	Agreed correc-tive action
1. Cash					
2. Projects					
3. Natural Resource Management					
4. Administration					

The community should discuss finances (and other matters) regularly (Box 16.10). A useful guideline is to hold a general meeting monthly to discuss community finances for the three to four months after the community receives the income, when most transactions are occurring. Once most of the money is spent, the frequency can be reduced to quarterly. Obviously, a single annual report to the community (which is the norm in many projects) is inadequate for preventing problems from occurring, or for correcting them in a timely manner.

BOX 16.10 RECOMMENDED FINANCIAL CYCLE FOR MANAGING COMMUNITY FINANCES

The following procedures should be standardised within the community:

1. Monthly:

    a) The community treasurer reconciles his or her books, and compiles a financial report.
    b) The community facilitator inspects the books and financial reports, and provides a statement to the community about their accuracy. This is an inexpensive way of providing an external audit.
    c) Monthly/quarterly general meeting – three short reports are provided to the community:

        i)   The financial variance analysis (Box 16.9, top part of Figure 16.2).
        ii) Reconciliation statement ('balance' table in Figure 16.2).
        iii) 'Audit' report.

Based on these three reports, the community approves or disapproves the monthly financial report, records its observations, and agrees any corrective action.

2. Annually – if not too expensive, the AGM should incorporate an independent financial audit.

Financial transparency and accountability

Effective and transparent financial management is critical for building impersonal governance systems in CBNRM, and for strengthening social capital and trust. It also internalises the value of a wild resource into community decisions, and allocates money efficiently between alternative uses. By contrast, suspicion about financial misappropriation and misuse quickly undermines community trust and rots the fabric of social capital.

In summary, the financial system described above is designed to ensure that the community understands the source and amount of income, and makes the link between protection of wild resources and community well-being. Participatory, activity-based budgeting democratises financial accountability by ensuring that the community understands its finances, makes well-considered financial decisions, and certifies that these decisions are implemented. To achieve social accountability, financial reports are presented to the community in ways that are easily understood on a monthly/quarterly basis, so that the community can approve or disapprove of the financial statement. This is supported by external auditing, which is often a weak mechanism of accountability in the absence of social scrutiny.

While these social processes are invaluable for building social capital, communities need basic skills in keeping a bank account, petty cash register, and a set of books, and providing simple income and expenditure statements and reconciliation statements on a monthly basis.

The importance of cash and choice

The importance and role of discretionary cash management in community development is often misunderstood. The way that wildlife income is handled is critical, and often makes or breaks a CBNRM programme. Moreover, as we saw in the case studies, discretionary choice over all the cash is symbolic of both democratic processes and proprietorship, and is a prerequisite for sustainable natural resource management. Annual cash dividends, and the process of sharing these, provide an unmatched mechanism for institutionalising participatory governance and discretionary economic choice. As equal shareholders of the wildlife resource, each member has the right to their full share of wildlife income as private income in the form of a cash dividend. These dividends are the rightful payment for the effective management of wild resources owned by people. They are in no way cash 'handouts' because the community has invested in wildlife production to get these benefits, including bearing the opportunity cost of leaving land untouched by farming and livestock. Cash satisfies important household needs like food, clothing, and school fees. These often have higher utility than community projects, especially for members with no other sources of cash income. Moreover, when managed properly, cash payments often result in more projects, not less (Figure 12.3). Thus, the processes and learning that happen around cash are more important than the cash itself, and certainly more important than the projects the cash can build.

Nonetheless, officials and NGO staff often misunderstand the importance of cash and choice, and argue that it is wasteful to allow communities to get cash from their wildlife. The historical treatment of wildlife as a public good means that it is often difficult for officials to accept that the people who live with wildlife, and bear the costs of doing so, should have the right to benefit directly and in cash. Spending money on behalf of the community may satisfy administrators or NGOs, but it provides few of the benefits listed above. The early WINDFALL programme (Chapter 11) showed how ineffective this is as a conservation or development mechanism. Building projects on behalf of the community, or even handing out cash, is retrogressive because it perpetuates the sense that people are the recipients of philanthropy and, like giving alms to beggars, is humiliating and disempowering. It is a serious mistake to rush in and build projects for the community.

Genuine development flows from empowering communities with the resources, systems, confidence, and pride to do things for themselves – building a classroom, or a clinic, employing and managing village scouts, managing money, making decisions, and so on. This can be achieved through carefully structured fiscal devolution, where we should be measuring improvements in community self-esteem, resource stewardship, and effective, transparent, accountable, and equitable community governance as much as we count the number of concrete projects that are built. As conservationists, we also needed to be at pains to link money to wildlife production, so that wildlife income symbolises both ownership and productivity. Therefore, we need to reconceptualise wildlife, and wildlife cash dividends, as a private good that can then be taxed by mutual agreement for community projects (Box 16.11).

BOX 16.11 COMMUNITY CASH AS TAX

To be consistent with the CBNRM objectives of promoting inclusive governance, and converting wildlife from a public into a private-community good, the nature of the money that a community gets from wildlife also needs to be defined correctly.

Community members are joint shareholders in their wildlife. Therefore, the cash represents a shareholder dividend from the village company. Company dividends represent individual benefit from joint investments. To signify this, communities require full choice in allocating their dividends, especially the option of keeping it all as cash. Lesser rights (i.e. constrained choices) will devalue wildlife in the eyes of individual members.

When communities then set aside money for community projects, this is the direct equivalent of a self-imposed community tax. The benefits of projects can be significant, but individual benefits may be more valuable. The importance of the participatory budgeting process described in this chapter is that the crucial trade-off between individual and collective benefits becomes the subject of intensive discussion, and the option of cash is healthy because it sharpens these choices.

Thus, a well-designed revenue distribution process (1) maximises the sense of proprietorship over the resource (because people get cash and full discretion in the use of this cash); (2) empowers people by involving them in discretionary choices; and (3) optimises the allocation of scarce cash, including tradeoffs between private and collective expenditure. Using community money to implement a project on their behalf, and even imposing percentages to follow in the allocation of money, indicates that the essential importance of rebuilding social capital is not understood.

Managing CBNRM governance and compliance

The management of a CBNRM programme is systematic and rigorous. All things being equal, it begins by institutionalising face-to-face inclusive governance, preferably as a set of written principles and procedures. I favour incorporating these principles and procedures into an individual bill of rights (six pages) and reinforcing these by duplicating them in a community constitution (20 pages). These documents describe the annual management cycle, and include the formats for key reports and information so that the community knows what to expect and demand.

CBNRM is then implemented through an annual management cycle. At the core of this is the annual general meeting and the monthly/quarterly general meetings, together with their associated requirements for information and training, reporting, and decision-making (Box 16.2). The mechanism of the extraordinary general meeting is an essential governance safety value. This should be codified in the bill of rights and constitution to allow ordinary community members to call a meeting independently of the committee to resolve problems, including requiring information from the committee and holding them to account in other ways, including fair dismissal.

The CBNRM management team then needs two semi-independent control systems: a system for monitoring and managing procedural compliance with the principles of inclusive

BOX 16.12 MONITORING FORM FOR CBNRM GENERAL MEETINGS

CBNRM General Meeting Form
(to be compiled by the community facilitator)

Name of community:
……………….

	Day 1	Day 2	Total
Men			
Women			
Total			

Dates of meeting:
…………………

Facilitator/s:
………………………..

Status of community register: complete/incomplete. Comment (…)

Report back on finances:
1. Were finances presented properly?
2. How did actual expenditure compare to budget?
3. Was external audit presented?
4. Was there any suspicion of misappropriation or misuse?
5. Was the financial report approved by the community?

Financial Allocations:

Total income for VAG:	
Number of members:	
Dividend:	

Decision-making:
1. Was information presented properly to support decision-making?
2. Did people, including women, participate fully in decisions?
3. Did people have full choice in using their money, including cash?

Summary of budget allocation:

Name of project or activity	Initial budget	Agreed budget	Contribution by each member
1. Cash			
2. Project/s			
3. NRM			
4. Admin			

Elections:
Were these held?   Yes / No / partly (explain ……………………………………..)
Were they free and fair?

Names of incoming and outgoing committee: (table of outgoing, incoming officers, plus others who stood in elections and votes received)

Hot issues:

Comments by facilitator:

governance and CBNRM (Box 16.12), and a simple 'client satisfaction' survey of ordinary members of the community to track their understanding of and satisfaction with the programme.

Compliance

Perhaps the most important role of facilitating agency is to ensure that the community complies with a set of procedures that track if CBNRM principles are being followed. This is theoretically a government function, but is often done by an agency acting on behalf of government.

Box 16.13 provides a governance compliance checklist to assess if members are meeting regularly, are participating in decisions, and are checking if their decisions are implemented. Compliance management requires that facilitators attend all general meetings and complete a two-page assessment form which is illustrated in Box 16.12 (note: spaces for comments etc. have been removed). The data in this form, when entered into a database, provide a tracking system that is invaluable for planning community support activities in monthly or quarterly planning meetings. The second set of information is the community financial system described above. Here, the primary role of facilitators is to 'audit' community finances monthly while they train communities to keep their books, and to ensure that community treasurers compile the variance analysis and financial reconciliation statements illustrated above.

This checklist becomes a powerful tool where full compliance is a necessary condition for the release of the next tranche of community benefits. In Luangwa, for example, community income was paid directly in a bank account with community and project signatures. The project had no rights to withdraw this money, but the community could only access its annual dividend when the project signed off the compliance monitoring form and the bank release form.

BOX 16.13 CHECKLIST OF GOVERNANCE COMPLIANCE CRITERIA FOR INCLUSIVE CBNRM

Procedural requirements	Means of verification
1. AGM: At least 60% of members attend, which follows the standard agenda (Box 16.2).	Meeting monitoring form
2. Monthly/quarterly general meetings: At least 60% of members attend three additional general meetings, which follow the agenda.	Meeting monitoring form
3. Financial decision-making	Meeting monitoring form
• At least 60% of members (or a minimum of 60 people) are involved in deciding how to allocate the year's wildlife income in a participatory way (and no decision is forced on them);	
• People have full choice of the use of their money, including household dividends (cash), projects, and activities.	

(Continued)

(Cont.)

Procedural requirements	Means of verification
4. Community decisions are recorded in the format of an activity-based budget (Box 16.8).	Budget form submitted
5. Quarterly financial 'audit'	Audit of accounts by community facilitator
• Local audit shows accounts are properly kept and expenditure follows budget;	
• The reconciliation statement (Figure 16.2) shows that all money is accounted for.	
6. Participatory financial review (monthly/quarterly)	Meeting form
• All finances are properly presented to the community using an activity-based variance analysis (Box 16.9);	
• The community reviews and approves expenditure compared to the budget (and work plan);	
• Any variance above 5%, or misappropriation, is dealt with properly.	
7. An annual (external) financial audit verifies the accounts and compares expenditure with budget (Figure 16.2).	External audit
8. People elect a new committee each year and consider this process to be free and fair.	Meeting form
9. The membership list is updated and approved by the community before each AGM.	Meeting form

Client satisfaction

It is easy to forget that community members are the primary 'client' of CBNRM activities. Livelihood surveys (every five years) and the governance dashboard process (every two years) are useful tools to remind the support agency of this focus (Chapter 15). However, it is invaluable for managers of a CBNRM project to track community understanding and satisfaction with a very simple survey called a client orientated monitoring system (Box 16.14). If community facilitators survey three to five people in each community each quarter with a half page form, this provides an annual sample of $n =$ 240 for a programme with only 20 communities. Like the compliance data (Box 16.12), this information is invaluable at the team's quarterly meeting, highlighting weak communities (e.g. where only one or two communities score low on a particular variable) or weak support or systems (e.g. where all communities score low on the same variable). Box 16.12 lists broad questions that track the village members' understanding of issues, and their opinions about their governance. These can be scored from 0 (no understanding) to 10 (perfect understanding).

BOX 16.14 A SIMPLE CLIENT ORIENTATED MONITORING SYSTEM

0 = no understanding; 10 = perfect understanding

Name of enumerator: Name of village: Name of person: (in confidence)	Score
Does the interviewee	
1. Know how much animals are worth?	
2. Know how much the community earned this year from wildlife?	
3. Know what the money is used for?	
4. Understand key principles in the constitution well?	
5. Understand what projects are being implemented, and who is responsible for them?	
In the opinion of the villager	
6. The CBNRM project is working well for me	
7. The committee is doing a good job	
8. The leaders can be trusted with the community's money	
9. Community projects are being implemented effectively	
10. Wildlife conservation is important to my future	

While the politics of getting the enabling environment right can be challenging and time consuming, managing CBNRM properly at the field level is relatively straightforward, although it requires careful planning, a performance monitoring system, and reliable and tenacious follow through. Micro-governance can be greatly improved with a set of reasonably standard skills.

The most technical challenge is formalising CBNRM principles as I have explained here for the 'rules of the money', and elsewhere for the individual bill of rights and constitution (Child & Wojcik, 2014). This includes clarifying the roles and responsibilities of community members, committee members, community employees, and support agencies, and reinforcing this on a regular basis. It is useful to complement formal documents (e.g. constitution) with mini-manuals outlining, for example, how to run the annual general meeting, how to manage projects, and the roles and responsibilities of office bearers.

The second task is to build these principles into the community management cycle through an effective general meeting with a clear agenda and well-defined expectations and rights for informed decision-making. This includes training the community to expect the monthly/quarterly financial variance analysis in a standard format, knowing how to approve or not approve this, and procedures for taking corrective action when things go amiss.

The third task is to provide technical training to support the management cycle. This involves training leaders and followers in bookkeeping, financial management, financial reconciliation, and the presentation and interpretation of monthly/quarterly financial statements.

Fourth, and critical, is compliance monitoring and enforcement. Linking compliance to the annual release of money (or quotas) provides strong motivation for the community to institutionalise impersonal decision systems and to develop supportive technical skills.

Finally, experience is needed to negotiate and resolve conflicts and the inevitable bumps in the road that occur.

This is an adequate framework for community governance. In most programmes, it is also valuable to train communities to implement projects, like clinics and schools, or community activities. It is rare that communities do not have the skills to build a clinic or construct a borehole. What is missing is clarity of purpose, and clarity of responsibility. Again, it is relatively straightforward to introduce simple work plans to communities and, even more importantly, regular peer review of progress using standard forms that also provide the CBNRM support agency project with a full record of community projects.

Finally, natural resource management requirements need to be integrated into the system using the same processes of experiential training, co-development of systems, and monitoring and adaptive management. For a single resource like wildlife, CAMPFIRE and Namibia alone have already developed methods and manuals for counting animals, monitoring hunting, monitoring and managing human–wildlife conflict, community patrolling and resource protection, negotiating and monitoring commercial hunting and tourism partnerships with the private sector, managing community campsites, producing and selling crafts, land use planning, fisheries management, and so on.[2]

With training and supervision, locally recruited school leavers can undertake most of the community training and support, with one person able to support roughly five village communities. On a monthly basis this will require the field-based para-professional to train communities to run meetings, manage finances, manage projects, and manage some aspects of natural resources, plus the responsibility for accurately monitoring the performance of these activities – meetings, finances, projects, wildlife management – using standardised checklists. It is much cheaper and more effective if these community facilitators live in the field, as locating them centrally distances them from their work and results in high transportation costs. A small support staff is necessary to manage databases and for supervision. Light touch high-level skills are essential to design new systems, train and supervise staff, resolve conflicts, to provide pro-active leadership, and to manage external factors. Especially in new programmes there is a strong case for locating this high-level expertise in the field, although the temptation is usually to build a head office team to support mid-level field staff.

Conclusions

This chapter has provided practical tools for institutionalising participatory community governance. Participatory institutions cannot be developed by dropping in and out of the community, or by providing training courses to the leaders (which exacerbates power asymmetries and increases the risk of elite capture). Capacity-building requires focused attention towards ongoing experiential learning processes, especially the tracking of compliance and performance using simple but robust monitoring systems. The implementation of CBNRM governance procedures needs to be systematic, with disciplined performance and compliance monitoring. Combining monitoring with a regular monthly or quarterly review process is key to the integrity and efficiency of the supporting management systems. It identifies areas of weakness, either in the system as a whole or in individual communities, which can be corrected by specific technical training that always involves both the leadership and followership. The process described in this chapter is quite manageable. It can be implemented effectively by young facilitators, often recruited from the community, provided quality skills are available to design and adhere to the administration systems described above, and to mentor and train community facilitators to apply these methodically.

However, CBNRM does not happen only within a community. Capacity-building also requires clarifying the roles of external agencies to support inclusive local systems through (1) institutional capacity building; (2) monitoring and ensuring conformance with governance principles; and, once effective governance systems are in place, (3) building systems and skills for economic development and natural resource management. There is inevitably the need to fight for and protect community rights, and to convince both government administrators and CBNRM support agencies of the importance of inclusive community-based governance.

Notes

1   I have developed a set of training manuals that include the following, and are available on request. This includes:

- A Member's Bill of Rights. This ranges from a two-page document that lays out the basic principles, to a nine-page document that also describes the annual management cycle, the purpose and agenda of each meeting, and templates of reporting documents, so that members know what information they can rightly demand.
- CBNRM rules and principles and/or a constitution. This can be presented as both a short two to three-page document (rules of the money), and/or elaborated as a community constitution, with the latter including a detailed description of most procedures. Although constitutions usually stretch over 20 pages, and are not used regularly, having a well-crafted constitution to fall back on in the case of procedural issues and conflicts is invaluable.
- A set of simple operations and procedures manual(s) that describes key procedures such as the AGM, general meetings, revenue distribution, financial management, participatory project planning and management, and so on.
- A set of mini-manuals that describe the roles of community members, the chair, the treasurer, the secretary, and so on. These include and cross-reference the above procedures, and trade comprehensiveness for being a bit long (8–15 pages).

2   Here I refer the reader to the WWF-Southern African Regional Programme Office's excellent Wildlife Management Series and also to the website of the Namibian Association of CBNRM Support Providers (NACSO) www.nacso.org.na/resources/training-manual.

17

CONCLUSIONS

Overview

Communities living on the Mozambican border of Kruger National Park epitomise the tragedy of the ungoverned spaces we introduced in Chapter 4. People live in pole and mud homes, and many of these families have no food in the current drought. Only by chopping down trees to make charcoal could they eke out some food, sometimes supplemented by stealing batteries from water pumps or neighbours. The poverty is heart wrenching. Little children die of completely avoidable diseases. There has been at least one suicide by a near-abandoned elderly person, while child-headed families live on the very cusp of desperation and survival. In an area heavily influenced by slavery in the early 1800s, and more recently by a brutal civil war, people don't trust each other, but the levels of trust in authority are even lower. Rhino poachers drive high-end vehicles conspicuously through the community. Criminality and social tension is rife, and people have no reliable law enforcement to turn to for protection. Ordinary people, especially women, are exceedingly uncomfortable with this criminality, but, without law enforcement, alternatives are few and the temptations extraordinary, especially for young men with no other options. Even a thoughtful and regal older man asks if he is really a man if he allows his family to starve in a drought, when for the risk of a few dangerous nights of poaching rhino in Kruger he could save them. Anti-poaching helicopters fly over, spending in a single hour what it would cost to feed many of these communities for a month.

Yet within 50 kilometres, a similar landscape supports a thriving wildlife economy, based on the soul-changing tranquillity of the African bush, and the wonder of its wildlife. Beneath this natural beauty beats the engine of a vigorous global economy, providing nearly 35 000 jobs, and well over half a billion dollars in economic activity each year (Chidakel & Child, in preparation). The Mozambican scenario is typical of large areas of the planet where poverty, social distress, and environmental destruction are interwoven. The real tragedy is that all this is avoidable. This raises the question of what we need to know and do to transform these systems because, clearly, the differential outcomes are human constructed.

Bad outcomes that are avoidable are as common as the dysfunctionality of the underlying rules that cause them. Chapter 2 used the broad sweep of history to show just how much the trajectory of society is influenced by the underlying rules – or institutions – that govern human cooperation. The Mozambican community is representative of ungoverned spaces locked into low-value production, which are replacing high-value wild species on a global scale (Chapter 3). The institutions to leverage these values include rights to land and wild resources, markets, and social capacity. These have been lost through a process of 'deinstitutionalisation', creating the configurations opposite to those underpinning the age of prosperity – the protection of person and property in inclusive societies, and the freedom to choose and trade (Chapter 2).

At the end of the 19th century, global leaders like Theodore Roosevelt established systems of public governance for wild resources to combat the massive depletion of wildlife, forests, and other wild resources (Chapter 7). However, treating wildlife as a public good on private (or community) land has left it economically defenceless in the face of the massive new threat of the expansion of low-value agriculture into forests and drylands. Reversing the decline of wildlife requires massive public financing of wildlife as a public asset, or an economic approach that combines proprietorship and appropriate actions, or preferably both in careful combination. What is does not require is more of the current inconsistent, incoherent, and over-centralised political governance.

Wildlife is misclassified as a public good

Chapter 5 defined public goods as being non-rival and non-excludable. In today's world of scarcity, wild resources are easily used up (i.e. they are rival goods, because use by one person reduces use by others). New knowledge and technologies allow potential owners of wild resources to exclude others, and to know how much is taken or polluted. This implies that many wild resources are private or private-community goods, with some common-pool characteristics.

Southern African's wildlife administrators were the first to put two-and-two together, and invent a different way of governing wildlife conservation (Chapter 8). They recognised that wildlife would not survive as a low-value public good on private land. They wagered that wild animals had ecological and economic advantages over farming monocultures, and that people would nurture wildlife if they owned it and it contributed meaningfully to their livelihoods. Consequently, they did the opposite of Roosevelt. They denationalised wildlife by devolving the rights to use, manage, and benefit from it to landholders and communities. They encouraged markets for wildlife, seeking to make it as valuable as possible. Then they combined privatisation and collective action in ways that were highly complementary to managing both the private and common-pool characteristics of wildlife (Chapter 8). Along the way, they made the case that wildlife was a viable form of land use (Chapter 9), and conceptualised these lessons as the sustainable governance approach (Chapter 10).

The rewilding of private land in southern Africa is, alongside national parks, one of the most successful conservation actions in history, bringing wildlife back from the brink of extinction as the foundation of a rapidly growing wildlife economy. However, private land is rare in Africa, and the next challenge was to adapt the proprietorship-price governance model to Africa's extensive communal lands and ungoverned spaces through what is now known as CBNRM. The CBNRM model was validated where implementation was done properly, providing lessons for within-community governance, managing scale, and social processes for building institutional and technical capital and capacities (Chapter 11). Perhaps the most important lesson is that there is a profound difference between inclusive (community-based) governance and representational (committee-based) community governance, even if the systems look remarkably alike to the uninitiated (Chapter 12).

In response to demography, poverty, environmental problems, and weaknesses in top-down governance, community conservation began to emerge globally in the late 1970s, spawning a considerable scholarship but no all-embracing theory (Chapter 13). To fill this gap, I have developed an operational model of CBNRM (Chapter 13) and a much more detailed model of micro-governance (Chapter 14). The latter stands on the intellectual shoulders of Ostrom, Murphree, and others, but advances theory and practice by describing the process by which these principles emerge, or can be encouraged to emerge, in circumstances that are less than ideal (Chapter 14). Finally, Chapter 15 has suggested a practical approach for implementing CBNRM, while Chapter 16 provides additional guidelines for implementing participatory governance in practice.

Lessons of economic history and the sustainable governance approach

There are surprising parallels between the historical transformation from feudalism and serfdom to the modern age of liberalism, free-markets and democracy, and the conditions for the economic rewilding of ungoverned spaces. John Locke's inclusive economic and political institutions and rights ('Lives, Liberties, and Estates') and Adam Smith's discretionary exchange allowed Englishmen, and then Europeans, to emerge from the drudgery and disease of the European Dark Ages. These institutional conditions, and this transformation, has bypassed the drylands and forests of African, Asia, and Latin America. Communities that coexist with wild life and wild lands do not share these freedoms. They have weak rights to use and benefit from wild products, markets for wild products are restricted and distorted, and people are vulnerable to the expropriation of their land (Murombedzi, 2014) and even violence. These conditions persist, allowing exploitation that fuels a free-riding modern, urban sector. They are so deeply entrenched that they are unthinkingly accepted as normal. Despite evidence of massive problems – less than 1% of Africans who live in forests have formal rights to their ancestral homes and Africa has lost as much as 90% of its wildlife – there is surprisingly little serious questioning of the institutions of open-access 'communal lands' and the top-down public governance of wild resources. These quasi-feudal conditions dispossessed and disempowered local communities in the colonial period. They are being perpetuated as the phenomena I have labelled 'ungoverned spaces', locking forests, wildlife, and people in negative environmental, economic, and political feedback loops. This raises the intriguing question of whether institutions that provide security of person and property are the missing ingredients in the economic and political transformation of communities that depend on wild resources. Case studies where communities have received some of these rights certainly suggest that this is the case.

Rules governing wildlife are outdated

The institutions (or rules) governing wild life are outdated, and no longer fit for purpose. They were created in an earlier age when wildlife was plentiful and cheap, and agriculture was the order of the day. Consequently, they greatly favour domestic plants and animals, and tilt the economic playing field strongly against wild species, creating dualism and dysfunction. This plays out through a relatively prosperous, modern urban sector extracting resources from rural people in communal lands who struggle to feed themselves despite living with valuable resources. It also plays out in the political allocation of wild resources, with the centre imposing so-called solutions on rural people, and paying far more attention to special interest than the democratic rights and participation of local people.

The underlying problem is that of differential ownership. Rich people own land, but poor people do not. Domestic resources (i.e. domestic crops and livestock) are private, replacing wild resources that are not, so rural people have come to rely on low value livestock and crops because this is all they own regardless of their underlying economic value. They are prohibited from owning or utilising age-old wild resources, which are therefore neglected, slashed down, replaced, and traded 'illegally'. The institutions governing wild resources do not offer a sustainable path out of poverty – they undermine the ability of wild resources to compete for space, and they also undermine the capacities, rights, and justice of the people who live with these resources.

The sustainable governance approach

Forests and drylands and systems with similar characteristics (e.g. mountains, fisheries) are the subject of a global tragedy. The juxtaposition of rich biodiversity and poor people is jarring, and unsustainable. As we pass the centenary of the public governance of wild resources, it is clear that the rules and institutions that we are most comfortable with are not coping, and are often counter-productive. They create ungoverned spaces, with symptoms that include poverty, vulnerability, violence, economic inefficiency, and environmental destruction. Like a playgroup with no rules, conflict and chaos are avoidable if we are bold enough to apply sensible rules (before these systems are overwhelmed by the demographics of poor people).

Southern Africa's surging wildlife economy is far from accidental. It is the anticipated outcome of deliberate policies put in place by a small cohort of prescient wildlife administrators in the 1960s, and of a considerable effort, experimentation, and learning since then. The wisdom, and even the morality, of strategies that place the future of wildlife in the economic hands of the people who live with it is not widely appreciated. I have used this experience to make the case for the 'sustainable governance approach', a set of rules for governing wildlife (outside parks) and other wild resources on a crowded planet. The entitlement of communities with stronger rights to use, manage, and protect their resources is gaining traction as an essential prerequisite for sustainable forest or wildlife management (Libecap, 2009). Democratisation, similarly, is an essential ingredient of shared prosperity (North et al., 2009) and of virtuous political and economic processes that allow ordinary people to shape the rules that affect them at national or local level (Ostrom, 1990; Acemoglu & Robinson, 2012).

If the conservation literature neglects the centrality of property rights in valuing, allocating, and democratising wild resources, even more so does it miscomprehend and oppose the role of markets in promoting the value, sustainability, and the spread of wild resources. For example, many CBNRM programmes flounder despite good intentions because they do not grasp the importance of making wild resources as valuable as possible, or ensuring that 100% of value is retained locally. The knee-jerk acceptance of local subsistence uses, and the aversion to trade (e.g. 'demand reduction'), also suggest a profound misunderstanding of the process of wealth creation and development. We don't need to fiddle tentatively with utilising wild resources at the margins. We need to take on the challenge of making them the foundation of a sustainable bio-economy.

Starting with Theodore Roosevelt, most conservationists vilify the economic use of wild resources, even more so in a time when markets are gamed by elites to generate unfair profits (Stiglitz, 2002). In laying the blame on greedy exploiters, we have misdiagnosed the problem, which starts with the importance of property rights and exclusion for economic efficiency and social equity. Moreover, the misuse of markets should not distract us from the fact that markets, when properly designed, are the best, fairest, and most democratic way that humankind has found to allocate scarce resources. Thus, the sustainable governance approach emphasises the importance of trade or exchange for valorising wild resources, without forgetting that this is only effective when combined with rights of exclusion that are designed so that benefits accrue to the people who live with the resources. Without proprietorship and exchange, rural people, like the serfs of Middle Ages Europe, are locked into a small-scale world of vulnerability, drudgery, and poverty. Without proprietorship, and without easy access to markets, they produce many things but few efficiently. Romantic as it may seem, restricting local people to subsistence hunting and gathering (as opposed to legal rights and markets for wild life) is a bad thing. It forces rural people to consume a lot of wildlife in low-value uses, or to replace wildlife with domestic species.

This is basic economics. Adam Smith showed that wealth is created when people apply their labour and initiative to turning raw materials into more highly ordered goods and services that people want (Beinhocker, 2006). Productivity increases when individuals specialise in producing what they are good at, but specialisation requires trade. Just as Adams Smith's pin makers exchange pins for bread, fish, and implements, communities can only prosper from wildlife by trading high-value hunting, tourism, or even wildlife products, for food and school fees (rather than relying on low-value meat).

However, free markets as defined by Adam Smith need to be carefully designed, and laissez-faire economists and libertarians are in error when they assume that they simply emerge. Thus, my advice to bodies that govern wildlife use and trade, like national governments and CITES, is simple (if controversial). Automatically encourage use and trade if wildlife is owned by landholders and communities (and they get 100% of the benefits), and view use and trade outside these conditions with scepticism.

Internalising the costs and benefits of land use, and moving beyond markets and states, requires polyvalent governance (Wall, 2017) through a combination of privatisation and collective action. With proprietorship and price as the overarching principles (i.e. privatisation), systems of community self-regulation are also necessary to internalise and account for the externalities and idiosyncrasies of managing fugitive wild resources (Chapter 9, Child and Child, 2015). The new rules that we need for a crowded planet should (1) devolve proprietorship to landholders and communities; (2) encourage trade where this is the case; and (3) establish conditions that encourage collective action and self-regulation at local and sector level. Clear, simple rules are always preferable. However, the rules also need to be crafted carefully to encourage specific outcomes, such as large wild landscapes rather than semi-domestication of wild systems.[1]

The sustainable governance approach and CBNRM are bottom-up, market-based, and democratic approaches to the governance of conservation. They recognise that the way humankind interacts with nature is shaped by rules-of-use, including proprietorship, markets, and collective action. This is the new knowledge that we need to reform the system. The sciences of habitat and wildlife management are not transformative, but the demand for these skills will rise as more and more landholders switch away from commodity agriculture to intact habitats and the bio-experience economy. Training institutions have not yet adjusted to these needs, and conservation scientists often lack institutional and economic skills.

Community-based natural resource management

CBNRM is a broad concept (Kellert et al., 2000; Hulme & Murphree, 2001; Brosius et al., 2005; Berkes, 2007) that seeks to reconcile conservation with development, involve local people in natural resource conservation and benefit, devolve power to decentralised organisations, and promote indigenous values and knowledge (Shackleton et al., 2010). CBNRM emerged spontaneously across the globe as a response to demographic and economic pressure, poverty, and weaknesses in the top-down management of forestry, fisheries, and wildlife (Pomeroy, 1995). It emerged in different forms, and as a result of different socio-economic processes, with change being initiated at different levels of the economy (as defined by Williamson, 2000; Box 4.1), through social movements, administrative leadership, and pilot examples (Chapter 13). Forest communities in Mexico acquired land rights with the Mexican Revolution in the early 1900s, but have only recently gained commercial rights to their forests (Bray et al., 2006). In the Amazon, local and indigenous communities gained territorial rights, arising out of extensive social movements in response to threats to their land by in-migration (Schmink & Wood, 1992; Cronkleton et al., 2008). However, their economic rights to use fish, wildlife, and forest products are significantly truncated, reflecting the Rooseveltian aversion to market hunting, and in marked contrast to community wildlife management in southern Africa. In an excellent study, Pinheiro (2018) shows that stocks of the pirarucu, one of the world's largest freshwater fish, often improved under local management, with the important finding that the emergence of Ostrom-type rules was directly related to social capital. However, the evolution of local natural resource governance was constrained by arbitrary decisions at higher levels of bureaucracy, including restrictions on markets. Other well-known examples of CBNRM are community forestry in South Asia (Arnold, 2001; Birch et al., 2014) and fisheries in South East Asia (Allison et al., 2012; Cinnera et al., 2012). Interestingly, excellent examples of community-based management in developed countries, such as forest and game management in Norway, Spain, and Germany (personal communication with Ralph Baldus, Sandy McDonald, and Thor Larsen), are not prominent in the literature. While there are a lot of commonalities in CBNRM across continents and resources, CBNRM in southern Africa has the peculiarities that it was initiated by government wildlife administrators, and had a far greater emphasis on markets and economics than its sister programmes.

The literature contains many deep insights (Agrawal & Gibson, 1999), but despite considerable investment in community conservation globally, the concept is not well defined, ranging from park outreach, to co-management, to CBNRM itself (Barrow & Murphree, 2001). In practice, CBNRM has not fulfilled its theoretical potential because these underlying principles are reflected more in good intention and rhetoric than in real changes in institutions and controls. An enthusiastic beginning was followed by a literature critical of CBNRM, including scholarship opposed to combining economy with conservation (Reid, 2016). Agreeing with Murphree (2000) that '[CBNRM] has not been tried and found wanting; it has been found difficult and rarely tried', Reid suggests that many 'community-based' approaches were externally initiated and a veneer through which top-down management maintained control (Reid, 2016). This view that the credibility of CBNRM is being smeared by poor, or even fraudulent, application of the term to initiatives that in no way reflect its principles is shared by others (Ribot et al., 2010).

So what are these principles, and how do we know if an initiative can be genuinely defined as deserving of the label CBNRM? Working from the operational model of CBNRM provided in Chapter 13 (Figures 13.2 and 13.3), this suggests four questions, the first two reflecting the sustainable governance approach, and the latter two specifically concerned with micro-governance, scale, and adaptive learning:

1. Have the rights to wild resources been devolved to well-defined communities at the right (very local) scale? Do communities have a full suite of rights, including the rights of access, benefit, management, exclusion, and sale?
2. Have sufficient measures been taken to ensure that the value of wild resources are maximised and returned to the producer community? Have market restrictions been removed, and markets created, and do they get close to 100% of the economic value of the resource? Critically, do they get 100% of what wild resources are sold for financially?
3. Do the communities themselves practise inclusive governance at a face-to-face scale within a supportive enabling environment? Are the people affected by rules and decisions participating in making them?
4. Are social, institutional and organisational capital, and natural resource management systems, being strengthened through well-facilitated, participatory, and adaptive learning processes?

The second half of the book focuses on the third question – the theory and practice of micro-governance. It suggests that the principles and implementation of effective micro-governance are ubiquitous, because small rural communities the world over are surprisingly alike, reflecting our evolution as bands of hunter-gatherers. It also emphasises that social learning processes are critical to the adaptive development of local rules and practices (Figures 14.1 and 14.2), but does not delve too deeply into the underlying theory. This discussion on micro-governance draws heavily on the scholarship of Ostrom and Murphree. Moving beyond scholarship, it tackles the challenge of implementing CBNRM in less-than-ideal conditions where institutions are weak, centralised, and extractive, where local communities are economically and politically marginalised, and where social capital is low. This leads to a recategorisation of Ostrom's eight principles as being foundational (principles 1, 3, 7) and emergent (principles 2, 4, 5, 6), with principles 7 and 8 concerning cross-scale linkages. The model specifically adds a ninth (foundational) principle relating to getting the prices of wild resources right (Figure 14.1).

Externally, and following the sustainable governance approach, CBNRM establishes private-community ownership, with the full devolution of wild resources, coupled with measures to maximise the value of wild resources and return these in full to producer communities. Internally, effective micro-governance requires that (1) communities and their members (shareholders) are clearly defined at the right scale, and (2) that they practice face-to-face inclusive governance. Inclusive citizenship is elemental to CBNRM. An important contribution of this book is the differentiation between inclusive (community-based) and representational (committee-based) governance. I have provided empirical evidence that inclusive governance delivers more public goods (especially informed participation and equitable benefit sharing) by an order of magnitude, and more reliably, than representational forms of governance (Chapter 12).[2] Representative governance (i.e. centralisation at the local level) can certainly provide environmental gains and generate significant revenues. However, it perpetuates authoritarian practices and seldom avoids the scourge of elite capture, posing serious questions about social equity and long-term sustainability.

Zimbabwe's CAMPFIRE programme was pioneering because it transferred the sustainable governance approach from private land to rural communities. This provided many new lessons (Chapter 11). The revenue distribution processes revealed profound differences between treating the income from wildlife publically (i.e. only funding projects) or privately (i.e. allowing full choice, including cash). Similarly, CAMPFIRE's best performing communities invariably followed face-to-face modes of self-governance, rather than representational forms of administration. In Luangwa, participatory governance delivered more benefits, cleaner and more responsive decisions, and greater participation, by an order of magnitude, than representational governance, a conclusion supported by data from across the region (Chapter 12). This should not be surprising. Acemoglu and Robinson (2012) go so far as to suggest that the differences between successful and unsuccessful societies can be predicted on whether their governance is extractive or inclusive. Indeed, the differences between participatory and representational forms of governance are so large that participatory and representational CBNRM should be considered completely different animals.

I cannot help thinking that, having evolved in small groups for most of our existence, we are hardwired to work at this scale with high levels of internal accountability. It is when we lose this scale that we flounder as societies until we are able to develop the rich institutions and the checks and balances associated with Ancient Greece and the modern age of prosperity. Ostrom reflects this same conclusion when she states that 'most individuals affected by the operational rules can participate in modifying the operational rules', emphasising the rights of citizens to influence the governance of their institutions, the same conditions that Figure 2.1 suggests underlie modern prosperity. It is certainly true that poverty and environmental decline are associated with the absence of these institutional conditions in the ungoverned spaces of forests and drylands. The obvious suggestion is that CBNRM can play a major role in the reinstitutionalisation of ungoverned spaces, by replacing top-down and extractive governance with inclusive, local, impersonal rules. Without delving too deeply into the theory of state building, I nevertheless wonder if CBNRM might not have some equivalency to town hall democracy in weaving a deep norm of participation and accountability into the American psyche that bubbles up when formal and representational institutions falter.

No one is naïve enough to believe that inclusive communities will emerge organically, and certainly not at the pace that we need to cope with current rates of poverty and environmental destruction. Therefore, CBNRM invariably requires the external design, facilitation and imposition of rules for inclusive governance, plus monitoring and sanctions to ensure compliance with these rules. Imposing rules on a system predicated on participation is not as contradictory as it seems. CBNRM rules are specifically designed to foster inclusivity and to protect the rights of marginalised groups, including women, to participate in self-governance, against the strong headwinds of extractive and patriarchal norms. This, of course, is a controversial conclusion, and the norm is either to expect democratic systems to somehow bubble up in idyllic communities, or to respect 'traditional' cultures and norms of governance. My argument is that there is a deep legacy of authoritarianism and even abuse that needs to be proactively reversed.

Nascent CBNRM communities lack rules of use, adaptive monitoring, graduated sanctions, and even conflict resolution (Ostrom's principles 2, 4, 5, 6), which are emergent properties of inclusive governance. Therefore, CBNRM also requires quality learning processes to rebuild institutional, social, and managerial capital. The key to this lies in experiential learning, which can also develop technical skills and capacities for managing resources and enterprises, and create resourcefulness and community self-confidence. Thus, the quality and tone of CBNRM support services and facilitation, and an emphasis on enabling communities to decide and do things for themselves, are essential. As noted, Pinheiro (2018) measured a direct relationship between social capital, the application of Ostrom's principles, and the recovery of the pirarucu fishery in the Brazilian Amazon. By contrast, Musavengane and Simatele (2016) blame the poor performance of CBNRM in parts of South Africa on low levels of social capital. The rapid early progress of CAMPFIRE reflects not only sound policy but a well-managed processes of social learning (Chapter 11), a feature shared by the highly successful Namibian CBNRM programme which, for 25 years, benefitted from high-quality leadership and facilitation, excellent monitoring, and a deep commitment to local empowerment. Unfortunately, the capacity for facilitative learning and adaptive management is not an attribute for which most natural resource administrators in government and NGOs are known, nor is it a recognised quality of the project cycle.

At the field level there is a high probability that CBNRM can be successful if it follows the right principles (encapsulated, largely, by Ostrom's first six principles, as elaborated above). Therefore, we should be extremely intolerant of non-performance, and highly critical of projects and project managers that let communities down. The poor record of accomplishment of CBNRM is a symptom of a system that accepts shoddy management and design. CBNRM, like rocket science, follows a rigorous and conceptually principled process. Without high quality management that follows the laws of physics, most rockets would crash and burn. A shoddy rocket-building process doesn't mean that rockets don't work, only that the process of building the rocket has not met the necessary manufacturing standards. We can apply the same arguments to CBNRM.

That said, a significant challenge to CBNRM lies in the enabling environment and the devolution of rights and powers, which can be influenced but not controlled by CBNRM practitioners. Conceptually, these conditions are less well articulated, and are also more context specific, than principles internal to the community. Indeed, there is considerable resistance to devolution, hence the preponderance of CBNRM initiatives beset by 'aborted devolution' (Hulme & Murphree, 2001). While there seems to be a lot of confusion about what exactly is meant by decentralisation and CBNRM (Ribot et al., 2010), perhaps this ambiguity deliberately hides the necessity of fully devolving rights for wildlife to the people who live with it. CBNRM is simply privatisation at the community level, yet we deny communities' rights that we take for granted.

The objective of CBNRM is to locate the origin of rights, governance, benefits, and choice in the citizens that live with wildlife. These ideas are reflected in the principles of subsidiarity (Handy, 1994), and by statements such as 'think global, act local', and 'never globalise a problem if it can possibly be solved locally' (Hardin, 2001). Meadows (2008) advises us that 'hierarchies exist to serve the bottom layers, not the top' and then warns us that 'economic examples of over-control from the top, from companies to nations, are the causes of some of the great catastrophes of history'. Following these observations, Chapter 13 calls on Murphree (2000) to provide theoretical guidance for allocating functions across scale in hierarchical systems. He identified sequencing to be the key, first devolving proprietorship of wild resources to individual landholders and communities, with nested institutions then emerging through the process of delegated aggregation.

Scale and the roles of meso- and macro-governance

Having dealt extensively with devolution and micro-governance, I will briefly describe the functions of meso- and macro-level governance (Table 17.1). While, theoretically, the rights and functions of micro- and macro-governance are reasonably clear, there is no such theory for defining the roles of the messy-middle (meso-governance) where top-down and bottom-up systems often meet. Here, we draw on practical reality, to outline three primary activities that occur at the meso-level.

Chapter 8 described the potential for collective self-regulation by landholder units, giving the example of large wildlife conservancies and Zimbabwe's well-crafted Intensive Conservation Areas (ICAs) in southern Africa. In addition, land units often work collectively to build scale (for markets and landscapes) and to facilitate learning and political advocacy.

Third-party enforcement is also operationalised at the meso-level function, including the monitoring and sanctioning of governance compliance, and monitoring to ensure that land units are fulfilling their environmental responsibilities (e.g. environmental inspectorates).

Finally, this is the level at which extension and capacity-building links grounded action to centralised research. Especially for CBNRM, external assistance and support to design governance systems, build capacity, and provide extension and financing is usually located in field offices at the meso-level. We need to challenge the norm that as people gain skills and experience they should be steadily promoted upwards from the field to head office. Certainly, farmers live on the farm, and rocket laboratories should not be denied the highest quality engineers. Similarly, CBNRM is not best served through a traditional skills hierarchy. Inventing new technical and organisational processes, and fostering the trust and relationships through which innovation occurs, requires direct long-term partnerships between high-level professionals and communities, and challenges the norm that the 'cleverest' people should be in head office; they are desperately needed in the field, preferably full time. Moreover, it is good practice for policy-makers and head office administrators to have field responsibility for at least one project to keep them grounded and realistic. I would not exclude the desk officers for major donors from this conclusion.

TABLE 17.1 Practical guidance for allocation functions across hierarchical scale

Level	Organisations	Roles
Level 1 (Micro)	Land units: • private landholders • private-community landholders	• Resource proprietorship – the rights to access, use, manage, sell, and benefit from the resources, and to exclude others from doing so. • Inclusive governance within communities
Level 2 (Meso)	• Landholder associations • Government agencies • NGOs	Functions delegated upwards by land units including: • Collective action for self-regulation between land units • Collective action for building economies or ecologies of scale such as groups of land units operating together as a 'conservancy' • Cross-scale learning and political representation by land units and people who represent them (e.g. landholders associations) External regulation: • Monitoring and sanctioning of governance compliance • Inspection of environmental sustainability Support functions: • Capacity-building, support, and financing • Design of inclusive community governance
Level 3 (Macro)	• Government agencies • NGOs • Universities • Landholder associations	• Legitimate state functions – justice, policing, legislation, and some service delivery • Functions that are centralised because they have significant economies of scale – research and education, programme monitoring, and adaptive management • Political representation and advocacy by landholders and communities

There are also three sets of functions at the macro-level (Table 17.1). Following political theory, the state provides public goods including justice, policing, policy, and legislation. Inclusive governance requires that landholders and communities are represented at this level, and have the capacity to influence the rules and policies that affect them (e.g. through CBNRM Associations). Finally, there are considerable economies of scale associated with research and the monitoring and adaptive management of national CBNRM programmes, through associations of CBNRM support providers and universities.

Is CBNRM onerous?

CBNRM seeks to address poverty and environmental degradation in complex ecosystems, where it is difficult to define ecosystem processes let alone value them, and where the political and economic institutions are weak and dysfunctional. Developing new systems of devolved governance and local collective action can be daunting, and working at the local level can be frustrating, when the romantic notion of local participation is exposed to include petty squabbling and power plays. It has been said that the conditions for CBNRM are onerous (Shackleton et al., 2010), and the examples of success are certainly out-numbered by cases that are struggling to overcome these numerous challenges. So, why do we persist with CBNRM?

First, there really is no palatable alternative. The world's poorest people really do live in our most bio-diverse ecosystems, and we have no option but to create new alternatives for conservation and development.

Second, CBNRM is hard, especially if it is tackled in the wrong way. It deals simultaneously with two of humankind's most intractable problems: how to reduce poverty, and how to live in a sustainable balance within our environmental limits. It deals with natural resources that are mobile and difficult to define, measure, or manage, often spread over large areas and characterised by uncertain, complex, and non-linear processes. It deals with poor people, often with weak or dislocated social structures and lacking basic services such as clean water, healthcare, and education. It deals with the need for radical institutional change in marginal areas, where the overarching conditions for institutional transformation are often hostile. Further, the best and the brightest are usually concentrated in capital cities making money, rather than resolving tough problems like poverty and environmental decline.

Third, the fact that something is onerous does not mean that it is not worthwhile (though it does imply that the quick-fix strategy associated with development assistance projects are inappropriate). The conditions for democracy are challenging and onerous (Dahl, 1998). Yet millions of people who were exploited by previous systems characterised by slavery, authoritarianism, child labour, ill health, and early death even into the 20th century are doubtless better off today, with greater freedoms and rights and more economic opportunity. And free people with opportunity work harder and produce more. The statistics bear this out as Douglass North and his colleagues argue so lucidly: the 30 richest countries in the world are all democracies with the exception of four small oil nations and Singapore (North et al., 2009).

Fourth, where CBNRM is carefully designed and properly applied, albeit a situation that is far less frequent than we would like, we see definite and often remarkable progress in conservation, rural governance, and even poverty reduction. Effective CBNRM catalyses the collective energies of rural people in the way promoted by Amartyr Sen in his powerful argument in *Development as Freedom* (Sen, 1999) and as modern, successful organisations do (Peters & Waterman, 1982). We accept that these situations are rare, and that success is often partial, for instance emphasising conservation more than democratic governance. And we also note with concern the backsliding of CBNRM initiatives and community democracy, even when people have experienced an alternative and better future (Rihoy et al., 2007; Lubilo & Child, 2010). This is often a reflection of the recentralisation of authority and benefit ('aborted devolution'), and suggests that communities, on their own, cannot protect the conditions that enable the emergence of effective CBNRM.

Nonetheless, perhaps we overestimate the challenges of implementing CBNRM. Exam-ples of successful CBNRM show that it is not particularly onerous if the enabling conditions and design are right. CBNRM, done properly, can be rapid and rewarding, with substantial gains across several axes – democracy, livelihoods, environmental management, and social capital (Chapters 11 and 12). The CAMPFIRE programme shows that, once the enabling conditions were in place, progress was extremely rapid on a national scale. The Namibian example demonstrates the wisdom of quality technical leadership and long-term commitment. It is not enough to transform a broken system without recognising that new systems require long-term and even permanent support. In Namibia, this cost about $10 million a year, but within 12 years a system that had been broken was covering its costs, and after 25 years the economic return on investments was high and increasing. This implies that, globally, it will cost a lot to repair and recapitalise ungoverned spaces, but that the return on investment may be significant if this is done properly.

Thus, fifth, these early successes, and even more so the many failures, suggest that there is a set of principles and practices that, provided they are followed, ensure a reasonable or even a high probability of success. Where we come unstuck is that CBNRM is multi-faceted and involves natural resource use, pricing theory, property rights and collective action, governance, learning systems and adaptive management, political ecology, and so on. Each of these facets is, in and of itself, a complex and captivating subject into which the literature is delving ever more deeply. It is easy to become entrapped by this complexity, which undermines our ability to design balanced, workable systems. I have therefore provided a theory of change model for CBNRM, including suggestions of where to start, and what to emphasise.

However, CBNRM is not a 'wicked problem' that is difficult or impossible to solve because of contradictory and complex requirements. It only becomes a wicked problem if approached from the wrong starting assumptions. Indeed, there is an underlying simplicity to CBNRM based around a core principle: that governance, rights, and choice should originate in the people who coexist with wildlife, forests, fish, and biodiversity. Effective CBNRM requires that:

- Rights to resources are fully devolved to communities at the right scale.
- Wild resources deliver their true value to landholders and communities, including through global trade and the development of missing markets.
- Community rights are governed following the principles that all people affected by decisions participate in making them, and are well informed.
- Institutional, managerial, social, and technical capital are cultivated through a well-facilitated adaptive learning process.

Building systems upwards from this origin through a process of delegated aggregation (a crucial concept developed by Marshall Murphree) provides a high likelihood of alignment between many of the facets that contribute to CBNRM, and avoids the scale mismatches that lie at the heart of so many of our challenges in managing natural resources and our political economy. As noted in Chapter 2, individual discretionary choice and property rights lie at the root of the free market economy, and also at the root of the democratic process. This is why I have emphasised the similarities between CBNRM, free market conservation, and environmental justice.

Changing the game

Centralised conservation emerged when the greatest threat to wild life was frontier capitalism, leading to a reliance – and a lock-in – around public and non-market solutions that now extends to international agencies and NGOs. But how appropriate are these late-19th-century solutions to 21st-century threats? Can we afford to entrust complex environmental problems to centralised institutions, when the selective pressures acting on them are their ability to raise money, not audited field results? The examples of top-down governance we see every day do not give us confidence (Meadows, 2008). The US Congress is gridlocked in partisan rancour. We see a similar loss of democratic accountability and quality deliberation in global conservation institutions, where local people are absent but single-issue special interests are not. We are not going to solve these massive problems by spinning old ideas as innovations at the political centre. Solving the problems of poverty and environment requires the serious application of knowledge that we already have, complemented by the evolution of new approaches through experiential learning. Evolution, in turn, requires that powerful selective pressures (e.g. rigorous performance assessment) act upon a diversity of approaches. The global conservation hegemony unfortunately lacks all these attributes, including performance accountability and a willingness to experiment with bold decentralisation.

The centralised governance of wild resources that is so ingrained in our psyches is taking us down a false trail that is perpetuated through ideology and our inability to think slowly, dispassionately, and in systems about emotional issues like wildlife conservation. We are addicted to strategies that don't work, yet don't refresh themselves (Table 17.2). For example, the early game laws lacked economic coherence, and often lacked social legitimacy outside of the ruling class. Without social legitimacy, the long arm of the law was unsuccessful in protecting wildlife. The primary exception is North America's model of socialised conservation that builds on a deep history of public benefits and public agency accountability. Without economically coherent policies that internalise the values of wildlife to landholders who are invariably deterministic of land use outcomes, high-value wildlife will continue to be replaced by low-value domestic species, and a green bio-experience economy will not supplant an unsustainable agro-extractive one.

A caveat and a challenge

My personal experience is with the reinstitutionalisation of wildlife in southern Africa, now in its seventh decade of providing lessons and learning experiences linked to the returning of rights to landholders and communities, the development of markets for wildlife, and the design of governance structures at all levels. I have used this example because of its demonstrated success, and also its reversals which provide lessons that are just as important.

My focus on wildlife and African communities should not detract from the value of these lessons for community fisheries or forestry governance. While the management of natural resources is specific, principles relating to governance, economics, and people are in all likelihood transferable. Indeed, having worked with Bantu, Bushmen, and European farmers, and advised students from Asia and Latin America, I am increasingly convinced that at the local level people are people and the principles and practice of micro-governance are widely applicable – after all, we were all living in bands of hunter gathers only 10 000 years ago. By contrast, each country has a unique culture, history, and politics. This implies that the principles of micro-governance are ubiquitous, but the process of establishing supportive macro-governance (and perhaps even meso-governance) is highly contextual.

TABLE 17.2 Conservation addictions and their antidotes

	Public	*Devolved*
Rules of the game	• Centralised management (public trust doctrine) • Paternalistic and even authoritarian	• Highly democratic • Local proprietorship • Bottom-up democratic governance and self-regulation (privatisation and CBNRM)
Innovation	• Vested interests prevent introspection and change	• Thousands of landholders and communities driving innovation
Mode of thinking	• Doctrinal rather than reflective • Blue-print planning	• Decentralised policy experimentation • Collaborative adaptive management
Value	• Public • Subsistence, low-value uses	• Private and local • Value increased through trade and globalisation
Outcomes	• Does little to address poverty • Wild species replaced by domestic species • Over-exploited because of weak, open-access property regimes	• Creates synergies between poverty reduction and conservation • Wildlife is 'produced' because its values are internalised • Promotes economic efficiency by allocating resources to their highest valued uses

When telling this story, I am regularly challenged by the observation that African wildlife is a charismatic and valuable resource that provides a powerful economic tool for transforming local governance and livelihoods. It is true that wildlife is a particularly apt tool for leveraging change, not least because it was a public resource with few locally entrenched interests blocking the introduction of innovative systems of community governance. It is also true that it is now valuable. This was not always the case. To those people who suggest that CBNRM succeeded in southern Africa only because of the high value of wildlife, and that the approaches offered within this book are therefore not applicable, I issue this challenge. When we embarked on the pathway that eventually led to CBNRM in the 1950s, it was far from a done deal that wildlife had value. Indeed, livestock was replacing wildlife for the very reason that it was worthless. Like today's forests and wetlands, we were convinced that intact land under wildlife was more valuable, economically speaking, than the alternative. Nonetheless, wildlife did not automatically become a competitive land use. This took nearly seven decades of proactive policy reform and thoughtful economic innovation, such as gradually easing up the price of hunting and changing the norm that hunting was a right to acceptance that it was a privilege that had to be paid for (Child, 1995). Not least, new policies deliberately unlocked the entrepreneurial juices of well over 10 000 private enterprises and landholders who, to this day, are inventing new ways, new products, and new experiences that make more money out of less wildlife in the way the public and NGO sectors can never do. Areas with charismatic species are commercially viable in southern Africa, but not in West Africa, while places like Java that still have elephants, rhinos, and tigers do not have a commercial model to sustain them. It is no accident that the range and quality of wildlife products and hospitality services in southern Africa exceeds those in, say, Kenya, or India, or Indonesia, and even those in North America. Indeed, southern African entrepreneurs are gradually introducing products like high quality tented camps and guiding services to countries as widely dispersed as Congo, Egypt, India, and the USA. This raises the question of whether non-African wildlife, other species like crocodiles and turtles, tropical forests, carbon and ecosystem services, and so on might not also become economically competitive if similar processes were applied to them.

Finally, it is my hope that the examples and principles in this book will assist and inspire those who care about rural people, wildlife, and their environments. I hope these ideas will increase the probability of their success in an environment that is often politically and ideologically weighted against rural people and sustainable use but, where, nevertheless, there are many moments of inspiration and breakthrough. I also hope that, in putting down some of these principles and practices, I will hold to account those who approach CBNRM with naivety, insufficient preparation, cynicism, or even as charlatans seeking selfish gain. If I reduce the space for badly designed and non-performing projects, and for efforts that sell themselves as CBNRM but give CBNRM a bad name, these efforts will be worthwhile. CBNRM is not for the faint hearted, but neither is it unreachable. There are no quick fixes but, properly done, effective CBNRM is attainable and worth fighting for. A planet where rural people govern themselves and their wildlife efficiently and sustainably will be a better place for us all.

Notes

1   Chapter 8 compares the Zimbabwean and South African legislation. By using collective action to manage ownership and externalities Zimbabwe built large wild landscapes, whereas South Africa's reliance on game fencing has resulted in landscape fragmentation. Likewise, subtle differences in the rules for governing the rhino trade (if we ever get there) will have massive implications for rhinos and landscapes. The financial and psychological costs of protecting rhinos is now so high that they are unlikely to survive except where landholders are financed to cover the high costs of protection, or can trade rhinos to cover these costs (Rubino & Pienaar, 2018; personal observation). Because philanthropy is fickle and unsustainable, the survival of rhinos probably depends on trade. However, subtle differences in the rules for trading rhinos will have major implications for landscapes. Simply allowing trade is highly likely to conserve rhinos, but through intensive rhino farming which is hardly the goal. By contrast, well-crafted policy could deliberately leverage the high value of rhinos to transform and rewild landscapes if, for example, the rights to trade them are made particularly easy if they are free-ranging and wild on areas exceeding 10 000 hectares (Child, 2012). Similarly, the rhino trade is so lucrative that it could be deliberately designed to benefit poor people. Rhinos are like the sheep with the Golden Fleece that keeps on giving; clipping off the horn every few years as it regrows is no different from shearing a sheep and, literally, provides a substance worth more than gold. A coherent wildlife economy would apply this gold to the protection of rhinos, poverty reduction, and the rewilding of Africa.

2   While this is a small dataset, to me it is very convincing. Nonetheless, comparative studies of inclusive versus representational community governance and scale require more academic attention.

REFERENCES

Acemoglu, D. & Robinson, J. 2012. *Why Nations Fail: The Origins of Power, Prosperity, and Poverty*, New York, Crown Business.

Agrawal, A. 2005a. *Environmentality: Technologies of Government and the Making of Subjects*, Durham, NC, Duke University Press Books.

Agrawal, A. 2005b. *'Environmentality': Community, Intimate Government, and the Making of Environmental Subjects in Kumaon, India*, Durham, NC, Duke University Press.

Agrawal, A. & Gibson, C. 1999. Enchantment and disenchantment: The role of community in natural resource conservation. *World Development*, 27, 629–649.

Alden Wily, L. 2009. Africa's big question. Can the continent find solutions to its colonial landownership legacy? *Tropical Forest Update*, 19, 10–12.

Allison, E. H., Ratner, B. D., Asgard, B., Willmann, R., Pomeroy, R. & Kurien, J. 2012. Rights-based fisheries governance: From fishing rights to human rights. *Fish and Fisheries*, 13(1), 14–29.

Anderson, J., Mehta, S. & Schwarz, J. 2013. The rich get richer and the poor get committees? Institutional arrangements and requirements for local forest management in Africa. *Nature & Faune*, 27, 26–30.

Anderson, T. L. & Leal, D. 1997. *Enviro-Capitalism. Doing Good by Doing Well*, Plymouth, Rowman & Littlefield.

Anon. 2011. A global perspective on the anthropocene. *Science*, 334, 34–35.

Aragonès, E. & Sánchez-Pagés, S. 2009. A theory of participatory democracy based on the real case of Porto Alegre. *European Economic Review*, 53, 56–72.

Arnett, E. B. & Southwick, R. 2015. Economic and social benefits of hunting in North America. *International Journal of Environmental Studies*, 72, 732–745.

Arnold, J. E. M. 2001. *25 Years of Community Forestry*, Rome, Food and Agricultural Organization of the United Nations.

Arnstein, S. R. 1969. A ladder of citizen participation. *American Institute of Planners Journal*, 35, 216–224.

Aryal, A., Dhakal, M., Panthi, S., Prasad Yadav, B., Shrestha, U., Bencini, R., Raubenheimer, D. & Ji, W. 2015. Is trophy hunting of bharal (blue sheep) and Himalayan tahr contributing to their conservation in Nepal? *Hystrix. The Italian Journal of Mammology*, 26(2), 85–88.

Astle, W. L. 1999. *A History of Wildlife Conservation and Management in the Mid-Luangwa Valley, Zambia*, Bristol, British Empire and Commonwealth Museum.

Balakrishna, M. & Ndlovu, D. 1992. Wildlife utilization and local people: A case study in Upper Lupande Game Management area, Zambia. *Environmental Conservation*, 19, 135–144.

Barnes, G. & Child, B. (eds.) 2014. *Adaptive Cross-Scalar Governance of Natural Resources*, London, Earthscan.

Barnes, J. & Jones, B. 2009. Game Ranching in Namibia. In: H. Suich & B. Child (eds.) *Evolution & Innovation in Wildlife Conservation*, London, Earthscan, 113–126.

Barrett, C. B. & Arcese, P. 1995. Are integrated conservation-development projects (ICDPs) sustainable? On the conservation of large mammals in sub-Saharan Africa. *World Development*, 23, 1073–1084.

Barrow, E. & Murphree, M. 2001. Community conservation. From concept to practice. In: D. Hulme & M. Murphree (eds.) *African Wildlife and Livelihoods. The Promise and Performance of Community Conservation*, Oxford, James Currey, 24–37.

Batterbury, S. P. J. & Fernando, J. L. 2006. Rescaling governance and the impacts of political and environmental decentralization: An introduction. *World Development*, 34, 1851–1863.

Beadle, C. J. & Macdonald, J. A. 1969. *R. v. Moresby-White*. Criminal law – Wild life conservation act [Chapter 199]–S. 41 – Construction of right of private landowner to kill animals on property discussed. *Rhodesian Law Reports*, 484–490.

Beinart, W. 1984. Soil erosion, conservationism and ideas about development: A Southern African exploration 1900–1960. *Journal of South African Studies*, 11, 52–83.

Beinhocker, E. D. 2006. *The Origin of Wealth. Evolution, Complexity and the Radical Remaking of Economics*, Boston, Harvard Business School Press.

Belay, M. 2012. *Participatory Mapping, Learning and Change in the Context of Biocultural Diversity and Resilience*. Ph.D., Rhodes University.

Berkes, F. 2007. Community-based conservation in a globalized world. *PNAS*, 104(39), 15188–15193.

Besley, T., Pande, R. & Rao, V. 2005. Participatory democracy in action: Survey evidence from South India. *Journal of the European Economic Association*, 3, 648–657.

Bethell, T. 1998. *The Noblest Triumph. Property and Prosperity through the Ages*, New York, St. Martin's Press.

Bigalke, R. C. 1966. Some thoughts on game farming. *Proceedings of the Annual Congresses of the Grassland Society of Southern Africa*, 1, 95–102.

Birch, J. C., Thapa, I., Balmford, A., Bradbury, R. B., Brown, C., Butchart, S. H. M., Gurung, H., Hughes, F. M. R., Mulligan, M., Pandeya, B., Peh, K. S.-H., Stattersfield, A. J., Walpole, M. & Thoma, S. H. L. 2014. What benefits do community forests provide, and to whom? A rapid assessment of ecosystem services from a Himalayan Forest, Nepal. *Ecosystem Services*, 8, 118–127.

Blair, H. 2000. Participation and accountability at the periphery: Democratic local governance in six countries. *World Development*, 28, 21–39.

Bond, I. 1994. *Importance of Elephant Hunting to CAMPFIRE Revenue in Zimbabwe*, Harare, WWF- Southern African Regional Programme Office.

Bond, I. 2001. CAMPFIRE and the incentives for institutional change. In: D. Hulme & M. W. Murphree (eds.) *African Wildlife & Livelihoods. The Promise and Performance of Community Conservation*, Oxford, James Currey, 227–243.

Bond, I. & Cumming, D. H. M. 2006. Wildlife research and development. In: M. Rukuni, P. Tawonezvi & C. Eicher (eds.) *Zimbabwea Agricultural Revolution Revisited*, Harare, University of Zimbabwe Publications, 465–496.

Bond, I., Grieg-Gran, M., Wertz-Kanounnikoff, S., Hazelwood, P., Wunder, S. & Angelsen, A. 2009. *Incentives to Sustain Forest Ecosystem Services: A Review and Lessons for REDD*, London, International Institute for Environment and Development.

Booth, V. 2002. *Analysis of Wildlife Markets (Sport Hunting and Tourism)*, Harare, WWF Southarn African regional Programme Office.

Booth, V. R. 2009. *A Comparison of the Prices of Hunting Tourism in Southern and Eastern Africa*, Budapest, CIC – International Council for Game and Wildlife Conservation and FAO – Food and Agriculture Organization of the United Nations.

Borgerhoff Mulder, M. & Coppolillo, P. 2005. *Conservation. Linking Ecology, Economics, and Culture*, Princeton, Princeton University Press.

Börner, J., Wunder, S., Wertz-Kanounnikoff, S., Tito, M. R., Pereira, L. & Nascimento, N. 2010. Direct conservation payments in the Brazilian Amazon: Scope and equity implications. *Ecological Economics*, 69, 1272–1282.

Borrini-Feyerabend, G., Dudley, N., Jaeger, T., Lassen, B., Pathak Broome, N., Phillips, A. & Sandwith, T. 2013. *Governance of Protected Areas. From Understanding to Action*, Gland, Switzerland, IUCN.

Boulding, C. & Wampler, B. 2010. Voice, votes, and resources: Evaluating the effect of participatory democracy on well-being. *World Development*, 38, 125–135.

Bowles, S. & Choi, J.-K. 2013. Coevolution of farming and private property during the early Holocene. *Proceedings of the National Academy of Sciences of the United States of America*, 110, 8830–8835.

Brandon, K. E. & Wells, M. 1992. Planning for people and parks: Design dilemmas. *World Development*, 20(4), 557–570.

Bray, D. B., Antinori, C. & Torres-Rojo, J. M. 2006. The Mexican model of community forest management: The role of agrarian policy, forest policy and entrepreneurial organization. *Forest Policy and Economics*, 8, 470–484.

Bray, D. B., Merino-Perez, L. & Barry, D. (eds.). 2005. *The Community Forestry of Mexico. Managing for Sustainable Landscapes*, Austin, TX, University of Texas Press.

Brinkley, D. 2009. *Wilderness Warrior. Theodore Roosevelt and the Crusade for America*, New York, HarperCollins.

Brockington, D. & Igoe, J. 2006. Eviction for conservation: A global overview. *Conservation and Society*, 4, 424–470.

Bromwich, M. 2014. *National Parks and Wildlife Management. Rhodesia and Zimbabwe 1928–1990. An Historical and Anecdotal Account by Those Who Served*, China, Michael Bromwich, SA Media Services South Africa and WKT Co. Ltd.

Brooks, J. S., Waylen, K. A. & Mulder, M. B. 2012. How national context, project design, and local community characteristics influence success in community-based conservation projects. *PNAS*, 109 (52), 21265–21270.

Brosius, J. P., Tsing, A. L. & Zerner, C. (eds.). 2005. *Communities and Conservation. Histories and Politics of Community-Based Natural Resource Management*, Lanham MD, AltaMira Press.

Brown, K. & Rosendo, S. 2000. The institutional architecture of extractive reserves in Rondijnia, Brazil. *The Geographical Journal*, 166, 35–48.

Bruntland, G. H. 1987. *Our Common Future, World Commission on Environment and Development*, Oxford, Oxford University Press.

Bukamuri, B. B., Manjengwa, J. M. & Anstey, S. (eds.). 2009. *Beyond proprietorship: Murphree's Laws on Community-Based Natural Resource Management in Southern Africa*, Harare, Weaver Press.

Burnes, B. 2004. Kurt Lewin and the planned approach to change: A re-appraisal. *Journal of Management Studies*, 41, 977–1002.

BWPA. 2005. *The Botswana Game Ranching Handbook*, Gabarone, Botswana Wildlife Producers' Association.

Byamugisha, F. F. K. 2013. *Securing Africa's Land for Shared Prosperity*, Washington, DC, World Bank.

Campbell, A. & Child, G. 1971. The impact of man on the environment of Botswana. *Botswana Notes and Records*, 3, 91–110.

CAMPFIRE-Association. 2016. *The Role of Trophy Hunting of Elephant in Support of the Zimbabwe Campfire Program*, Harare, CAMPFIRE Association.

Carruthers, J. 1989. Creating a National Park 1910–1926. *Journal of Southern African Studies*, 15, 188–216.

Carruthers, J. 1995. *The Kruger National Park. A Social and Political History*, Pietermaritzburg, University of Natal Press.

Carruthers, J. 2008. 'Wilding the farm or farming the wild?' The evolution of scientific game ranching in South Africa from the 1960s to the present. *Transactions of the Royal Society of South Africa*, 63, 160–181.

Castello, L., Viana, J. O. P., Watkins, G., Pinedo-Vasquez, M. & Luzadis, V. A. 2009. Lessons from integrating fishers of Arapaima in small-scale fisheries management at the Mamiraua' Reserve, Amazon. *Environmental Management*, 43, 197–209. Castello et al. 2009 Fisheries Mamiraua Amazon.

CBD. 2004. *Addis Ababa Principles and Guidelines for the Sustainable Use of Biodiversity*, Montreal, Secretariat of the Convention on Biological Diversity.

Chabal, P. & Daloz, J. P. 1999. *Africa Works: The Political Instumentalization of Disorder*, Oxford, James Currey.

Chadwick, D. H. 1996. A place for parks in the new South Africa. *National Geographic*, 190, 2–41.

Chambers, R. 1983. *Rural Development: Putting the Last First*, Oxford, Pearson Education Limited.

Chambers, R. 1994. The origins and practice of participatory rural appraisal. *World Development*, 22, 953–969.

Charnley, S. & Poe, M. R. 2007. Community forestry in theory and practice: Where are we now? *Annual Review of Anthropology*, 36(1), 301–336.

Chase, A. 1987. *Playing God in Yellowstone: The Destruction of America's First National Park*, Orlando, FL, Harcourt Brace & Co.

Chayes, S. 2017. Kleptocracy in America. Corruption is reshaping governments everywhere. *Foreign Affairs*, 96, 142–150.

Chidakel, A. & Child, B. in preparation. *The economic impact of Kruger National Park and the surrounding reserves: a policy brief*, Gainesville, University of Florida.

Chidakel, A. & Child, B. (in review). Evaluating the economic impacts of park-based tourism using a grounded approach for South Luangwa National Park in Zambia.

Child, B. 1988. *The Role of Wildlife Utilization in the Sustainable Economic Development of Semi-Arid Rangelands in Zimbabwe*. D.Phil., University of Oxford.

Child, B. 1995. *A Summary of the Marketing of Trophy Quotas in CAMPFIRE Areas 1990-1993*, Harare, Department of National Parks and Wildlife Management, Zimbabwe.

Child, B. (ed.). 2004. *Parks in Transition. Biodiversity, Rural Development and the Bottom Line*, London, Earthscan.

Child, B. 2012. The sustainable use approach could save South Africa's rhinos. *South African Journal of Science*, 108, 1–4.

Child, B., Gorsevski, V., Chidakel, A. & Souza, T. in preparation. *Assessing the Socio-Economic Impacts of GEF-Supported Terrestrial Protected Areas. A STAP Guidance Document*, Washington, DC, Scientific and Technical Advisory Panel of the Global Environmental Facility.

Child, B. & Jones, B. 2006. *Practical tools for community conservation in southern Africa: Special edition. Participatory Learning and Action*, Participatory Learning and Action 55, London, International Institute for Environment and Development.

Child, B., Jones, B., Mazamban, I., Mlalazi, A. & Moinuddin, H. 2003. *Final Evaluation Report: Zimbabwe Natural Resources Management Program – USAID/Zimbabwe Strategic Objective No. 1. CAMPFIRE Communal Areas Management Programme for Indigenous Resources*, Harare, USAID.

Child, B., Mupeta, P., Muyengwa, S. & Lubilo, R. 2014. Community-based natural resource management: Micro-governance and face-to-face participatory democracy. In: R. W. Merle Sowman (ed.) *Governance for Justice and Environmental Sustainability. Lessons across Natural Resource Sectors in Sub-Saharan Africa*, London, Earthscan, 156–179.

Child, B. & Murphree, M. 2004. *Principles and Criteria for Evaluating the Effectiveness of Community Institutions and Capacity for Managing Natural Resources at an Ecosystem Level*, Building Community Institutions and Capacity for Managing Natural Resources at an Ecosystem Level (CICENRM), Washington DC, World Bank.

Child, B., Musengezi, J., Parent, G. & Child, G. 2012. The economics and institutional economics of wildlife on private land in Africa. *Pastoralism Journal*, 2(18), 1–32.

Child, B., Muyengwa, S., Lubilo, R. & Mupeta-Muyamwa, P. 2014. Using the governance dashboard to measure, understand and change micro-governance. In: G. Barnes & B. Child (eds.) *Adaptive Cross-Scale Governance of Natural Resources*, London, Earthscan, 203–237.

Child, B. & Peterson, J. 1991. CAMPFIRE in rural development. The Beitbridge experience. Branch of terrestrial ecology, Department of National Parks and Wildlife Management & Centre for Applied Social Studies, University of Zimbabwe.

Child, B., Ward, S. & Tavengwa, T. 1997. *Zimbabwe's CAMPFIRE Programme: Natural Resource Management by the People*, Harare, Canon Press.

Child, B. & Weaver, C. 2006. Marketing hunting and tourism joint ventures in community areas. *Participatory Learning and Action*, 55, 37–44.

Child, B. & Wojcik, D. 2014. *Developing Capacity for Community Governance of Natural Resources: Theory & Practice*, Bloomington, AuthorHouse.

Child, G. 1968. *Report to the Government of Botswana on an Ecological Survey of North-Eastern Botswana*, Rome, Italy, FAO.

Child, G. 1995. *Wildlife and People: The Zimbabwean Success. How the Conflict between Animals and People Became Progress for Both*, Harare, Wisdom Foundation.

Child, G. 2009. The growth of park conservation in Botswana. In: H. Suich & B. Child (eds.), *Evolution and Innovation in Wildlife Conservation*, London, Earthscan, 51–66. Click here to enter

Child, G. & Child, B. 2015. The conservation movement in Zimbabwe: An early experiment in devolved community based regulation. *Southern African Journal of Wildlife Research*, 45(1), 1–16.

Child, G. F. T. & Riney, T. 1987. Tsetse control hunting in Zimbabwe, 1919–1958. *Zambezia*, 14, 11–72.

Churchill, W. 1908. *My African Journey*, London, Hodder and Stoughton.

Cinnera, J. E., Basurtob, X., Fidelmana, P., Kuangec, J., Laharic, R. & Mukminin, A. 2012. Institutional designs of customary fisheries management arrangements in Indonesia, Papua New Guinea, and Mexico. *Marine Policy*, 36, 278–285.

Cleaver, F. 2005. The inequality of social capital and the reproduction of chronic poverty. *World Development*, 33, 893–906.

Coase, R. H. 1937. The nature of the firm. *Economica*, 4(16),386–405.

Coase, R. H. 1960. The problem of social cost. *The Journal of Law and Economics*, 3(1), 1–44.

Coe, M. J., Cumming, D. H. M. & Phillipson, J. 1976. Biomass and production of large African herbivores in relation to rainfall and primary production. *Oecologia*, 22(4), 341–354.

Cooney, R., Roe, D., Dublin, H., Phelps, J., Wilkie, D., Keane, A., Travers, H., Skinner, D., Challender, D. W. S., Allan, J. R. & Biggs, D. 2017. From poachers to protectors: Engaging local communities in solutions to illegal wildlife trade. *Conservation Letters*, 10, 367–374.

Costanza, R., D'arge, R., Groot, R. D., Farber, S., Grasso, M., Hannon, B., Limburg, K., Naeem, S., O'neill, R. V., Paruelo, J., Raskin, R. G., Sutton, P. & Belt, M. V. D. 1997. The value of the world's ecosystem services and natural capital. *Nature*, 387, 253–260.

Costanza, R., D'arge, R., Groot, R. D., Farber, S., Grasso, M., Hannon, B., Limburg, K., Naeem, S., O'Neill, R. V., Paruelo, J., Raskin, R. G., Sutton, P. & Belt, M. V. D. 1998. The value of the world's ecosystem services and natural capital. *Ecological Economics*, 25, 3–15.

Cousins, J. A., Sadler, J. P. & Evans, J. 2008. Exploring the role of private wildlife ranching as a conservation tool in South Africa: Stakeholder perspectives. *Ecology and Society*, 13, 43 online.

Covey, S. R. 2004. *The 7 Habits of Highly Effective People*, New York, Free Press.

Cox, M, Arnold, G. & Tomás, S. V. 2010. A review of design principles for community-based natural resource management. *Ecology and Society*, 15, www.ecologyandsociety.org/vol15/iss4/art38/.

Craigie, I. D., Baillie, J. E. M., Balmford, A., Carbone, C., Collen, B., Green, R. E. & Hutton, J. M. 2010. Large mammal population declines in Africa's protected areas. *Biological Conservation*, 143, 2221–2228.

Cronkleton, P., Taylor, P. L., Barry, D., Stone-Jovicich, S. & Schmink, M. 2008. *Environmental Governance and the Emergence of Forest-Based Social Movements*, Bogor, Indonesia, Center for International Forestry Research.

Crook, R. C. 2003. Decentralisation and poverty reduction in Africa: The politics of local-central relations. *Public Administration and Development*, 23, 77–88.

Crook, R. C. & Manor, J. 1998. *Democracy and Decentralisation in South Asia and West Africa: Participation, Accountability and Performance*, Cambridge, Cambridge University Press.

Cumming, D. H. M. 1982. The influence of large herbivores on savanna structure in Africa. In: B. J. Huntley & B. H. Walker (eds.) *Ecology of Tropical Savannas*, Berlin Heidelberg New York, Springer, 217–245.

Cumming, D. H. M. & Bond, I. 1991. Animal production in southern Africa: Present practices and opportunities for peasant farmers in arid lands. *Multispecies Animal Production Systems Project Project Paper No. 22*, Harare, WWF Multispecies Project.

Cumming, D. H. M., Fenton, M. B., Rautenbach, I. L., Taylor, R. D., Cumming, G. S., Cumming, M. S., Dunlop, J. M., Ford, G. S., Hovorka, M. D., Johnston, D. S., Kalcounis, M. C., Mahlanga, Z. & Portfors, C. V. 1997. Elephants, woodlands and biodiversity in Miombo woodland in southern Africa. *South African Journal of Science*, 93, 231–236.

Dahl, R. A. 1989. *Democracy and Its Critics*, New Haven, CT, Yale University Press.

Dahl, R. 1998. *On Democracy*, New Haven, CT, Yale University Press.

Dalal-Clayton, B. & Child, B. 2003. *Lessons from Luangwa. The Story of the Luangwa Integrated Resource Development Project, Zambia*, London, International Institute for Environment and Development.

Daly, H. 2005. Economics in a full world. *Scientific American*, 293, 100–107.

Dasmann, R. F. & Mossman, A. S. 1961. *Commercial Utilization of Game Animals on a Rhodesian Ranch*, Salisbury, National Museums.

Davies, J., Poulsen, L., Schulte-Herbrüggen, B., Mackinnon, K., Crawhall, N., Henwood, W. D., Dudley, N., Smith, J. & Gudka, M. 2012. *Conserving Dryland Biodiversity*, Nairobi, Kenya, International Union for Conservation of Nature and Natural Resources, United Nations Environment Programme- World Conservation Monitoring Programme (UNEP-WCMC), and United Nations Convention to Combat Desertification (UNCCD).

de Soto, H. 2000. *The Mystery of Capital. Why Capitalism Triumphs in the West and Fails Everywhere Else*, New York, Basic Books.

DEA. 2015. *Situation Analysis of Four Selected Sub-Sectors of the Biodiversity and Conservation Sector in South Africa, and Transformation Framework*, Pretoria, Department of Environmental Affairs, South Africa.

Deepak, R. 2016. The elasticity of global cropland with respect to crop production and its implications for peak cropland. *Environmental Research Letters*, 11, 114016.

Demsetz, H. 1967. Towards a theory of property. *American Economic Review*, 57, 347–359.

Dennis A. Rondinelli, G. S. C. (ed.) 2003. *Reinventing Government for the Twenty-First Century. State Capacity in a Globalizing Society*, Bloomfield, CT, Kumarian Press.

Department-for-Community-Development. 2006. *Contemporary Literature on Capacity Building and the Strengths Perspective and Good Practice Wisdom for the Capacity Building Strategic Framework 2005 to 2007*, Perth, Government of Western Australia, Department for Community Development.

Derman, W. 1990. *The Unsettling of the Zambezi Valley: An Examination of the Mid-Zambezi Rural Development Project*, Centre for Applied Social Studies, University of Zimbabwe Working Paper.

Diamond, J. 2002. Evolution, consequences and future of plant and animal domestication. *Nature*, 418, 700.

Diamond, J. 2005. *Collapse: How Societies Choose to Fail or Succeed*, New York, Viking Press.

Dickson, B., Hutton, J. & Adams, W. M. (eds.). 2009. *Recreational Hunting, Conservation and Rural Livelihoods*, Oxford, Wiley-Blackwell.

Dillon, P. 2006. *The Last Revolution: 1688 and the Creation of the Modern World*, London, Thistle Publishing.

Dinda, S. 2004. Environmental Kuznets curve hypothesis: A survey. *Ecological Economics*, 49, 431–455.

Ding, H., Veit, P., Gray, E., Reytar, K., Altamirano, J., Blackman, A. & Hodgdon, B. 2016. *Climate Benefits, Tenure Costs. The Economic Case for Securing Indigenous Land Rights in the Amazon*, Washington, World Resources Institute.

DNPWLM. 1991. *Guidelines for CAMPFIRE*, Harare, Department of National Parks and Wildlife Management, Zimbabwe.

Donnall, T. E. 2010. *How Sportsmen Saved the World: The Unsung Conservation Efforts of Hunters and Anglers*, Guilford, CT, Globe Pequot Press.

Dressler, W., Scher, B. B., Schoon, M., Brockington, D., Hayes, T., Kul, C. A., Mccarthy, J. & Shrestha, K. 2010. From hope to crisis and back again? A critical history of the global CBNRM narrative. *Environmental Conservation*, 37, 5–15.

Drucker, P. 1973. *Management: Tasks, Responsibilities, Practices*, New York, HarperCollins Publishers.

Dry, G. 2010. Why game farming should be taken seriously. *Farmer's Weekly*, 14 May, 5–6.

Dry, G. 2011. Commerical wildlife ranching's contribution to a resource efficient, low carbon, pro employment green economy. *7th International Wildlife Ranching Symposium*, Kimberly, South Africa.

Duchelle, A. E., Cromberg, M., Gebara, M. F., Guerra, R., Melo, T., Larson, A., Cronkleton, P., Börner, J., Sills, E., Wunder, S., Bauch, S., May, P., Selaya, G. & Sunderlin, W. D. 2014. Linking forest tenure reform, environmental compliance, and incentives: Lessons from REDD+ initiatives in the Brazilian Amazon. *World Development*, 55, 53–67.

Duffy, R. 2000. *Killing for Conservation. Wildlife Policy in Zimbabwe*, Oxford, James Curry.

Duffy, R. 2016. War, by conservation. *Geoforum*, 69, 238–248.

Dunbar, R. I. M. 1993. Coevolution of neocortical size, group size and language in humans. *Behavioral and Brain Sciences*, 16, 681–735.

Dunbar, R. I. M. 1998. The social brain hypothesis. *Evolutionary Anthropology*, 6(5), 178–190.

du Toit, J. G. 2007. *Role of the Private Sector in the Wildlife Industry*, Pretoria, Wildlife Ranching SA.

du Toit, J. T. & Cumming, D. H. M. 1999. Functional significance of ungulate diversity in African savannas and the ecological implications of the spread of pastoralism. *Biodiversity and Conservation*, 8, 1643–1661.

Eguren, A. & Sprague, A. 2014. *Baseline Livelihood Report for Makuleke Community, South Africa*. Masters in Sustainable Development Practice Practicum, University of Florida.

Ellis, R. C. 2011. Anthropogenic tranformation of the terrestrial biosphere. *Philosophical Transactions of the Royal Society*, 369, 1010–1035.

Emerton, L. 1999. The nature of benefits and the benefits of nature: Why wildlife conservation has not economically benefitted communities in Africa. *Community Conservation Research in Africa: Principles and Comparative Practice*, Manchester, Institute for Development Policy and Management, University of Manchester.

Fabricius, C., Kock, E., Magome, H. & Turner, S. 2004. *Rights, Resources & Rural Development. Community-Based Natural Resource Management in Southern Africa*, London, Earthscan.

FAO. 2009. *FAOSTAT Agriculture*. www.fao.org/corp/statistics/en/.

FAO. 2016. *Global Forest Resources Assessment 2015. How Are the World's Forests Changing?* Rome, Food and Agriculture Organization of the United Nations (FAO).

Feeny, D., Berkes, F., Mccay, B. J. & Acheson, J. M. 1990. The tragedy of the commons: Twenty-two years later. *Human Ecology*, 18, 1–19.

Feld, L. P. & Kirchgässner, G. 2000. Direct democracy, political culture, and the outcome of economic policy: A report on the Swiss experience. *European Journal of Political Economy*, 16, 287–306.

Fitter, R. S. R. & Scott, S. P. 1978. *The Penitent Butchers. The Fauna Preservation Society, 1903–1978*, Reading, Fauna Preservation Society.

Foley, J. A., Ramankutty, N., Brauman, K. A., Cassidy, E. S., Gerber, J. S., Johnston, M., Mueller, N. D., O'Connell, C., Ray, D. K., West, P. C., Balzer, C., Bennett, E. M., Carpenter, S. R., Hill, J., Monfreda, C., Polasky, S., Rockström, J., Sheehan, J., Siebert, S., Tilman, D. & Zaks, D. P. M. 2011. Solutions for a cultivated planet. *Nature*, 478, 337.

Franks, P. & Small, R. 2016. *Social Assessment for Protected Areas (SAPA). Methodology Manual for SAPA Facilitators*, London, IIED.

Frizen, S. A. 2007. Can the design of community-driven development reduce the risk of elite capture? Evidence from Indonesia. *World Development*, 35, 1359–1375.

Frost, P. G. H. & Bond, I. 2008. The CAMPFIRE programme in Zimbabwe: Payments for wildlife services. *Ecological Economics*, 65, 776–787.

Fuglie, K. O., Macdonald, J. M. & Ball, E. 2007. *Productivity Growth in U.S. Agriculture*, Economic Brief No. 9, Washington, DC, U.S. Dept. of Agriculture (USDA), Economic Research Service.

Fukuyama, F. 1992. *The End of History and the Last Man*, New York, Free press.

Fukuyama, F. 2011. *The Origins of Political Order: From Prehuman Times to the French Revolution*, New York, Farrar, Straus & Giroux.

Gambiza, J. & Nyama, C. 2006. *Country Parture/Forage Report, Zimbabwe*, Rome, FAO.

Gari, L. 2006. A history of the hima conservation system. *Environment and History*, 12, 213–228.

Gee, A. 2018. This land is your land. Rotting cabins, closed trails: Why we're shining a light on US National Parks. *The Guardian, International Edition*, 29 January.

Gerson, M. 2015. Myths, meaning and Homo sapiens. *The Washington Post*, 11 June.

Gibson, C. C. 1999. *Politicians and Poachers. The Political Economy of Wildlife Policy in Africa*, Cambridge, Cambridge University Press.

Gibson, C. C. & Marks, S. A. 1995. Transforming rural hunters into conservationists: An assessment of community-based wildlife management programs in Africa. *World Development*, 23, 941–957.

Goldewijk, K. K., Beusen, A. & Janssen, P. 2010. Long-term dynamic modeling of global population and built-up area in a spatially explicit way: HYDE 3.1. *Holocene*, 20, 565–573.

Goredema, L., Bond, I. & Taylor, R. 2006. Building capacity for local-level management through participatory technology development. *Participatory Learning and Action*, 55, 30–36.

Grainger, J. & Llewellyn, O. 1994. Sustainable use: Lessons from a cultural tradition in Saudi Arabia, *Parks: The International Journal for Protected Areas*.

Grindle, M. S. 2007. *Going Local. Decentralization, Democratization and the Promise of Good Governance*, Princeton, NJ, Princeton University Press.

Grindle, M. S. & Thomas, J. W. 1991. *Public Choices and Policy Change. The Political Economy of Reform in Developing Countries*, Baltimore and London, The Johns Hopkins University Press.

Gruber, J. S. 2010. Key principles of community-based natural resource management: A synthesis and interpretation of identified effective approaches for managing the commons. *Environmental Management*, 45, 52–66.

Guhrs, T., Rihoy, L. & Guhrs, M. 2006. Using theatre in participatory environmental policy making. *Participatory Learning and Action*, 55, 87–93.

Gunderson, L. H. & Holling, C. S. 2002. *Panarchy. Understanding Transformations in Human and Natural Systems*, Washington, DC, Island Press.

Hallmann, C. A., Sorg, M., Jongejans, E., Siepel, H., Hofland, N., Schwan, H., Stenmans, W., Müller, A., Sumser, H., Hörren, T., Goulson, D. & de Kroon, H. 2017. More than 75 percent decline over 27 years in total flying insect biomass in protected areas. *PLOS One*.

Handy, C. 1994. *The Empty Raincoat. Making Sense of the Future*, London, Arrow Books Limited.

Hanks, J. 2006. *Mitigation of Human-Elephant Conflict in the Kavango-Zambezi Transfrontier Area, with Particular Reference to the Use of Chilli Peppers*, Cape Town, Conservation International.

Harari, Y. 2014. *Sapiens: A Brief History of Humankind*, London, Random House.

Harari, Y. 2014. *Sapiens: A Brief History of Humankind*, London, Vintage.

Hardin, G. 2001. There is no global population problem. *The Social Contract*, 7(1), 48–57.

Hardin, G. J. 1968. The tragedy of the commons. *Science*, 162, 1243–1248.

Hardin, G. J. 1971. The tragedy of the commons. *Science*, 162, 1243–1248.

Hatcher, J. & Bailey, L. 2009. *Tropical Forest Tenure Assessment. Trends, Challenges and Opportunities*, ITTO Technical Series #37, Washington, DC, International Tropical Timber Organization Rights and Resources Initiative.

Hayek, F. A. 1944. *The Road to Serfdom*, Chicago, University of Chicago Press.

Heffelfinger, J. R., Geist, V. & Wishart, W. 2015. The role of hunting in North American wildlife conservation. *International Journal of Environmental Studies*, 70, 399–413.

Heijnsbergen, V. P. 1997. *International Legal Protection of Wild Fauna and Flora*, Amsterdam, OIS Press.

Hess, C. & Ostrom, E. 2007. *Understanding Knowledge as a Commons: From Theory to Practice*, Cambridge, MA, MIT Press.

Hill, K. A. 1996. Zimbabwe's wildlife utilization programs: Grassroots democracy or an extension of state power? *African Studies Review*, 39, 103–123.

Hobbes, T. 1651. *Leviathan*, England.

Hochschild, A. 1999. *King Leopold's Ghost: A Story of Greed, Terror, and Heroism in Colonial Africa*, New York, Houghton Mifflin Harcourt.

Holling, C. S. 1973. Resilience and stability of ecological systems. *Annual Review of Ecology and Systematics*, 4, 1–23.

Hulme, D. & Murphree, M. 2001a. *African Wildlife & Livelihoods. The Promise and Performance of Community Conservation*, Oxford, James Currey.

Hulme, D. & Murphree, M. 2001b. Community conservation in Africa. An introduction. In: D. Hulme & M. Murphree (eds.) *African Wildlife & Livelihoods. The Promise and Performance of Community Conservation*, Oxford, James Currey, 1–37.

Huntley, B. J. & Redford, K. H. 2014. *Mainstreaming biodiversity in pratice. STAP Advisory Document*, Washington, Global Environment Facility.

HURID. 2002. *Policy and Legislation Review of the Fisheries, Forestry, Wildlife and Water Sectors vis-à-vis Community Based Natural Resource Management*. The Institute of Human Rights, Intellectual Property and Development Trust (HURID) for CONASA (Community-based Natural Resource management and Sustainable Agriculture Project, USAID, Zambia).

Hutton, J., Adams, W. & Murombedzi, J. 2005. Back to the barriers? Changing narratives in biodiversity conservation. *Forum for Development Studies*, 2, 341–370.

Hutton, J. & Dickson, B. 2000. *Endangered Species, Threatened Convention: The Past, Present and Future of CITES*, London, Earthscan.

Hutton, J. & Webb, G. 2003. Crocodiles: Legal trade snaps back. In: S. Oldfield (ed.) *The Trade in Wildlife. Regulation for Conservation*, London, Earthscan, 108–120.

Hyden, G. 2006. *African Politics in Comparative Perspective*, Cambridge, Cambridge University Press.

Hyden, G. & Court, J. 2002. Governance and development. *World Governance Survey Discussion Paper 1*, United Nations University.

Hyden, G., Court, J. & Mease, K. 2003. Governance and development. *World Governance Survey Discussion Paper 1*. London, ODI. www.odi.org/publications/3135-making-sense-governance-need-involving-local-stakeholders.

IPBES. 2018. Worsening worldwide land degradation now 'critical', undermining well-being of 3.2 billion people.

IPBES. 2018. Media Release: Worsening worldwide land degradation now 'critical', undermining well-being of 3.2 billion people. *Issued by the IPBES secretariat on 23 March*, www.ipbes.net/news/media-release-worsening-worldwide-land-degradation-now-%E2%80%98critical%E2%80%99-undermining-well-being-32.

IUCN. (ed.). 1963. *Conservation of Nature and Natural Resources in Modern African States*, Morges, Switzerland, International Union for the Conservation of Nature and Natural Resources.

IUCN. 1980. *The World Conservation Strategy*, Geneva, International Union for Conservation of Nature and Natural Resources, United Nations Environment Programme, World Wildlife Fund.

Jacobsen, T. 2012. Rhino and Vicuna: A parallel, unpublished MS.

Jansen, D. J., Bond, I. & Child, B. 1992. *Cattle, Wildlife, Both or Neither? A Survey of Commercial Ranches in the Semi-Arid Regions of Zimbabwe*, Harare, WWF Multispecies Animal Production Project.

Johnstone, P. A. 1971. Evaluation of a Rhodesian game Ranch. Part 1: Operations, productivity and economic appraisal over six years 1967–72. *International Symposium on the Behaviour of Ungulates and its Relation to Management*, University of Calgary, Alberta, IUCN Publications.

Jones, B. & Murphree, M. 2001. The evolution of policy on community conservation in Namibia and Zimbabwe. In: D. Hulme & M. Murphree (eds.) *African Wildlife & Livelihoods. The Promise and Performacne of Community Conservation*, Oxford, James Curry, 38–58.

Jones, B. & Weaver, C. 2009. CBNRM in Namibia: Growth, trends, lessons and constraints. In: H. Suich & B. Child (eds.) *Evolution & Innovation in Wildlife Conservation*, London, Earthscan, 223–242.

Jones, B. T. B. 1999. Policy lessons from the evolution of a community-based approach to wildlife management, Kunene Region, Namibia. *Journal of International Development*, 11, 295–304.

Kabiri, N. 2010. Historical and contemporary struggles for a local wildlife governance regime in Kenya. In: F. Nelson (ed.) *Community Rights, Conservation & Contested Land. The Politics of Natural Resource Governance in Africa*, London, Earthscan, 121–144.

Kachel, S. M., McCarthy, K. P., McCarthy, T. M. & Oshurmamadov, N. 2016. Investigating the potential impact of trophy hunting of wild ungulates on snow leopard Panthera uncia conservation in Tajikistan. *Oryx*, 51(4), 597–604.

Kahneman, D. 2011. *Thinking, Fast and Slow*, New York, Farrar, Straus and Giroux.

Kellert, S. R., Mehta, J. N., Ebbin, S. A. & Lichtenfeld, L. L. 2000. Community natural resource management: Promise, rhetoric, and reality. *Society and Natural Resources*, 13, 705–715.

Korten, D. C. 1980. Community organization and rural development: A learning process approach. *Public Administration Review*, 40, 480–511.

Kotter, J. 1996. *Leading Change*, Boston, MA, Harvard Business School Press.

Koziell, I. & Inoue, C. Y. A. 2006. Mamiraua sustainable development reserve. Lessons learnt in integrating conservation with poverty reduction. In: *Biodiversity and Livelihoods Issues No. 7*, London, IIED.

Krausman, P. & Mahoney, S. P. 2015. How the Boone and Crocket Club (B&C) shared North American conservation. *International Journal of Environmental Studies*, 72, 746–755.

Krutilla, J. V. 1967. Conservation reconsidered. *The American Economic Review*, 57, 777–786.

Lang, D. J., Wiek, A., Bergmann, M., Stauffacher, M., Martens, P., Moll, P., Swilling, M. & Thomas, C. J. 2012. Transdisciplinary research in sustainability science: Practice, principles, and challenges. *Sustainability Science*, 7, 25–43.

Lange, G.-M., Barnes, J. I. & Motinga, D. J. 1997. *Cattle Numbers, Biomass, Productivity, and Land Degradation in the Commercial Farming Sector of Namibia, 1915 to 1995*, Windhoek, Namibia, Directorate of Environmental Affairs, Ministry of Environment and Tourism, research discussion paper, number 17.

Langholz, J. A. & Kerley, G. I. H. 2006. *Combining Conservation and Development on Private Lands: An Assessment of Ecotourism-Based Private Game Reserves in the Eastern Cape*. Center for African Conservation Ecology, Nelson Mandela Metropolitan University.

Larson, A. & Ribot, J. 2004. Democratic decentralisation through a natural resource lens: An introduction. *The European Journal of Development Research*, 16, 1–1.

Larson, A. M. & Soto, F. 2008. Decentralization of natural resource governance regimes. *Annual Review of Environment and Resources*, 33, 213–239.

Larsson, G. 1991. *Land Registration and Cadastral Systems: Tools for Land Information Management*, New York, John Wiley.

Lewin, K. 1958. Group decision and social change. In: E. E. Maccoby, T. M. Newcomb & E. L. Hartley (eds.) New York, Holt, Rinehart and Winston, 197–211.

Libecap, G. D. 2009. The tragedy of the commons: Property rights and markets as solutions to resource and environmental problems. *Australian Journal of Agricultural and Resource Economics*, 53, 129–144.

Lindsey, P., Du Toit, R., Pole, A. & Romanach, S. 2009. Save valley conservancy: A large-scale African experiment in cooperative wildlife management. In: H. Such & B. Child (eds.) *Evolution & Innovation in Wildlife Conservation*, London, Earthscan, 163–184.

Lindsey, P., Havemann, C. P., Lines, R. M., Price, A. E., Retief, T. A., Rhebergen, T., Waal, C. V. D. & Romanach, S. S. 2011. Benefits of wildlife-based land uses on private lands in Namibia and limitations affecting their development. *Oryx*, 47, 41–53.

Lindsey, P. A., Barnes, J., Nyirenda, V., Tambling, C. & Taylor, W. A. 2012. *The Zambian Game Ranching Industry: Scale, Associated Benefits, and Limitations Affecting Its Development*, Lusaka: A study commissioned by the Wildlife Producers Association of Zambia and funded by Livestock Services Cooperative Society Limited.

Lindsey, P. A., Nyirenda, V. R., Barnes, J. I., Becker, M. S., McRobb, R., Tambling, C. J., Taylor, W. A., Watson, F. G. & T'sas-Rolfes, M. 2014. Underperformance of African protected area networks and the case for new conservation models: Insights from Zambia. *PLoS ONE*, 9.

Lomborg, B. 2001. *The Skeptical Environmentalist*, Cambridge, Cambridge University Press.

Lubilo, R. 2007. A comprehensive report on the status of governance and participation in Lupande Game Management area of South Luangwa area Management unit in Zambia.

Lubilo, R. 2011. *An Assessment of the Impact of Devolved Governance in Natural Resources Management: A Case Study of Wuparo Conservancy, Namibia*, MSc thesis, University of Kent, UK.

Lubilo, R. & Child, B. (eds.). 2010. *The Rise and Fall of Community-Based Natural Resource Management in Zambia's Luangwa Valley: An Illustration of Micro- and Macro-Governance Issues*, London, Earthscan.

Lymbery, P. 2017. *Dead Zone: Where the Wild Things Were*, London, Bloomsbury.

MacArthur, R. H. & Wilson, E. O. 1967. *The Theory of Island Biogeography*, Princeton, Princeton University Press.

Mahoney, S. P. & Jackson, J. J. I. 2015. Enshrining hunting as a foundation for conservation: The North American model. *International Journal of Environmental Studies*, 70, 448–459.

Malenga, G. 2004. *Audit of 8 Community Resource Boards in Mumbwa, Namwala and Kafue Flats Game Management Areas. Period Covering 1 January 2000 to 31 August 2004*, Lusaka, Zambia, CBNRM-Mumbwa project (Danida).

Maliaoa, R. J., Pomeroy, R. S. & Turingana, R. G. 2009. Performance of community-based coastal resource management (CBCRM) programs in the Philippines: A meta-analysis. *Marine Policy*, 33, 818–825.

Mapedza, E. & Bond, I. 2006. Political deadlock and devolved wildlife management in Zimbabwe. The case of Nenyunga ward. *The Journal of Environment and Development*, 15, 407–427.

Marquetti, A., Da Silva, C. E. S. & Campbell, A. 2011. Participatory economic democracy in action: Participatory budgeting in Porto Alegre, 1989–2004. *Review of Radical Political Economics*, 44, 62–81.

Martin, G. 2012. *Game Changer. Animal Rights and the Fate of Africa's Wildlife*, Berkeley, Los Angeles, University of California Press.

Martin, R. 1981. Sebungwe Region: a planning framework. Harare, Zimbabwe, Joint Report. Department of National Parks and Wildlife Management, Department of Physical Planning.

Martin, R. 1989. *The Status of Projects Involving Wildlife in Rural Development in Zimbabwe 1989 Report*, Harare, Zimbabwe, Department of National Parks and Wildlife Management.

Martin, R. 1999. Adaptive management. The only tool for decentralised systems. *Paper presented to the Conference on the Ecosystems Approach for Sustainable Use of Biological Diversity*, Trondheim, Norway.

Martin, R. 2009a. *From Sustainable Use to Sustainable Development. Evolving Concepts of Natural Resource Management*, IUCN – Southern African Sustainable Use Specialist Group, Harare, 55.

Martin, R. 2009b. Murphree's laws, principles, rules & definitions. In: B. B. Mukamuri, J. M. Manjengwa & S. Anstey (eds.) *Beyond Proprietorship. Murphree's Laws on Community-Based Natural Resource Management in Southern Africa*, Harare, Weaver Press, 7–28.

Martin, R. B. 1984. *Communal Areas Management Programme for Indigenous Resources (CAMPFIRE)*, Harare, Department of National Parks and Wildlife Management, Zimbabwe.

Martin, R. B. 1986. *Communal Areas Management Programme for Indigenous Resorces (CAMPFIRE)*, Harare: Branch of Terrestrial Ecology, Department of National Parks and Wildlife Management, Zimbabwe.

Martin, R. B. 1994. *Resolving the Conflict between People and Parks in Zimbabwe*, Harare: Branch of Terrestrial Ecology, Department of National Parks and Wildlife Management.

Martin, R. B. 2003. Conditions for effective, stable and equitable conservation at the national level in Southern Africa. *World Parks Congress Workshop 'Local Communities, Equity and Protected Areas'*, Durban, South Africa. IUCN Commission on Environmental, Economic and Social Policy (CEESP), 23.

Martin, R. B., Craig, G. C., Booth, V. R. & Conybeare, A. M. G. 1992. *Elephant Mangement in Zimbabwe. A Review Compiled by Department of National Parks and Wild Life Management, Zimbabwe*, Harare, Zimbabwe, Department of National Parks and Wild Life Management.

Maslow, A. H. 1943. A theory of human motivation. *Psychological Review*, 50, 350–396.

Matema, S. & Andersson, J. A. 2015. Why are lions killing us? Human–Wildlife conflict and social discontent in Mbire District, northern Zimbabwe. *The Journal of Modern African Studies*, 53, 93–120.

Matheka, R. M. 2008. Decolonisation and wildlife conservation in Kenya, 1958–68. *The Journal of Imperial and Commonwealth History*, 36, 615–639.

Mavah, G. A. 2015. *Integrating Rural People Into Natural Resources and Wildlife Management in the Republic of the Congo*. Ph.D, University of Florida.

Mazambani, D. & Dembetembe, P. 2010. Community based natural resource management stocktaking assessment. Zimbabwe Profile. United States Agency for International Development.

Mbaiwa, J. E. 2017. Effects of the safari hunting tourism ban on rural livelihoods and wildlife conservation in Northern Botswana. *South African Geographical Journal*, DOI: http://dx.doi.org/10.1080/03736245.2017.1299639.

McCarty, C., Killworth, P. D., Bernard, H. R., Johnsen, E. C. & Shelley, G. A. 2001. Comparing two methods for estimating network size. *Human Organization*, 60, 28–39.

McGregor, D. 1960. *The Human Side of Enterprise*, New York, McGraw-Hill.

Mckean, M. A. 2000. Common property: What is it, what is it good for, and what makes it work? In: C. C. Gibson, M. A. Mckean & E. Ostrom (eds.) *People and Forests: Communities, Institutions, and Governance*, Cambridge, MA, The MIT Press, 27–55.

Mckean, M. A. 2000. Common property: What is it, what is it good for, and what makes it work? In: C. C. Gibson, M. A. Mckean & E. Ostrom (eds.) *People and Forests: Communities, Institutions, and Governance*, Cambridge, MA, The MIT Press, 75–101.

McShane, T. O. & Wells, M. P. 2004. *Getting Biodiversity Projects to Work. Towards More Effective Conservation and Development*, New York, Columbia University Press.

MEA. 2005. *Millennium Ecosystem Assessment. Ecosystems and Their Services*, Washington, DC, Island Press.

Meadows, D. H. 2008. *Thinking Is Systems. A Primer*, London, Earthscan.

Measham, T. G. & Lumbasi, J. A. 2013. Success factors for community-based natural resource management (CBNRM): Lessons from Kenya and Australia. *Environmental Management*, 52, 649–659.

Menard, C. & Shirley, M. M. 2011. *The Contribution of Douglass North to New Institutional Economics.* halshs-00624297.

Menard, C. & Shirley, M. M. 2014. The contribution of Douglass North to new institutional economics. In: S. Galani & I. Sened (eds.), *Institutions, Property Rights and Economic Growth: The Legacy of Douglass North*, Cambridge, Cambridge University Press, 11–29.

Merz, L. 2014. *Situational Analysis of Mangalane, Mozambique for a Community Based Natural Resource Management Program*. Masters, University of Florida.

Metcalfe, S. 1993a. CAMPFIRE - Zimbabwe's communal areas management programme for indigenous resources. In: D. Western, M. Wright & S. Strum (eds.) *Natural Connections: Perspectives in Community-Based Conservation*, Washington, DC, Island Press, 161–192.

Metcalfe, S. 1993b. Rural development and biodiversity: Prospects for wildlife habitat on communal land in Zimbabwe's Zambezi Valley. *Biodiversity in Practice Symposium held by South African Wildlife Management Association*, Port Elizabeth, South Africa.

Micklethwait, J. & Wooldridge, A. 1996. *The Witch Doctors: Making Sense of the Management Gurus*, London, William Heinemann.

Milledge, S. 2007. *Forestry, Governance and National Development: Lessons Learned from a Logging Boom in Southern Tanzania*, Dar es Salaam, Tanzania, TRAFFIC East/Southern Africa/Ministry of Natural Resources and Tourism/Tanzania Development Partners' Group.

Minister-of-Environment-and-Tourism. 1992. *Policy for Wild Life*, Zimbabwe, Department of National Parks and Wildlife Management, Ministry of Environment and Tourism.

Monbiot, G. 2017. Insectageddon: Farming is more catastrophic than climate breakdown. *Guardian*, 20 October.

Monke, E. A. & Pearson, S. R. 1989. *The Policy Analysis Matrix for Agricultural Development*, Ithaka, Cornell University Press.

Morgan-Brown, T. 2014. *Governance and Incentive Structures for Reducing Emissions from Deforestation and Degradation (REDD) in Tanzania*. Ph.D., University of Florida.

Mossman, A. S. & Mossman, S. L. 1976. *Wildlife Utilization and Game Ranching: Report on a Study of Recent Progress in This Field in Southern Africa*, Morges, Switzerland, National Union for Conservation of Nature and Natural Resources.

Muir, J. 1912. *The Yosemite*, New York, The Century Co.

Mulindahabi, F. 2017. *Assessment of the Impacts of the Conservation of Protected Areas to the Improvement of Livelihoods of Adjacent Communities of the Nyungwe National Park, Rwanda*. Masters, University of Florida.

Mupeta, P. 2008. Decentralization, devolution & democratic governance in community based natural resource management (CBNRM): Exploring the concepts, principles and practices. Independent Study with Professor Goran Hyden, Department of Political Sciences, University of Florida.

Murombedzi, J. 1991. Decentralizing common property resources management: A case study of the Nyaminyami District council of Zimbabwe's Wildlife Management Programme. *Paper prepared for the Common Property Conference*, University of Manitoba-Winnipeg, September 26-29, 1991.

Murombedzi, J. 2014. National and transnational land grabs in Africa: Implications for local resource governance. In: G. Barnes & B. Child (eds.) *Adaptive Cross-Scale Governance of Natural Resources*, Abingdon, UK, Earthscan from Routledge, 75–101.

Murphree, M. 1991. *Communities as Institutions for Resource Management*, Harare, Zimbabwe, Centre for Applied Social Sciences, University of Zimbabwe.

Murphree, M. 1993. Communities as resource management institutions. *International Institute for Environment and Development, GATEKEEPER SERIES*, 14p.

Murphree, M. 1994a. Communities as resource management institutions. *International Institute for Environment and Development, Gatekeeper Series*.

Murphree, M. 1994b. The role of institutions in community-based conservation. In: D. Western, R. M. Wright & S. C. Strum (eds.) *Natural Connections. Perspectives in Community-Based Conservation*, Washington, D.C., Island Press.

Murphree, M. 1995. Optimal principles and pragmatic strategies: Creating an enabling politico legal environment for community based natural resource management. In: E. Rihoy (ed.) *The Commons without the Tragedy. Strategies for Community Based Natural Resource Management in Southern Africa. Proceedings of the Regional Natural Resource Management Programme Annual Conference*, Mowana Lodge, Kasane, SADC Wildlife technical Coordination Unit, 47–52.

Murphree, M. 1997a. Articulating voices from the commons, interpretation, translation and facilitation: Role and modes for common property scholarship. *Society and Natural Resources*, 10, 415–421.

Murphree, M. 1997b. *Congruent Objectives, Competing Interests and Strategic Compromise. Concept and Process in the Evolution of Zimbabwe's CAMPFIRE Programme*, Manchester, U.K, Institute for Development Policy and Management (IDPM), University of Manchester.

Murphree, M. 1999. Enhacing sustainable use. Incentives, politics and science. *Berkeley Workshop on Environmental Politics, Working Paper 99-2*, Berkeley, CA, Institute of International Studies, University of California, Berkeley, 12.

Murphree, M. 1999a. *Enhacing Sustainable Use*, Incentives, Politics and Science.

Murphree, M. 2000. Constituting the commons: Crafting sustainable commons in the new millennium. *'Multiple Boundaries, Borders and Scale' at the Eighth Biennial Conference of the International Association for the Study of Common Property (IASCP)*, Bloomington, Indiana, U.S.A.

Murphree, M. 2001. Experiments with the future. *A prologue to a seminar on an interdisciplinary, longitudinal and interactive methodology to explore environmental and institutional sustainability in the human use of nature*, University of California at Berkeley.

Murphree, M. 2005. Congruent objectives, competing interests, and strategic compromise: Concept and process in the evolution of Zimbabwe's CAMPFIRE, 1984-1996. In: J. P. Brosius, A. L. Tsing & C. Zerner (eds.) *Communities and Conservation. Histories and Politicis of Community-Based Natural Resource Management*, Oxford UK, Rowman & Littlefield, 105–148.

Murphree, M. W. 1989. *Research on the Institutional Contexts of Wildlife Utilization in Communal Areas of Eastern and Southern Africa*, Harare, Centre for Applied Social Studies, University of Zimbabwe.

Murphree, M. W. 2000. Community-based conservation: Old ways, new myths and enduring challenges. *Conference on African Wildlife Management in the New Millenium*, College of African Wildlife Management, Mweka, Tanzania, 13–15 December 2000.

Murphree, M. W. 2002. Protected areas and the commons. *The Common Property Resource Digest*, 60, 1–3.

Murphree, M. W. 2004. Communal approaches to natural resource management in Africa: From whence to where? *Breslauer Symposium on Natural resource Issues in Africa*, University of California, Berkeley.

Murphree, M. W. 2009. The strategic pillars of communal natural resource management: Benefit, empowerment and conservation. *Biodiversity and Conservation*, 18, 2551–2562.

Murrell, P. 2017. Design and evolution in institutional development: The insignificance of the English bill of rights. *Journal of Comparative Economics*, 45, 36–55.

Musavengane, R. & Simatele, D. M. 2016. Community-based natural resource management: The role of social capital in collaborative environmental management of tribal resources in KwaZulu-Natal, South Africa. *Development Southern Africa*, 33(6), 806–821.

Musengezi, J. 2010. *Wildlife Utilization on Private Land: Understanding the Economics of Game Ranching in South Africa*. Ph.D., University of Florida.

Mutandwa, E. & Gadzirayi, C. 2007. Impact of community-based approaches to wildlife management: Case study of the CAMPFIRE programme in Zimbabwe. *International Journal of Sustainable Development & World Ecology*, 14, 336–344.

Muyengwa, S. & Child, B. 2017. Re-assertion of elite control in Masoka's wildlife program, Zimbabwe. *Journal of Sustainable Development*, 10(6), 28–40.

Muyengwa, S., Child, B. & Mupeta, P. 2014. *Baseline Study for the Kasonso and Mulobezi Game Management Areas (GMAS) in Zambia*, Lusaka, Zambia, The Nature Conservancy (TNC) and Zambia Community Based Natural Resource Management Forum (ZCBNRM).

Muyengwa, S., Child, B., Mupeta, P. & Clough, L. 2014. *Baseline Study for the Kasonso and Mulobezi Game Management Areas (Gmas) in Zambia*, Lusaka, Zambia, The Nature Conservancy (TNC) and Zambia Community Based Natural Resource Management Forum (ZCBNRM).

Myers, N., Mittermeier, R. A., Mittermeier, C. G., Fonseca, G. A. B. D. & Kent, J. 2000. Biodiversity hotspots for conservation priorities. *Nature*, 403, 853–858.

NACSO. 2015. *The State of Community Conservation in Namibia – A Review of Communal Conservancies, Community Forests and Other CBNRM Initiatives (2015 Annual Report)*, Windhoek, Namibian Association of CBNRM Support Organisations (NACSO).

NACSO. 2016. *The State of Community Conservation in Namibia*, Windhoek, Namibia, Namibian Association of CBNRM SUpport Organizations.

NACSO. 2018. *The State of Community Conservation in Namibia. Annual Report 2017*, Windhoek, Namibian Association of CBNRM Support Providers, 84.

Naidoo, R., Weaver, C., Diggle, R. W., Matongo, G., Stuart-Hill, G. & Thouless, C. 2016. Complementary benefits of tourism and hunting to communal conservancies in Namibia. *Conservation Biology*, 30, 628–638.

Nakashima, D. J., Galloway Mclean, K., Thulstrup, H. D., Ramos Castillo, A. & Rubis, J. T. 2012. *Weathering Uncertainty: Traditional Knowledge for Climate Change Assessment and Adaptation*, Paris, France and Darwin, Australia, UNESCO and UNU.

Nelson, F. (ed.). 2010. *Community Rights, Conservation and Contested Land: The Politics of Natural Resource Governance in Africa*, London, Earthscan.

Nelson, F. & Agrawal, A. 2008. Patronage or participation? Community-based natural resource management reform in sub-Saharan Africa. *Development and Change*, 39(4), 557–585.

Nelson, F. & Blomley, T. 2009. Peasants' forests and the king's game? Institutional divergence and convergence in Tanzania's forestry and wildlife sectors. In: F. Nelson (ed.), *Community Rights, Conservation and Contested Land. The Politics of Natural Resource Governance in Africa*, London, New York, Earthscan, 79–106.

Nepstad, D., Schwartzman, S., Bamberger, B., Santilli, M., Ray, D., Schlesinger, P., Lefebvre, P., Alencar, A., Prinz, E., Fiske, G. & Rolla, A. 2006. Inhibition of Amazon deforestation and fire by parks and indigenous lands. *Conservation Biology*, 20, 65–73.

Nickerson, B. J. 1994. The environmental laws of Zimbabwe: A unique approach to management of the environment. *Boston College Third World Law Journal*, 14, 189–230.

North, D. C. 1990. *Institutions, Institutional Change and Economic Performance*, Cambridge, Cambridge University Press.

North, D. C. 1992. *Transaction Costs, Institutions, and Economic Performance. Occasional Papers No 30*. San Francisco, California, International Center for Economic Growth.

North, D. C. 1995. The new institutional economics and third world development. In: J. Harriss, J. Hunter & C. M. Lewis (eds.) *The New Institutional Economics and Third World Development*, London, Routledge, 17–26.

North, D. C. 2003a. *The Role of Institutions in Economic Development*, Geneva, Switzerland, United Nations Economic Commission For Europe.

North, D. C. 2003b. Understanding the process of economic change. In: *Forum Series on the Role of Institutions in Promoting Economic Growth*, Washington, DC, Mercatus Center at George Mason University and The IRIS Center, 23.

North, D. C. 2005. *Understanding the Process of Economic Change*, Princeton, NJ, Princetone University Press.

North, D. C. & Thomas, R. P. 1973. *The Rise of the Western World: A New Economic History*, Cambridge, UK, Cambridge University Press.

North, D. C., Wallis, J. J. & Weingast, B. R. 2009. *Violence and Social Orders. A Conceptual Framework for Interpreting Recorded Human History*, Cambridge, Cambridge University Press.

Nuulimba, K. & Taylor, J. J. 2015. 25 years of CBNRM in Namibia: A retrospective on accomplishments, contestation and contemporary challenges. *Journal of Namibian Studies History Politics Culture*, 18, 89–110.

OECD. 2016. *Agricultural Policy Monitoring and Evaluation 2016*, Paris, OECD Publishing.

Ogutu, J. O., Owen-Smith, N., Piepho, H.-P. & Said, M. Y. 2011. Continuing wildlife population declines and range contraction in the Mara region of Kenya during 1977–2009. *Journal of Zoology*.

Ogutu, J. O., Piepho, H.-P., Said, M. Y., Ojwang, G. O., Njino, L. W., Kifugo, S. C. & Wargute, P. W. 2016. Extreme wildlife declines and concurrent increase in livestock numbers in Kenya: What are the causes? *PLoS ONE*.

Oldfield, S. (ed.). 2003. *The Trade in Wildlife. Regulation for Conservation*, London, Earthscan.

Olson, M. 1965. *The Logic of Collective Action: Public Goods and the Theory of Groups*, Cambridge, MA, Harvard University Press.

Olsen, M. 2000a. *Power and Prosperity*, New York, Basic Books.

Olson, M. 2000b. *Power and Prosperity. Outgrowing Communist and Capitalist Dictatorships*, New York, Basic Books.

Onishi, N. 2015. A hunting ban saps a village's livelihood. *New York Times*, September 12.

Ostrom, E. 1990. *Governing the Commons: The Evolution of Institutions for Collective Action*, Cambridge, Cambridge University Press.

Ostrom, E. 2000a. Design principles of robust property-rights institutions: What have we learned? In: K. G. Ingram & Y.-H. Hong (eds.) *Property Rights and Land Policies*, Cambridge MA, Lincoln Institute of Land Policy, 25–51.

Ostrom, E. 2000b. Collective action and the evolution of social norms. *The Journal of Economic Perspectives*, 14, 137–158.

Ostrom, E. 2007. A diagnostic approach for going beyond panaceas. *Proceedings of the National Academy of Sciences of the United States of America*, 104, 15181–15187.

Ostrom, E. 2009a. Design principles of robust property-rights institutions: What have we learned? In: E. Ostrom, K. G. Ingram & Y.-H. Hong (eds.) *Property Rights and Land Policies*, Cambridge, MA, Lincoln Institute of Land Policy.

Ostrom, E. 2009b. A general framework for analysing sustainability of social-ecological systems. *Science*, 325, 419–422.

Ostrom, E. 2009c. Nobel prize lecture. http://nobelprize.org/mediaplayer/index.php?id=122.

Ostrom, E. 2009d. Beyond markets and states: Polycentric governance of complex economic systems. *Nobel Prize Lecture*.

Ostrom, E. & Hess, C. 2007. Private and common property rights. *Bloomington, Workshop in Political Theory and Policy Analysis*, Indiana University, 116.

Ostrom, E. & Nagendra, H. 2006. Insights on linking forests, trees, and people from the air, on the ground, and in the laboratory. *Proceedings of the National Academy of Science*, 103, 19224–19231.

Overdevest, C. 2000. Participatory democracy, representative democracy, and the nature of diffuse and concentrated interests: A case study of public involvement on a national forest district. *Society and Natural Resources*, 13, 685–696.

Parker, I. 2004. *What I Tell You Three Times Is True. Conservation, Ivory, History & Politics*, Kinloss, Librario.

Parris, R. & Child, G. 1973. The importance of pans to wildlife in the Kalahari and the effects of human settlement on these areas. *Journal of Southern African Wildlife Management Association*, 3, 1–8.

Pearce, D. W. & Turner, R. K. 1990. *Economics of Natural Resources and the Environment*, Baltimore, The Johns Hopkins Hniversity Press.

Pearce, F. 2016. *Common Ground: Securing Land Rights and Safeguarding the Earth*, Oxfam, International Land Coalition, Rights and Resources Initiative.

Peters, T. & Waterman, R. H. 1982. *In Search of Excellence. Lessons from America's Best-Run Companies*, London, HarperCollinsBusiness.

Phillips, A. 2007. *A Short History of the International System of Protected Areas Management Categories*, Andalusia, Spain, IUCN World Commission on Protected Areas Task Force: IUCN Protected Area Categories.

Phimister, I. R. 1978. Meat and monopolies: Beef cattle in Southern Rhodesia, 1890–1938. *The Journal of African History*, 19(3), 391–414.

Pimbert, M. P. & Pretty, J. N. 1994. *Participation, People and the Management of National Parks and Protected Areas: Past Failures and Future Promise*, London, United Nations Research Institute for Social Development, IIED and WWF.

Pimbert, M. P. & Pretty, J. N. 1997. Parks, people and professionals: Putting 'participation'into protected area management. *Social Change and Conservation*, 16, 297–330.

Pinheiro, P. S. 2018. *Co-Management of Natural Resources in the Lowe Jurua Extractive Reserve, Central-West Brazilian Amazon*. Ph.D., University of Florida.

Pinker, S. 2012. *The Better Angels of Our Nature: Why Violence Has Declined*, New York, Viking.

Poffenberger, M. 2006. People in the forest: Community forestry experiences from Southeast Asia. *International Journal of Environment and Sustainable Development*, 5, 57–69.

Polski, M. M. & Ostrom, E. 1999. *An Institutional Framework for Policy Analysis and Design*, Department of Political Science, Indiana University.

Pomeroy, R. S. 1995. Community-based and co-management institutions for sustainable coastal fisheries management in Southeast Asia. *Ocean & Coastal Management*, 27, 143–162.

Porter-Bolland, L., Ellis, E. A., Guariguata, M. R., Ruiz-Mallén, I., Negrete-Yankelevich, S. & Reyes-García, V. 2012. Community managed forests and forest protected areas: An assessment of their conservation effectiveness across the tropics. *Forest Ecology and Management*, 268, 7–17.

Porter-Bolland, L., Ruiz-Mallén, I., Camacho-Benavides, C. & McCandless, S. R. (eds.). 2013. *Community Action for Conservation. Mexican Experiences*, New York: Springer.

Poulsen, J. R., Clark, C. J., Mavah, G. & Elkan, P. W. 2009. Bushmeat supply and consumption in a tropical logging concession in Northern Congo. *Conservation Biology*, 23, 1597–1608.

Prager, K. & Vanclay, F. 2010. Landcare in Australia and Germany: Comparing structures and policies for community engagement in natural resource management. *Ecological Management and Restoration*, 11, 187–193.

Pricewaterhouse. 1994. *The Lowveld Conservancies: New Opportunities for Production and Sustainable Land-Use*, Harare, Zimbabwe, Published by Save, Bubiana and Chiridzi River Conservancies.

Randall, A. 1983. The problem of market failure. *Natural Resources Journal*, 23, 131–148.

Reader, J. 1999. *Africa: A Biography of A Continent*, New York, Vintage.

Reed, M. S. 2008. Stakeholder participation for environmental management: A literature review. *Biological Conservation*, 141, 2417–2431.

Reed, T. 2002. *The Function and Structure of Protected Area Authorities. Considerations for Financial and Organizational Management*, Washington, DC, World Bank.

Reid, H. 2016. Ecosystem- and community-based adaptation: Learning from community-based natural resource management. *Climate and Development*, 8. DOI: 10.1080/17565529.2015.1034233.

Rekacewicz, P. 2005. *World Atlas of Desertification*, Arendal, Norway GRID-Arendal. A Centre Collaborating with UN Environment: www.grida.no/resources/6338.

Ribot, J. C. 2002. *Democratic Decentralization of Natural Resources. Institutionalizing Popular Participation*, Washington, DC, World Resources Institute.

Ribot, J. C. 2003. Democratic decentralisation of natural resources: Institutional choice and discretionary power transfers in sub-Saharan Africa. *Public Administration and Development*, 23, 56–65.

Ribot, J. C. 2007. Representation, citizenship and the public domain in democratic decentralization. *Development*, 50, 43–49.

Ribot, J. C. 2008. *Building Local Democracy through Natural Resource Interventions. An Environmentalist's Responsibility*, Washington, DC, World Resources Institute.

Ribot, J. C., Lund, J. F. & Treue, T. 2010. Democratic decentralization in sub-Saharan Africa: Its contribution to forest management, livelihoods, and enfranchisement. *Environmental Conservation*, 37, 35–44.

Ridley, M. 2010. *The Rational Optimist: How Prosperity Evolves*, New York, HarperCollins.

Rigava, N., Taylor, R. & Goredema, L. 2006. Participatory wildlife quota setting. *Participatory Learning and Action*, 55, 62–69.

Rihoy, E., Chirozva, C. & Anstey, S. 2007. *'People are Not Happy' – Speaking up for Adaptive Natural Resource Governance in Mahenye*, Cape Town, Programme for Land and Agrarian Studies, University of the Western Cape.

Riney, T. 1964. Development of the wildlife resource in Africa. *Unasylva*, 15(2), 76–80.

Riney, T. 1967. *Conservation and Management of African Wildlife*, Rome, FAO.

Riney, T. 1982. *Study and Management of Large Mammals*, Chichester, John Wiley.

Riney, T. & Hill, P. 1967. *Conservation and Management of African Wildlife*, Rome, FAO.

Ripple, W. J., Newsone, T. M., Wolf, C., Dirzo, R., Everatt, K. T., Galetti, M., Hayward, M. W., Kerley, G. I. H., Levi, T., Lindsey, P. A., Macdonald, D. W., Malhi, Y., Painter, L. E., Sandom, C. J., Terborgh, J. & van Valkenburgh, B. 2015a. Collapse of the world's largest herbivores. *American Association for the Advancement of Science*, 12.

Ripple, W. J., Newsone, T. M., Wolf, C., Dirzo, R., Everatt, K. T., Galetti, M., Hayward, M. W., Kerley, G. I. H., Levi, T., Lindsey, P. A., Macdonald, D. W., Malhi, Y., Painter, L. E., Sandom, C. J., Terborgh, J. & Van Valkenburgh, B. 2015b. Collapse of the world's largest herbivores. *Science Advances*, 1(4), e1400103.

Roberts, R. G., Flannery, T. F., Ayliffe, L. K., Yoshida, H., Olley, J. M., Prideaux, G. J., Laslett, G. M., Baynes, A., Smith, M. A., Jones, R. & Smith, B. L. 2001. New ages for the last Australian Megafauna: Continent-wide extinction about 46,000 years ago. *Science*, 292, 1888–1892.

Rockström, J., Steffen, W., Noone, K., Persson, Å., Chapin, S., Lambin, E. F., Lenton, T. M., Scheffer, M., Folke, C., Schellnhuber, H. J., Nykvist, B., Wit, C. A. D., Hughes, T., Leeuw, S. V. D., Rodhe, H., Sörlin, S., Snyder, P. K., Costanza, R., Svedin, U., Falkenmark, M., Karlberg, L., Corell, R. W., Fabry, V. J., Hansen, J., Walker, B., Liverman, D., Richardson, K., Crutzen, P. & Foley, J. A. 2009. A safe operating space for humanity. *Nature*, 461, 472–475.

Rodriguez, G. 1985. The economic implications of the beef pricing policy in Zimbabawe. International Livestock Centre for Africa.

Rondinelli, D., Mccullough, J. S. & Johnson, R. W. 1989. Analysing decentralization policies in developing countries: A political-economy framework. *Development and Change*, 20, 57–87.

Ropke, I. 2004. The early history of modern ecological economics. *Ecological Economics*, 50, 293–314.

Rubino, E. C. & Pienaar, E. F. 2018. Rhinoceros ownership and attitudes towards legalization of global horn trade within South Africa's private wildlife sector. *Oryx*, 52, 175–185.

Rubino, E. C. & Pienaar, E. F. 2018. Understanding South African private landowners' decision to manage rhinoceros. *Human Dimensions of Wildlife*, 23(2), 160–175.

Runte, A. 1979. *National Parks. The American Experience*, Lincoln & London, University of Nebraska Press.

SADC. 1999. *Protocol on Wildlife Conservation and Law Enforcement*, Maputo, Southern African Development Community.

Safriel, U., Adeel, Z., Niemeijer, D., Puigdefabregas, J., White, R., Lal, R., Winslow, M., Ziedler, J., Prince, S., Archer, E., King, C., Shapiro, B., Wessels, K., Nielsen, T., Portnov, B., Becker-Reshef, I., Thonnell, J., Lachman, E. & McNab, D. 2005. Dryland systems, ecosystems and human well-being: Current state and trends. In R. Hassan & R. Scholes (eds.), *Washington Millennnium Ecosystem Assessment: Ecosystems and Human Well-Being*, Washington, DC, Island Press, 623–662.

Sample, I. 2019. Strongest opponents of GM foods know the least but think they know the most. *The Guardian*.

Sanchez, P. A. 2015. En route to plentiful food production in Africa. *Nature Plants*, 1.

Sandford, S. 1983. *Management of Pastoral Development in the Third World*, Chichester, Wiley.

Sandom, C., Faurby, S., Sandel, B. & Svenning, J.-C. 2014. Global late quaternary Megafauna extinctions linked to humans, not climate change. *Proceedings of the Royal Society B: Biological Sciences*, 281.

SASUSG. 1996. *Sustainable Use Issues and Principles*, Southern Africa Sustainable Use Specialist Group, IUCN Species Survival Commission, Harare.

SASUSG. 2003. *Principles of Sustainable Use*, Windhoek, Namibia Nature Foundation.

Save Valley Conservancy. 2002. Directed by Taylor, S. Zimbabwe.

Schalkwyk, D. L. V., Mcmillin, K. W., Witthuhn, R. C. & Hoffman, L. C. 2010. The contribution of wildlife to sustainable natural resource utilization in Namibia: A review. *Sustainability*, 2, 3479–3499.

Schlager, E. & Ostrom, E. 1992. Property-rights regimes and natural resources: A conceptual analysis. *Land Economics*, 68, 249–262.

Schmink, M. & Wood, C. 1992. *Contested Frontiers in Amazonia*, New York, Columbia University Press.

Schoones, I. 1994. *Living with Uncertainty: New Directions in Pastoral Development in Africa*, London, Intermediate Technology Publications.

Scott, J. C. 1985. *Weapons of the Weak: Everyday Forms of Peasant Resistance*, New Haven and London, Yale University Press.

Scott, J. C. 1999. *Seeing Like a State: How Certain Schemes to Improve the Human Condition Have Failed*, New Haven, CT, Yale University Press.

Scriabine, R. 1983. Third world national parks conference. *The Environmentalist*, 3, 75–77.

Sen, A. 1999. *Development as Freedom*, New York, Anchor Books.

Senge, P. 1990. *The Fifth Discipline: The Art and Practice of the Learning Organization*, New York, Doubleday.

Shackleton, C. M., Shackleton, S. E., Buiten, E. & Bird, N. 2007. The importance of dry woodlands and forests in rural livelihoods and poverty alleviation in South Africa. *Forest Policy and Economics*, 9, 558–577.

Shackleton, C. M., Willis, T. J., Brown, K. & Polunin, N. V. C. 2010. Reflecting on the next generation of models for community-based natural resources management. *Enviromental Conservation*, 37, 1–5.

Sharp, R. 2017. Five giants of conservation pass on. SULiNews 11, August 2017. Gland, Switzerland, Sustainable Use and Livelihoods Initiative, IUCN Commission on Environmental, Economic and Social Policy.

Shelhas, J. 2001. The USA national parks in international perspective: Have we learned the wrong lesson? *Environmental Conservation*, 28, 200–304.

Skyer, P. & Saruchera, M. 2004. Community conservancies in Namibia: An effective institutional model for commons management? In: *Policy Brief. Debating Land Reform and Rural Development No 14*, Cape Town, Programme for Land and Afrarian Studies, University of the Western Cape, 4.

Sloman, S. & Fernbach, P. 2019. Human expertise: Its not what you know, it who ... *The Guardian*, Sunday 9 April 2017.

Smil, V. 2002. *The Earth's Biosphere. Evolution, Dynamics and Change*, Cambridge, MA, MIT Press.

Smil, V. 2011. Harvesting the biosphere: The human impact. *Population and Development Review*, 37, 613–636.

Smith, A. 1776. *An Inquiry into the Nature and Causes of the Wealth of Nations*, London, W. Straham and T. Cadell.

Sobrevila, C. 2008. *The Role of Indigenous Peoples in Biodiversity Conservation: The Natural but Often Forgotten Partners*, Washington, DC, World Bank.

Souza, T., Chidakel, A., Child, B. & Chang, W. in review. *Tourism Economic Model for Protected Areas, TEMPA. Estimating the Economic Impact of Visitor Spending in Developing Country Protected Areas*, Washington, DC, Scientific and Technical Advisory Panel, Global Environmental Facility.

Souza, T. D. V. S. B. & Thapa, B. 2018. Tourism demand analysis of the federal protected areas of Brazil. *Journal of Park and Recreation Administration*, 36, 1–21.

Spenceley, A. (ed.). 2009. *Responsible Tourism. Critical Issues for Conservation and Development*, London, Routledge.

Steffen, W., Broadgate, W., Deutsch, L., Gaffney, O. & Ludwig, C. 2015. The trajectory of the anthropocene: The great acceleration. *The Anthropocene Review*, 2, 81–98.

Stiglitz, J. 2002. *Globalization and Its Discontents*, London, Penguin Books.

Stocking, M. 1985. Soil conservation policy in colonial Africa. *Agricultural History*, 59, 148–161.

Stroup, R. & Baden, J. 1983. *Natural Resource Economics. Bureaucratic Myths and Environmental Management*, Cambridge, MA, Ballinger Publishing Company.

Stuart-Hill, G., Diggle, R., Munali, B., Tagg, J. & Ward, D. 2005. The event book system: A community-based natural resource monitoring system from Namibia. *Biodiversity and Conservation*, 14, 2611–2631.

Stuart-Hill, G., Diggle, R., Munali, B., Tagg, J. & Ward, D. 2007. The event book system: Community-based monitoring in Namibia. *Participatory Learning and Action*, 55, 70–78.

Stynes, D. 2005. Economic significance of recreational uses of National Parks and other public lands. *Social Science Research Review*, 5, 36.

Stynes, D., Propst, D., Chang, W. & Sun, Y. 2000. *Estimating National Protected Area Visitor Spending and Economic Impacts; the MGM2 Model*, Michigan, Department of Park, Recreation and Tourism Resources, Michigan State University.

Suich, H. & Child, B. (eds.). 2009. *Evolution & Innovation in Wildlife Conservation. Parks and Game Ranches to Transfrontier Conservation Areas*, London, Earthscan.

Tacconi, L. 2007. Decentralization, forests and livelihoods: Theory and narrative. *Global Environmental Change*, 17, 338–348.

Talbot, L. M., Ledger, H. P. & Payne, W. J. A. 1961. The possibility of using wild animals for animal production on East African rangelands based on a comparison of ecological requirements and efficiency of range utilization by domestic livestock and wild animals. *8th International Congress of Animal Production*, Hamburg.

Talbot, L. M., Payne, W. A., Ledger, H. P. & Talbot, M. H. 1965a. The meat production potential of wild animals in Africa. *Commonwealth Agriculture Bureau Technical Communication*, No. 16.

Talbot, L. M., Payne, W. J. A., Ledger, H. P., Verdcourt, L. D. & Talbot, M. H. 1965b. The meat production potential of wild animals in Africa. A review of the biological knowledge. *Technical Communication No. 16*, Edinburgh: Commonwealth Bureau Animal Breeding and Genetics.

Taylor, A., Lindsey, P. & Davies-Mostert, H. 2016. *An Assessment of the Economic, Social and Conservation Value of the Wildlife Ranching Industry and Its Potential to Support the Green Economy in South Africa*, Johannesburg, The Endangered Wildlife Trust.

Taylor, R. 2009. Community based natural resource management in Zimbabwe: The experience of CAMPFIRE. *Biodiversity and Conservation*, 18, 2563–2583.

Taylor, R., Bond, I. & Rigava, N. 1997. Quota setting manual. WWF-SARPO Wildlife Management Series. Harare, WWF, Southern African Regional Programme Office, 43.

Taylor, R. & Murphree, M. W. 2007. *Zimbabwe: Masoka and Gairezi. Case Studies on Successful Southern African NRM Initiatives and Their Impacts on Poverty and Goverance*, Washington, DC, USAID.

Taylor, R. D. & Child, B. 1991. Does wildlife offer comparative advantages over cattle? The buffalo range case study: Ecological considerations. *Meeting Rangelands Challenges in Southern Africa in the 1990s*, CSIR Conference Center, Pretoria.

Taylor, R. D. & Murphree, M. 2007. *Case Studies on Successful Southern African NRM Initiatives and Their Impacts on Poverty and Governance*, Zimbabwe, Masoka and Gairesi. USAID-FRAME & IUCN.

Taylor, R. D. & Walker, B. H. 1978. Comparison of vegetation use and herbivore biomass on a Rhodesian game and cattle ranch. *Journal of Applied Ecology*, 15, 565–581.

Tchawa, P. 2009. Land and development. *Tropical Forest Update*, 19, 8–9.

TEEB. 2009. The economics of ecosystems and biodiversity: Climate issues update. The Economics of Ecosystems and Biodiversity. www.teebweb.org/media/2009/09/.

Terborgh, J. 1999. *Requiem for Nature*, Washington, DC, Island Press.

Themba, M. 2017. *Dashboard Survey Report 2017, Mangalane Community, Mozambique, Moamba District*, Southern African Wildlife College.

Thomas, C. C., Huber, C. & Koontz, L. 2014. *2013 National Park Visitor Spending Effects. Economic Contributions to Local Communities, States, and the Nation*, Natural Resource Report NPS/NRSS/EQD/NRR—2014/824, Fort Collins, CO, U.S. Department of the Interior, National Park Service.

Tietenberg, T. H. 2006. *Emissions Trading*, New York, Routledge.

Tilley, C. 2007. *Democracy*, Cambridge, Cambridge University Press.

Tilly, C. 2009. Grudging consent. *Social Science Research Council*, 27 May.

Tocqueville, A. 1994. *Democracy in America*, New York, Alfred A. Knopf.

Tocqueville, A. 2000. *Democracy in America*, Chicago and London, University of Chicago Press.

Tole, L. 2010. Reforms from the ground up: A review of community-based forest management in tropical developing countries. *Environmental Management*, 45, 1312–1331.

UNEP-WCMC. 2012. *Protected Planet Report 2012. Tracking Progress Towards Global Targets for Protected Areas*, Cambridge, UK, United Nations Environment Programme World Conservation Monitoring Centre (UNEP-WCMC).

United Nations. 2011. *Global Drylands: A UN System Wide Response*, Geneva, United Nations Environment Management Group, UNEP.

USFWS. 2017. *2016 National Survey of Fishing, Hunting, and Wildlife-Associated Recreation*, Washington, DC, U.S. Fish & Wildlife Service, U.S. Department of the Interior.

van der Reit, M. 2008. Participatory research and the philosophy of social science: Beyond the moral imperative. *Qualitative Inquiry*, 14, 546–565.

van Hoven, W. 2015. Private game reserves in southern Africa. In: R. van der Duim, M. Lamers & J. van Wijk (eds.) *Institutional Arrangements for Conservation, Development and Tourism in Eastern and Southern Africa. A Dynamic Perspective*, London, Springer, 101–118.

Vatn, A. 2015. *Environmental Governance. Institutions, Policies and Actions*, Northhampton, MA, Edward Elgar Publishing, Inc.

Vincent, V. & Thomas, R. G. 1961. *An Agricultural Survey of Southern Rhodesia*, Salisbury, Federation of Fhodesia and Nyasaland.

Vitousek, P. M., Ehrlich, P. R., Ehrlich, A. H. & Matson, P. A. 1986. Human appropriation of the products of photosynthesis. *BioScience*, 36, 368–373.

Vogel, G. 2017. Where have all the insects gone? *Science*, 356, 576–579.

Wade, R. 1987. The management of common property resources: Collective action as an alternative to privatisation or state regulation. *Cambridge Journal of Economics*, 11, 95–106.

Wade, R. 1987b. *Village Republics. Economic Conditiosn for Collective Action in South India*, Cambridge, Cambridge University Press.

Wainright, C. 1996. Evaluating community based natural resource management: A case study of the Luangwa Integrated Resource Development Project (LIRDP), Zambia. MSc Dissertation, University of kent.

Walker, B., Holling, C. S., Carpenter, S. R. & Kinzig, A. 2004. Resilience, adaptability and transformability in social-ecological systems. *Ecology and Society*, 9, 2–10.

Walker, B. H. 1979. Game ranching in Africa. In: B. H. Walker (ed.) *Management of Semi-Arid Ecosystems*, Oxford, Elsevier Scientific, 55–81.

Wall, D. 2017. *Elinor Ostrom's Rules for Radicals. Cooperative Alternatives beyond Markets and States*, London, Pluto Press.

Warren, M. E. 1993. Can participatory democracy produce better selves? Psychological dimensions of Habermas's discursive model of democracy. *Political Psychology*, 14, 209–234.

Wells, M. 1992. Biodiversity conservation, affluence and poverty: Mismatched costs and benefits and efforts to remedy them. *Ambio*, 21, 237–243.

Whande, W., Kepe, T. & Murphree, M. 2003. *Local Communities, Equity and Conservation in Southern Africa. A Synthesis of Lessons Learnt and Recommendations from a Southern African Technical Workshop*, Cape Town, Programme for Land and Agrarian studues (PLAAS), African Resources Trust (ART) and Theme on Indigenous and Local communities, Equity and Protected Areas (TILCEPA).

White, R. P., Tunstall, D. & Henninger, N. 2002. *An Ecosystem Approach to Drylands: Building Support for New Development Policies, Information Policy Brief No.* Washington, DC, World Resources Institute.

Whitlow, R. 1988. Potential versus actual erosion in Zimbabwe. *Applied Geography*, 8, 87–100.

Wild, H. 1968. Bechuanaland protectorate. In: L. A. O. Hedberg (ed.) *Conservation of Vegetation in Africa South of the Sahara*, Uppsala, Almqvist & Wiksblls Boktryckbri Ab, 198–202.

Williams, T. O. 1993. Livestock pricing policy in sub-Saharan Africa: Objectives, instruments and impact in five countries. *Agriculture Economics*, 8, 139–159.

Williamson, O. E. 2000. The new institutional economics: Taking stock, looking ahead. *Journal of Economic Literature*, 38, 595–613.

Wilshusen, R. 2008. Shades of social capital: Elite persistence and the everyday politics of community forestry in southeastern Mexico. *Environment and Planning*, 41, 389–406.

Wilson, E. O. 2016. *Half-Earth: Our Planet's Fight for Life*, London, Liveright.

Wunder, S., Wertz-Kanounnikoff, S. & Ferraro, P. 2010. *Payments for Environmental Services and the Global Environment Facility: A STAP Advisory Document*, Washington, DC, Scientific and Technical Advisory Pane, Global Environment Facility.

WWF. 2018. *Living Planet Report – 2018: Aiming Higher*, Gland, Switzerland, WWF.

WWF, London, Z. S. O., Network, G. F. & Agency, E. S. 2012. *Living Planet Report 2012. Biodiversity, Biocapacity and Better Choices*, Gland, WWF International.

WWF-SARPO. 2017. *Marketing Wildlife Leases*, Harare, WWF-World Wide Fund for Nature (formerly World Wildlife Fund) Programme Office, Zimbabwe.

INDEX

activity-based budgeting 212–217, 232–233, 327–328
Adamson, Joy and George 13
adaptive management 195–197, 266, 304, 306–308
Africa 134–156; evolution of man 15–16; farming 38, 41; land rights 21, 57–58; species diversity 36–37, 43; wildlife economy 48, 65, 125–132, 137, 355–356; wildlife loss 118, 125, 135–136; *see also* southern Africa; *individual countries*
Africa Special Project 147
Agricultural Revolution 12, 18, 19–20
agriculture *see* farming
agro-extractive economy *179*
'alternative livelihoods' 109
Ancient Egypt 19, 20, 21–22
Ancient Greece 22, 228
Ancient Rome 22
Anthropocene 34, 48–49
anthropomorphisation 13, 30, 124
Asia 254–256, 346
Australia 36
Australopithicus afarens 15

banking 330
Big Man rule 25, 28, 63, 67
bill of rights 280, 284, 317, 318, 334
biodiversity: conservation 89, 121–122, 154, 185, 263; loss and recovery 2, 5, 29, 33, 46, 62, 99, 131; sustainable governance 36, 182, 279, 344, 353
bio-experience economy *179*
bipedalism 15
bookkeeping 330
Botswana **176**

bottom-up community-based management 236–238
Brazil 254
budget-funded institutions 53–54
Buffalo Range Ranch, Zimbabwe 166–168
bureaucracy 26–27, 149–150
bushmeat trade 108–110

cadastres 19, 20, 21–22, 80
CAMPFIRE 9, 200–224; core principles 175, 208–209; enabling environment 209–210, 285, 297, 298, 306, 349; history 201–207; outcomes and successes 217–224, 242, 280, 348, 353; price and proprietorship 77, 188–189; tools and processes 210–217
capacity building 279–282, 308–313
carbon payments 48
cascaded institutions 278–279
cash management 333–334
catchment communities 113
cattle 169–171
CBNRM (community-based natural resource management) 259–267; background and emergence 251–253, 259–260; case for 351–353; comparison of programmes **242**; defining features 93–94, 252, 263–266, 305–306, 347; feasibility assessment 289–298; foundational conditions and principles 7, 9–10, 10, 195, 262–263, 273–276, 306–308, 349–350, 353; management, governance and compliance *283*, 285, 323, 325–328, 330–337, 348–349; monitoring and evaluation 337–338; outcomes and impacts 41–42, 127–128, 286, **306–307**, 342–343; and property rights 71, 74, 76–77; roles and responsibilities 323–325; scope 261–262; stages of implementation 287–313;

support for programmes 306–308; theory of
change 282–286; *see also* CAMPFIRE; Luanga
Valley, Zambia; Namibia, CBNRM
programme
Cecil the Lion 13, 30, 124
change, theory of 282–286
Child, Graham 129, 138, 142–146, 203
China 20, 26
CITES (Convention for the International Trade
in Endangered Species) 79, 91–92, 129
classical period 22
Coase theorem 104
'cognitive revolution' 16, 101
collective action 89, *90*, 284–285
common property/resources **75**, 76, 83–86
communal lands 41, 57–59, 175; *see also* 'tragedy
of the commons'
community citizenship 316–317
community conservation 250–253
community meetings 320
community membership 321–322
comparative advantage 103
conflict resolution 278
conservation: contemporary issues 67, **355**;
economic analysis 101–104, 112–113; founding
ideologies 4, 98; North American Model 62;
and state agencies 55–56; *see also* community
conservation; public conservation
constitution, for CBNRM 230–231, 232, 274,
280, 284, 310, 317, 318, 323, 334, 338
constitutional governance 23, 63, 65
corruption 29
costs and benefits: Buffalo Range Ranch case
study 166–169; CAMPFIRE programme 204;
in CBNRM 246, 261, 263, 275, 284, 285, 286;
Coase theorem 104; free markets 110, 112;
Murphree's principles 271, 272, 277;
proprietorship and price 73, 75–76, 79, 80–89,
113, 114, 345; sustainable governance approach
181, 183, 191–192; ungoverned spaces
98–100, 103
cross-scale governance *194*

Dark Ages 3
decentralisation 79, 93–94, 223, 227–228, 251,
255, 297
deforestation 45, 48
deinstitutionalisation 56–57, 61, 100, 261, 342
delegated aggregation 191–192
'demand reduction' 153–154
democracy/democratisation 2, 22–23, 27, 276,
344; *see also* inclusive governance; liberal
democracies
demography 4–5
differential taxation 110–111, 220–221
disease 39
divine right of kings 4, 12, 19, 23

domestic species 45, 73, 189–191
drylands 42–44, 45–46, 47–48
Dunbar's number 12, 17, 18, 295

economic development 33, *34*, 63–66
economic efficiency 6, 73–74
economic justice 6, 74
economic man 101
economic pricing/analysis 111–112, 169,
290–291
economy, and environment 98, 158–179
elections 329
elite capture 268–269
Enclosure movement 21
England 25
English Bill of Rights 24, 63, 156
Enlightenment, the 3, 21, 24, 65
environmentalism 29, 30
Europe 20, 21, 25, 53
evolution of man 14–36
exclusion, right of 72, 76–77, 86, 104
extinction 36–37
extractive governance 25–26, 62–63, 113–114,
268–269, **317**

farming 37–42
Fertile Crescent 19
feudal systems 21, 22
financial management 323, 325–328, 330–334
financial pricing/analysis 111–112, 168, **290**
fiscal devolution 317–319
fisheries 255–256
food production 19–20, 37–42
forests 42–43, 44–46, 47–48, 130–131, 253–255
Frazer, Archie 138
free markets 27–28, 64, 66–67, 82, 106–108,
113–114
frontier economy 53–56, **117**, 118–119, 186–187
fugitive resources 80–82

'game ranching' 161–165
Glorious Revolution 3, 7, 10, 12–13, 23–25, 63,
67–68
governance 13, 52, 117–119; *see also* extractive
governance; inclusive governance; institutions
'governance compliance' 282
governance dashboard process 302–305
graduated sanctions 277–278
Great Acceleration 33

habitat replacement 8
'half-earth' 6; *see also* ungoverned spaces
Hobbes, Thomas 21, 26, 105, 285
Holocene 33, 37
Homo erectus 15
Homo habilis 15

Homo sapiens: evolution of 14–36; impact on the planet 33–50; language, reciprocity and exchange 16, 17–18; norms and institutions 2, 12–14; and species extinction 36–37; ways of thinking 12–14
hunter-gatherers 33, 36–37
hunting 31, 54–55, 129–130, 165–166, 178

inclusive governance: in CBNRM 261, 263, 268–269; core principles 322; enabling environment 279–282, 284–285, 295–298; *cf.* extractive governance 25–26, 62–63, **317**; Luanga Valley case study 226–249; and prosperity 28–29; *cf.* representational governance 239–240, 243–244, *247*; sustainable governance 181, 279–282; tools for developing 315–340; *see also* CBNRM
India 20
Industrial Revolution 3–4, 12, 24, 33, 63–64, 118
institutional change 64–66
Institutional Revolution 12
institutions 2–4, 13–14, 20–36, 52–62, 343–344; *see also* governance; weak institutions
Intensive Conservation Areas (ICAS), Zimbabwe 140–141

land tenure 20–22
land-use trends 152–153, 174–175
Latin America 65
learning communities *197*
Leviathan authority 82
liability 79
liberal democracies 20–21, 24, 28–29, 63, 65, 105, 223
livelihood survey **299–300**, 301–302
Living Planet Index 33
local commons 52, 72–73, 108
Locke, John 21, 26, 61, 343
London Convention 125–126
Luanga Valley, Zambia 229–240, 297

macro-governance 350–351
Mahenye community, Zimbabwe 205–206
markets/market failure 111–112, 175, 294, 344–345
Martin, Rowan 202–204
meritocracy 26
meso-governance 350–351
Mexico 253–254
micro-governance 268–286, 347–348
Millennium Ecosystem Assessment (MEA) 46
missing markets **291**, 294–295
MOMS (Management Orientated Monitoring System) 280–281, 310
monitoring 277, 278, 280–281, 284, 311
Mozambique 302, 304–305, 319–321, 341–342

Muir, John 98
Murphree, Marshall 77, 149, 191, 198–199, 204–205, 269–272, 350, 353
myths 18–19

Namibia, CBNRM programme 174–175, **176**, 257–259, 280–281, 294, 298, 308, 349, 353
Napoleon 22
nationalisation 29, 119, 125–132
national parks 4, 89, 125–128, 135, 185
nation states 27
Natural Resources Board (NRB), Zimbabwe 140–141
negotiation 312
Nepal 255
new institutionalism 3
NGOs 297, 308
North America 36, 118
North American Model (conservation) 62, 119–125, **184**

open access property **75**, 76
organisations, *cf.* institutions 13
Ostrom, Elinor 7, 9–10, 140, 192, 227, 228, 260, 269–272

Parks and Wild Life Act (1975) 79, 139, 140, 142, 145, 162, 192, **193**, 203
participatory adaptive management 181
participatory governance *see* inclusive governance
participatory rural appraisal 301
policy analysis matrix (PAM) 171–174
polyvalent governance *194*
Porto Alegre, Brazil 229
poverty: benefits of CBNRM 229; and biodiversity 2, 263, 352; economic justice 6, 74; and environmental loss 4–5, 6, 8, 20, 41, 286; extractive governance 25, 113; farming 38, 40–41; and hunting 129–130; and national parks 126; property rights 74, 344; and ungoverned spaces 12, 34–35, 41, 46, 56, 68, 98, 344
pre-modern economy **117**, 118
'preservation' movement 98
price 104–114, 181, 263
private goods/property 75, 83–86
product development **291**, 294–295
Progressive Era 29
proprietorship (property rights) 20–22, 71–96; case study: Americas 253–254; history of 24; Murphree's principles 271; and price 104–105; regulation 89–92; sustainable governance approach/CBNRM 181, 263, 284, 286, **290–293**, 345; types 75–76
proprietorship–price hypothesis 7, 183–189, 282, 342

prosperity 3, 27–28; *see also* wealth
proto-humans 15
public conservation 53–56
public goods/property **75**, 76, 83–86, 246, 342
public governance, of wild resources 2–4, 29, 53, 89, **117**, 119, **184**, *186*, 188

recentralisation 239
REDD (reducing emissions from deforestation and degradation) 48, 253, 256–257, 260, 286, 321
Reformation, the 12
regulation 79, 89–92, 110–111
representational governance 239–240, 243–244, *247*, 348
resource allocation 104
revenue distribution ceremony 329–330
rewilding 9, 342
rhinos 91–92, 187
rights: of the individual 4, 10, 23–24; to land 57–59; to wild resources 59–60; *see also* propriertorship
Riney, Thane 138
Roosevelt, Theodore 53, 54–55, 62, 98, 127, 342, 345
rule of law 25, **317**
rule of man **317**
rules *see* institutions

SASUSG (Southern Africa Sustainable Use Specialist Group) 148
savannas 44, 45–46
scaling 189–197, 285, **292**, 350–351
Scientific Revolution 3, 12, 24
scouts 282
security 285–286
self-regulation 140–142
service institutions 53–54
Smith, Adam 4, 63–64, 106, 343, 345
Smithers, Reay 138
social capital 61, 282, 349
South Africa 175, **176**
South America 36
southern Africa 8–9, 91–92, 152–153, 174–175, **176**, 342, 344, 346
Spain 25
state, the 26
state agencies 55–56
Stockil, Clive 205
subsidiarity 189–191

subsidiary governance 181
subsistence farming 40–42
subsistence hunting 109
supply and demand 106–107
sustainability 6–7, 101
sustainable governance 180–199, 344–346; *see also* CBNRM
Switzerland 228

Tanzania 15, 147, 253, 256–257, 321
taxation, differential 110–111
thinking, patterns and systems of 12–14, 30–31, 354
top-down community-based management 236–238
total economic value 46, 47, 176–177
tourism 178
'tragedy of the commons' 41, 56–57, 72–73, 118, 130, 260, 298
trauma 61

ungoverned spaces 6, 8, 29, 34–36, 42–50, 52–62, 98–100, 344

value estimates, of wild resources 46–48
Village Action Group (VAG) 229, 235, 324–325
Village Company 77
vision building 298–305
visualisation 281

warfare 25
weak institutions 6–7, 41, 48, 67, 73, 99, 286
wealth 33, 36, 39, 64, 106; *see also* prosperity
Weber, Max 26–27
wildlife: loss 48–49; pricing 49–50, 101–103; production 161–165; profitability **163**, **164**; as public good 342–343
wildlife trade 31, 108–110
wild resources 5–6; future of 353–356; rights, ownership and governance 2, 3, 4, 8, 59–60, 86–89, 277–282; values and markets 46–48, 60, 98–115, 181–183
WINDFALL 202–203, 216

Yellowstone Model 125, 185

Zambia **176**, 229–240
Zimbabwe 79, 135, 137–140, 158–179, 191–193; *see also* CAMPFIRE